Environmental Protection: Techniques and Sustainability Measures

Environmental Protection: Techniques and Sustainability Measures

Editor: Kane Harlow

Callisto Reference
www.callistoreference.com

Callisto Reference,
118-35 Queens Blvd., Suite 400,
Forest Hills, NY 11375, USA

Visit us on the World Wide Web at:
www.callistoreference.com

ISBN: 978-1-63239-875-8 (Hardback)

Printed in China.

Cataloging-in-Publication Data

Environmental protection : techniques and sustainability measures / edited by Kane Harlow.
p. cm.
Includes bibliographical references and index.
ISBN 978-1-63239-875-8
1. Environmental protection. 2. Environmental protection--Technological innovations.
3. Environmental engineering. 4. Sustainability. 5. Environmental quality. I. Harlow, Kane.
TD170 .E58 2017
363.7--dc23

Table of Contents

Preface

Over the recent decade, advancements and applications have progressed exponentially. This has led to the increased interest in this field and projects are being conducted to enhance knowledge. The main objective of this book is to present some of the critical challenges and provide insights into possible solutions. This book will answer the varied questions that arise in the field and also provide an increased scope for furthering studies.

Environmental protection is defined as a set of processes that protect the quality of the environment in the face of pollution and harmful contaminants. This book on environmental protection deals with the various approaches that forefront environmental conservation and ecosystem monitoring. Environmental protection uses methods such as computational analysis and population dynamics assessment for its research. This book unfolds the innovative aspects of environmental protection which will be crucial for the progress of this field in the future. This book elucidates new techniques and their applications in a multidisciplinary approach. From theories to research to practical applications, case studies related to all contemporary topics of relevance to this field have been included in this book. It will provide comprehensive knowledge to the readers.

I hope that this book, with its visionary approach, will be a valuable addition and will promote interest among readers. Each of the authors has provided their extraordinary competence in their specific fields by providing different perspectives as they come from diverse nations and regions. I thank them for their contributions.

Editor

Continuous process design model simulation for the anaerobic digestion of vegetable oil wastewater

Nweke Chinenyenwa Nkeiruka*, Nwabanne Joseph Tagbo

Department of Chemical Engineering, Nnamdi Azikiwe University, PMB 5025, Awka, Nigeria

Email address:
chidarl2000@yahoo.com (Nweke C. N.), joe_nwabanne@yahoo.com (Nwabanne J. T.)

Abstract: Simulation of a mathematical model for the design of steady-state continuous reactors required for the anaerobic digestion of wastewaters was carried out. This was obtained so as to make the design of continuous stirred tank reactor (CSTR) that would be used for the anaerobic digestion of vegetable oil wastewater (VOW) possible. The model considered the kinetics of the process which has been previously determined during the batch digestion of the wastewater. The hydraulic retention time (HRT) and solid retention time (SRT) which are important parameters required for the design were estimated by varying different model parameters such as efficiency of reactor, influent substrate concentration (S_o) and biomass concentration (X) at various food-microorganism ratios (F/M). The HRT was calculated for different reactor efficiency (*E*) varying in the range of 70-95% at a constant influent substrate concentration. It was observed that the HRT increased with increase in efficiency ranging between 1.10 to 7.06 days. HRT also increased as the biomass concentration increased. SRT at constant influent substrate concentration and biomass concentration were observed to increase as the reactor efficiency increased. The effect of different substrate concentration on HRT to attain a targeted efficiency in the CSTR under steady state condition was also studied. It was observed that HRT linearly increased with increase in influent substrate concentration at constant reactor efficiency and biomass concentration. HRT also increased with decrease in biomass concentration and constant reactor efficiency for a particular substrate concentration. There was also an increase in HRT as reactor efficiency increased at constant biomass and influent substrate concentrations. SRT calculated at E1 (70%) and E2 (80%) efficiencies were observed to decrease with increase in influent substrate concentration. As the efficiencies of the reactor increased, SRT was also observed to increase. With the information obtained from the model simulation, a continuous stirred tank reactor that would treat VOW by anaerobic digestion could be designed.

Keywords: Anaerobic Digestion, Continuous Stirred Tank Reactor (CSTR), Process Design Model, Model Simulation, Vegetable Oil Wastewater (VOW)

1. Introduction

Nigeria has established industries like petroleum refinery, soap and detergent, breweries, textiles, food and beverage industries. The various effluents produced by these industries are discharged into the environment. Treatment of all kinds of wastewater before discharging into the environment is essential to prevent the pollution of water bodies where it is most likely to be discharged. These toxicans and pathogens when not sufficiently reduced cause a lot of diseases and premature deaths. Among the highly productive industries in Nigeria is the vegetable oil industry [1].

In order to obtain refined vegetable oil from seeds, many technological processes which include pre-treatment of oil seeds, manufacturing, refining and modification of oils are carried out. By-products and wastes are formed during these processes which when improperly disposed can cause pollution to the environment. These wastes include pesticides used in vegetable oil seed growing, chemical compounds like phenol, heavy metals from bleaching earth, catalysts used in the hydrogenation process, oxidizable substances and fats and oils [2].

There is growing interest in alternative renewable energy sources as the cost of available fuel rises [3]. The effect of the negative environmental impact caused by industrialization can be minimized and energy can be obtained as well by the means of anaerobic digestion. Anaerobic conversion of organic materials and pollutants is an established technology for environmental protection through the treatment of wastes and wastewater [4].The end product is biogas. This product is a useful renewable energy source as it is composed of a mixture of methane and carbon dioxide. Anaerobic digestion is a technologically simple process with a low energy requirement used to convert organic material from a wide range of wastewater types, solid wastes and biomass into methane [4]. It consists of the decomposition of organic material in the absence of free oxygen thereby producing biogas rich in methane, carbon dioxide, ammonia and traces of other gases including volatile fatty acids (VFA) within the reactor [5]. A much wider application of the technology is desirable in the current endeavors towards sustainable development and renewable energy production. Currently, the production of methane from wastes is receiving renewed attention as it can potentially reduce carbon dioxide (CO_2) emissions via the production of renewable energy and limit the emission of the greenhouse gas from especially animal manure [4]. The authors stated that this trend is supported by the growing market demand for 'green' energy and by the substantial optimization of anaerobic digestion technologies in the past decades, especially the development of modern 'high rate' and co-digestion systems. Anaerobic digestion when compared to other treatment methods have shown to be the principal method of an advanced technology for environment safeguards [6]. This is due to the fact that it handles the most important goals required in wastewater treatment which are lowering the amount of pollution generated and the reduction of dependency in fossil fuels through the production of bio-energy [5].

In choosing a reactor for wastewater treatment, certain factors are considered. Industries prefer technology that requires less capital, area and a higher reliability when compared to other options. The attractiveness of an anaerobic digestion system requires that it is able to run at high organic and hydraulic loading rates with a reduced operation and maintenance requirement [3]. The factors that affect these requirements include the amount of active biomass that can be retained by a reactor per unit volume, the contact opportunity between the retained biomass and the incoming wastewater; the diffusion of substrate within the biomass [3]; and optimum environmental conditions [5]. There are various types of reactors in use today and each design is related to the quality of the waste to be digested, with capital investment factors, and with the principal function of digestion. The different groups of reactors designs commonly used for the anaerobic digestion (AD) of wastes are batch, continuous one-stage and continuous two-stage reactors, tubular reactor, up-flow anaerobic sludge blanket (UASB), anaerobic sequencing batch reactor (ASBR), anaerobic filter (AF), and the plug flow reactor (PFR) [7]. These reactors and how they operate are also discussed in [7] and [8].

In a batch reactor, substrate is treated and discharged before the cycle is repeated [9]. The digester is filled once at the start of the process, with or without the addition of seed material, and allowed to go through all the degradation steps sequentially leading to the formation of biogas. A minimum amount of water is added and the leachate can be re-circulated to the top of the fermenting waste [10]. The liquid contents of the reactor are mixed completely during the process [9]. Anaerobic batch digestion is useful because it can be performed with simple, inexpensive equipment and with waste with total solids concentration as high as 90%, e.g, staw. It is also useful in laboratory scale studies in assessing the rate at which a material can be digested and in determining the yields of biogas obtainable. The batch reactor is a useful tool to improve the understanding of the outcome of anaerobic digestion processes [11] and is mainly used to blend chemicals or dilute concentrated chemicals [9]. The major disadvantages of batch systems are their large footprint, a possible need for a bulking agent and a lower biogas yield caused by impairment of the percolation process due to channeling or clogging due to compaction [10]. Another disadvantage of the system is that a separate influent tank and effluent tank are needed [4].

For plug flow reactors, fluid particles are passed through the reactors with little or no longitudinal mixing and leave in the same sequence in which they entered [9]. Plug flow reactor is found to be superior to the conventional processes due to low concentrations of VFA in the effluent, a high degree of sludge retention and stable reactor performance [3]. The disadvantage of this treatment process is in its restrictions due to the nature of wastewater to be treated, nature of reaction; whether homogeneous or heterogeneous, reaction kinetics for the treatment process, process performance requirements and the conditions of the environment where the treatment process would be carried out. Where the high rate hybrid filters provide high sludge retention times by means of sludge recycling which results in maximum efficiency at short hydraulic times, however, their performance with concentrated wastewater has been shown to be poor [9]. A common problem encountered in the industrial anaerobic plants is biomass washout which can be addressed by the use of membranes coupled with the anaerobic digester for biomass retention [3]. However, membrane separation has been considered for anaerobic reactors but the technology is still in a developmental stage [9].

In the continuous stirred tank reactor, it is assumed that complete mixing occurs instantaneously and uniformly throughout the reactor as fluid particles enter the reactor. Fluid particles leave the reactor in proportion to their statistical population. The time required to achieve complete mixing depends on power input and reactor geometry [9]. The content in the reactor is a mixture of fresh material and fully digested material. Examples of CSTR systems are sewage sludge digesters or well mixed anaerobic treatment pond. Some fresh water is fed into the digester on daily basis with an equal quantity of digested material being discharged. Degradation of the material is obtained when a sufficiently long hydraulic retention time (HRT) is observed [8].

According to [12], dairy wastewater has been treated using anaerobic digestion in CSTRs with the incorporation of a biofilm support in some of them. It was observed that 20%

better improvement in methane yield was obtained when compared to CSTRs operated without biofilm support systems. This showed that an improvement in the efficiency of digesters was obtained without biomass recycling. The authors also reported that anaerobic digestion of a mixture of pig manure, fish oil waste and waste from bentonite of edible oil filtration process for biogas has been carried out by the use of a laboratory scale CSTR operated at 30°C. During the experiment, an average methane content of 65% was obtained and after a HRT of 15 days, maximum methane production of 74% was obtained in the digester biogas. An investigation was carried out to obtain the continuous biogas production from fodder beet silage without manure addition using a CSTR. The content of methane in the biogas produced ranged between 62% and 64% [12]. The authors also reported an investigation on the effects of OLR, HRT and pH on conversion of food waste to hydrogen were investigated using a thermophilic CSTR. It was observed that after a 5 day HRT at a loading rate of 8g VS/L/d and a pH of 5.5, maximum operational conditions for continuous hydrogen production could be attained. Reference [6] studied the use of CSTR in the anaerobic digestion of petrochemical wastewater treatment by raising the organic loading rate. Results showed that highest COD elimination efficiency was 98%±0.5% at 7.5g COD/Ld OLR and 4d HRT, and the average methane produced was 65.49%. Anaerobic digestion of VOW in a CSTR is uncommon.

The anaerobic degradation of complex organic material involves sequence of subsequent steps which are carried out by different kinds of bacteria in a mixed population [13]. These steps or metabolic stages are explained in [5]. The performance of the bacteria in degrading the organic matter can be described by kinetics. Further, kinetics is an important tool to understand the anaerobic degradation, reactor design and operation. Its analysis is an accepted route for describing the performance of biological treatment systems and for predicting their performance. Generally, the results of kinetic studies obtained from experimental studies can be used for estimating treatment efficiencies of full scale reactors with the same operational conditions [13]. Simulation makes the design of full scale reactors easier without pre-experimentation on a pilot scale of the reactor to be designed. To ensure effective wastewater treatments in biological systems, Solids Retention Time (SRT) and Hydraulic Retention Time (HRT) need to be controlled.

Solids Retention Time (SRT) is the estimated time the activated-sludge solids are to be retained in the reactor. The growth rate of microorganism is controlled by SRT [14]. It plays a vital role in the removal of pollutants and in the reduction of quantity of unrestrained sludge. Microorganisms in the wastewater are not optimized for the current growth conditions when SRT is not controlled. One of the various methods used to make waste flow adjustments and control SRT include food-to-microorganism ratio (F/M) control also known as organic load [15]. This method requires TSS and BOD/COD estimations [14]. Hydraulic retention time (HRT) is the period in which a specific volume of liquid is retained in the working volume of a reactor. The inverse of HRT is F/M ratio [16]. The F/M ratio and the volume of the reactor determine the HRT of the wastewater treatment. The determination of required HRT will prevent unfavorable metabolic activities of the microbes and maintain a control of the biomass concentration [15, 16].

Simulation of a mathematical model in which HRT and SRT were estimated for the design of a continuous reactor for anaerobic digestion of wastes was developed by [17]. The model was developed to treat MSW in a continuous flow reactor unit under homogeneous steady state conditions. The simulation was carried out using the kinetic parameters obtained from [18] where the kinetics of batch anaerobic digestion of MSW was evaluated. The simulation model was applied in this work for the design of a CSTR for the anaerobic digestion of VOW using the kinetic parameters obtained by considering Monod's kinetics of biomass growth and substrate utilization in [19] while treating VOW in a batch anaerobic digester. CSTR has been applied to treat a lot of effluents but its use in the treatment of VOW is scarce in literature.

2. Materials and Methods

2.1. Batch Experimental Procedure

The batch digestion procedure was carried out as stated by [19]. During the batch digestion period, chemical oxygen demand (COD) and total suspended solids (TSS) were tested after every five-day interval from the point of charging to the end of the retention period. These parameters were analyzed in order to obtain some kinetic values which were evaluated from a simplistic design model applied by [18] and [20].

2.2. Kinetics of Batch Process Anaerobic Digestion

As described in [18] and [20], limited substrate consumption is a first order reaction which can be expressed as:

$$\frac{-dS}{dt} = KS \tag{1}$$

Where K = the rate constant
S = Substrate concentration
t = hydraulic retention time (HRT)

The above equation is characterized as exponential growth and the substrate concentration profile with respect to hydraulic retention time (HRT) as follows:

$$S_e = S_0 \exp(-Kt) \tag{2}$$

Where S_o is the influent substrate concentration (mg/l), S_e is the effluent substrate concentration (mg/l) and t is the hydraulic retention time (days).

Equation (2) shows the exponential growth of the organism as the substrate is utilized. Rearranging and taking natural logarithm of both sides of (2) gives (3) [13,18,20]:

$$\ln\left(\frac{S_e}{S_o}\right) = -Kt \quad (3)$$

Where K is the first order inactivation rate coefficient (l/day).

A linear plot of -ln (S_e/S_o) against t is obtained with the regression coefficient showing the order of reaction.

The rate of substrate utilization (U) is related to effluent substrate concentration (S_e) in (4) according to [18,20].

$$\frac{1}{U} = \frac{K_S}{K}\frac{1}{S} + \frac{1}{K} \quad (4)$$

From (4), a linear plot of $1/U$ against $1/S_e$ gave $1/U$ as intercept and K_s/K as slope from where K and K_s were obtained.

Where K_s = half-velocity constant/ saturation constant (mg/l)

K = maximum rate of substrate utilization (day^{-1})

$U = \frac{dS/dt}{X} =$ Specific rate of substrate utilization (mg COD/L/day)

$X =$ Average total suspended solid (biomass concentration) (mg/l)

K is used to determine the volume of biological reactors. The greater the value of 'K', smaller will be the size of the reactor [21]. K also shows whether inoculation is required for better performance [18,20]. This makes the design of bioreactors easier. K_s give an idea about the change in the specific growth rate of bacteria with a change in the concentration of the growth limiting substrate [21]. The specific rate of substrate utilization is related to mean cell residence time [18,20] in (5)

$$\frac{1}{\theta} = YU - K_d \quad (5)$$

Where $\theta = \frac{X}{dX/dt} =$ the mean cell residence time (day) = SRT = sludge age

U = the specific rate of substrate utilization

Y = the biomass yield/microbial growth yield (mg/mg)

K_d = endogenous decay coefficient (day^{-1})

From (5), biomass yield (Y) and endogenous decay coefficient (K_d) were obtained from the plot of $1/\theta$ versus U. Y is used to estimate the total amount of sludge produced as a result of wastewater treatment. K_d is used to determine the net amount of sludge to be handled. Hence the size and cost of the sludge handling facilities can be determined from this information since the higher the value of Y and K_d, the smaller the sludge production and sludge handling facilities [21].

The maximum specific growth rate of microorganism, μ_{max} is evaluated from (6).

$$K = \frac{\mu_{max}}{Y} \quad (6)$$

Where K = maximum rate of substrate utilization (day^{-1})

μ_{max} = maximum specific growth rate of microorganisms (day^{-1})

Also from [22] and the Monod equation,

$$\mu_{net} = \mu_{max}\frac{S}{K_s + s}[X] - K_d[X] \quad (7)$$

Dividing through by X,

$$\mu_{net} = \mu_{max}\frac{S}{K_s + s} - K_d \quad (8)$$

μ_{net} = net specific growth rate of microorganisms (day^{-1})

2.3. Continuous Process Design Model for Anaerobic Digestion

Reference [17] proposed a kinetic model considering the case of continuous mode homogeneous reactor under steady state condition.

The HRT at any desired efficiency, E (in fraction) of the reactor and food to microorganism can be evaluated from eq. (9) can be expressed as

$$t = \frac{ES_O[K_S + S_O(1-E)]}{KXS_O(1-E)} \quad (9)$$

where $X = F/M$ (from M = 0.7 – 1.0) = S_o/M

F/M = food to microorganism ratio

As soon as the hydraulic retention time is evaluated, the volume of the reactor can be determined as follows

$$V = Qt \quad (10)$$

Where V = volume of the reactor (m^3)

Q = the desired flow rate (m^3/day)

The solid retention time, SRT (days) required for the anaerobic digestion system, can be determined as

$$\theta = \frac{X}{r_g - K_d X} = \frac{X}{-\frac{YdS}{dt} - K_d X} \quad (11)$$

Where r_g = volumetric cell production rate (gVSSm^{-3}day^{-1})

X = F/M ratio

Again, according to [17] and [23], the continuous reactor under steady state condition, (11) can be modified as

$$\theta = \frac{X}{Y\frac{S_0 - S}{t}K_d X} = \frac{X}{-Y\frac{ES_0}{t} - K_d X} \quad (12)$$

Table 1. *Kinetic parameters used for the study*

Parameters	Values
K	0.8215day^{-1}
K_s	163.31mg/l
K_d	0.0838day^{-1}
Y	0.2813mg/mg

The retention period (hydraulic retention time, HRT, days and solid retention time, SRT, days) for the continuous reactor is determined after K_s, K, S_e and S_o have been evaluated from batch experiment. The kinetic parameters obtained from [19] after the batch digestion of VOW is shown in Table 1.

2.4. Simulation Methodology

The important CSTR process design parameters for the anaerobic digestion of VOW are influent and effluent substrate concentration, biomass concentration, hydraulic retention time (HRT), solid retention time (SRT) and the kinetic constants [17]. The HRT and SRT are obtained from (9) and (12). Since the vegetable oil wastewater being considered for treatment is that analyzed by [19] from batch digestion of VOW, hence the kinetic parameters, the substrate concentrations and biomass concentration are already known.

In the present study, the biodegradable organics were represented as influent COD (640mg/l) which was initially analyzed by [19]. Hence, the values of HRT and SRT which are required in the design of a CSTR for the anaerobic digestion of VOW can be obtained.

3. Results and Discussion

For the design of a steady state continuous stirred tank reactor (CSTR) that would be used to treat VOW wastewater by anaerobic digestion, the hydraulic retention time (HRT) at which the reactor would operate is required. The model simulation was evaluated for different reactor efficiencies (*E*) of 70, 75, 80, 85, 90 and 95% with F/M ratios of 0.7, 0.75, 0.8, 0.85, 0.9, 0.95 and 1.0 respectively. These were considered at a constant influent COD of 640mg/l. The biomass concentrations calculated for F/M ratios between 0.7-1.0 were determined to be 914.3, 853.3, 800, 753, 711.7, 673.7 and 640mg/l respectively. It was observed that the biomass concentration decreased as the F/M ratio increased, although the rate at which the biomass concentration decreased was not proportional. The biomass concentration decreased by 61, 53.3, 47, 41.9, 37.8, 33.3, mg/l as the F/M ratio increased from 0.7-1.0 respectively. This showed that the rate at which the biomass concentration decreased became higher as microorganisms increased. The variation of HRT for the various efficiencies of anaerobic digestion of VOW is shown graphically in Fig. 1 and the values are given in Table 2. It was observed that the HRT increased with increase in the efficiency of the reactor. This was observed to have a hyperbolic curve. Also, HRT increased with increase in F/M ratio. For the reactor to operate at 95% efficiency, HRT of 7 days at F/M ratio of 1.0 is required. This is lesser than the time applied in the batch treatment of VOW [19]. Hence treatment using CSTR is less time consuming therefore less expensive.

The SRT required for the anaerobic digestion of VOW in a CSTR under steady state was calculated for different removal efficiencies (*E*) of COD varying in the range of 70, 75, 80, 85, 90 % at a constant influent COD of 640mg/l and F/M ratio of 0.7. The biomass concentration was not varied while varying efficiency at a fixed influent COD for the determination of SRT as there was no effect in the values obtained [17]. The calculated values of SRTs at different treatment efficiencies are represented graphically in Fig. 2 and the values are given in Table 2. The SRT was observed to increase as the digestion efficiency increased.

The HRT required for targeted treatment efficiency at various substrate concentrations of the wastewater under steady state condition was determined. The influent substrate concentration was varied between 530.28-914.3 mg/l. The biomass concentration was taken as 914.3 and 640 mg/l corresponding to F/M ratio of 0.8 and 1.0 respectively. The HRT was calculated for 70, 80 and 90% reactor efficiencies in response to F/M ratio of 0.8 and 1.0. It was observed that the HRT required for anaerobic digestion of the wastewater increased with increase in influent substrate concentration at steady reactor efficiency and biomass concentration. HRT also increased as biomass concentration decreased for particular reactor efficiency. Again, at a constant biomass concentration, the HRT would increase if the reactor efficiency is increased. The variation of HRT as influent substrate concentration increased has been shown in Fig. 3 and the values are provided in Table 3.

The effect of different substrate concentration on SRT value to attain a targeted efficiency under steady state condition was also studied on VOW digestion. The influent concentration was varied in the range of 530.28-914.3 mg/l. The biomass concentration was taken as 914.3 and 640 mg/l corresponding to F/M ratio of 0.8 and 1.0. The SRT was calculated for 70 and 80% reactor efficiency in response to F/M ratio of 0.8 and 1.0 respectively. The variation of SRT at different reactor efficiencies as influent substrate concentration increased is shown in Fig. 4and Table 3. It was observed that SRT decreased with increase in influent substrate concentration. Increasing the reactor efficiency at a constant substrate concentration however brought about an increase in SRT. The profile of SRT also indicated a hyperbolic nature and does not dependent on the biomass concentration [17].

Table 2. *Variation of HRT and SRT with variation of efficiency of anaerobic digestion of vegetable oil wastewater*

Efficiency (%)	Influent COD, S_0 (mg/l)	Biomass Concentration X (mg/l)							HRT (day)							SRT (day)
		F/M =0.7	F/M =0.75	F/M =0.8	F/M =0.85	F/M =0.9	F/M =0.95	F/M =1.0	F/M =0.7	F/M =0.75	F/M =0.8	F/M =0.85	F/M =0.9	F/M =0.95	F/M =1.0	
70	640	914.3	853.3	800	753	711.1	673.7	640	1.10	1.18	1.26	1.34	1.42	1.50	1.58	10.53
75	640	914.3	853.3	800	753	711.1	673.7	640	1.29	1.38	1.48	1.57	1.66	1.75	1.84	11.67
80	640	914.3	853.3	800	753	711.1	673.7	640	1.55	1.66	1.77	1.88	1.99	2.11	2.22	13.39
85	640	914.3	853.3	800	753	711.1	673.7	640	1.96	2.10	2.24	2.38	2.52	2.66	2.79	16.45
90	640	914.3	853.3	800	753	711.1	673.7	640	2.72	2.91	3.11	3.31	3.50	3.70	3.89	23.01
95	640	914.3	853.3	800	753	711.1	673.7	640	4.94	5.29	5.65	6.00	6.35	6.71	7.06	49.74

Table 3. Variation of HRT and SRT with variation of influent substrate concentration of anaerobic digestion of VOW

Influent COD, S_0	Reactor Efficiency (%)			Biomass Concentration, X (mg/l)		HRT (day)						SRT (day)	
(mg/l)	E1	E2	E3	F/M=0.8X1	F/M=1.0X2	E1 +X1	E1 + X2	E2 + X1	E2 + X2	E3 + X1	E3 + X2	E1	E2
530.28	70	80	90	914.3	640	1.00	1.43	1.44	2.05	2.59	3.70	17.84	23.71
585.14	70	80	90	914.3	640	1.05	1.50	1.49	2.13	2.66	3.80	16.78	21.95
640.0	70	80	90	914.3	640	1.10	1.58	1.55	2.22	2.72	3.89	16.04	20.65
694.86	70	80	90	914.3	640	1.15	1.65	1.61	2.30	2.79	3.99	15.31	19.48
749.72	70	80	90	914.3	640	1.21	1.72	1.67	2.38	2.86	4.08	14.70	18.50
804.58	70	80	90	914.3	640	1.26	1.80	1.73	2.47	2.92	4.17	14.27	17.76
859.44	70	80	90	914.3	640	1.31	1.87	1.95	2.55	2.99	4.27	13.81	17.04
914.3	70	80	90	914.3	640	1.36	1.94	1.84	2.63	3.05	4.36	13.41	16.41

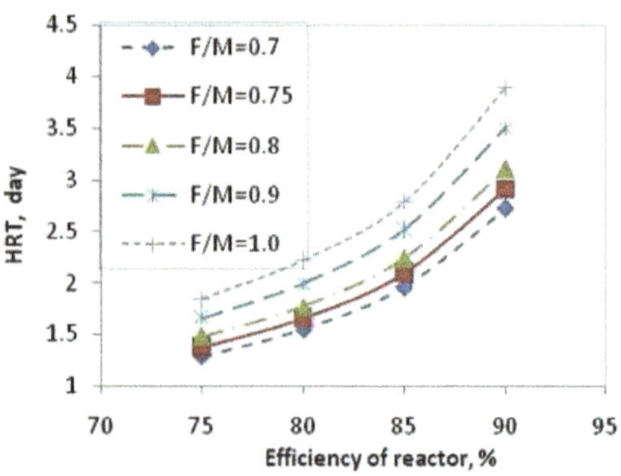

Fig 1. Variation of HRT with efficiency of anaerobic digestion of VOW

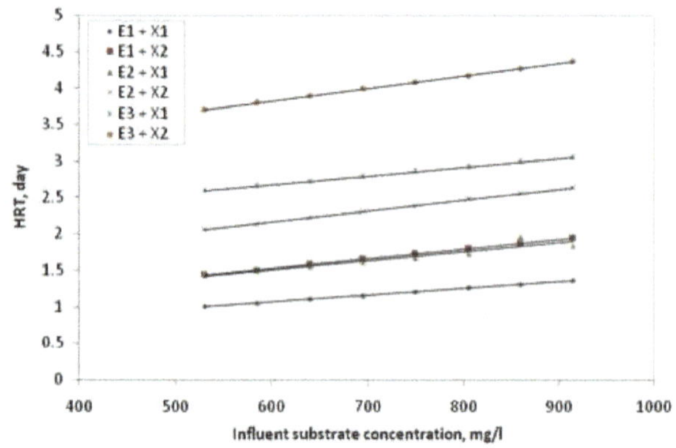

Fig 2. Variation of SRT with efficiency of anaerobic digestion of VOW

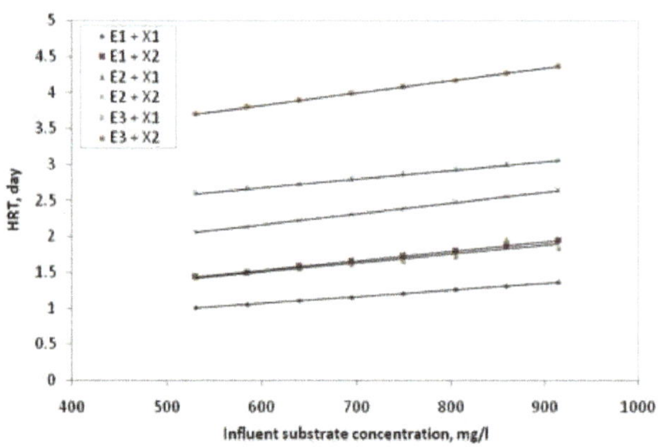

Fig 3. Variation of HRT of anaerobic reactor for VOW digestion with influent substrate concentration (E1=70%, E2=80%, E3=90%, XI=914.3mg/l, X2=640mg/l)

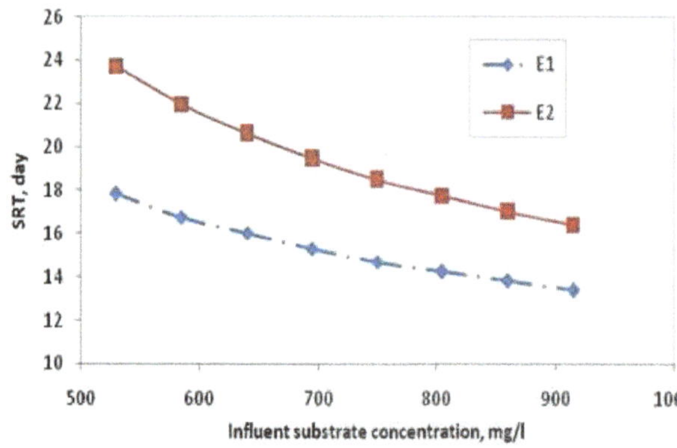

Fig 4. Variation of SRT of anaerobic reactor for VOW digestion with influent substrate concentration

4. Conclusion

The kinetic parameters obtained from the batch process anaerobic digestion of VOW were used to simulate a model for the design of continuous reactors for the same treatment method and wastewater. This was done by obtaining the HRT and SRT that would be required for the treatment at any given reactor efficiency. With these data, the volume and flow rate of the reactor can be calculated. While varying the HRT and SRT with variation of efficiency, the reactor operation had an upward shift as the efficiency increased from 70-95%. Variation of HRT and SRT with influent substrate concentration for continuous process indicated that the HRT increased with increase in varying substrate concentration. The SRT on the other hand decreased as influent substrate concentration increased in a hyperbolic pattern. It could be concluded that a simplistic process design model for anaerobic digestion of VOW is possible to be developed. The developed mathematical model can be employed to predict the process control parameters like the HRT and SRT of both batch and continuous flow type homogenous reactors under steady state condition.

Nomenclature

AD	Anaerobic Digestion
AF	Anaerobic Filter
ASBR	Anaerobic Sequencing Batch Reactor
BOD	Biological Oxygen Demand
COD	Chemical Oxygen Demand
COD/Ld	Chemical Oxygen Demand per litre per day
CO_2	Carbon dioxide

CSTR	Continuous Stirred Tank Reactor
E	Efficiency (in fraction) of the reactor
F/M	Food to Microorganism ratio
HRT	Hydraulic Retention Time (day)
K	First Order Inactivation Rate Coefficient (l/day)
K	Maximum Rate of Substrate Utilization (day^{-1})
K_d	Endogenous Decay Coefficient (day^{-1})
K_s	Half-velocity constant, mg/l
MSW	Municipal Solid Waste
OLR	Organic Loading Rate
PFR	Plug Flow Reactor
Q	Flowrate of the Reactor (m^3/day),
r_g	Volumetric cell production rate ($gVSSm^{-3}day^{-1}$)
S_0	Initial value of COD before the onset of the experiment (mg/l)
S_e	COD value after every five days on charging the digester (mg/l)
SRT	Solid Retention Time (day)
t	Time for batch digestion, day
t	Hydraulic Retention Time (day) for continuous reactor
t_c	Solid Retention Time (day) for the continuous reactor
TSS	Total Suspended Solid
$U = \frac{ds/dt}{x}$	Rate of Substrate Utilization (mg COD/L/day)
V	Volume of the Reactor (m^3)
VFA	Volatile Fatty Acids
VOW	Vegetable Oil Wastewater
VS/L/d	Volatile Solids per litre day
WHO	World Health Organization
X	Average Total Suspended Solid (biomass concentration) (mg/l)
Y	Biomass yield (mg/mg)
$\theta = \frac{x}{dx/dt}$	Mean cell residence time (day)
μ_{max}	Maximum Specific Growth Rate of Microorganisms (day^{-1})
μ_{net}	Net Specific Growth Rate of Microorganisms (day^{-1})

References

[1] Olaniyi, I., Odoh, R. and Nwadiogbu, J.O. (2012) Effect of Industrial Effluent on the Surrounding Environment. Available at Scholars Research Library in Archives of Applied Science Research, 2012, 4(1):406-413.

[2] Chipasa, K.B. (2001) Limits of Physicochemical Treatment of Wastewater in the Vegetable Oil Refining Industry. Polish Journal of Environmental Studies. 10(3):141-147.

[3] Rajeshwari, K.V., Balakrishnan, M., Kansal, A., Lata, K. and Kishore, V.V.N. (2000) Renewable and Sustainable Energy Reviews, 4(2000): 135-156.

[4] De Mez, T.Z.D., Stams, A.J.M., Reith, J.H. and Zeeman, G. (2003) Methane Production by Anaerobic Digestion of Wastewater and Solid Wastes. Dutch Biological Hydrogen Foundation, C/O Energy Research Centre of the Netherlands.

[5] Randa, M.O. (2014) Anaerobic Fermentation of Industrial Wastewater (Review Article), Open Journal of Chemical Engineering and Science, 1(1):50-78.

[6] Siddique, M.N.I. and Zularisam, A.W. (2012) Continuous Stirred Tank Reactor Application for Renewable Methane from Anaerobic Biodegradation of Petrochemical Wastewater. Engineering e-Transaction, 7(1):23-27.

[7] Nasir, I., Ghazi, T.I.M. and Omar, R. (2012) Anaerobic Digestion Technology in Livestock Manure Treatment for biogas production: A Review. Engineering Life Sciences, 12(3):258-269.

[8] Stewart, S.W. (2004) Co-Generation Opportunities Utilizing Sugar Industry Wastewater Through the Use of Biological Treatment Systems. Electricity Supply Industry in Transition: Issues and Prospect for Asia, 14-16:43-51.

[9] Krishna, R.H. (2013) Review of Research on Bio Reactors used in Wastewater Treatment for production of Bio Hydrogen: Future Fuel. International Journal of Science Inventions Today, 2(4):302-310.

[10] Parawira, W. (2004) Anaerobic Treatment of Agricultural Residues and Wastewater: Application of High-Rate Reactors. Department of Biotechnology, Lund University, Sweden.

[11] Pandey, K.P. and Soupir, L.M. (2011) Escherichia Coli Inactivation Kinetics in Anaerobic Digestion of Dairy Manure under Moderate, Mesophilic and Thermophilic Temperatures. Springopen Journal: AMB Express, 1(1):18.

[12] Demirel, B., Scherer P., Yenigun, O. and Onay, T.T. (2010) Production of Methane and Hydrogen from Biomass through Conventional and High-Rate Anaerobic Digestion Processes. Critical Reviews in Environmental Science and Technology, 40:116-146.

[13] Mullai, P., Huu-Hao, N., Sabarathinam, P. (2011) Substrate Removal Kinetics of an Anaerobic Hybrid Reactor Treating Pharmaceutical Wastewater. Journal of Water Sustainability. 1(3):301-312.

[14] Greenwood, S., Anderson, C., Rieth, M. and Stein, T. (2002) A Constant SRT Calculated from Liquid Flows Improves BNR Activated Sludge Performance. Central States Waster Environment Association, Illinois. www.cswea.org/papers/2002Radebaugh.pfd

[15] Aline, F.V., Geraldo, F. S Jr. (2008) Is Hydraulic Retention Time an Essential Parameter for MBR Performance? Elsevier Journal of Hazardous Materials, 150(1):185-186.

[16] Parviz, M., Shaliza, I., Mohamad, S. M. A., Shahin, G., Sabaratnam, V., Ali, A. Z. (2012) Influences of Environmental and Operational Factors on Dark Fermentative Hydrogen Production: A Review. Clean-Soil, Air, Water, 40(11):1297-1305.

[17] Asok, A., Debabrata M. and Pratip, B. (2011) Simulation of a Process Design Model for Anaerobic Digestion of Municipal Solid Wastes. International Journal of Civil and Environmental Engineering. (3)3:177-182.

[18] Nwabanne, J.T., Onukwuli, O.D. and Ifeakandu, C.M. (2009) Biokinetics of Anaerobic Digestion of Municipal Waste. International Journal of Environmental Resources. 3(4):551-556.

[19] Nweke, C.N., Nwabanne, J.T. and Igbokwe, P.K. (2014) Kinetics of Batch Anaerobic Digestion of Vegetable Oil Wastewater. In press.

[20] Nwabanne, J.T., Okoye, A.C. and Ezedinma, H.C. (2012) Kinetics of Anaerobic Digestion of Palm Oil Mill Effluent. Canadian Journal of Pure & Applied Sciences. 6(1):1877-1881.

[21] Haydar, S. and Aziz, J.A. (2009) Kinetic Coefficients for the Biological Treatment of Tannery Wastewater Using Activated Sludge Process, Pak. J. Engg. & Appl. Sci. 5:39-43.

[22] Igoni, A.H., Abowei, M.F.N., Ayotamuno, M.J. and Eze, C.L (2008) Comparative Evaluation of Batch and Continuous Anaerobic Digesters in Biogas Production from Municipal Solid Waste, Agricultural Engineering International: CIGR EJOURNAL. Volume X:1-12.

[23] Maharajh, N. (2010) Effect of Feed Rate and Solid Retention Time (SRT) on Effluent Quality and Sludge Characteristics in Activated Sludge Systems using Sequencing Batch Reactors, Environmental Science and Engineering, State University, Blacksburg, Virginia.

Effect of coffee processing plant effluent on the physicochemical properties of receiving water bodies, Jimma zone Ethiopia

Dejen Yemane Tekle[1,*], Abebe Beyene Hailu[2], Taffere Addis Wassie[3], Azeb Gebresilassie Tesema[1]

[1]Mekelle University, College of Health Science, Department of Public Health, Mekelle, Ethiopia
[2]Jimma University, College of Public Health and Medical Sciences, Department of Environmental Health Science and Technology, Jimma, Ethiopia
[3]Addis Ababa University, Ethiopian Institute of Water Resource, Addis Ababa, Ethiopia

Email address:
yemaned@yahoo.com (D. Y. Tekle), abebe.beyene@ju.edu.et (A. B. Hailu), taffere.addis@yahoo.com (T. A. Wassie), azeb18@gmail.com (A. G. Tesema)

Abstract: Although the coffee wastewater emanating from the traditional coffee processing plants in Jimma zone is a valuable resource, it is disposed off to the nearby water course without any treatment. As a result, it becomes a severe threat to the aquatic ecosystem and downstream users. To tackle this problem, understanding the nature of the coffee processing wastewater is fundamental for the design and operation of appropriate and effective treatment technologies. Thus, the main objective of this study was to assess the effect of coffee processing plant effluent on the physicochemical properties of receiving water bodies of Jimma zone Ethiopia. Based on the results of the physicochemical parameters, it was proved that the coffee effluent has a remarkable polluting potential during the wet coffee-processing season. The concentrations of the physicochemical parameters were significantly ($p<0.05$) increased following effluent discharge except TSS and temperature, when downstream or impacted (L) compared with upstream or non-impacted (U) sites. If business-as-usual scenario is followed, the economic gains accrued as a result of coffee export will be worthless due to the alarming water quality degradation and aquatic ecosystem disturbance. Therefore, urgent intervention in the area of coffee factory for effluent management options should be dealt with top priority to avoid further needless damage to the environment.

Keywords: Coffee Processing Plant, Coffee Wastewater, Physicochemical Properties, Water Bodies, Jimma Zone

1. Introduction

Environmental pollution because of improper waste management is an alarming challenge for developing countries to meet the millennium development goals. Several studies reported that untreated waste from traditional and modern industries is threatening surface waters worldwide, and it is severe in developing countries (1). Based on the type of industry, various levels and quantity of pollutants can be discharged into the environment directly or indirectly.

Jimma zone the major and famous coffee producer region in Ethiopia; have a number of small-scale coffee pulping industries situated along the banks of rivers and/or streams with a varying degree of hydraulic gradients, due to the fact that washed coffee requires considerable amount of water during processing for receiving the cherries, transporting them hydraulically through the pulping machine, removing the pulp, and sorting and re-passing any cherries with residual pulp adhering to them (2). The rise in the number of wet processing plants has therefore resulted in the generation of large amounts of processing by-products mainly coffee pulp and effluents which are discharged unwisely into nearby natural water way which flow into rivers and/or infiltrate into ground water and become main threat to the surface and ground water qualities.

Coffee processing industries produces very high pollution load wastewater because coffee processing industries are one of the significant consumers of water and produces large amount of wastewater that contains high concentrations of organic matter, nutrients, suspended matter and highly acidic wastewater. Thus, considering the volume generated and the pollutants released through the wastewaters, the coffee processing industry represents one of the main contributors to the severe pollution problems and it is reported that they don't have any effluent treatment plants. They directly discharge untreated colored and acidic effluent into the nearby water bodies, streams and open land (3, 4).

Legesse *et al.* (2003), reported that the Pollution problems from coffee processing emanates from the fact that rivers and streams are location factors for coffee factories and effluent disposal pits are constructed very close to watercourses mostly on sloppy grounds. This represents a potential threat to the ecosystem since the effluent discharged from the coffee factories can easily reach the rivers when there is overloading of seepage pits during the peak season or run off during the heavy rains (5).

According to Kansal et al. (1998) these discharges are of environmental concern due to the decrease in the dissolved oxygen concentration in the water which brings great impact over the biotic environment, and can be fatal to fish and other aquatic animals, as well as originate odoriferous products resulting from an anaerobic processes and decreased suitability of surface water for different purposes (6).

Few if any studies have investigated this issue in Ethiopia to assess the extent of the problem and to suggest solutions and recommendations accordingly. Therefore, it is with this ground that this study will be very important to characterize the coffee processing wastewater and assess the effect of coffee processing plant effluent on the physicochemical properties of receiving water bodies of Jimma zone Ethiopia.

2. Materials and Methods

2.1. Study Area

The study was conducted in Gomma and Mana districts of Jimma Zone (Figure 1). Jimma zone is located 390 km southwest of Addis Ababa the capital city of Ethiopia, lying between Latitude of 6^0 and 9^0 North and Longitude of 34^0 and 38^0 East. The altitude ranges from 880m to 3340m above sea level.

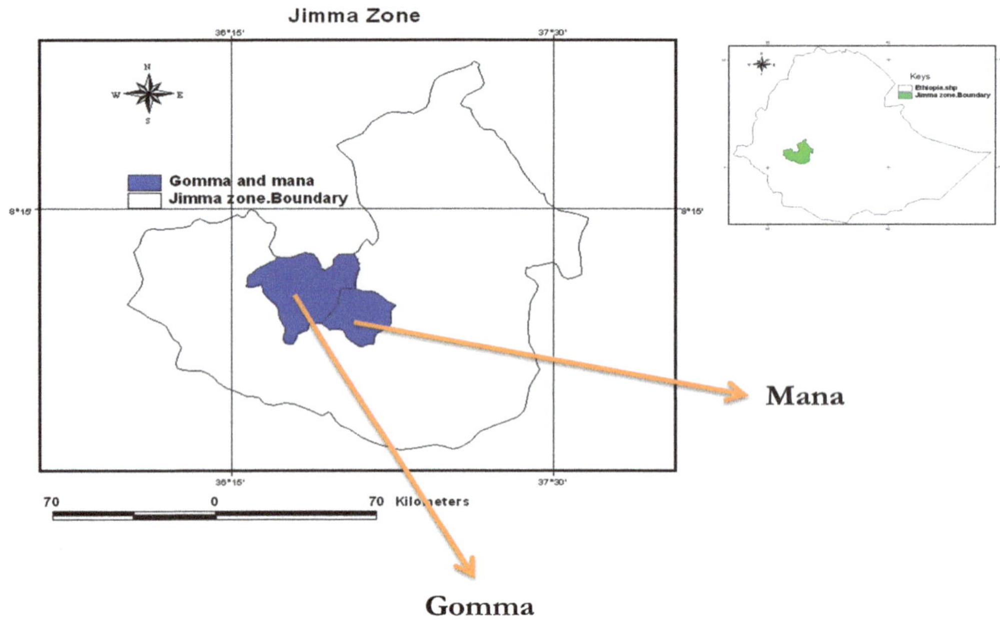

Figure 1. *Digitized map showing sampling sites in manna and gomma districts, Jimma zone (7).*

2.2. Wastewater Sampling and Sample Analysis

Wet coffee processing in Jimma Zone usually begins at the end of August and proceeds until mid-December. Consequently, we collected samples in November 2010 (the month of the year for peak wet coffee processing). Although all the coffee processing plants are following the same wet coffee processing method and expected to produce wastewater of similar composition, we randomly selected six coffee processing plants to make the samples more representative.

In order to assess the influence of coffee processing wastewater being discharged, water samples were taken from the rivers that receive wastewater from coffee processing at upstream and downstream of the discharge points (minimum of 100 m above and billow the coffee processing plant). Samples were also taken from the effluent released from the coffee processing plant for wastewater characterization. In order to have a representative composite sample, the wastewater samples were collected at the peak hours of coffee processing for three days of a week from the chosen sampling points. All samples were collected using polyethylene plastic bottle by the principal investigator.

On-site measurements were taken for pH, temperature, turbidity and electric conductivity using digital-portable probes. The collected samples for BOD_5, COD, TN, NH_4^+, NO_3^-, TN, PO_4^{3-}, TP, VSS, TDS and TSS were kept in thoroughly washed plastic containers to minimize contamination. These samples were properly and carefully labeled, sealed and transported to the laboratory of the Department of Environmental Health Sciences, Jimma University. Cold storage was maintained throughout the process till analysis.

2.3. Data Quality Management

Standard methods were applied for all procedure of the set of experiments (8). All the chemical reagents used are of analytical grade and their expiry dates were checked. For each test, triplicate sample analysis was made to maintain accuracy. Blank and control experiments were run. The results of all tests were honestly and cautiously recorded on a prepared data registration format.

2.4. Data Analysis and Presentation

The duplicate data that were collected for analysis from different sites of the coffee processing industry for characterization, were recorded, organized and summarized in descriptive statistics methods using Microsoft Excel program and Statistica version 7.0 software packages. Since environmental variables are not normally distributed, non-parametric Mann–Whitney U test and krusical-wallis ANOVA test were done and the results were presented in tables (mean value and percentage), graphs and findings of the study were discussed and compared with available literatures.

2.5. Ethical Consideration

Ethical clearance was obtained from the Ethical Committee of the College of Public Health and Medical Sciences, Jimma University. The laboratory analysis was followed scientific procedures and the results were recorded honestly in data collection formats. In addition Authors of books and journals that are used were cited properly. Scholars, individuals and organizations contributed for the successful completion of this study were acknowledged.

3. Results

3.1. Physicochemical Characterization of Coffee Processing Wastewater

A detailed characterization of coffee processing plant wastewater and downstream water samples were carried out to determine the downstream pollution loads on the nearby rivers. The results obtained from the wastewater streams revealed that there were variations in physicochemical parameters along the course of the rivers and among the rivers following the effluent discharge (Table 1) and indicating the increasing impact of the effluents on the nearby downstream water body.

Table 1. Physico-chemical characteristics of water samples of rivers at the upstream (U) and downstream (L) from the coffee processing plant discharge points (units in mg/l or otherwise stated).

Parameters	Sampling Site (mean values)											
	Yebu		Urgessa		Chiseche		Sunde		Gurracho		Funtule	
	U	L	U	L	U	L	U	L	U	L	U	L
BOD_5	12.6	1241	11.5	2967	18	1739	1.9	128	9.2	601	4.7	379
COD	29.5	2841	123	3244	28	1935	74	849	18.5	831.2	32.7	1247
NH_4^+	0.80	81.3	3.25	129.3	0.44	57.87	0.76	8.55	0.18	5.54	0.32	16.45
TDS	246	3051	113	9034	233	2147	73	709	118	1148	135	349
TSS	55	789	73	1105	85	2504	18	47	31	57	26	78
NO_3^-	1.52	11.9	0.96	13.2	0.37	8.86	3.07	19.3	2.97	27.1	1.1	32.7
PO_4^{-3}	0.08	12.1	5.72	18.5	1.4	13.12	0.59	7.6	0.26	10.84	0.05	9
VSS	62	753	41	867	47	605	16	294	23	92	26	561
TP	0.37	17.9	1.35	11.3	1.02	12.9	0.75	9.2	1.2	8	0.63	3.2
TN	30	101	31	155	12.9	97	7.8	63	10.2	78	13	84
Temp. (^{0}C)	17.9	21.8	15.46	23.1	19.3	18.67	17.5	20.01	19.4	18.3	16.6	19.9
pH	6.93	3.14	7.71	2.51	6.59	3.73	7.06	5.8	7.58	4.37	6.77	4.24
EC (µS/cm)	312	1240	110	930	45	1106	90	520	70	197	160	710
Turb.(NTU)	36	279	19	347	25	741	12	183	11	253	7	68

Higher concentrations were observed at downstream sites than upstream sites based on the values presented in Table 1. However, Unlike the other parameters, the amount of pH was found to be high in the upstream (7.11) and reduced pH values were recorded at downstream locations of most rivers (as low as 3.24). The lower site of the rivers, which are impacted by discharge of untreated coffee effluents, like Yebu, Urgessa, and Chiseche were very acidic (Table 1). The lowest level recorded (2.51) was at the lower course of Urgessa River while the upper course of most of the rivers was above 6.59 (Table 1). Similar analysis was conducted using the non parametric Mann–Whitney U test to verify the difference between coffee processing plants, but no significant difference were found between the sampling sites($p>0.05$).

From the Mann-Whitney U Test all the physicochemical parameters were altered significantly by the discharge of untreated coffee waste except TSS and Temperature (Table 2).

Figure 2 above compares the value of water quality parameters using box and whisker plots. During the study period the physicochemical parameters displayed significant variation at locations above and below effluent discharge points. Mann–Whitney U test, a powerful non-parametric test to verify the significant difference of the water quality parameters between the two groups of upper and lower sampling sites of the coffee processing plants show only TSS and temperature were found not significantly different ($p > 0.05$) between the two sample groups (impacted (L) versus non-impacted (U)) see Table 2.

The characteristics of coffee effluent taken at the discharge points just before its entrance into the effluent disposal pits, from six pulping stations, revealed that the coffee effluent on average consists of very high amount of BOD_5, COD, NH_4^+, PO_4^-, TP, TN, NO_3, TSS, TDS and VSS (Tables 3). However, measurements of these physicochemical parameters after getting into the river bodies were decreased (Table 1).

The pollution profiles for COD along the water bodies downstream ranged from (831.2-3244mg/l) and BOD_5 concentration (128-2967 mg/l), respectively (Table 1). Compared to maximum concentration at the discharge points (8079 mg/l COD and 5861 mg/l BOD_5) (Tables 3), the passage of the effluent through the disposal pits and due to the dilution of river water was found to reduce chemical and biological oxygen demanding substances.

Table 2. *The immediate impact of coffee waste on the physicochemical parameters of the nearby rivers as it is compared upstream (U) with lower stream (L).*

Mann-Whitney U Test (Spreadsheet1), By variable Parameter, Marked tests are significant at p <.05000

Parameters	Rank Sum (U)	Rank Sum (L)	U	Z	P-level	Z adjested	P-level	Valid N (U)	Valid N (L)	2*1sided exact p
BOD_5	21.00000	57.00000	0.000000	-2.80224	0.005075	-2.80224	0.005075	6	6	0.002165
COD	21.00000	57.00000	0.000000	-2.80224	0.005075	-2.80224	0.005075	6	6	0.002165
NH_4^+	21.00000	57.00000	0.000000	-2.80224	0.005075	-2.80224	0.005075	6	6	0.002165
TDS	21.00000	57.00000	0.000000	-2.80224	0.005075	-2.80224	0.005075	6	6	0.002165
TSS	27.00000	51.00000	6.000000	-1.84147	0.065553	-1.84147	0.065553	6	6	0.064935
NO_3^-	21.00000	57.00000	0.000000	-2.80224	0.005075	-2.80224	0.005075	6	6	0.002165
PO_4^{-3}	21.00000	57.00000	0.000000	-2.80224	0.005075	-2.80224	0.005075	6	6	0.002165
VSS	21.00000	57.00000	0.000000	-2.80224	0.005075	-2.80224	0.005075	6	6	0.002165
TP	21.00000	57.00000	0.000000	-2.80224	0.005075	-2.80224	0.005075	6	6	0.002165
TN	21.00000	57.00000	0.000000	-2.80224	0.005075	-2.80224	0.005075	6	6	0.002165
Temp.	27.00000	51.00000	6.000000	-1.84147	0.065553	-1.84147	0.065553	6	6	0.064935
pH	57.00000	21.00000	0.000000	-2.80224	0.005075	-2.80224	0.005075	6	6	0.002165
EC	22.00000	56.00000	1.000000	-2.64211	0.008239	-2.64211	0.008239	6	6	0.004329
Turbidity	21.00000	57.00000	0.000000	-2.80224	0.005075	-2.80224	0.005075	6	6	0.002165

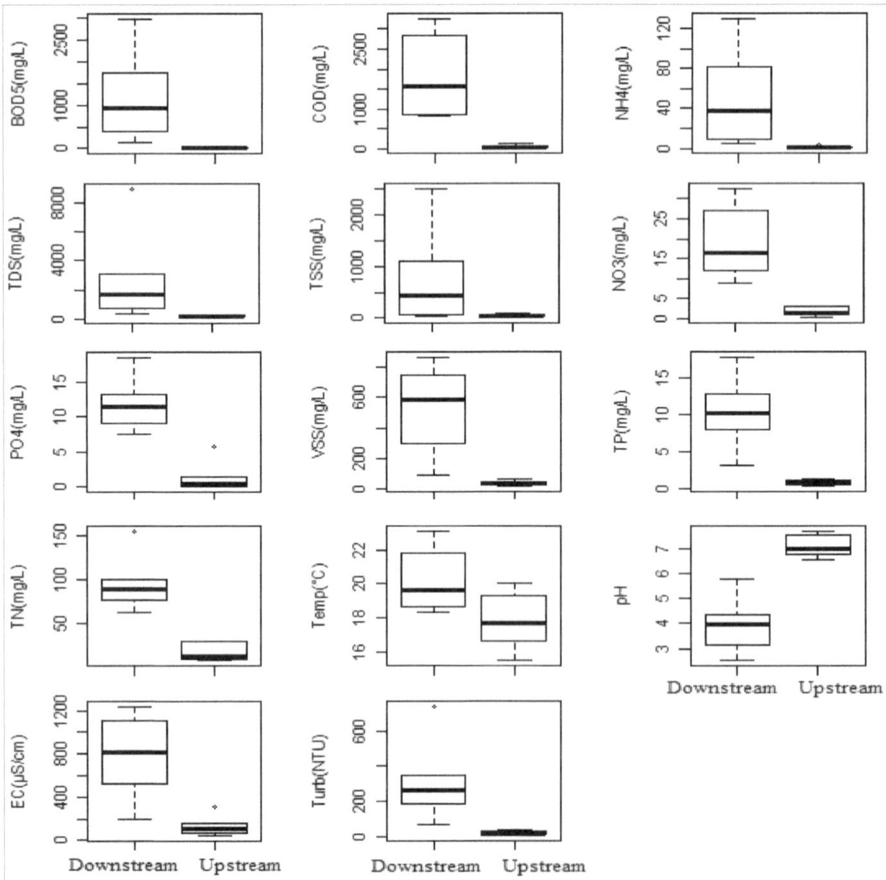

Figure 2. *Box and whisker plot (using median and standard deviation) of water quality variables of the river at the upstream and downstream of the coffee processing plants.*

Table 3. *A profile showing average characteristics of the physico-chemical parameters of samples at discharge points of coffee processing plants (Units in mg/l or otherwise stated).*

Parameter	Sampling sites (mean value)					
	Asefa B.	Bedilalo A.	Mulege	Haji A.	Getu	Bedeyu
BOD_5	4820	8242	3367	7391	4571	6104
COD	7903	10,524	6685	8909	7427	9082
NH_4^+	340.8	209.62	253.34	238.8	272.63	192
TDS	8638	5666	5808	4768	7503	4765
TSS	4452	950	1939	3492	5434	4217
NO_3^-	95.6	59.49	38.72	51.65	81.07	44.83
PO_4^{3-}	22.69	17.45	12.84	9.67	20.1	23.43
VSS	1415	689	1109	914	1084	1763
TP	11.3	5.84	16.5	4.41	21.7	8.16
TN	186	144.5	111	152	210	193
Temp.(^{0}C)	22.1	19.7	21.5	18.5	19.3	19.6
pH	3.1	5.15	3.78	3.96	4.22	3.57
EC(µS/cm)	1008	1852	3137	1813	1706	3799
Turb.(NTU)	573	430	568	531	712	662

4. Discussion

The coffee processing industries uses large quantity of water (an average of 147 m^3 /day) for pulping, fermentation and washing of the coffee cherry with no recirculation. Consequently, the wet coffee processing stations are generating large amounts of high strength wastewater and discharge directly into the pits that are intended to serve as wastewater stabilization are neither properly constructed nor the right dimension to accommodate the generated waste during peak processing time leading to overflow of raw effluents into natural watercourses and aggravate river pollution of the area. The non parametric Kursical wallis test to verify the difference of effluent discharged between coffee processing plants also indicated the coffee processing industries discharge homogenous wastes.

It was evident from Table 1 that the wastewater was heavily polluted with organic load, nutrients and suspended matter. As a result the polluting potential of the factories is enormous at locations below effluent discharge points even after stabilized in a disposal pits, indicating deterioration of water quality in downstream locations where coffee effluent is seeped in to the water course.

Relatively lower concentration of physicochemical parameters were observed in the downstream of Sunde, Gurracho and Funtule rivers (table 1) which have relatively large size and located at a reasonable far distance (> 100 m) from the rivers. Although the measured pollution parameters downstream along the sampling sites were highly reduced compared to the raw wastewater, most of the pollutant concentrations were still very high to meet the provisional discharge limits set by the WHO (9).

Various researchers reported high pollution from wet processing (3, 4, 10, and 11). As a result the polluting potential of the coffee processing factories is enormous as shown by the BOD_5 and COD content of coffee effluent reaching up to 2967 and 3244 mg/l even after stabilized in a pit respectively at downstream of Urgessa river (See Table 1) and differ significantly ($P <0.05$). This BOD_5 and COD content of coffee wastewater was much lower than the previous work undertaken by Hadis and Devi (2007) reported BOD_5 and COD values up to 10,800 and 15,780 mg/l in Jimma zone (11).

WHO standard for effluent discharges on land for irrigation and to receiving water has a limit value of (300 mg/l) COD and (100 mg/l) BOD_5, the maximum effluent concentration obtained from this analysis were higher than the acceptable limit and the reference samples (9) (Table 1) respectively indicating the pollution strength of the wastewaters. This indicates that large amount of chemical and biological oxygen demanding substances in the effluent are released from the coffee processing wastewater into the river. They also indicate that there could be low oxygen available for living organisms in the wastewater when utilizing the organic matter present.

The TSS of the water samples among the upstream and downstream sites generally showed that no significant difference ($P > 0.05$) (Figure 2), and results below the permissible limit (Table 1) were also observed at three impacted sites, which have large size pits to accommodate the generated waste during peak processing time. Based on the standard discharge limit value, the suspended solids in the Yebu, Urgessa and Chiseche Rivers may adversely affect the use of water for various purposes (Table 1) by exacerbating the dissolved oxygen problem by sedimentation and forming oxygen demanding sludge deposit, which cause turbidity in the receiving water and may alter the habitat of aquatic microorganisms.

The total dissolved solid (TDS) profile of the upstream and receiving water body samples vary significantly ($P < 0.05$) and ranged from 73-246 mg/l for upstream and 349-9034 mg/l for downstream respectively and was high all along the sampling points (Table 1) as compared to the provisional discharge limits set by the WHO (9). And showed increment as compared to the reference samples (Table 1) indicating the increasing impact of the effluents on the downstream water bodies. The relatively higher amount of TDS at the downstream site might be attributed by the high mucilage coming out from coffee processing stations. Moreover, Von

Enden and Calvert (2002) described the high suspended material (especially the digested mucilage) when precipitated out of the solution builds a crust on the surface, clogging up water ways and further contributing to anaerobic condition (4). These TDS concentrations automatically influenced the quality of the received water bodies. Elevated TDS can be toxic to freshwater animals by causing osmotic stress and affecting the osmoregulatory capability of the organisms and can give rise to obnoxious odours from the decomposition of organic matter and foul smelling.

The nitrate concentration level in the downstream of river water was found within range of (8.86-32.7 mg/l) (Table 1) and changed significantly ($P < 0.05$). Based on the WHO (1995) standard, the nitrate concentrations along downstream sites were higher than the acceptable limit and as compared with the reference the value was 11 times greater, which indicated the pollution of the rivers (9). This finding is in consistent with similar study done in Jimma zone by Hadis and Devi (2007) reported the nitrate concentration of 10.5 mg/l (11). High nitrate concentrations at downstream of Gurracho and Funtale rivers (Table 1) may occur as a result of the deamination of ammonium nitrogen from nitrogenous materials raw wastes that can be oxidized to nitrate by the action of microbiological agents (12). This would allow for growth of algae which would lower the dissolved oxygen levels and it is important to note that high nitrate level in the downstream could be a source of eutrophication for the receiving water bodies as the values obtained exceeded the recommended limits for no risk of 5 mg/l (9). Lower nitrate and higher ammonium during the peak coffee processing season in downstream site might be due to the fact that the high suspended material (especially the digested mucilage) when precipitated out of the solution contributing to anaerobic condition (4). This might have suppressed the production of nitrate while producing high levels of ammonium as a result of fermentation and decomposition of coffee mucilage. At lower pH of the sampling sites along the downstream of Yebu, Urgessa and Chiseche rivers, the ammonium concentrations were also high this was likely to be used by microorganisms in the anaerobic conditions as a nutrient source (13) and this can pose severe health problems among the residents of nearby community who uses the downstream water for different purposes and this is in agreement with the previous study by Hadis and Devi (2007) in Jimma zone (11). In addition to the presence of high dissolved solids the reasons for the bad odour that can be sensed along the river courses during the field sampling could be also due to presence of high ammonium. With regard to TN content of the sampling sites, there is a general pattern of its increment at downstream sites compared to the reference sampling sites. High concentration of total nitrogen could indicate pollution of a water body that is rapidly converted to ammonia and become toxic to aquatic life. The concentration of NH_4^+ and TN (Table 1) was higher at Urgessa River than the other sites. Animal and human waste, decaying organic matter and agricultural runoff can contribute to total nitrogen and ammonium enrichment of water and the Site (Urgessa River) is highly exposed to these practises.

In contrast, the increases in PO_4 (a necessary nutrient for plant growth and generally is limiting in the freshwater environment) in the processing water are of concern. Phosphorus concentrations in downstream sites were higher than the acceptable provisional discharge limits set by the WHO (9) and the values obtained from the reference samples (Table 1). Comparison of the result obtained in this present study from the receiving water bodies, e.g. Hadis and Devi (2007) (4.6 mg/l) showed lower phosphate concentrations than that obtained in this study (7.6-18.5 mg/l) (11). In general nitrogenous and phosphorus compounds are essential nutrients to plant life, but when found in excessive quantities; they can stimulate excessive and undesirable plant growth such as algal blooms and changing the types of plants and animals that live in a stream and this could adversely affect the use of rivers for different purposes as the covering of large areas by macrophytes could prevent access to waterways and could cause unsightly and malodorous scum which could lead to the growth of blue-green algae and release toxic substances (cyanotoxins) into the water systems. These substances are well known to cause the death of farm livestock (14) and this must be a matter of concern in the study area as these receiving water bodies are used for drinking by the farm livestock.

Water temperature was not significantly different between impacted and unimpacted sites all along the sampling points. The downstream water temperature in particular had temperature ranges of 15.46-23.1°C were below 25 °C, which is the recommended limit for no risk according to the WHO quality guidelines for discharging of effluents for irrigation purpose and allowed to enter into natural surface water bodies (9). Based on this guideline, the temperature of the effluent does not appear to pose any threat to the homeostatic balance of the receiving water bodies. These results are in agreement with previous work undertaken by Hadis and Devi (2007) in Jimma zone in which the water temperature of coffee processing wastewater was reported as 22°C (11). The insignificant difference of water temperature might be due to lower temperature during sampling period. Therefore, samples taken around morning are expected to have a relatively lower temperature.

The electrical conductivities of the water samples generally varied significantly ($P < 0.05$) throughout the study period (figure 2). Higher conductivities were observed downstream of the discharged points and based on WHO guideline (9); the effluent quality does not appear to be compliant with the regulation for electrical conductivity. Thus, the parameter does indicate that the water is unsuitable for direct domestic use. The electrical conductivity in the samples showed, however, the presence of a relatively large amount of ions in solution concentration, which makes the carrying out of a treatment necessary, if there is a desire to introduce these waters in to water bodies without incurring eutrofization risks of the water bodies.

Various researchers reported lower pH and acidic water

from wet processing (3, 4, and 15). However, the pH value recorded at the lower course of Urgessa, Chiseche and Gurrachu Rivers (Table 1), was even lower than that reported by Hadis and Devi (2007) and Kassahun *et al.* (2007) in Bilida River (5.15) and Boru River (4.63) of Jimma zone respectively, suggesting that there was fermentation of sugars in the mucilage in the presence of yeasts to alcohol and CO_2 (7, 11, 16). It is also known that as the organic wastes oxidize, CO_2 is released and increase the acidic characteristics of the water decreasing the pH value below the range of 6.5 - 8.5; which is WHO standard for any source of water for human use (9). Then the pH of the water may influence the species composition of an aquatic environment and affect the availability of nutrients and the relative toxicity of many trace elements (17). Under these acid conditions, higher plants and animals will hardly survive.

As Kassahun *et al.* (2007) reported this observation highlights thc fact that poorly designed and constructed pits do not prevent pollution of water bodies and the resulting threat to aquatic life unless well-designed waste treatment technologies are provided for the coffee waste in addition to adopting sound environmental practices (7).

5. Conclusion and Recommendation

The traditional wet coffee processing industries are generating huge amounts of organic and nutrient rich and acidic wastewater. This huge untreated wastewater is discharged directly into the nearby pits that are intended to serve as wastewater stabilization but are neither properly constructed nor the right dimension to accommodate the generated waste during peak processing time. This leads to overflow of raw effluents into natural watercourses and severely damaged the surface waters and aquatic life. Such an alarming pollution in the region calls for an urgent action and seeks a sound effluent management option in order to ensure sustainability of coffee production and to avoid irreversible environmental damage. So that, individual coffee processers should be informed about the importance of waste minimization, fate of the wastes after they are disposed and usefulness of coffee wastewater as a valuable resource. Besides, the regional EPA in collaboration with Jimma University should think and take mitigation measures so as to address this serious pollution problem

Authors' Contributions

DY: designed the study, participated in the data collection, performed analysis an interpretation of data and drafted the paper and prepared the manuscript for publication. AB: Involved in proposal writing, designing, analysis and write-up and in all stages of the project implementation. TA: Involved in proposal writing, designing, analysis and write-up of the project. AG: Involved in analysis and write-up and manuscript preparation. Finally all authors have been read and approved the manuscript.

Acknowledgements

The authors would like to thank Jimma Zone traditional coffee processing cooperatives for their collaboration during coffee wastewater sampling. We are also grateful to Jimma University, College of Public Health and Medical Sciences, for financial and logistic support. We also thank the anonymous reviewers for their great help, and the editor for his constant support.

References

[1] Beyene A, Kassahun Y, Addis T, Assefa F, Amsalu A, Legesse W, Kloos H, Triest L (2011). The impacts of traditional coffee processing on river water quality in Ethiopia and the urgency of adopting sound environmental practices. J Environ Monit Assess 184:7053–7063

[2] Coste, R. (1992). Coffee: thc plant and thc product. Macmillan press limited, London.

[3] Mburu, J.K., Thuo, J.T., Marder, R.C., (1994). The characterization of coffee waste water from coffee processing factories in Kenya. Kenya Coffee 59, 1756- 1763.

[4] Von Enden, J.C. and Calvert, K.C. (2002). Limit Environmental Damage by basic knowledge of coffee wastewaters. GTZ – PPP Project "Improvement of coffee quality and sustainability of coffee production in Vietnam".

[5] Legesse W., W/Senbet B., Endris S., Kassahun Y. and Kebede T. (2003). Effect of coffee processing on ecological integrity of river systems in Jmma Zone, Southwestern Ethiopia.

[6] Kansal A., Rajeshwari K V., Balakrishnan M., Lata K., Kishore V V N. (1998). Anaerobic digestion technologies for energy recovery from industrial wastewater - a study in Indian context. TERI Information Monitor on Environmental Science 3(2): 67–75

[7] Kassahun Y., Beyene A., Addis T., Assefa F., Amsalu A., Legesse W., Kloos H., Triest L. (2007). The impact of traditional coffee processing on river water quality in Ethiopia and the urgency of adopting sound environmental practices M.sc. thesis, unpublished.

[8] Eaton A.D., Clesceri L.S., and Greenberg A.E. (1995). Standard Method For Examination of Water and Wastewater, 19th Edn. Town

[9] WHO (1995), Guideline for Discharge of Industrial Effluent Characteristics, vol.3, pp. 231–236.

[10] De Matos, T, A., Lo Monaco, P.A., A.B. Pinto, R. Fia, and D.C. Fukunaga. (2001). Pollutant Potential of Wastewater of the Coffee Fruits Processing. Federal University of Viçosa, Department of Agricultural Engineering, Viçosa-MG, Brazil.

[11] Haddis A. and Devi R. (2008). Effect of effluent generated from coffee processing plant on the water bodies and human health in its vicinity. Journal of Hazardous Materials. 152:259-262.

[12] Morrison, G., Fatoki, O S., Persson, L. and Ekberg, A. (2001). Assessment of the Impact of Point Source Pollution from the Keiskammahoek Sewage Treatment plant on the Keiskamma River- pH, Electrical Conductivity, Oxygen -Demanding Substance (COD) and Nutrients. Water SA 27(4): 475-480.

[13] Gerardi Michael H., (2003). The Microbiology of Anaerobic Digesters, third Edition.

[14] Holdsworth, R., (1991). New health consideration in water treatment. Avebury Technical Publications, Aldershot.

[15] Murthy, K.V.N., A. D'Sa, and G. Kapur. (2004). An effluent treatment-cum-electricity generation option at coffee estates: is it financially feasible? Draft version. International Energy Initiative, Bangalore.

[16] Hue N. V., Bittenbender H. C., and Ortiz-Escobar M. E., (2004). Managing coffee processing water in Hawaii, Department of Tropical Plant and Soil Sciences, College of Tropical Agriculture and Human Resources, University of Hawaii, Manoa, Honolulu, HI 96822, USA

[17] Canada Environmental Protection Agency. (1994). monitoring surface water quality: A Guide for Citizens, Students and Communities in Atlantic Canada. http://www.ns.ec.gc.ca/community/acap/pdf/waterquality_e.pdf

3

The characteristics of ambient gases pollutant, South Korea

Amin Kalantarifard, Eun-Song Byeon, Jo Gwanggon, Go Su Yang*

Department of Environmental Engineering, Chonbuk National University, Jeonju, South Korea

Email address:
Aminkalantarifard@gmail.com (A. Kalantarifard), Gsyang@jbnu.ac.kr (G. S. Yang)

Abstract: In this study, DPNH-sampling technique, indophenol method, gas chromatography-flame ionization detector (GC-FID) and High performance liquid chromatography were applied to conduct three days continuous measurements in areas of Gunsan city, South Korea to detect the concentration of formaldehyde, acetaldehyde, styrene and ammonia emissions level during 21th to 23rd of October 2012. A total 288 samples were collected at six different locations included residential and industrial areas of the city. The sampling was run in the period of 6 hr for three days. With the combination of local meteorological data, including wind direction and speed, temperature and humidity, the result showed that the concentration of ammonia were higher during the afternoon when temperature increases while the values of formaldehyde, acetaldehyde and styrene were high during the night and early morning. The major source of ammonia, aldehydes and styrene were mainly industries chimney with high average concentration. Highest level of ammonia concentration was detected as 125.8 ppb in the atmosphere during the third day 12 pm when temperature raised to 20.1°C. The results indicate that in addition the formaldehyde, acetaldehyde and styrene average levels ranged from 3 to 11 ppb, 1.2 to 5.6 ppb and 4 to 16.2 ppb during sampling time. This pilot study was undertaken to provide reliable information on distribution patterns of gaseous pollutants in ambient air in Gunsan. The results indicated that the values of emitted pollutants concentration investigated in the present study were below the limits that are mandatory in the South Korea guidelines.

Keywords: Gases Emission Level, Local Industrial Park, South Korea

1. Introduction

Globally, there is a concern related to the effects of air pollution on the environmental system. The growth of the economy, industrialization and urbanization are identified as major sources of air pollution emissions in developing countries. Several volatile organic compounds (VOCs), including carbonyl compounds (formaldehyde and acetaldehyde), styrene and ammonia have carcinogenic and mutagenic effects on living organisms and the health of humans, and also contribute to promoting photochemical smog. Annually, governments invest huge budgets to measure and monitor the air pollutant emissions to prevent the consequences of these pollutants on the environment and human life.

Formaldehyde and acetaldehyde compounds are ubiquitous in the environment, present in the gaseous, liquid, and particulate phases of the atmosphere [1]. These complex chemical compositions play an important role in the atmosphere, physically and chemically, and also on climate processes or even human health [2]. These compounds are mostly unwanted by-products of many industrial processes especially waste wood combustion. Formaldehyde is the most abundant compound among the emitted aldehydes from wood combustion processes with an emission factor of 422 mg/kg of wood while acetaldehyde has lower emission factors of 86.3 mg/kg of wood [3]. These compounds are stable and lipophilic and can be transported over long distances from their original source when they are released into the atmosphere. The increase of concentration levels of these compounds has become one of the major causes of the degrading air quality in developing countries. These two compounds are known to be irritants of the eye and upper airways, especially the nasal cavity. In addition, formaldehyde can act as a precursor to organic aerosol formation in urban air.

Styrene (C_8H_8) is a clear, colorless liquid hydrocarbon which is easily polymerized and therefore widely used in industrial processes such as the manufacturing of plastics, packaging, synthetic rubbers, resins, insulators and also in the production of protective surface coatings [4]. Chronic exposure to styrene emissions can affect the central nervous system causing symptoms such as depression, headache, and fatigue and can cause minor effects on kidney functions and the blood [5]. Styrene is classified by the World Health Organizations International Agency for Research on Cancer (IARC) as a Group 2B carcinogen that is possibly carcinogenic to humans [6,7]. Styrene has been found to be metabolized to styrene-7,8-oxide [8] which is an active metabolite of styrene and classified by the IARC as a Group 2A carcinogen that is possibly carcinogenic to humans as well [9].

The emission of ammonia (NH_3) as a primary gaseous base in the atmosphere is a concern because of its direct contribution to secondary photo- or smog air pollutants, and for its adverse influences on various ecosystems, such as bodies of water and soil [10]. As a component of atmospheric nitrogen (N) deposition, NH_3 also plays a role in the cycling of N in terrestrial and aquatic ecosystems. Some sources of NH_3 emissions include chemical industries, food processing, fertilizers, and industrial and publicly owned treatment works. One of the main sources of ammonia identified is uric acid from poultry, which under moist conditions is quickly broken down to urea and then to ammonium-N from which ammonia is released. Ammonia is very soluble in water and readily reacts with other substances in the atmosphere to form ammonium (NH_4^+) compounds such as ammonium sulphate which can have profound effects on natural ecosystems. Previous studies indicate that increases of NH_4^+ concentration level in the atmosphere can cause environmental stresses such as soil acidification [11], deforestation [12], and th eutrophication of surface waters [13,14]. Therefore, it is necessary to characterize the magnitude and spatiotemporal variability of atmospheric NH_3 and NH_4^+ concentrations.

1.1. Sampling-Site Description

Gunsan city, with an area of approximately 390.20 km^2 and population of 275,298 (2012), is located in the North Jeolla Province of South Korea. The large scale industrialization in the last decade had a major effect on increasing the percentage of emissions formed of different gases in this city. The Gunsan industrial park is located in the northwest with an area of 28.402 km^2. Waste wood combustion, chemical materials & products, plastic manufacturing, food production & processing, concrete and construction materials production are examples of the current industrial plants in this city. The waste wood combustion power plant located in Gunsan was established in 1990 with an area of 171.816 m^2 and represents one of the biggest combustion power plants in the region. Medium density fiberboard (MDF) is a main product by this company. Existing combustion chambers has a maximum burning capacity of 200 tons/day of wood waste. Figure 1 shows the percent distribution of industries in Gunsan city. Over the years, setting the plants in residential neighborhoods of this city has caused a number of problems and extra expenses such as high taxes, local political involvement, and complaints about gaseous pollutants. Since there was no statistical data available for aldehydes, styrene and ammonia emission levels in the area of this city, this study focuses to estimate the concentration level of these pollutants in the atmosphere of Gunsan.

Industrial share overall (Gunsan)

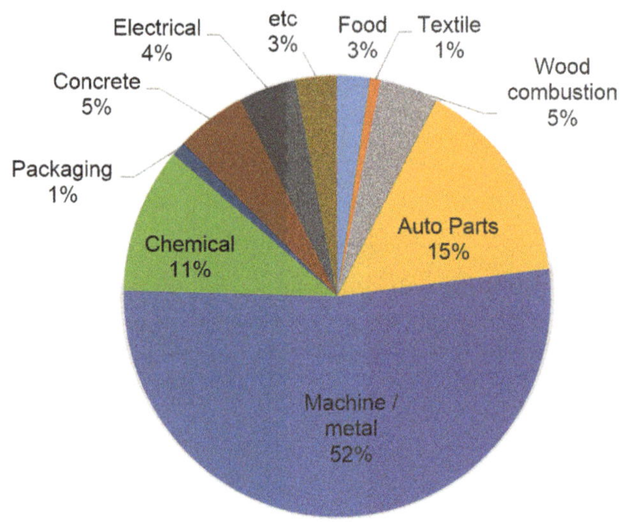

Figure 1. *Percent distribution of industries at Gunsan city.*

The sampling points were sited to be representative of the air quality in a surrounded area. For this particular research, six total points were selected in the area of Gunsan to provide data within zones and agglomerations where the highest and lowest concentration of pollutants occur during the day and night. The number of sampling points for measurement calculated the base on accounted emission densities, likely distribution patterns of ambient air pollution, and potential exposure. The aldehydes, ammonia and styrene measurements were intially done near the industrial area which were called points A-1 and A-2. The other points were selected in residential area, schools, and the center of the city. A total of 288 samples were collected in the period from the 21st to the 23rd of October 2012. The sampling was run in the period of 6 hr for three days. Figure 2 shows the photographical image of the Gunsan industrial and residential areas and the location of selected points.

2. Materials and Methods

2.1. Formaldehyde and Acetaldehyde Analysis

Versapak-bonded C18 cartridges (360 mg) purchased from a Top-trading Company in South Korea was used to determine the aldehydes concentration in the atmosphere. The procedure involved the collection of aldehydes samples in C18 cartridges coated with 2,4-dinitrophenylhydrazine (DPNH) and quantified them using HPLC coupled with a

UV-Vis detection system at a wavelength of around 360 nm. In the collection procedure, aldehydes were collected from the atmosphere by pumping air into connected cartridges using a personal sampling pump (KEMIK, SARA 5100). The optimum flow rate kept in order to operate the pump stably for at least 10 min was 2 L/min, which resulted in a total of ~20 L of air. Before reaching the cartridges, the air sample passed through the ozone scrubber consisting of a 40mm holder containing 1.5 grams of high purity potassium iodide to prevent the oxidation of DPNH by the ozone. After sampling, the end of the cartridge column was closed with a plastic cap, and the cartridges were wrapped individually in aluminum foil and placed in hermetically closed plastic bags and stored in a refrigerator until analysis time. The collection system for aldehydes with the ozone scrubber is shown in Figure 3. The DNPH compound found in the cartridge was slowly eluted with acetonitrile (ACN). The obtained mixed solution (approximately 5mL) of ACN and aldehydes compounds were kept in the refrigerator for nearly one day to ensure the stability of samples.

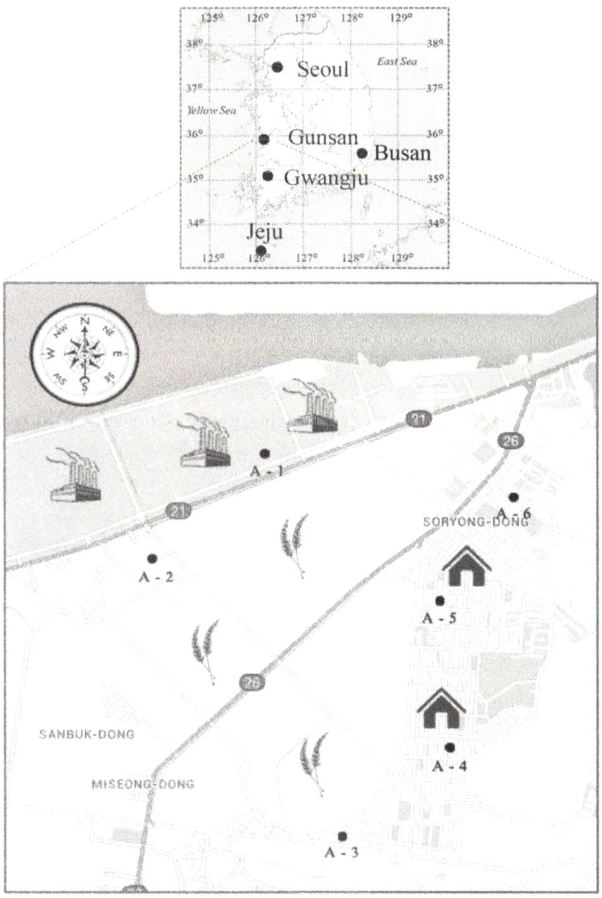

Figure 2. The geographical location of industrial and residential areas at Gunsan and selected air pollution sampling points, A-1 to A-6.

The quantification of aldehyde compounds carried out by HPLC in an Agilent 1100 series capillary liquid chromatography system. The systems consisted of a binary pump, a degasser, an automated injector, a thermo stated column compartment, and a UV-DAD detector. The analytical conditions were as follows: HPLC column dimensions (250mm× 4.6mm×5μm), injection volumes of 10μg, a temperature of 35°C and mobile phases composed of binary mixtures of ACN and a water flow rate of 1.2 mL/min. An equilibration time of 1 min between successive HPLC runs was always adopted. All samples were detected in 360 nm. Aldehydes were identified by retention times, elution order and their absorption spectra. The analytical precision was determined with injections of true standards of formaldehyde and acetaldehyde acquired from (Supelco, USA) an HPLC system to observe retention times and absorption spectras. Collected data from Supelco Company shows the calibration curve for a solution containing true standard formaldehyde and acetaldehyde in concentrations of 14.73 to 14.88 μg/ml. These results were compared with chromatogram curves from collected samples to identify the concentration of formaldehyde and acetaldehydes.

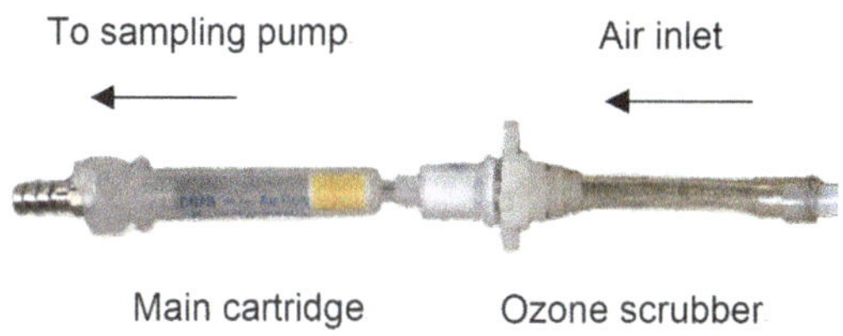

Figure 3. Collection system for aldehydes with ozone scrubber.

2.2. Styrene

Styrene in the atmosphere is determined by a capillary column gas chromatography (GC) with a flame ionization detector (FID). Air samples are collected by drawing air using a personal vacuum pump with a flow rate of 0.05 L/min for 15 minutes through glass sampling tubes packed with two sections of coconut shell charcoal that has been coated with p-tert-butylcatechol (TBC), which is 10% by weight. The front section contains 110 mg and the back section contains 55 mg of TBC-coated coconut shell charcoal [15]. After 15 minutes sampling, the samples were removed and tube was sealed with a plastic end cap and sent to the Chonbuk National University laboratory for analysis. In the laboratory, samples are desorbed with toluene and analyzed by GC using a flame ionization detector. The detailed information about GC-FID experiments, sampling conditions, standard calibration methods and curves can be found in the OSHA analytical manual and NIOSH recommended standards [16].

2.3. Ammonia

An ammonia trap system consisted of boric acid bottles in order to trap the emitted ammonia, a variable area flow meter to regulate the flow rate of the sweep-air, a vacuum pump to pull air through the system and a bottle containing silica gel to adsorb the humidity (Figure 4). This method involves capturing ammonia in a boric acid solution and quantifying according to the indophenols-UV method [17]. In our experiment, ammonia absorption in boric acid occurred by vacuuming the ambient air through two bottles containing 25 milliliters of boric acid (0.5%). The vacuum pump was adjusted with a flow rate of 5.0 L/min for 10 minutes. The collected samples were buffered at a pH of 9.5 with a borate buffer in order to decrease the hydrolysis of cyanates and organic nitrogen compounds then finally distilled into a solution of boric acid. Alkaline phenol and hypochlorite react with ammonia to form an indophenol blue that is proportional to the ammonia concentration (EPA Method 350.1) The blue color formed was intensified with sodium nitroprusside and measured by the OPTIZEN 1412V spectrophotometer at a wavelength of 640 nm. The calculation of the actual ammonia concentrations in the samples was done using a calibration curve based on an ammonia standard solution.

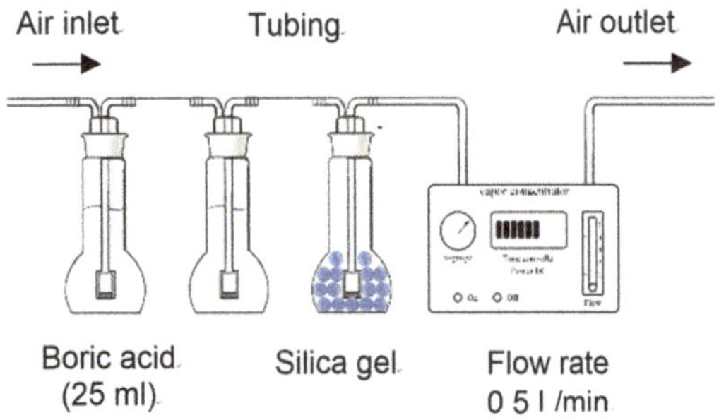

Figure 4. Schematic of ammonia absorption in boric acid.

3. Results and Discussion

3.1. Diurnal Variation of Meteorological Parameters

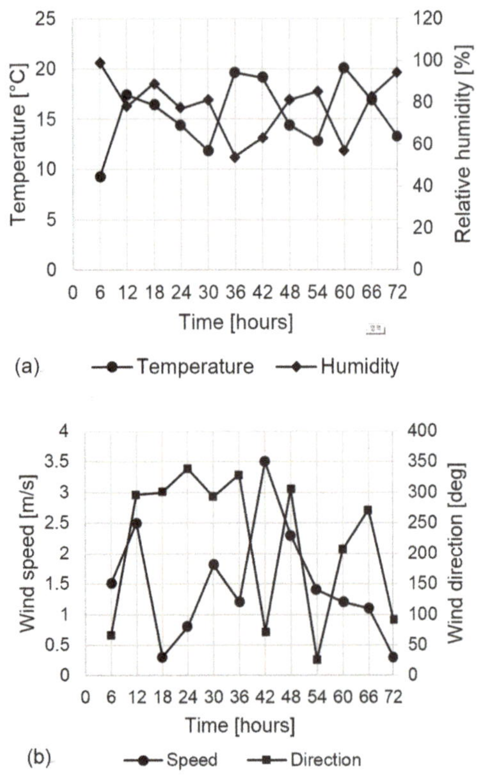

Figure 5. Meteorological conditions recorded during sampling campaign from 21th to 23rd October 2012, the data are 6-h average values. a) relative humidity and temperature, b) wind direction, wind speed.

The spring and summer climate in Gunsan is hot and humid whereas the fall and winter is cold and dry with local variations due to altitude differences. In 2012, the seasonal average temperature was 13.5°C in spring, 24.1°C in summer, 15.3°C in fall and 11°C in winter. During the sampling period, the highest temperature obtained was on the 23rd with 20.1°C and the lowest was 9.3°C on the 21st. The prevailing wind was a steady 3.5 m/s^{-1} with a slight angle across the sea. The wind direction was from the north and northwest in the first and second days and from west to southwest and north to south on the third day of sampling. The meteorological parameters including wind direction, wind speed, relative humidity and temperature were recorded continuously and represented in Figure 5a and b.

3.2. Diurnal Variation of Formaldehyde, Acetaldehyde and Styrene

Six hours of average concentrations of pollutants were measured from the 21st to the 23rd of October 2012. The sampling campaign started at 6 am to 12 am each day. The average concentrations of formaldehyde, acetaldehyde and styrene in each location are shown in Table 1. During the studied period, concentrations of acetaldehyde were generally very low and some samples were under the limit of detection while the formaldehyde concentration was higher than the acetaldehyde in all selected areas. The average formaldehyde concentrations remained above 4.3 ppb at point A-1 and A-2 through all three sampling days. There is a consistent underestimation of formaldehyde at 12 pm each day which can probably be attributed to a moisture and temperature effect. The ambient temperature range observed during the sampling period was 9.3-12.8°C in the early morning whereas it is increased to 17.4-20.1°C at noon. The concentration of formaldehyde generally decreases by increasing the temperature. The average variation of formaldehyde concentrations ranged from 3 to 11 ppb and acetaldehyde concentrations from 1.2 to 5.6 ppb (Table 1).

Table 1. The average gases concentrations (ppb) in each location during sampling period

Locations		Day 1	Day 2	Day 3
	formaldehyde	11	8.3	4.4
A-1	acetaldehyde	5.6	3.2	2.6
	styrene	16.2	11.9	6.5
	formaldehyde	8	5	4.3
A-2	acetaldehyde	3.2	3	2.4
	styrene	12.5	9.3	7
	formaldehyde	4.5	3	3.8
A-3	acetaldehyde	3	1.2	2.1
	styrene	5.6	4.7	4
	formaldehyde	5.2	4.4	3.3
A-4	acetaldehyde	3.3	2.4	2.7
	styrene	8.2	7.9	5.05
	formaldehyde	4.8	4.7	3.9
A-5	acetaldehyde	2.4	2.4	2.2
	styrene	8.5	8.2	6.2
	formaldehyde	6.2	5.1	4
A-6	acetaldehyde	3.5	3.8	2.2
	styrene	8.8	6.5	6.1

The highest concentration level of formaldehyde was obtained during the first day at night when the temperature and humidity was recorded as 14.4°C and 77% while the highest level of acetaldehyde was obtained during the first morning when the humidity was recorded as 99%. Figure 6 shows the continuously measured 6 h average values of formaldehyde, acetaldehyde and styrene in all selected locations. The lower concentrations of formaldehyde and acetaldehyde at points A-3 to A-6 could be the result of a longer distance from industrial sections which possibly decrease the pollutants emission level. In general, based on the time of sampling, similar patterns for the concentration level of formaldehyde and acetaldehyde were observed each day. During this monitoring period, the highest average styrene concentration observed for day one was 16.2 ppb and the lowest instantaneous level was 4 ppb on the third day. The highest recorded concentrations of airborne styrene were measured at point A-1 and A-2, which are at a very close proximity to the industrial area. The residential area located in the south and southwest was only impacted slightly by the styrene emissions and only when the winds were from the southwest.

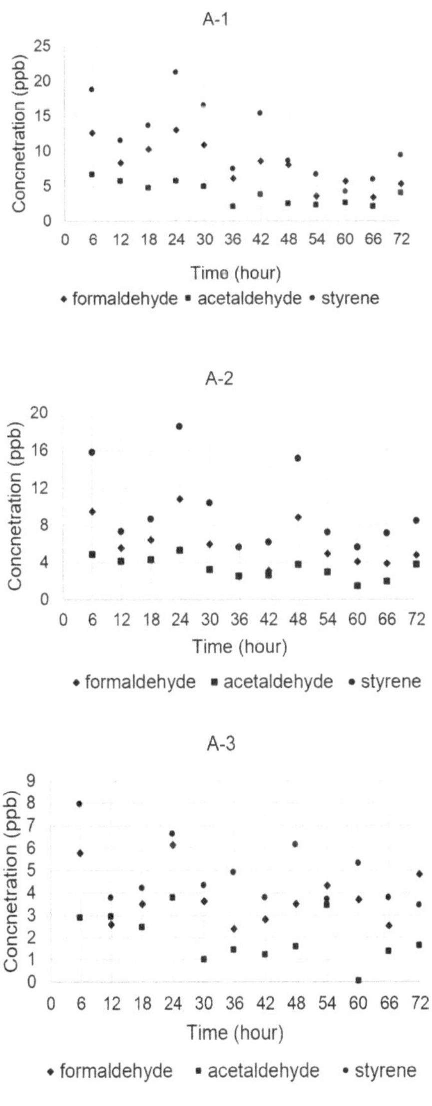

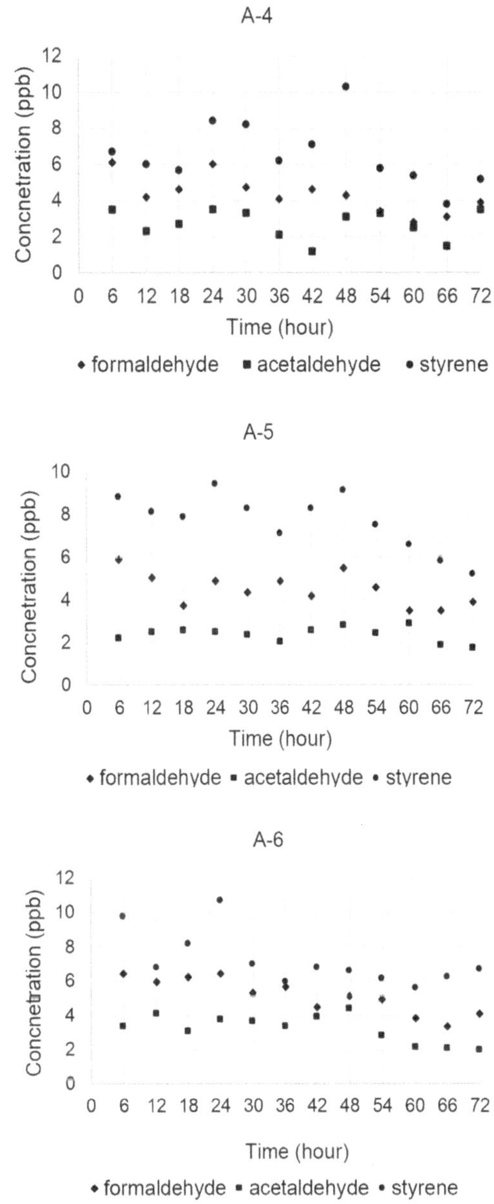

Figure 6. *Continuously measured 6-h average values of formaldehyde acetaldehyde and styrene in all selected locations from 21th to 23rd October 2012.*

3.3. Diurnal Variation of Ammonia

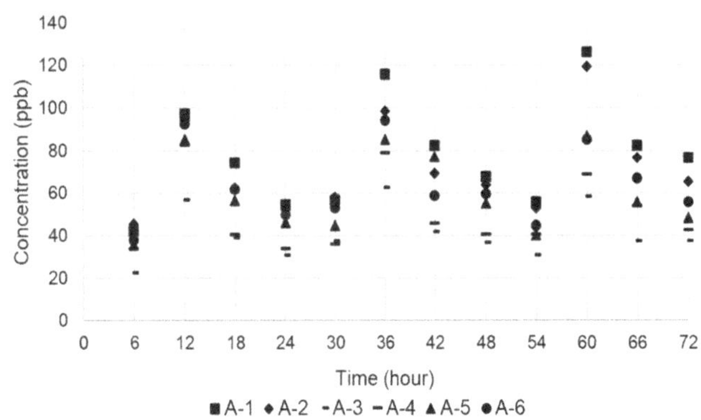

Figure 7. *Daily variation of ambient ammonia concentrations in all selected locations at Gunsan.*

Figure 7 represents the daily variation of ambient NH_3 concentrations at the six sampling locations. In Gunsan, the relative mean temperature level at 12 pm was approximately 1.3 times higher than the mean level for 12 am as is showed in Figure 5a. The concentration of ammonia generally increased when temperature increased. Figure 7 shows that the mean NH_3 levels in Gunsan ranged from 22.5 to 57.8 (mean = 40.1 ppb) at 6 am, 56.6 to 125.8 ppb (mean = 91.2 ppb) at 12 pm, 37.4 to 82.3 ppb (mean = 59.8 ppb) at 6 pm, and from 30.4 to 76.5 (mean = 53.4 ppb) at 12 am, respectively. For all the sampling durations, the highest level of 125.8 ppb was obtained on the third day at 12 pm at point A-1. To discern the difference between sampling times, the mean average NH_3 concentrations at the highest level at 12 pm was divided by the mean NH_3 concentration at 6 am. The mean ratios were 2.2, 2.0, 1.9, 2.1, 2.14 and 2.0 in location A-1 to A-6 respectively. However, the ratio of 2.2 in locations A-1 is the biggest number which indicates that there is either an increased emission under the same sources, or there is a new emission source during the daytime. At this stage, one explanation for the higher emission level of NH_3 in the location A-1 is a close distance to chemical industries which has a significant impact on the overall elevation of ammonia in the area.

3.4. Influence of Surrounding Areas and Wind Direction

The industrial sites in Gunsan are located about 500-1000 meters away from the sea in the northwest. High concentrations of primary aldehydes and ammonia pollutants generated from these manufacturing processes during the day and night is a main cause of the increasing concentration levels of harmful gases in Gunsan. These pollutants are usually transported to the residential areas with the aid of the sea breeze. Therefore, a distribution of pollutants by sea and land breezes significantly affects the actual pollutant concentration levels in selected sampling locations in this study. Taking a close look at the effect of wind direction on formaldehyde, acetaldehyde and styrene dispersion data during the campaign, the wind direction was from north and northwest (from the sea) during the first two days (Fig. 5b) while on the 3rd day, the wind was blowing from the west and northeast. Weak

Previous research has reported a seasonal cycle in ambient NH_3 concentrations [18,19] with maximum concentrations occurring during warm days. This seasonal cycle in ambient concentrations is in agreement with the temperature dependence of aqueous-phase partitioning between NH_3 and NH_4^+, as well as the equilibrium between aqueous-and gas phase NH_3 as predicted by Henry's law, which results in increasing ammonia emissions from the emission sources [20,21]. Recorded results shows the concentrations of NH_3 reach a maximum at 12 pm on the third day when the temperature rose to 20.1°C at all sites whereas the minimum concentration was observed during the first day morning of sampling when the temperature was 9.3°C. winds with a speed of 0.3 m/s and 0.8 m/s were during the evening of the 1st day, which increased to more than 3 m/s during the 2nd day. Formaldehyde and acetaldehyde concentration levels increased with the decreased wind speed. On the first day, the concentration level of formaldehyde, acetaldehyde and styrene were recorded as 12.9, 5.7 and 21.2 ppb at 12 am in point A-1 when the wind speed dropped to 0.8 m/s. In addition, Figure 5a shows that the temperature dropped from 17.4 at 12 pm to 14.4 at 12 am. The decreased temperature at midnight is another factor which caused an increase of these pollutant levels during the sampling time.

The highest pollutant levels were expected for all sampling points at 6 am during the first day since the lowest temperature, highest humidity and lowest wind speed had been recorded. However, as is shown in Figure 5b, wind direction was from the northeast to the southwest during this time which significantly reduced the emission concentration levels of all selected sampling points. At point A-1, we measured the concentration of formaldehyde, acetaldehyde and styrene as 12.51 ppb, 6.63 ppb and 18.8 ppb at 6 am that we believe is lower than the actual concentration level of pollutants at this time.

4. Conclusion

The importance, in a South Korean context, to estimate and assess the concentration levels of gaseous pollutants from industrial power plants, agriculture and livestock as an environmental impact has become an increasing concern to public and private sectors. The acceptable levels of acetaldehyde, ammonia and styrene in South Korea have been continuously reduced in the past decade. South Korea's ministry of environment (MEV) announced in January 2010 that it had settled the suit by agreeing to implement performance standard limits for acetaldehyde as 0.05 ppm, ammonia as 1 ppm and styrene as 0.4 ppm for a 24 hour average value in residential areas. According to this agreement there is no standard limit for formaldehyde in the atmosphere but there is an indoor standard limit value as 0.12 mg/m^3.

The overall results of this continuous measurement of gaseous pollutants shows that the industrial area of Gunsan is clearly a source for generated criteria pollutants. The correlation between meteorological data, especially air direction and pollutant levels, shows during the time that air direction is from the west and northwest to the east and southeast (from industrial area to residential), higher concentration levels at selected points have been recorded. In general, a higher concentration of ammonia was recorded at above 125 ppb during the third day (12 pm) at point A-1 when the average temperature rose to 20.1°C. The results indicate that the formaldehyde, acetaldehyde and styrene levels ranged from 3 to 11 ppb, 1.2 to 5.6 ppb, and 4 to 16.2 ppb during sampling times.

The obtained result in this research showed that the highest emission values of pollutants during sampling in the residential areas of Gunsan are still much lower than standard limits. Measurements of pollutants such as aldehydes and ammonia are often limited by the non-availability of feasible techniques for online analysis. In this study, measurements of pollutants have been done during October, which is considered autumn in South Korea. We believe seasonal experiments should be considered in the future to obtain the representative data of pollutant concentrations in this city.

Acknowledgements

This work was supported financially by the climate change specialization graduate school of Chonbuk National University.

References

[1] Oliveira, F.S., E.L. Sousa, and J.B. Aderade, 2007: Sensitive flow analysis system for the fluorimetric determination of low levels of formaldehyde in alcoholic beverages. J. Talenta, 73, 561-566.

[2] Beiner, K., A. plewka, S. Haferkorn, Y. linuma, W. Engewald, and H. Hermann, 2009: Quantification of organic acids in particulate matter by coupling of thermally assisted hydrolysis and methylation with thermodesorption-gas chromatography–mass spectrometry. J. Chromatography A, 1216, 6642-6650.

[3] Hedberg, E., A. Kristensson, M. Ohlsson, C. Johansson, P.A. Johansson, E. Swietlicki, V. Vesely, U. Wideqvist, and R. Westerholm, 2002: Chemical and physical characterization of emissions from birch wood combustion in a wood stove. J. Atmospheric Environment, 36, 4823-4837.

[4] Chen, Q. F., R. K. Milburn, and N.S. Karellas, 2006: Real time monitoring of hazardous airborne chemicals: A styrene investigation. J. Hazardous Materilas, 132, 261-268.

[5] U.S. Environmental Protection Agency, 2006: Health and Environmental Effects Profile for Styrene. Washington, DC, ECAO/CIN. 103pp.

[6] IARC (International Agency for Research on Cancer), 2002: Styrene. World Health Organization, Lyon. IARC monographs on the evaluation of the carcinogenic risks to humans. 82, 437-550pp.

[7] ATSDR (Agency for Toxic Substances and Disease Registry), 2007: Toxicological profile for styrene (draft for Public Comment). US Department of Public Health and Human Services, Public Health Service. Atlanta, GA. 140.Available: http://www.atsdr.cdc.gov/ToxProfiles /tp53.pdf.

[8] Guengerich, F.P., D.H. Kim, and M. Iwasaki, 1991: Role of human cytochrome P450 IIE1 in the oxidation of many low molecular weight cancer suspects. J. Chemical Research in Toxicology, 4, 168-179.

[9] IARC (International Agency for Research on Cancer), 1994: IARC monographs on the evaluation of carcinogenic risk to humans. Some Industrial Chemicals. 60, 233-320pp.

[10] Hsieh, L. T., and T. C. Chen, 2010: Characteristics of ambient ammonia levels measured in three different industrial parks in Southern Taiwan. J. Aerosol and Air Quality Research, 10, 596-608.

[11] Roelofs, J. G. M., A. J. Kempers, A. L. F. M. Houdijk, and J. Jansen, 1985: The effect of airborne ammonium sulphate on Pinusnigra var. maritima in the The Netherlands. J. Plant and Soil, 84, 45-56.

[12] Nihlgard, B., 1985: The ammonium hypothesis-an additional explanation to the forest dieback in Europe, J. Ambio, 14, 2-8.

[13] Paerl, H.W., 1995: Coastal eutrophication in relation toatmospheric nitrogen deposition: current perspectives. J. Ophelia, 41, 237–259.

[14] Paerl, H.W., and D. R. Whitall, 1999: Anthropogenically derived atmospheric nitrogen deposition, marine eutrophication and harmful algal bloom expansion: is there a link?. J. Ambio, 28, 307-311.

[15] OSHA (Analytical Methods Manual), 1990: U.S. Department of Labor, Occupational Safety and Health Administration. Method 9 – Styrene, American Conference or Government Industrial Hygienists (ACGIH); Cincinnati, OH-No. 4542.

[16] NIOSH, Criteria for a Recommended Standard, 1983: Occupational Exposure to Styrene, U.S. Department of Health and Human Services, Public Health Service, Center for Disease Control, National Institute for Occupational Safety and Health, Cincinnati, OH-No. 83-119.

[17] Kawashima, S., and S. Yonemura, 2001: Measuring ammonia concentration over a grassland near livestock facilities using a semiconductor ammonia sensor. J. Atmo Envir, 35, 3831-3839.

[18] Robarge, W. P., J. T. Walker, R. B. McCulloch, and G. Murray, 2002: Atmospheric concentrations of ammonia and ammonium at an agricultural site in the southeast United States. J. Atmospheric Environment, 36, 1661-1674.

[19] Roelofs, J. G. M., A. J. Kempers, A. L. F. M. Houdijk, and J. Jansen, 1985: The effect of airborne ammonium sulphate on Pinusnigra var. maritima in the The Netherlands. J. Plant and Soil, 84, 45-56.

[20] Asman, W. A. H., M. A. Sutton, and J. K. Schjorring, 1998: Ammonia: emission, atmospheric transport, and deposition. J. New Phytologist, 139, 27-48.

[21] Ni, J., 1999: Mechanistic models of ammonia release from liquid manure: a review. J. Agricultural Engineering Research, 72, 1-17.

4

Suppressing Hydrogen Sulfide Generation by Nitrate or Oxygen Addition in Column Percolation Experiment

Shumpei Kitazaki[1], Kei Nakagawa[1], Tomonori Kindaichi[2], Hiroshi Asakura[1, *]

[1]Graduate School of Fisheries Science and Environmental Studies, Nagasaki University, Nagasaki, Japan
[2]Graduate School of Engineering, Hiroshima University, Hiroshima University, Higashihiroshima, Japan

Email address:
asakura_hiroshi@yahoo.co.jp (H. Asakura)

Abstract: H_2S generation suppression by the addition of NO_3 from reagent (NR) or compost (NC), and O_2 supply (OX) was quantitatively estimated in column percolation experiments. The cumulative amounts of oxygen added exceeded 300 mmol-O/L in NR and NC, and approximately 2 mmol-O/L in OX on day 75. The cumulative amounts of H_2S generated on day 60 in control, NR, NC, and OX were 69, 26, 71, and 31 mmol-H_2S/L, respectively. The amounts in NR and OX were half of that in control, whereas the amount in NC was almost the same as that in control. The H_2S generation rates from day 50 to day 60 in control, NR, NC, and OX were 0.86, 0.019, 0.75, and 0.041 mmol-H_2S/(L·d), respectively, and the rates in NR, NC, and OX were 0.02, 0.9, and 0.05 times that in control. In this way, the H_2S generation suppression effects of NO_3 reagent addition were observed. In NR, although NO_3 was injected from the top of the column, the suppression effect reached the deep layer.

Keywords: Landfill Site, Waste Gypsum Board, Hydrogen Sulfide, Generation Suppression, Nitrate

1. Introduction

Hydrogen sulfide (H_2S) gas is generated at landfill sites globally [1–6]. The inhalation of H_2S gas at 750–1000 ppm causes abrupt physical collapse [7] (hereafter, H_2S exceeding 1000 ppm is called H_2S_{high}). In 1999, three workers at an inert solid waste landfill site (a type of landfill site in Japan where waste plastic, rubber, metal, glass, ceramic, and construction and demolition debris are acceptable, as well as waste gypsum board without protective paper sheet at the time of the accident. No gas venting pipes are required.) died of exposure to 15000 ppm H_2S gas. H_2S gas is produced at a landfill when sulfate-reducing bacteria (SRB) consume organic matter and convert sulfate (SO_4) under anaerobic conditions [1, 8, 9]. The source of SO_4 in construction and demolition debris is predominantly gypsum ($CaSO_4$) [10]. Suggested measures for preventing H_2S generation or diffusion at a landfill site include the exclusion of anaerobic conditions (by water exclusion and/or oxygen (O_2) supply) [11], and the trapping of generated H_2S gas (by conversion into iron sulfide using iron-containing material [12, 13], and by adsorption to activated carbon [14] or zeolite [15]). The Ministry of the Environment in Japan recommended the appropriate installation of cover soil (to prevent rainwater percolation and trap generated H_2S gas by iron in soil) and the use of gas venting pipes (O_2 supply). However, in regard to the gas venting pipe, a large cost is necessary for the installation and the identification of H_2S generation spots as the installation points. In addition, oxygen supply to the waste layer via gas venting pipes is not possible at coastal landfill sites because a major portion of the waste layer is found below the surface of inner leachate. It is also a fact that trapping by iron does not prevent H_2S generation per se. Therefore, it is necessary to develop a method to suppress H_2S generation.

In sewage systems, H_2S generation suppression by nitrate (NO_3) addition has been reported [16, 17]. This has been attributed to the increase in the oxidation-reduction potential (ORP) caused by the presence of NO_3. As the Gibbs free energies of the oxidation of organic matter (CH_2O) with dissolved O_2, NO_3, and sulfate (SO_4) as the electron acceptor are −125, −119, and −25.4 kJ/mol [18], respectively, the reaction is said to occur in the order of respiration by aerobic bacteria > denitrification by denitrifying bacteria > SO_4 reduction by SRB (H_2S generation) [18]. Therefore, if dissolved O_2 or NO_3 exists, SO_4 reduction hardly occurs.

If the NO_3 addition method were applicable to the waste layer as well, that method would be a realistic H_2S generation suppression method at an actual landfill site. NO_3 reagent is used in sewage systems [17]. Compost made from waste organic matter also contains NO_3 [19, 20]. Therefore, if useless NO_3 in the market could be used at landfill sites, low cost and effective use of resources would be accomplished.

The authors quantitatively estimated H_2S generation suppression by NO_3 (reagent and compost-derived) addition [21]. The experiment was conducted on a beaker scale and in a complete mixing system. However, column percolation experiments are required to study the applicability of the NO_3 addition method at an actual landfill site. In regard to the addition of compost-derived NO_3, not only the injection of extract derived from compost into the waste layer, but also the utilization of rainfall infiltration through compost on top of the waste layer is assumed. Furthermore, as NO_3 addition is an alternative method for supplying oxygen, H_2S generation suppression by NO_3 addition should be compared with that by O_2 supply.

In this study, H_2S generation suppression by the addition of NO_3 from reagent and compost, and O_2 supply was quantitatively estimated by conducting column percolation experiments.

2. Materials and Methods

2.1. Outline of Experiments

H_2S was generated in an airtight acrylic column, and H_2S generation suppression by NO_3 or O_2 addition was evaluated. Specifically, the column was filled with gypsum, organic matter solution was introduced from the top, and an anaerobic condition was preserved. Then, the reproducibility of H_2S_{high} generation was confirmed. H_2S in liquid and gas phases was removed by bubbling nitrogen (N_2) gas. NO_3 or O_2 was added into the column, and the concentration of re-generated H_2S was measured.

2.2. Materials

Table 1. Details of samples.

Component	Product name, company, type
Organic matter	Dried Yeast Extract D-3, Wako Pure Chemical Industries Lactose Broth DAIGO, Wako Pure Chemical Industries (equivalently mixed)
SO_4	Waste gypsum board (crushed into <10 mm particles)
Gravel	<10 mm, washed with water
Nutrient	BOD Nutrient Buffer Pillows for 6 L sample, APHA Formulation, HACH COMPANY (1 pillow mixed with 500 mL pure water)
Seeding	Black sediment in an ornamental pond (3 g mixed with 500 mL pure water)
NO_3 (reagent)	Sodium Nitrate, 1st grade, Wako Pure Chemical Industries
NO_3 (compost)	Sewage sludge compost
O_2	Aerated water or air bubbling

Details of the samples are shown in Table 1. Medium for microbial culture and waste gypsum board ($CaSO_4{\cdot}2H_2O$) were used as sources of organic matter and SO_4, respectively. Sodium nitrate ($NaNO_3$) reagent and sewage sludge compost were used as sources of NO_3, and air was used as the source of O_2. NO_3 content in the compost was 2.8 g-NO_3/100 g-compost.

2.3. Experimental Setup

As shown in Figure 1, an airtight column was made by connecting the acrylic column (10 cm inner diameter, 40 cm height, and 3.1 L inner volume) to a 100 mL plastic syringe. The syringe was filled with 50 mL of N_2 gas. The column was filled with solid sample (1.8 kg of a mixture of gypsum particles and gravel in 1:10 weight ratio; called waste layer hereafter). The height of the waste layer was 35 cm. Then, 400 mL of a mixture of nutrient and seeding was added. Tap water was introduced to a height of 30 cm. Four columns were prepared, i.e., columns for control, NO_3 (reagent) addition, NO_3 (compost) addition, and O_2 addition. Hereafter, those columns are called control, NR, NC, and OX, respectively.

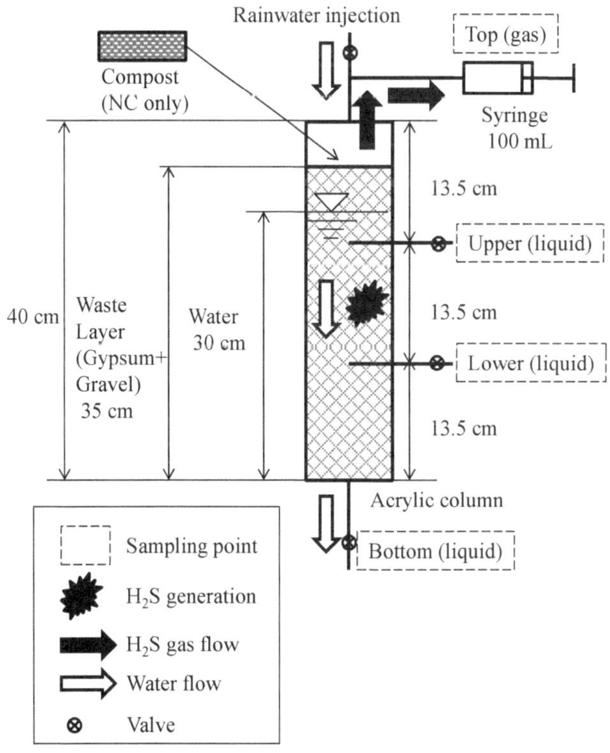

Figure 1. Schematic diagram of column.

2.4. Experimental Methods

At first, the stable generation of H_2S_{high} is necessary prior to the H_2S generation suppression experiment. The conditions for H_2S_{high} generation were as follows: maintenance of (1) an appropriate environment for SRB multiplication (anaerobic condition, temperature for mesophilic bacterial growth), (2) high concentration of organic matter, and (3) presence of SO_4.

To maintain condition (1), the column head space was filled with N_2. Valves shown in Figure 1 were shut except for rainwater injection or sampling. Most of the waste layer was filled with rainwater, and the column was set in an incubator (35 ^{o}C). To maintain condition (2), organic matter was added to rainwater. To maintain condition (3), SO_4 was introduced into the column in an amount equivalent to approximately 20 times that required for the saturation of $CaSO_4{\cdot}2H_2O$ (approximately 1700 mg-SO_4/L, 17.6 mmol-SO_4/L [22]).

Assuming rainfall at an actual landfill site, artificial rainwater (tap water) was injected with a syringe from the top of the column once every three days. The injected volume was 110 mL every three days, which corresponded to the annual precipitation in Japan, i.e., approximately 1700 mm/y. Organic matter was added to the rainwater (13.6 g-C/L). H_2S in the liquid and gas phases was removed by bubbling N_2 gas when H_2S_{high} was observed. H_2S_{high} observation and H_2S removal were conducted three times as the acclimation stage, and then, the H_2S generation suppression test was performed.

In NR, NO_3 reagent was added to the rainwater (19.05 g-$NaNO_3$/L). In NC, the waste layer was overlaid with approximately 150 g of compost. NO_3 in the compost was leached and transported to the waste layer by rainfall. The compost was replaced once every six days. In OX, two methods of O_2 supply were adopted, i.e., rainwater was aerated to saturate O_2, and air was directly bubbled from the bottom of the column. In air bubbling, air in an empty plastic bottle was introduced into the column by a closed system in order to prevent leakage of H_2S from the column (Figure 2).

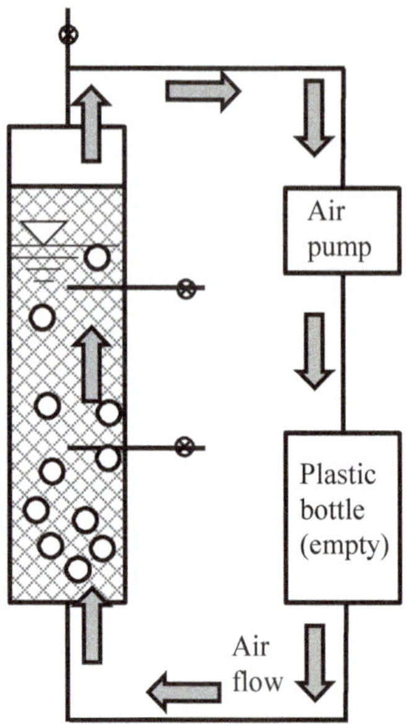

Figure 2. *Schematic diagram of air bubbling (OX).*

2.5. Analytical Procedure

The objects of measurement were the gas and liquid phases in the column. Measurement items for the gas phase were H_2S concentration and gas volume. Measurement items for the liquid phase were pH, electric conductivity (EC), ORP, dissolved oxygen (DO), TOC, NO_3 concentration, and liquid volume. Sampling was conducted once every three days.

In regard to the gas phase, H_2S was measured with a portable gas detector for worker safety (XS-2200, controlled potential electrolysis, NEW COSMOS ELECTRIC CO., LTD.). Gas volume in the plastic syringe was measured by reading the scale of the syringe.

In regard to the liquid phase, HORIBA B-212 (pH), HORIBA B-173 (EC), HORIBA-D-55 and 9300-10D (ORP), HORIBA-D-55 and 9520-10D (DO), Shimadzu TOC-V_{WS} (TOC), and Thermo Fisher Scientific ICS−1100 (NO_3) were used. Liquid volume was estimated by measuring the gross weight of the column. One hundred and ten milliliters of liquid was sampled from the liquid sampling points (Figure 1) with a plastic syringe. H_2S in the liquid phase was measured by head space sampling.

3. Results

The cumulative amounts of oxygen added (O of NO_3 or O_2) are shown in Figure 3. The unit is mmol-O/L, i.e., added oxygen/liquid volume in the column. The amount in NR was calculated from the amount of $NaNO_3$ added. The amount in NC was determined by calculating the difference between the original amount of compost NO_3 and the residual amount. The amount in OX was determined by calculating the difference of DO in water in the column between before and after aeration. The cumulative amounts of oxygen added exceeded 300 mmol-O/L in NR and NC on day 75, although it was only approximately 2 mmol-O/L in OX.

Changes in H_2S concentration and ORP in control, NR, NC, and OX at the lower point of the column are shown in Figures 4 to 7, respectively (* means N_2 gas bubbling). pH ranged from 5.2 to 8.9, showed an increase with time, and was around 8 after 40 days. EC ranged from 2.7 to 14.8 mS/cm, and was increased with time upon the addition of solution.

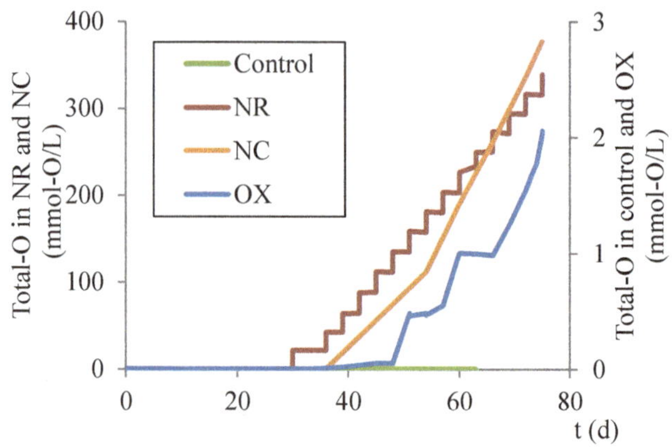

Figure 3. *Cumulative amounts of oxygen added.*

3.1. Control

ORP in control was below −300 mV (standard hydrogen electrode) after day 10 and H_2S_{high} was generated repeatedly (Figure 4). Thereafter, ORP was increased from approximately −500 to −400 mV, although the reason was unknown. The average ORP after 50 days was −356 mV. H_2S

was rapidly regenerated after removal by bubbling N_2 gas.

3.2. NR: Nitrate (Reagent) Addition

ORP in NR was increased upon the addition of NO_3 (Figure 5). H_2S_{high} was generated and ORP was below −200 mV from day 30 to day 52. After day 52, ORP exceeded −200 mV three times and no H_2S_{high} was generated. The average ORP after 50 days was −283 mV.

3.3. NC: Nitrate (Compost) Addition

ORP in NC was increased repeatedly upon the addition of compost (Figure 6). However, the average ORP after 50 days was −316 mV, i.e., almost the same as that of control. H_2S_{high} was generated repeatedly.

3.4. OX: Oxygen Addition

ORP in OX was increased drastically upon the addition of oxygen (Figure 7). H_2S_{high} was generated repeatedly from day 36 to day 51 during which O_2 was introduced by adding saturated water. However, the average ORP after 50 days was −166 mV. Thereafter, O_2 was added by air bubbling and H_2S generation was decreased.

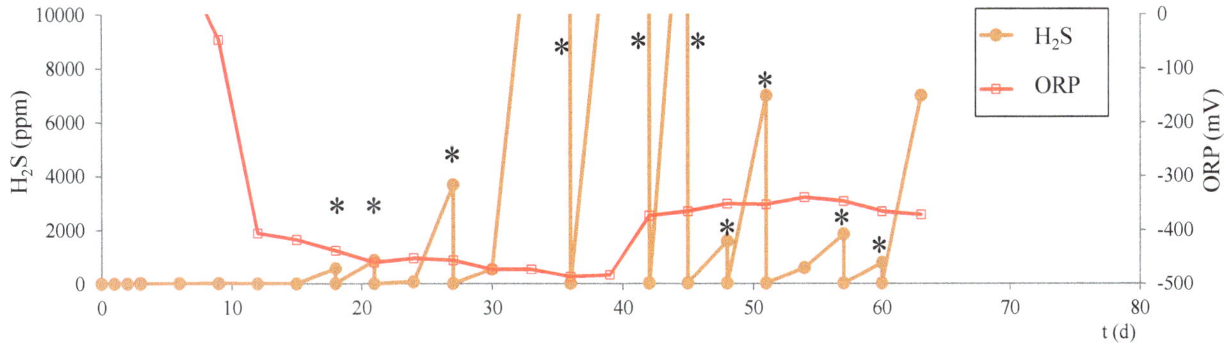

Figure 4. *ORP and H_2S concentration changes in control.*

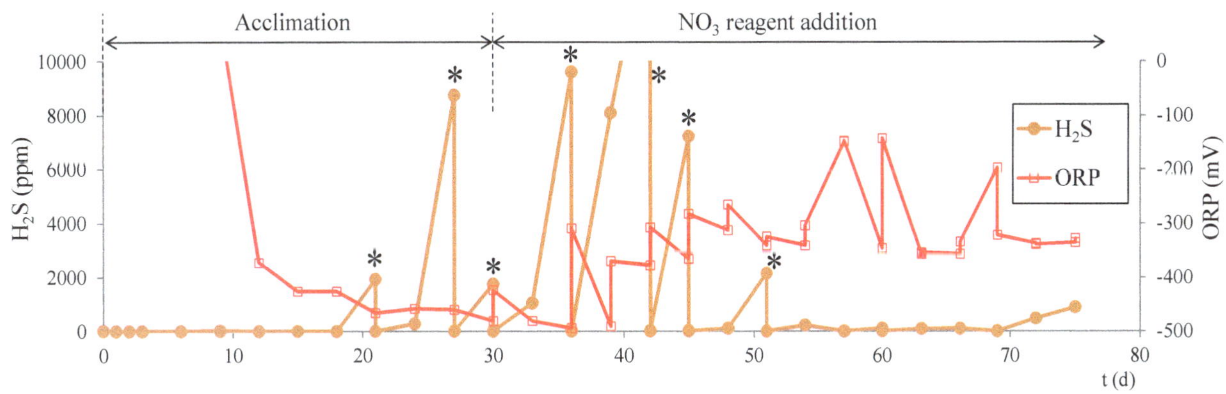

Figure 5. *ORP and H_2S concentration changes in control.*

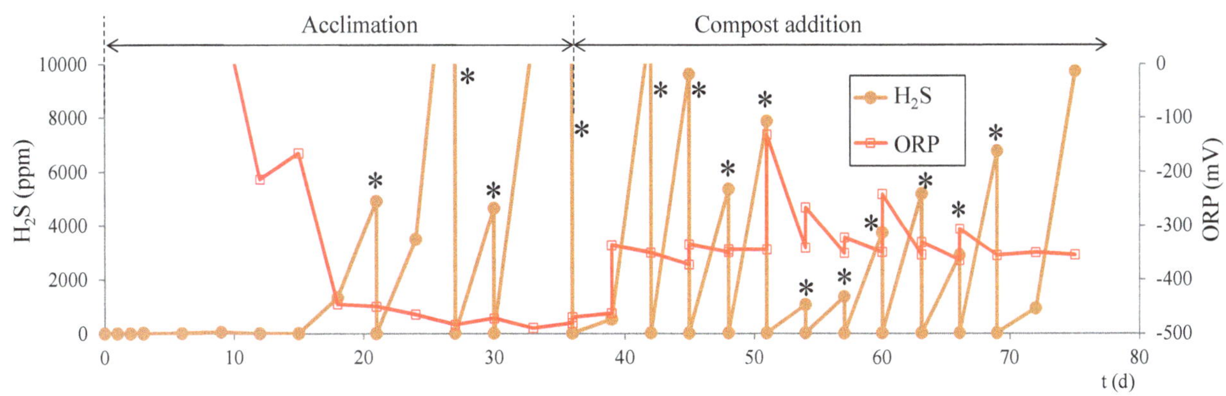

Figure 6. *ORP and H_2S concentration changes in control.*

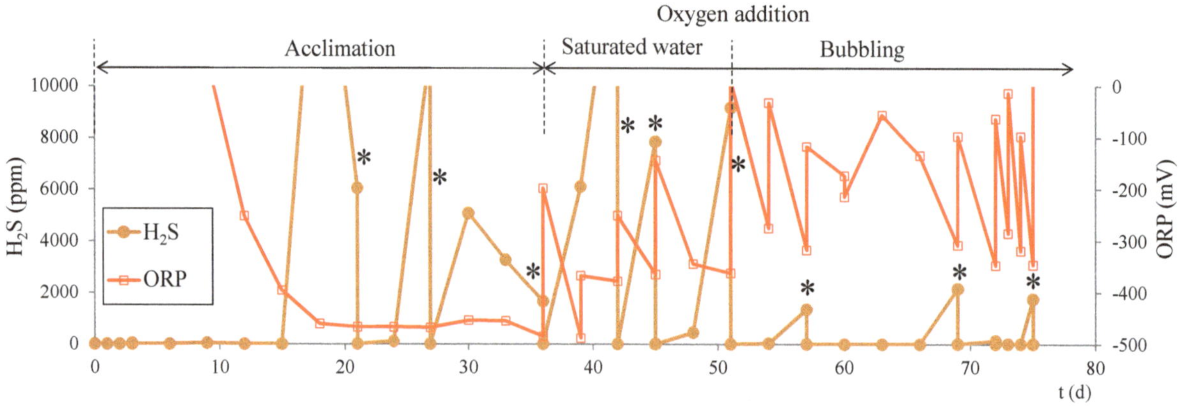

Figure 7. ORP and H_2S concentration changes in control.

4. Discussion

4.1. Quantitative Evaluation of H_2S Generation Suppression

The effects of NO_3 or O_2 addition on H_2S generation suppression were quantitatively estimated as follows. The cumulative amounts of H_2S generated are shown in Figure 8. The cumulative amount was determined from the total amount of H_2S in the column and the amount of H_2S removed by bubbling N_2 gas. It is known that H_2S exists in not only the gas phase but also the liquid phase. Therefore, the total amount of H_2S in the column was equivalent to the amounts in the gas and liquid phases. However, only the concentration of H_2S in the gas phase was measured. Using the concentration in the gas phase (determined by taking the average of the concentrations measured at the top, upper, lower, and bottom parts of the column), gas volume, liquid volume, and pH, concentration in the liquid phase was calculated according to Henry's law and an equilibrium relationship in the liquid phase [23]. As H_2S was generated from the liquid phase, the amount generated (mmol-H_2S) was converted into the amount per unit liquid volume (mmol-H_2S/L). The cumulative amounts on day 60 in control, NR, NC, and OX were 69, 26, 71, and 31 mmol-H_2S/L, respectively. The amounts in NR and OX were half that in control. However, the amount in NC was approximately the same as that in control. The effects of H_2S generation suppression were observed in NR after 45 days and in OX after 50 days. In NR, ORP exceeded −200 mV after 45 days and the suppression effect was observed. In OX, air bubbling was started after 50 days and the suppression effect was also observed. Therefore, the H_2S generation rates from days 50 to 60 were calculated. The generation rates in control, NR, NC, and OX were 0.86, 0.019, 0.75, and 0.041 mmol-H_2S/(L·d), respectively, and the rates in NR, NC, and OX were 0.02, 0.9, and 0.05 times that in control. In this way, the H_2S generation suppression effects of NO_3 reagent addition were observed.

The purpose of this study was to estimate suppression effects not in a complete mixing system, but in column percolation experiments. H_2S concentration distributions in the height direction (median value from days 50 to 70) are shown in Figure 9. In NR, although NO_3 was injected from the top of the column, H_2S concentrations at the lower and bottom parts of the column were not high compared with the top and upper parts, i.e., the suppression effect reached the deep layer. In NC, although no suppression effect was observed according to the cumulative amounts of H_2S generated mentioned above, H_2S concentrations at the top and upper parts were low (< 20 ppm), i.e., the suppression effect was observed in the shallow layer. In OX, H_2S concentrations at all the sampling points were low (< 20 ppm). The reason was probably because O_2 was supplied to all the column layers by injecting air from the bottom to the top.

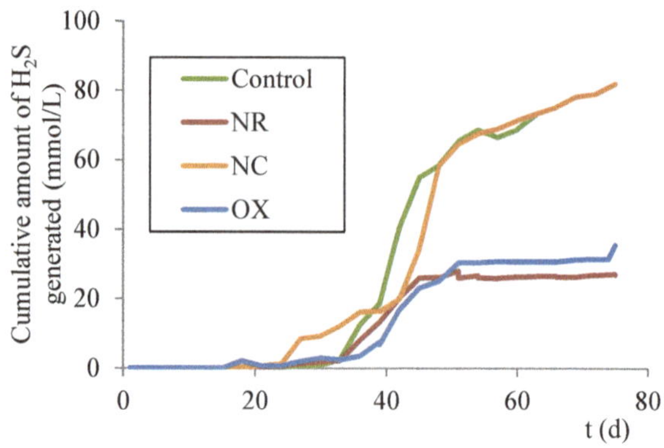

Figure 8. Cumulative amounts of H_2S generated.

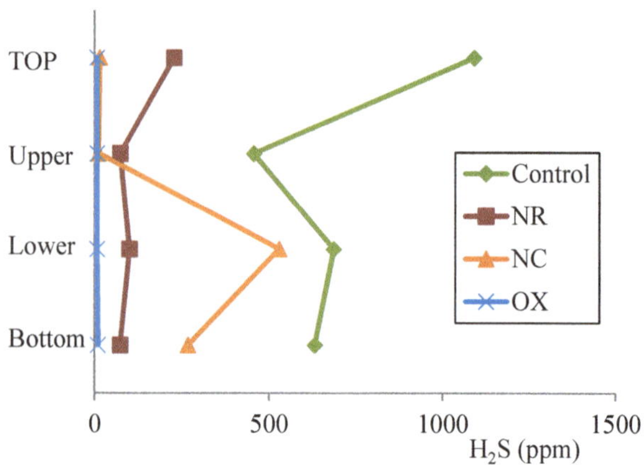

Figure 9. H_2S concentration distributions (median value from day 50 to day 70).

4.2. Problems Encountered, Application of Results, and Future Issues

In this study, H_2S generation suppression effects were estimated not in a complete mixing system, but in column percolation experiments. In our previous investigation, as we could control NO_3 concentration in the liquid phase in the complete mixing system, the relationship between H_2S generation suppression effect and NO_3 concentration (reagent and compost-derived) could be expressed [21]. In this study, as the concentration of H_2S or NO_3 varied with column depth, concentration control was difficult. Therefore, only the cumulative amounts of oxygen added and H_2S generated could be determined, and the suppression effects of NO_3 or O_2 addition were simply compared. Even though the amount of O_2 added was much smaller than that of NO_3 added, the suppression effect of O_2 was as high as that of NO_3. However, the suppression effect by injecting rainwater saturated with O_2 was low. Assuming a closed system, such as the column percolation experiment used in this study, the whole prepared air could be introduced by bubbling into the system. However, an actual landfill site is an open system. Assuming that air is bubbled into the H_2S generating zone in the waste layer, it would take a long time to dissolve O_2 in water and the air would diffuse out of the zone. Therefore, O_2 injection is difficult. It is difficult to compare the suppression effects of NO_3 and O_2 while taking the ease of injection into consideration.

The amount of NO_3 to be added to a waste layer of a landfill site where H_2S is generated could be estimated from the results of this study. Because the suppression effect of NO_3 injection from the top of the waste layer reached not only the shallow layer but also the deep layer, it is suggested that the NO_3 injection method from the top of the waste layer would be possible also at an actual landfill site. In other words, we could dispense with the NO_3 injection method by introducing pipes deep into the waste layer.

Future issues are as follows. In regard to NC, although the same amount of NO_3 was added as in NR, no suppression effect on the cumulative amount of H_2S generated was observed. Some substances contained in compost would hinder the suppression effect. Considering NO_3 injection into an actual landfill site, the utilization of not NO_3 reagent but useless NO_3 in the market should be considered. Therefore, it is necessary to investigate factors hindering the suppression effect and the possibility of utilizing useless NO_3 in the market, which has sufficient suppression effect. Environmental loading, such as eutrophication by application of the NO_3 addition method at a pilot scale or an actual landfill site, and cost reduction by using useless NO_3 in the market should be evaluated. As NO_3 addition is an alternative method for supplying oxygen, its advantages should be evaluated quantitatively. The advantages include no consumption of large amounts of NO_3 in the aerobic zone and consumption of organic matter by denitrification.

5. Conclusion

In this study, the H_2S generation suppression effects of NO_3 addition using reagent (NR) and compost (NC), and O_2 supply (OX) were quantitatively estimated in column percolation experiments.

The cumulative amounts of oxygen added exceeded 300 mmol-O/L in NR and NC, and approximately 2 mmol-O/L in OX on day 75. The cumulative amounts of H_2S generated on day 60 in control, NR, NC, and OX were 69, 26, 71, and 31 mmol-H_2S/L, respectively. The amounts in NR and OX were half that in control. However, the amount in NC was approximately the same as that in control. H_2S generation rates from day 50 to day 60 in control, NR, NC, and OX were 0.86, 0.019, 0.75, and 0.041 mmol-H_2S/(L·d), respectively, and the rates in NR, NC, and OX were 0.02, 0.9, and 0.05 times that in control. In this way, the H_2S generation suppression effects of NO_3 reagent addition were observed. In NR, although NO_3 was injected from the top of the column, the suppression effect reached the deep layer.

Acknowledgement

This research was supported by a Grant-in-Aid for Young Scientists B (25740039) from Japan Society for the Promotion of Science (JSPS).

References

[1] Lee, S., Xu, Q., Booth, M., Townsend, T.G., Chadik, P., Bitton, G., 2006. Reduced sulfur compounds in gas from construction and demolition debris landfills. Waste Management 26, 526–533.

[2] Eun, S., Reinhart, D.R., Cooper, C.D., Townsend, T.G., Faour, A., 2007. Hydrogen sulfide flux measurements from construction and demolition debris (C&D) landfills. Waste Management 27, 220–227.

[3] Xu, Q., Townsend, T., Reinhart, D., 2010. Attenuation of hydrogen sulfide at construction and demolition debris landfills using alternative cover materials. Waste Management 30(4), 660–666.

[4] Fang, J.J., Yang, N., Cen, D.Y., Shao, L.M., He, P.J., 2012. Odor compounds from different sources of landfill: Characterization and source identification. Waste Management 32, 1401–1410.

[5] López, A., Lobo, A., 2014. Emissions of C&D refuse in landfills: A European case. Waste Management 34, 1446–1454.

[6] Yue, E., Han, B., Sun, Y., Yang, T., 2014. Sulfide emissions from different areas of a municipal solid waste landfill in China. Waste Management 34, 1041–1044.

[7] Milby, T.H., Baselt, R.C., 1999. Hydrogen Sulfide Poisoning: Clarification of Some Controversial Issues. American Journal of Industrial Medicine 35, 192–195.

[8] Townsend, T., Tolaymat, T., Leo, K., Jambeck, J., 2004. Heavy metals in recovered fines from construction and demolition debris recycling facilities in Florida. Science of the Total Environment 332, 1–11.

[9] Townsend, T., Jambeck, J., Jang, Y., Plaza, C., Xu, Q., Clark, C., 2005. C&D Waste Landfill in Florida: Assessment of True Impact and Exploration of Innovative Control Techniques. Florida Center for Solid and Hazardous Waste Management, Gainesville, FL, USA.

[10] Jang, Y.C., Townsend, T., 2001. Sulfate leaching from recovered construction and demolition debris fines. Advances in Environmental Research 5(3), 203–217.

[11] Masamoto, H., Matsukiyo, A., Shigematsu, M., Matsufuji, Y., Yanase, R., 2012. Suppression of Hydrogen Sulfide Release from Waste Plasterboard Disposal in a Landfill-type Lysimeter. Journal of the Japan Society of Waste Management Experts 23(3), 144–153. (in Japanese)

[12] Naruoka, T., Ono, Y., 2004. Control measures against hydrogen sulfide production in final landfill sites. Journal of National Environmental Research 29(4), 14–19. (in Japanese)

[13] Bergersen, O., Haarstad, K., 2014. Treating landfill gas hydrogen sulphide with mineral wool waste (MWW) and rod mill waste (RMW). Waste Management 34, 141–147.

[14] Adib, F., Bagreev, A., Bandosz, T.J., 1999. Effect of Surface Characteristics of Wood-Based Activated Carbons on Adsorption of Hydrogen Sulfide. Journal of Colloid and Interface Science 214(2), 407–415.

[15] Karge, H.G., Raskó, J., 1978. Hydrogen sulfide adsorption on faujasite-type zeolites with systematically varied Si-Al ratios. Journal of Colloid and Interface Science 64(3), 522–532.

[16] Bentzen, G., Smit, A.T., Bennett, D., Webster, N.J., Reinholt, F., Sletholt, E., Hobsont, J., 1995. Controlled dosing of nitrate for prevention of H_2S in a sewer network and the effects on the subsequent treatment processes. Water Science and Technology 31(7), 293–302.

[17] Zhang, L., Schryver, P.D., Gusseme, B.D., Muynck, W.D., Boon, N., Verstraete, W., 2008. Chemical and biological technologies for hydrogen sulfide emission control in sewer systems: A review. Water Research 42, 1–12.

[18] Stumm, W., Morgan, J.J., 1996. Aquatic chemistry -Third Edition, John Wiley & Sons, New York.

[19] Farrell, M., Jones, D.L., 2010. Food waste composting: Its use as a peat replacement. Waste Management 30, 1495–1501.

[20] Hasan, K.M.M., Sarkar, G., Alamgir, M., Bari, Q.H., Haedrich, G., 2010. Study on the quality and stability of compost through a Demo Compost Plant. Waste Management 32, 2046–2055.

[21] Kitazaki, S., Xiao, K., Nakagawa, K., Kindaichi, T., Asakura. H., 2014. Hydrogen sulfide generation suppression by nitrate addition – application to solid waste landfill site. American Journal of Environmental Protection 3(5), 267–274.

[22] American Chemical Society, 2006. Reagent chemicals: specifications and procedures: American Chemical Society specifications, Oxford University Press. p. 242.

[23] Ono, Y., Tanaka, N., 2003. Preventing hazardous-level generation of hydrogen sulfide gas in landfill sites: Possibilities and administrative criteria. Journal of the Japan Society of Waste Management Experts 14(5), 248–257. (in Japanese).

5

Climate Change Adaptation: Services and Role of Information Communication Technology (ICT) in India

Anup Prakash Upadhyay, Arvind Bijalwan*

Indian Institute of Forest management, Nehru Nagar, Bhopal, Madhya Pradesh, India

Email address:
arvindbijalwan276@gmail.com (A. Bijalwan)

Abstract: Changing climate is one the most challenging issue that humankind is facing and its intensity seems to be more throbbing in the time to come. In order to overcome the alarming situation of climate change, mitigation and adaptation are two alternatives. On one hand, the mitigation is gradual and long term process however on other adaptation is usually based on quick responses to the changing scenario. Adaptation requires effective and timely delivery and exchange of all supporting mechanism and meteorological information, thus successful adaptation to climate change necessitate the Information Communication Technologies (ICT) to disseminate the information in multifarious ways. ICT not only reduce the time lag but also systemically transform the information through networked governance. ICTs can play a crucial role in supporting public outreach, building a data base, creating awareness for the adaptation against the impact of climate change and thus secure well being of biophysical entities.

Keywords: Climate Change, Information Communication Technologies (ICT), Adaptation, Mitigation, Climatic Aberration, Geography, Biophysical Entities

1. Introduction

Climate change, a comprehensive global environmental changes in the behaviour of climate in recent and distance past likely to have lethal effects on biophysical entities that can lead to critical mark on the survival of human being. The impact of climate change is undoubtedly being felt by the entire globe; however, the scenario in India seems to be comparatively pertinent due to varying geography and anthropogenic pressure. The situation, such as abnormal fluctuation in temperature (low and high), erratic rain fall (pattern and distribution), drought, flood etc are the witness of climatic aberration leading to series of problems in the survival and existence of the human being.

The consequences of climate change in India can be observed in all the sectors; however the fluctuation in time of flowering pattern in the plants (early and late), rising trend of unpredictable disease in animals including human being and shifting agricultural cropping pattern and sequences are some of them. Take an example of one of the prominent sector (agriculture) of India basically includes crop husbandry, animal husbandry, horticulture and fisheries, experiencing vagaries in climatic parameters particularly, rise in temperature and erratic rainfall pattern making agriculture to more vulnerable. The emerging situations like drought to flood in the agricultural fields and chilling to heat energies in crops and animals making the toppling state of Indian agriculture. It is very much evident that the major impacts of climate change will be on agricultural and food systems over the next few decades [1]. However application of crop modeling tools, have pointed out that climate change is likely to reduce food availability due to a reduction in agricultural production [2].

Now there is a unified voice from the scientific community, policy makers and common man of the country to develop proper adaptation and mitigation mechanism against changing climatic parameters. In order to overcome the alarming situation of climate change, adaptation and mitigation are two options. Adaptation is said to be older than Mitigation. Various strategies of adaptation to climate change have been employed throughout the history, while mitigation initiatives have evolved only when the scientific community reached on a conclusion and determined a possible interaction between human actions and climate. There is uncertainty in the exact intensity and impact of climate change because of complex interaction in ecosystem, improper feedback.

2. Climate Change Adaptation

Adaptation to climate change can be expressed as a fine-tuning with ecological, economic and social systems in response to observed or expected changes in climatic stimulus and their effects [3], moreover, adaptation can also explained as a set of approaches and actions taken by human in anticipation to the changing climate (acclimatization). Hence adaptation is a constant flow of activities, performance, choice and approaches that report to decisions about all aspects of life and that reflects existing social norms and processes [4]. Though, mitigation is equally important, however it is long term strategy and the real beneficiaries of mitigation will be the coming generations.

The adaptation is a great deal toward the minimizing the impact of climate change and as quick as needed in the present time. For example, remote sensing techniques have been used for environmental monitoring purposes, decision support tools and information and advisory system have been designed to guide users in the planning and implementing adaptation processes. In country like India the knowledge about adaptive ability of dry land farmers, forest dwellers, fisher folk and nomadic shepherds is very low [5] which directly hit such poor communities. There are two main practices of adaptation; autonomous and planned adaptation, the autonomous adaptation is mainly triggered by ecological changes in natural systems whereas planned adaptation is the result of a deliberate policy decision, creation of awareness and applying tools of adaptation [3]. There is an important need to compare the cost and benefits of adaptation strategies for making equilibrium among the society, further, if adaptation to climate change could be carried out at negligible cost then it may be less expensive, at least in the short-term, than any alternate strategy [6].

The task of adaptation to the impact of climate change in India requires wide spread knowledge, awareness and their communication not only to the scientific and administrative community but also to the common man's level. There are countless effects likely to be projected for climatic vagaries, however, these effects are weakly understood and our projections to climate change are still not very comprehensive and communicative. The factual impact of climate change in Indian is difficult to predict that too becomes more complex in diverse agro-ecological regions. Adaptation is the major corridor to pass through Indian situation for the sustenance in future. The sharing of information of adaptation to the common man is the success for fighting against climate change, yet the question is remaining same as how to disseminate the techniques and technology to the last man standing on the row. In this context, Information Communication Technologies (ICT) is being helpful to disseminate the knowledge of adaptation for fighting against climate change. TCT is a wide term which refers to all computer-based advanced technologies for managing and communicating information. In a literal sense it is broader than Information Technology (IT). It is the study, design, development, implementation, support or management of computer-based information systems, particularly software applications and computer hardware.

In the current scenario, ICTs not as such playing full-fledged role in the climate change sector however ICT has been applied in the study of the issue of climate change, but the application of ICTs in the adaptation and mitigation domains is yet to explore its full potential. As many international analysts noticed, ICT is greatly absent in this field, especially if we separate the systems based information management (such as Information Systems) from the technologies that are more oriented to communication services [7]. The World Telecommunication Development conferences, in 1998, 2002 and in 2006 [14-16] invited the Telecommunication Development Sector, in general, and the Telecommunication Development Bureau in particular, to assist developing countries in the use of information and communication technologies for the protection of the environment. For climate change there could be various reasons, but the major concern is man-made climate change as it is leading to gradual increase in warming of planet at an alarming rate.

It has been perceived that yesterdays backward nation were those who missed industrial revolution but tomorrows backward nation will those who will be missing communication revolution. ICT has revolutionised the whole world. Now when India is regarded as hub of the ICT every sector in the country is talking of using ICT. However efforts in this direction are state specific while a centralized approach is advocated to avoid duplication of efforts and to maintain uniformity and compatibility. ICT has shown immense potential in almost the entire prime sectors; be it medical, agriculture, Science and Technology etc., however, environmental issues also can't remain in isolation and untouched. ICT has vast potential and its potential could be utilised in combating the problems of environment and changing climate. It has untapped potential in this sector [8]. ICT has already been used in various countries in various forms ranging from cutting down the emissions to developing a data base of change occurring. It is being used in satellite communication to provide weather forecast, terrestrial systems that are also used for dissemination of information concerning different natural and man-made disasters (early warning), atmosphere composition data base, Ocean parameters etc. Till date use of ICT, climate change adaptation, mitigation and sustainable development has largely focused on ICT industry audience and its larger scope is still not much discussed. The ICTs have the potential to strengthen long-term adaptation strategies, decision-making processes, raising awareness, capacity building and participation, among other areas [9].

The situation at present is argued as ICTs provide generic support to the process of information gathering, decision making, implementation and evaluation for national level adaptation. Specific ICT applications have enough potential to enable delivery of particular adaptation actions for the threat being imposed by climate change. Moreover, knowledge and information can play a key role in overcoming constraints. Timely information and awareness are pivotal for building and strengthening the capacity of multiple stakeholders involved in adaptation strategies at the micro, meso and macro levels. ICTs, the Internet based applications, mobile phones, telecentres, community radio, etc those are increasingly available in developing regions like India posses an exceptional potential and provides opportunity to improve the creation, management, exchange and application of relevant climate change information and knowledge.

The climate change adaptation could be in two ways *viz.* planned and spontaneous. One possible assumption says it could be planned another is emergent. While the latter refers to spontaneous actions taken by people affected by climatic change or the events led by climate change, planned or policy driven adaptation processes involve the formulation of strategies that consist of a general plan of action, including policies and measures, for addressing climate change impacts within a given context. As many other developing countries, India is facing significant challenges linked to the increasing intensity and frequency of climate change related events. India with a rapidly expanding ICT industry and a advanced telecommunications sector, has been a pioneer in the exploration of ICTs as key tools towards the achievement of adaptation goals of climate change.

ICT potential as it has been discussed widely and yet many things to be done to harness the full potential of ICT in the field of climate change adaptation. There are certain issues which has major bearing on the role of ICT in climate change adaptation that too when this adaptation has to be a planned one [10]. Adaptation takes place at all levels in society may it be large-scale interventions which are generally driven by governments and regional organizations or the autonomous actions taken by threatened communities and individuals. In a study [11] mentioned that aappropriate adaptation in climate change requires:

- Access to current information and understanding of the physical and socio-economic changes along with the potential changes and effects which are or could occur as a result of climate change
- Appropriate adaptation calls for design and implementation of effective responses mechanism to the challenges posed by climate change or potential challenges and opportunities arising from climate change
- Coordinated action by all stakeholders at all levels in synchronous fashion, including those at the local level
- Reliable information and appropriate guidance on actions that may be or likely to be taken to increase the resilience of vulnerable.

Human skills are just as essential to the application of ICTs for adaptation as infrastructure, equipment and applications. ICTs can play a crucial role in supporting public outreach and building awareness of the impact of climate change and adaptation, as well as offering opportunities to address those challenges. To analyze the role that ICTs can play in supporting top-down and bottom-up approaches to adaptation, as well as the development of new arrangements for changing pattern of global governance and assessing the local and national responses to climate change, it is necessary to use a framework found in much of the current literature on the relationship between ICTs and the environment. In a study conducted about strategic use of ICT [12] in various sectors such as Social, rural development, Sustainable natural resource management the same is possible to apply with the three main functions of ICTs to the development of an adaptation strategy for climate change. First, data collection is important for constant monitoring of the climate system. Second, it is important to have tools for processing data and information which will be guiding the people in the analysis, planning and implementation of possible adaptation strategies. Finally, strategic communication tools are needed to manage the adaptation process and build on the knowledge which is generated within and outside the given system. Based on the current and similar kind of literature ICT application in climate change adaptation could significantly be categorized in three distinguished categories.

3. Direct Role of ICT

The use of ICTs can be for assessing the effect of climate change by developing a data base. This data base could be utilized for monitoring and measuring climate change, measuring the effects of climate change and controlling the level of interactions with the environment (e.g., the Internet of Things, monitoring and measurement of climate change impacts, the impact of weather/ climate on natural resources such as water sea levels, annual rainfall and weather-dependent economic sectors such as agriculture, horticulture, forestry, fisheries, and the potential impact of climate change on health related issue which could be of long term in nature).

Indirect Role of ICT

ICT is and can act as medium for mass communication. It could continue doing so in the field of climate change with a lot of efficiency. The ICTs could be used as a medium for increasing awareness and facilitating dialogue about the effects of climate change. It could be utilized for creating mass awareness and developing sensitivity among masses for climate change and responsible factors for climate change (e.g., Web 2.0 and social networking applications, early-warning systems concerned particularly with acute events resulting from climate variability and climate change).

4. System Based Role of ICT

The ICTs could find a new role in the changing economic and social organization as enabler for networked governance. The new forms of economic and social organization and decision-making that will likely be needed not only to adapt to climate change, but to achieve sustainable development (e.g., the kinds of Internet-based, open organizations and decision-making processes, transmission of information as well as guidance to and between local communities with reference to impacts that they may experience due to climate change, expert advice system which could tell people about actions that they could take to protect themselves against sudden weather-induced crises, and potential lifestyle along with behavioral changes that will help to secure lives and livelihoods in the longer term).

Understanding local situation the ecosystem itself and its social and cultural environment is very much important if adaptation interventions are to be successful. In general keeping other factors constant current uses of ICTs in the context of climate change may achieve the objectives of generating, organizing and communicating information about the risks resulting from climate change, climate variability and extreme climate events, as well as preparing for their effects on food security and water supply.

ICT for climate change adaptation may help in developing information systems within the food sector that are better able to reflect household access to food and food consumption. ICT could prove to be a great asset in developing early warning and hazard risk information systems to deal with the additional fire hazards associated with climate change and to enable integrated fire management; and developing ICT mapping tools to map vulnerable areas and provide spatial representations of climate change impacts.

These practical applications use a wide variety of technologies which are interconnected using existing communications networks and services, including:

- High-value remote monitoring equipment such as satellites
- Networks of remote sensors
- Global positioning and GIS applications
- Communications services such as the internet, mobile networks and SMS
- Handheld devices such as mobile phones and PDAs.

5. Challenges Ahead

ICT as it has been quoted in various literature, itself a major factor responsible for climate change. The intensity of ill effect of ICT is found lesser than the positives which could be derived by use of ICT for climate change adaptation. But the fact is above quoted statement lacks substantial evidence. It becomes itself a challenge for establishing ICT as a tool for climate change adaptation. Besides it is very much evident that building the capacity for integrating ICTs into adaptation policy development will have significant benefits, if particular attention is being paid to including more detailed application of ICTs in National Adaptation Plans. For a successful employment of ICT at the community level it is essential to frame strategy within Communication for Development this strategy which meant to be as the systematic application of processes, strategies and principles of Communication aimed at bringing positive social change itself is a big challenge in South Asia which have so much of diversity [13].

Engagement of the ICT private sector in climate change adaptation is less substantial and widespread phenomenon till today. It is more in use in climate change mitigation policy, where the private sector is directly involved in developing clean technologies, energy efficiency, dematerialization to reduce carbon footprints and various other areas of innovation. There is need to do more work to engage private ICT sectors/players strategically to review where privately motivated interests in applications would directly or indirectly enhance adaptation while also delivering commercial value. Identifying a public private partnership platform for integration mechanism to combat the issues of climate change and facilitate adaptation itself is a big challenge.

Another challenge in establishing and utilizing ICT potential for climate change adaptation is in making the use of ICTs embedded in the adaptation planning processes. This step is already being taken up at global level by most of the countries. Preparation for these should include a capacity building programme to review where and how investments in ICTs might be best integrated within these plans need to be decided. In India most of the ICT-related interventions that have taken place in last decade or so are either at a macro or a micro level. When we say macro level use of ICT for climate change adaptation it resembles with the intergovernmental dialogue, national strategies related with the use of ICT for climate change monitoring system, large-scale projects specifically meant for weather monitoring by satellite and large-scale sensor networks. Micro level ICT use for climate change adaptation programmes or projects aimed at increasing awareness and information for vulnerable communities and individuals. This awareness creation is done with assumption that it will help people identify and manage their own adaptation needs. There is a lack of the meso level use of ICT for climate change adaptation which is the need of the hour in bridging the gap between grand strategies having a national level perspective and local circumstances which are very much regional in nature.

6. Conclusion

ICTs as of today in the country like India have not yet become effectively integrated into adaptation planning when the issue of climate change adaptation is discussed. The potential of ICT-related interventions is insufficiently explored. When we say ICT for climate change adaptation we need reliable, high quality communications infrastructure which will ensure the effective use of ICTs. This ICT-enabled creative capacity is one of the keys to adapting to the consequences of climate change, particularly in the most vulnerable regions of India in which areas that are geographically, economically or socially marginal, and therefore tend to lie at the edges of mainstream concerns. Linking successful ICT pilots with farmer field schools may be one way of providing information and agro-advisory services. This could include those related to climate and building capacity on a wider scale. Such approaches enhance the understanding of farmers' needs as it has bottom up orientation and thus help redress and improve the existing situations for delivering agro-met services.

As it has been discussed that how ICT can play its role in climate change adaptation the point here to understand is technology can only deliver services and applications if that underlying infrastructure is available. It needs strengthening across the country in terms of infrastructure and capacity building. ICTs can play a crucial role in supporting public outreach, building a data base, creating awareness of the impact of climate change and adaptation, suggest possible interventions based on the data base for adaptive mechanism as well as ICT can offer opportunities to address those challenges.

Acknowledgement

The authors are thankful to the Director, IIFM, Bhopal, India for encouragement and the anonymous reviewers for improving the quality of this manuscript.

References

[1] Brown, Molly E. and Funk, Christopher C. (2008). Food Security under Climate Change. Science; Vol. 319, no. 5863, pp. 580-581.

[2] Lobell, David B.; Burke, Marshall B.; Tebaldi, Claudia; Mastrandrea, Michael D.; Falcon, Walter P.; Naylor, Rosamond L. (2008). Prioritizing Climate Change Adaptation Needs for Food Security in 2030. Science; Vol. 319, no. 5863, pp. 607-610.

[3] IPCC (2007). Fourth Assessment Report - Climate Change: Synthesis Report, 2007.

[4] Daw, T., Adger, W. N., Brown, K., and Badjeck, M. C. (2009). Climate change and capture fisheries: potential impacts, adaptation and mitigation. In: Climate change implications for fisheries and aquaculture overview of current scientific Knowledge, Cochrane, K., Young, C. De, Soto, D., & Bahri, T. (Eds). FAO Fisheries and Aquaculture Technical paper: No. 530, pp.107-150, FAO, Rome.

[5] Ravindranath, N. H. and Sathye, J. (2002). Climate Change and Developing Countries, Kluwer Academic Publisher, Dordrecht, Netherland.

[6] Sathaye, J., Shukl, P.R. and Ravindranath, N.H. (2006). Climate Change, Sustainable development and India: Global and national concerns. Current Science, 90 (3), 314-325.

[7] Sala, S. (2009). Information and communication technologies for climate change adaptation, with a focus on the agricultural sector. Thinkpiece for CGIAR Science Forum Workshop on "ICTs transforming agricultural science, research and technology generation," Wageningen, Netherlands. 16–17 June 2009.

[8] Parry, Jo-Ellen., Drexhage, J. and Gagnon-Lebrun, F. (2008). Adaptation in a Post-2012. Climate Regime," in John Drexhage, Deborah Murphy and Jenny Gleeson (eds), *A Way Forward: Canadian perspectives on post-2012 climate policy*, IISD. http://www.iisd.org/pdf/2007/a_way_forward.pdf. Accessed on 13th November, 2014.

[9] Heeks, R. (2010). 'Coherent ICT-for-Development Policies', paper presented at the workshop Policy Coherence in the Application of Information and Communication Technologies for Development, Organization for Economic Co-operation and Development (OECD) and info-Dev, Paris, France, 10-11 September. www.oecd.org/dataoecd/18/8/43762187.pdf Accessed on 19th December, 2015.

[10] Prasad, L., and Heeks, R. (2011). ICT Enabled Development of Capacity for Climate Change Adaptation, Centre for Development Informatics, University of Manchester http://www.niccd.org/PantHeeksClimateChangeAdaptationICTs.pdf. Accessed on 15th November, 2014

[11] Karanaisos, S. (2011). New and Emergent ICTs and Climate Change in Developing Countries Centre for Development Informatics, University of Manchester, Manchester. http://www.niccd.org/KaranasiosClimateChangeEmergentICTs.pdf. Accessed on 13th November, 2014.

[12] Mohammad, U.M., Abdur-rahman, M.D., Kazi A. and Rajatheva, M. A. P. (2005), ICT Based Sustainable Rural Business Opportunities in Developing Countries",In American Journal of Applied Sciences, Vol. 2, No. 8, pp. 1256-1260.

[13] UNFCCC (2007). Assessing Climate Change Impacts and Vulnerability: Making Informed Adaptation Decisions, United Nations Framework Convention on Climate change (UNFCCC), http://unfccc.int/files/adaptation/application/pdf/11unf051_nwp-was-web[1].pdf. Accessed on 22nd December, 2015

[14] World Telecommunication Development Conference, Valletta, Malta (1998). Resolution 8: "Telecommunication support for the protection of the environment." http://www.itu.int/ITU-D/conferences/wtdc/1998/VAP.html. Accessed on 2nd November, 2014.

[15] World Telecommunication Development Conference, Istanbul, Turkey (2002). Recommendation 7: "Role of telecommunication and information technologies in the protection of the environment." http://www.itu.int/ITU-D/cyb/app/docs/Rec007.pdf. Accessed on 2nd November, 2014.

[16] World Telecommunication Development Conference, Doha, Qatar (2006). Resolution 54: "Information and communication technology applications." http://www.itu.int/ITU-D/cyb/app/docs/wtdc_resolution_54.pdf. Accessed on 2nd November, 2014.

6

Germination and root nodule formation of soybean (*Glycine max* (L.) Merr.) in ridomil and chlorpyriphos treated soil

Ibiang Young Bassey[1,*], Ekanem Bassey Effiong[1,2], Usanga Daniel Archibong[1], Williams Uduak Ita[1]

[1]Department of Genetics and Biotechnology, University of Calabar, PMB 1115, Calabar, Nigeria

[2]Department of Science Technology, Akwa Ibom State Polytechnic Ikot Osurua, PMB 1200, Ikot Ekpene, Nigeria

Email address:

youngangale@yahoo.com (Ibiang Y. B.), ekanembassey@yahoo.com (Ekanem B. E.), danielusanga@yahoo.com (Usanga D. A.), favourwilliams29@gmail.com (Williams U. I.)

Abstract: Germination and root nodule formation of soybean (*Glycine max*) was evaluated in soil treated with Ridomil (R), Chlorpyriphos (CH), and a Combination of both (COM); using a completely randomized design, and a 3×3 factorial arrangement. Factor 1 was the pesticide type with 3 levels - R, CH, and COM; while factor 2 was the contamination rate (v/wt) with 3 levels - 0% (control), 2%, and 4%. Aqueous pesticides preparations (as recommended by manufacturers) were applied once, before seeds were planted. More than Ridomil and Combination, significantly ($P<0.05$) diminished percentage germination due to Chlorpyriphos was observed at 2, 3, and 4 weeks post application. The control had significantly higher germination than 2% and 4% groups, up until the 5th week. The Chlorpyriphos and pesticide Combination groups both had significantly lower root nodule number/plant, than Ridomil. Root nodule number/plant, and nodule biomass/plant were significantly reduced due to pesticide stress at 2% and 4%. Pesticides application at 2% and 4% negatively affected germination and root nodule formation; but there was no indication of an enhanced effect due to pesticides combination.

Keywords: Chlorpyriphos, Ridomil, Pesticide Interaction, Toxicity, Germination, Root Nodule, Soybean

1. Introduction

With the growing global concern about environmental sustainability, the impact of environmental stressors on the welfare of life forms and ecosystem processes remains a front burner in the intellectual and public domains. While environmental pollutants such as spilled crude oil, heavy metals, and toxic industrial effluents are widely regarded as stressors of living systems, Ashraf and Harris [1] describe plant "stress" to be any environmental factor that limits crop productivity or obtainable biomass, including drought, heat, cold, pathogens, salt, chemicals, etc [2]. Rapidly increasing human population, especially in developing countries, necessitates increased food and cash crop production, with attendant increase in utilization of farm inputs such as pesticides and fertilizers [3]. Despite the usefulness of pesticides in crop protection and food preservation, they have been associated with food poisoning, biological magnification, toxicity to non target species, and are agents of stress in the ecosystem. Further, pesticides use has not been without peculiar challenges such as high residues on food produce; the use of banned formulations; improper application procedures/precautions; inadequate agri-extension support and monitoring; soil and water pollution; amongst others [4, 5]. In Nigeria, cocoa farming contributes greatly to overall pesticides use, with formulations like Ridomil, deltamethrin, chlorpyriphos, etc, being popular with local farmers [6].

Aside toxicity to micro and meso-fauna [7], short and medium term environmental toxicity of pesticides have been illustrated in the form of oxidative stress in plant cells with reduced plant metabolism and growth, reduced nodulation, nitrogenase activity, as well as legume-rhizobium symbiosis, amongst others [8-11]. Other parameters such as germination, litter decomposition, soil enzyme levels, microbial counts/metabolic coefficients [12, 13], are influenced by chemical pesticides, and have a potential to be used as markers of environmental stress; although plants have generally not been utilized in environmental health monitoring as much as animals like fish and macroinvertebrate have. Soybean (*Glycine max* (L.) Merr.) is of ecological significance due to its role in enhancing nitrogen fixation by harbouring nitrogen fixing bacteria in its root nodules, as is typical of legumes [14].

While utilization of pesticides in line with manufacturer recommendations reduces the risk of harm due to their handling and application, chemical pesticides in use still remain an environmental hazard to be managed; and assessment within the scope of their local environmental releases/use patterns shed more light in this regard. Against the backdrop of pesticides drift and deposition, as well as dissolution in and adsorption to matter when they are applied, organisms could be exposed to more than one pesticide in the ecosystem, at any given locality. And interaction between two (or more) pesticides can occur, resulting in (enhanced) toxic effects much greater than is obtainable for equivalents of either pesticide alone [15]. Ridomil is a fungicide employed against the black pod disease, while chlorpyriphos is a broad spectrum insecticide used against mirids on cocoa. As a result, both pesticides do get to be used together in the same vicinity; taking into consideration the diversity of spraying regimen possible at farms. These form much of the underlying rationale for this study, using two ecologically significant processes; germination and root nodule formation, in the soybean plant. Having observed the paucity of literature reports on the combination of both pesticides, this study was carried out to evaluate the effect of Ridomil and Chlorpyriphos, singly and in combination, on germination as well as root nodule formation, in soybean; with a view to assessing the potential for enhanced effects, when both pesticides are used at rates recommended by the manufacturers.

2. Materials and Methods

2.1. Pesticides Preparations

Ridomil Gold® 66WP (6% metalaxyl; 60% copper I oxide) (Syngenta Agro AG) and Chlorpyriphos 20% EC (Perfect Killer®, Nantong Jinling Agric Chem Co. Ltd, Jiangsu, China) were purchased from Egoob Chemicals, Watt market Calabar. Manufacturer recommended preparation of Ridomil for use on cocoa was 50g in 15 L of water; while that for Chlorpyriphos was 1250 ml in 500 L of water. These were stated on the pesticide labels, and adopted for the study. 1 L aqueous pesticides preparations were made for both pesticides, in accordance with the manufacturer recommendations (i.e. 3.33 g L^{-1} for Ridomil; and 2.5 ml L^{-1} for Chlorpyriphos), and applied immediately to respective experimental units.

2.2. Experimental Design and Treatment Groupings

Employing a Completely randomized design, the experiment was laid out in a 3×3 factorial. Factor one was the pesticide type with three levels - Ridomil (R), Chlorpyriphos (CH), and Combination of both (COM); while factor two was the application/contamination rate (v/wt), also with three levels - 0% (control), 2% (20 ml of aqueous pesticide preparation kg^{-1} of soil), and 4% (40 ml of aqueous pesticide preparation kg^{-1} of soil). This gave a total of nine treatment combinations. 20 ml of pesticides combination was achieved thus: 10 ml of Ridomil preparation + 10 ml of Chlorpyriphos preparation kg^{-1} of soil; while 40 ml of pesticides combination was achieved in similar vein i.e. 20 ml of Ridomil preparation + 20 ml of Chlorpyriphos preparation kg^{-1} of soil). Three replicates were made for each treatment combination, giving a total of 27 experimental units.

2.3. Experimental Procedure

Top soil was collected 0-20 cm deep at Biological Sciences Experimental Farm, University of Calabar, Calabar. The soil (5.3% clay; 6.7% silt; 87.2% sand; pH 6.03) was sieved using a wire mesh (<5mm) to remove stones and other debris, mixed thoroughly, and kept for one week on a spread-out polythene sheet. 1 kg of soil sample was weighed into labeled plastic containers, to which corresponding pesticide preparations were applied, once. Distilled water was added to ensure even liquid content in all vessels (plus control), and approximately 70% of maximum water holding capacity [12]; and mixed. After pesticides treatment, experimental units were kept in the open, exposed to natural daylight but protected from rainfall. After 24 hours, soybean (*G. max*) seeds were planted. Seeds were obtained from the germplasm collection of Department of Genetics and Biotechnology, University of Calabar, Calabar, and were of the same cultivar type; TGX 728-ID. Prior to planting, seeds were soaked in tap water for 5 minutes, and floating ones discarded [16]. 15 seeds were planted in each container, at a depth of 2cm. After 72 hours, soybean seeds were recovered for estimation of percentage germination (based on unequivocal radical emergence) [17]. With the exception of pesticide treatment, which was done once, seed planting and percentage germination was determined weekly, for 7 more weeks. Only distilled water (up to 70% holding capacity) was repeatedly added to soil (with re-mixing) in each container, once a week, exactly 24 hours prior to seed planting. For the root nodule study, 5 kg of soil sample was weighed into polythene earth bags with holes and polluted with respective pesticides preparations. Soybean seeds were planted and later thinned to 2 stands per bag. After three months the earth bags were carefully destroyed with a knife in a bucket of water, and number and dry biomass of root nodules (oven dried at 50°C until constant weight) was determined. In the same vein, pesticide treatment was performed once, while weekly addition of water was done at the same volume in all the experimental units. A plastic tray was placed under each earth bag and collected water (drain) was returned into the same earth bag.

$$\% \text{ Germination} = \text{No. of germinated seeds/Total no. of seeds planted} \times (100) \quad (1)$$

2.4. Statistical Analysis

All data generated were subjected to two-way analysis of variance (ANOVA) to check for significant differences between the treatment groups, and conclusions drawn at 5% probability level. Least significant difference (LSD) test was performed to separate means, where significant differences where noticed.

3. Results

3.1. Germination Rate

Table 1 show the effect of the pesticide types (Ridomil, Chlorpyriphos, and their Combination) on percentage germination of soybean. In the first week, no significant (P>0.05) differences were observed in percentage germination, between the pesticide types, but Chlorpyriphos had the lowest germination rate (54.07%) while the combination had the highest (61.47%). In the second week, significant (P < 0.05) differences were observed in the germination rates, with chlorpyriphos having the lowest value (57.77%) which differed significantly from that of Ridomil (69.62%) and combination (69.58%). In the third week, Ridomil and Combination had no difference between them for soybean germination rate. But their values, however, differed significantly from that of Chlorpyriphos, which had the lowest value. In the fourth week, Ridomil had a significantly higher value (67.40%) than Chlorpyriphos (55.55%) and Combination (60.73%). In the fifth to eight weeks there were no significant differences in germination rate, between the pesticide types.

Table 1. *Effect of pesticide type on germination of soybeans (G. max).*

Week	R	CH	COM	LSD
1	57.03 ± 6.09	54.07 ± 6.23	61.47 ± 6.55	NS
2	69.62^a± 2.96	57.77^b ± 6.47	69.58^a ± 1.73	8.33
3	69.55^a ± 3.33	58.51^b ± 6.36	69.62^a ± 3.86	10.2
4	67.40^a ± 2.59	55.55^b ± 4.44	60.73^b ±3.22	6.00
5	64.44 ± 4.15	67.40 ± 4.77	67.40 ± 3.91	NS
6	65.92 ± 3.03	68.14 ± 3.81	68.14 ± 2.42	NS
7	64.44 ± 2.93	65.25 ± 3.03	65.51 ± 3.09	NS
8	66.47 ± 2.89	65.95 ± 1.62	64.77 ± 3.68	NS

Values are presented as Mean ± SEM. ^{ab}Values across the table with similar superscript are not significantly different at 5% based on ANOVA. NS = Not significant; R=Ridomil; CH = Chlorpyriphos; COM = Ridomil + Chlorpyriphos Combination.

Table 2 show the effect of pesticides concentration on germination of soybeans. In the first week, pesticide concentration showed dose dependent significant differences, with the control (0 ml kg^{-1}) having the highest percentage germination (75.55%), followed by 20 ml kg^{-1} (58.51%) and 40 ml kg^{-1} (38.51%), which had the lowest. All three concentrations differed significantly from one another. In the second and third week, the effect of pesticide concentration differed significantly in a dose-dependent manner, with the control having the highest germination rate and 40 ml kg^{-1} having the lowest. In the fourth week, the control had significantly higher germination (78.88%) than both 20 ml kg^{-1} (61.25%) and 40 ml kg^{-1} (55.55%). Again this was the case in the fifth week, with the control having a significantly higher value than the other groups. From week six to eight, there were no significant differences in germination rate between the control and the other treatment groups. Nevertheless, control had higher values than 20 ml kg^{-1} and 40 ml kg^{-1} treatment groups.

Table 2. *Effect of pesticide contamination rate on germination of soybean (G. max).*

Week	0 ml kg^{-1}	20 ml kg^{-1}	40 ml kg^{-1}	LSD
1	75.55^a± 2.93	58.51^b± 4.12	38.51^c± 3.29	10.47
2	73.33^a±1.92	68.36^a ±2.51	56.29^b ±6.19	8.33
3	79.99^a ± 3.84	64.44^b± 5.44	59.25^b ±3.41	10.20
4	78.88^a± 1.11	61.25^b± 3.03	55.55^b± 4.84	6.60
5	75.55^a ± 2.93	64.44^{ab}± 3.51	59.25^b ± 4.22	11.98
6	73.33 ± 1.92	65.81 ± 3.64	63.7 ± 2.51	NS
7	76.66 ± 1.92	69.29 ±2.51	69.25 ± 3.59	NS
8	74.44 ± 1.11	69.51 ± 2.89	69.25 ± 3.75	NS

Values are presented as Mean ± SEM. ^{abc}Values across the table with similar superscript are not significantly different at 5% based on ANOVA. NS = Not significant; 0ml kg^{-1} = control; 20 ml kg^{-1} = 2% (v/wt); 40 ml kg^{-1} = 4% (v/wt).

3.2. Root nodule Formation

There were significant differences in root nodule number/plant between the pesticide types, with Ridomil having the highest number (6.00) while Chlorpyriphos had the lowest (4.77) (see Table 3). There were no significant differences in root nodule biomass/plant between the pesticide types, but the highest biomass was recorded in the Ridomil group (0.39g), while the lowest was recorded in the chlorpyriphos group (0.26g). Pesticides concentration had significant effects on root nodule number/plant (Table 4), with the control having a significantly higher number of root nodules (6.00) than 40 ml kg^{-1} (4.44). 0 ml kg^{-1} group (control) had significantly higher root nodule biomass than both 20 ml kg^{-1} and 40 ml kg^{-1} groups.

Table 3. *Effect of pesticide type on root nodule formation in soybean (G. max).*

Parameter	R	CH	COM	LSD
Root nodule number/plant	6.00^a± 0.23	4.77^b±0.46	5.00^b±0.46	0.95
Root nodule biomass/plant (g)	0.39± 0.04	0.26± 0.06	0.28±0.06	NS

Values are presented as Mean ± SEM. ^{ab}Values across the table with similar superscript are not significantly different at 5% based on ANOVA. NS = Not significant; R=Ridomil; CH = Chlorpyriphos; COM = Ridomil + Chlorpyriphos Combination.

Table 4. Effect of pesticide contamination rate on root nodule formation in soybean (G. max).

Parameter	0 ml kg^{-1}	20 ml kg^{-1}	40 ml kg^{-1}	LSD
Root nodule number/ plant	6.00^{a}± 0.28	5.33^{ab}± 0.40	4.44^{b}± 0.37	0.95
Root nodule biomass/plant (g)	0.49^{a}± 0.04	0.24^{b}± 0.03	0.21^{b}±0.05	0.13

Values are presented as Mean ± SEM. ^{ab}Values across the table with similar superscript are not significantly different at 5% based on ANOVA. NS = Not significant; 0ml kg^{-1} = control; 20 ml kg^{-1} = 2% (v/wt); 40 ml kg^{-1} = 4% (v/wt).

4. Discussion

Results obtained showed that the pesticides inhibited seed germination of soybean, as well as development of root nodules. Between the pesticide types, chlorpyriphos had the lowest seed germination, while Ridomil and the Combination were comparables; with the significant effect occurring from weeks 2 to 4. This implies that as recommended for use on cocoa by the manufacturers, chlorpyriphos is a "higher" stressor to soybean seeds than Ridomil. This is in line with the position of Dubey and Fulekar [18] that chlorpyriphos contamination in soil reduces germination rate of some plants. No significant differences between the pesticide types were observed in the first week (Table 1). This perhaps, is because all pesticide types were nearly equally inhibitory to germination in the first week (very low germination rates in week one of Table 2, for both 20 ml kg^{-1} and 40 ml kg^{-1} supports this). But by the second – fourth weeks, chlorpyriphos retained the highest inhibitory effect (below 50%) on soybean seeds germination. Overall, the seeds in control soils had significantly higher germination rates than those in soils treated with 20 ml kg^{-1} and 40 ml kg^{-1} of pesticides (Table 2) up until the fifth week. This highlights the (short term) stressful effects both pesticides had on soybean seed germination, at 2% and 4% concentration (v/wt). Siddiqui and Ahmed [19] asserted that pesticide induced stress in plants could trigger the formation of phenolic compounds, such as the isoflavones - genistein and diadzein, phenolic acids and hydroxycinnamic acid derivatives such as hydroxybenzoic acids, ferulic acids, p-coumaric acids, which are all potential inhibitors of germination and plant growth [20-22]. Perhaps some of these phytochemicals were triggered in response to chlorpyriphos and Ridomil at both 2% and 4% pollution concentrations.

Legumes are important in nitrogen fixation in soils and have increasingly become models/markers of toxic effects of chemicals (such as pesticides) in the environment, on plants. Ahemad and Khan [23] reported that fipronil (an insecticide) was stressful to legumes including chickpea, pea, lentil, and green gram, and observed the negative effects of the insecticide on shoot dry biomass, chlorophyll content, root nodule numbers and biomass. Chlorpyriphos had a significantly lower root nodule number/plant than Ridomil. Though an interesting observation on its own, the absence of a corresponding significance in root nodule biomass/plant makes it rather difficult to conclusively declare that chlorpyriphos affected root nodule formation more than Ridomil (see Table 3). We observed that control plants had significantly higher root nodule number/plant and root nodule biomass/plant, than plants in pesticides treated soils (Table 4). These attest to the negative influence of both pesticides on these legume parameters. According to Anderson *et al.*, [24], pesticides could affect nodulation in legumes by; reducing carbohydrate supply to existing nodules; directly affecting rhizobial survival/activity in the soil; inhibiting biochemical signaling in plant roots that initiate/stimulate rhizobial attachment, inhibiting cell division of rhizobium, amongst others. Obviously, as described for use on cocoa, the pesticides had inhibitory effect on one or more of the above mentioned factors, in soybean.

The capacity of these pesticides as stressors, no doubt, owes much to their half lives in the environment. In this regard, while metalaxyl, the active ingredient in Ridomil has a half life of eight weeks (two months) and readily leaches in soil [25]; Sing *et al.*, [26] asserted that chlorpyriphos has a half life of 36-46 days. Our findings, in soil exposed to natural illumination, suggesting that toxicity to soybean seed germination abates by the sixth week (as no significant differences were observed due to the pesticides concentrations in the sixth week after initial application), are not far-fetched. Glesy *et al.*, [27] in a review on evidence for classification of chlorpyriphos and its metabolite - chlorpyriphos oxon as a: persistent organic pollutant (POP), and persistent bioaccumulative and toxic (PBT), concluded that there was not sufficient justification for the classification of chlorpyriphos as a POP or PBT. It appears our results on seed germination of soybean, add to, rather than subtract from their position; keeping in mind that manufacturer recommended preparations were used. On the other hand, root nodule formation, determined after three months of pesticides application, appeared a sensitive marker of chlorpyriphos and ridomil stresses, as nodule number and biomass were significantly reduced in the 2% and 4% treatment groups, as compared with the control.

Positive interaction between two pesticides should lead to a significantly higher response (by the marker), than is observed for equivalent single pesticides [15]. For both seed germination and root nodule formation, the effect (reduction) induced by a combination of both pesticides was not observed to be more than that due to chlorpyriphos or Ridomil alone. This implies a reduced likelihood of interaction between these two pesticides, for an enhanced stressful effect, when they are present together in the soil, on these legume parameters. One limitation of this study might be that the seeds were recovered after 72 hours for estimation of germination rate, rather than allowing them more time to perhaps germinate at their own pace. But a lag/delay in seed germination is itself a response to environmental stress [17].

5. Conclusion

This study investigated the effect of chlorpyriphos and Ridomil, singly and in combination on seed germination and root nodule formation in soybean, *Glycine max*. Pesticides were prepared in accordance with manufacturer recommendations for use on cocoa. More than Ridomil and Combination, significantly diminished percentage germination due to chlorpyriphos was observed at two, three, and four weeks after pesticides application. The control (0%) had significantly higher germination than 2% and 4% groups, up until the fifth week. Root nodule formation was also significantly reduced, due to pesticides stress at 2% and 4% (v/wt) pollution levels; as the control also had significantly higher root nodule number/plant, and root nodule biomass/plant than 2% and 4% groups. The chlorpyriphos and pesticide combination groups both had significantly lower root nodule number/plant, than Ridomil, but this was not reinforced by a corresponding difference in root nodule biomass/plant between the pesticide types. Pesticides application at 2% and 4% negatively affected germination and root nodule formation; but there was no indication of an enhanced effect due to pesticides combination, as chlorpyriphos negatively affected tested parameters more than the tested equivalents of Ridomil and their Combination.

Acknowledgement

We gratefully acknowledge the assistance of Mr. Paul Ekpo of Environmental Biotechnology Laboratory, Department of Genetics and Biotechnology, University of Calabar, in pesticides application and soil conditioning.

References

[1] Ashraf M., Harris P.J.C. 2004. Potential biochemical indicators of salinity tolerance in plants. *Plant Science*, 166, pp 3-16

[2] Gürsoy M., Balkan A., Ulukan H. 2012. Ecophysiological responses to stress in plants: a general approach. *Pakistan Journal of Biological Science*, 15(11), pp 506-516, doi:10.3923/pjbs.2012.506.516

[3] Singh T.B., Mukhopadhayay S.K., Sar T.K., Ganguly S. 2012 Acetamiprid Induces Toxicity in Mice under Experimental Conditions with Prominent Effect on the Hematobiochemical Parameters. *J. Drug Metab Toxicol*, 3:134. doi:10.4172/2157-7609.1000134

[4] Tijani, A. A. 2006. Pesticide Use Practices and Safety Issues: The case of Cocoa farmers in Ondo State Nigeria. *J. Hum. Eco.*, 19, pp 183-190.

[5] Asogwa, E.U., Dongo, L.N. 2009. Problems associated with pesticide usage and application in Nigerian cocoa production: A review. *African Journal of Agricultural Research*, 4, pp 675-683.

[6] Ibiang, Y.B., Ekaluo U.B., Nta A.I., Ikpeme E.V., Ekanem B.E., Erem, F.A. 2013. Effect Of Deltamethrin And Ridomil On Serum Biochemical Parameters In The Rat (*Rattus Norvegicus*). *European Journal of Toxicological Sciences*, 2013:5.

[7] Römbke J., Schmelz R.M., Knaebe S. 2009. Field studies for the assessment of with soil mesofauna, in particular enchytraeids, mites and nematodes: Design and first results. *Soil Organisms*, 81(2), pp 237-264

[8] Toscano N.C., Sances F.V., Johnson M.W., Laprel F. 1982. The effect of various pesticides on lettuce physiology and yield. *Journal of Economical Entomology*, 75, pp 738-741

[9] Jones V.P., Toscano N.C., Johnson M.W., Welter S.C., Youngman R.R. 1986. Pesticide effect on plant physiology: Integration into a pest management program. *Bulletin of the Entomological Society of America*, 32, pp 103-109.

[10] Dunfield K.E., Siciliano S.D., Germida J.J. 2000. The fungicides thiram and captan affect the phenotypic characteristics of Rhizobium leguminosarum strain C1 as determined by FAME and Biolog analyses. *Biology and Fertility of Soils*, 31 (3), pp 303-309

[11] Ahemad M. (2011). A Comparative analysis of Tebuconazole mediated phytotoxicity to legumes. *Journal of Environmental Science and Technology* 4(6), pp 630-637 10.3923/jest.2011.630.637

[12] Rasool, N., Reshi, Z. 2010. Effect of the fungicide Mancozeb at different application rates on enzyme activities in a silt loam soil of the Kashmir Himalaya India. *Tropical Ecology*, 51(2), pp 199-205

[13] Goswami M.R., Pati U.K., Chowdhury A., Mukhopadhyay A. 2013. Studies on the effect of cypermethrin on soil microbial biomass and its activity in an alluvial soil. *International Journal of Agriculture and Food Science*, 3(1), pp 1-9

[14] Raven P.H., Johnson G.B., Losos J.B., Singer S.R. 2005. *BIOLOGY*, 7th Edn, McGraw Hill, New York, p 741.

[15] Hodgson, E., Mailman B. R., Chambers J. E. (Eds) 1998. *Dictionary of Toxicology*, 2nd ed, Macmillan Reference Ltd, London, pp 186-190.

[16] Nta, A.I., Ibiang Y.B., Uyoh E.A., Edu N.E., Ekanem B.E., John Q.E. 2013. Insect pest damage to leaves of cowpea (Vigna unguiculata L. Walp): comparative effects of aqueous extracts of Piper guineensis, Allium sativum and Myristica fragrans. *IOSR Journal of Environmental Science, Toxicology and Food Technology*, 3 (2), pp 17-20

[17] O.M. Agbogidi (2011). Susceptibility of five cultivars of soyabean (Glycine max (L.) Merr) seeds to spent engine oil. *International Journal of Science and Nature*, 2(3), pp 447-451

[18] Dubey K. K., Fulekar M. H. 2011. Effect of pesticides on the seed germination of Cenchrus setigerus and Pennisetum pedicellatum as monocropping and co-cropping system: implications for rhizospheric bioremediation. *Romanian Biotechnological Letters*, 16 (1), pp 5909-5918.

[19] Siddiqui Z.S., Ahmed S. 2006. Combined effects of pesticide on growth and nutrient composition of soybean plants. *Pak. J. of Botany*, 38, pp721-733

[20] Einhellig, F.A., Muth M.S., Schon M.K. 1985. Effects of allelochemicals on plant-water relationship. In: Thomson A.C. (Ed.) *The chemistry of Allelopathy*. American Chemical Society, Washington.

[21] Macias, F.A., Galindo J.C.G., Massanot G.M. 1992. Potential allelopathic activity of several sesquterpene lactone models. *Phytochem.*, 31, pp 1969-1777.

[22] Mersie, W., Singh, M. 1993. Phenolic acids affect photosynthesis and protein synthesis by isolated leaf cells of velvet-leaf. *J. Chem. Ecol.,* 19, pp 1293-1301.

[23] Ahemad M. and Khan M. S. 2011. Comparative study of the growth parameters of legumes grown in fipronil-stressed soils. *Eurasia J. Biosci.* 5, pp 29-36.doi:10.5053/ejobios.2011.5.0.4

[24] Anderson A., Baldock J. A., Rogers S. L., Belloti W., Gill G. 2004. Influence of chlorsulfuron on rhizobial growth, nodule formation, and nitrogen fixation with chickpea. *Australian Journal of Agriculture Research*, 55, pp 1059-1070.

[25] Kimmel, E.C., Casida, J.E., Ruzo, L.O. 1986. Formamide Insecticide and chimoacetanide herbicides; disubstituted Anilines and Nitro benzenes as mammalian metabultes and bacterial metagens. *Journal of Agriculture and Food Chemistry*, 34, pp 157-161.

[26] Sing, B.K., Walker A., Wright D.J. 2002. Degradation of chlorpyrifos, fenamiphos and chlorothelonil alone and in combination and their effects on microbial activity. *Environmental Toxicology and Chemistry*, 21, pp 2600-2605.

[27] Glesy J.P., Solomon K.R., Mackay D., Anderson J. 2014. Evaluation of evidence that the organophosphorus insecticide chlorpyrifos is a potential persistent organic pollutant (POP) or persistent, bioaccumulative, and toxic (PBT). *Environmental Sciences Europe*, 26:29. doi:10.1186/s12302-014-0029-y

7

Analysis of Production, Transport and Lifecycle of Pellets

Ricardo Díaz Martín[1], Gaston Sanglier Contreras[2], Adib Guardiola Mouhaffel[3]

[1]Chairman building Engineering department, CEU San Pablo University, Madrid, Campus de Monte príncipe
[2]Associate professor building Engineering department, CEU San Pablo University, Madrid, Campus de Monte príncipe
[3]CEU San Pablo University, Madrid, Campus de Monte príncipe

Email address:
Ricadima@ceu.es (R. D. Martín), Sanglier.eps@ceu.es (G. S. Contreras), aa.Guardiola@usp.ceu.es (A. G. Mouhaffel)

Abstract: The forestry production industry in the province of Galicia, represents one of the most available areas of biomass production in Spain. The aim of this study is determine the carbon footprint score in forestry production and processing of pellets for subsequent transportation to a community with little locally available biomass. The methodology proposed for processing forestry production and consumption, taking actual data, actual emissions in the area of Pontevedra, and considering fundamental aspects from the environmental point of view provides the most accuracy. The performance of machinery and consumption in the forestry process, as well as distance to be covered, ultimately give us the impact of the carbon footprint for comparison with predetermined targets in emissions savings.

Keywords: Carbon Footprint, Forestry Treatments, Transport, Pellets

1. Introduction

This article aims to explain the methodology based specifically on the life cycle analysis (LCA),which has been taken into account in order to determine the carbon footprint (CFP) for forestry production and transportation of pine wood pellets used as biomass for energy production. A quantitatively determined amount and percentage of CFP in the production and transportation of pine wood pellets used as fuel for thermal purposes in the Canaries is suggested.

At present, however, the ACV still have uncertainties about the risks: there are numerous methodologies capable of being applied o each case, leading to results which are, on many occasions, diverse and even contradictory [1]. It is important to that efforts to implement measures within sectors that are in transition due to the implementation of ISO 14001 systems, such as transport, agriculture, tourism, the residential sector, and where knowledge of the impact of CFP are causing themto change their attitude towards management and development for responsible and lasting business [6].

So that we can accurately characterize CFP in regards to production and transport, raise awareness and make improvements in moving away from the policies associated with "Greenwashing", will jump on the train by looking at products and sources of renewable energy which lack sufficient information for determining the impact of the product [6].

2. Methodology

For the evaluation of production, a system of forestry production of medium difficulty in terms of accessibility [8], the pine which has an average density of 460 kg / m^3 with humidity at the time of collection of around 50% has been taken into account. Once dried and pelletized, it is a wood that reaches a density of around 670 kg/m^3, on ready for commercialization.

As the combustion of bio-fuels (from biomass) is considered not to generate GHG emissions - their emission factor is zero. Therefore, the emission factor of fuel formed by the mixture of fossil fuels and bio-fuels will be the result of subtracting from the first, the extent to which is the second. The consideration of the emission during the drying period will correspond to supplementary accessories in the trammel

Ground transportation to Vigo and subsequently by sea to the port in Gran Canaria, and finally the trip along the length of the island, will give us the results of emissions using the coefficients applied in each case.

The quantification system in kilos of emission of carbon dioxide (CO_2 Kg / T) will be given by the following methods, depending on the activity:

- Forestry production tasks will be determined by the equipment used and performance in obtaining a ton of wood [1].

- The drying treatment is carried out with a trammel fed with biomass fuel, so the emission rate will be compensated and only consider the purpose of issuing operating auxiliaries.
- The system is considered as pelletizing, pelletized at30 kW with a production rate of 0.5 tons/h, calculated by the emission power required by Tn [2].
- Land transport by mode of transport, loading and distance [2].
- Shipping, Kuehne + Nagel - the software will give us equates tonsduring the trip [4].

3. Results

3.1. Forestry Production

Forestry production during the maintenance of the forest environment requires a set of machines that maximize and ensure optimum functionality in the execution of tasks. The coefficients of work applied to the machinery required based on performance [1], will give the resulting emissions for obtaining pine wood.

Taking the density of the pine [8], we managed to obtain the total emissions of forestry production, which would be 98,65 kg CO_2/T; treatment and management of pelletized wood would be 25,08 kg CO_2 / T

Table 1. *CO_2 emissions forestry pine production process.*

Cut down trees sytem	Total Kg CO_2/m^3
Hand saw machine	7,88 kg CO_2/m^3
Jaw Crushing-wood chipper	14,20 kg CO_2/m^3
Transport	16,80 kg CO_2/m^3
Total	38,88 KgCO_2/m^3

Table2. *CO_2 emissions Factory process.*

Factory process	(Kw)	Pollutantemissions
Conveyorized	18,00	1,44 Kg CO_2/m^3
Wood chipper	15,00	1,20 Kg CO_2/m^3
Sieve shaker	11,00	0,88 Kg CO_2/m^3
Shredding machine	19,00	1,52 Kg CO_2/m^3
Rotary –trammel dryers	23,00	1,84 Kg CO_2/m^3
Pelletizing machine	30,00	2,40 Kg CO_2/m^3
Cooled and packed	9,40	0,72 Kg CO_2/m^3
Total		**10,03 Kg CO_2/m^3**

3.2. Transport

3.2.1. Maritime Transport

Using the software company Kuehne + Nagel [4], which calculates the carbon footprint of a load from the port of departure, Vigo, to the destination of Las Palmas, 22 tons of pellets could be loaded and including the weight of a foot container at 2.230 Kg.

From	To	Mode of transport	Distance in km	CO2 in kg	NOx in kg	S02 in kg	PM10 in kg
VIGO, SPAIN (ESVGO)	LAS PALMAS, SPAIN (ESLPA)	Sea	2,235	794.51	16.32	10.03	1.41
Total			**2,235**	**794.51**	**16.32**	**10.03**	**1.41**

From	To	Mode of transport	Distance in km	EC (TTW) in GJ	EC (WTW) in GJ	CO2e (TTW) in kg	CO2e (WTW) in kg
VIGO, SPAIN (ESVGO)	LAS PALMAS, SPAIN (ESLPA)	Sea	2,235	9.52	10.36	739.32	802.12
Total			**2,235**	**9.52**	**10.36**	**739.32**	**802.12**

Figure 1. *Footprint maritime transport, Vigo- Las Palmas.*

Fuente: EmpresaKuehne+Nagel

3.2.2. Land Transport

CO2 emissions of diesel is 2.47 kg CO2/l [2], considering the necessary vehicle to carry the burden of 24,3Tn pellet, plus the weight of a 20-foot container, requires a power of 340 CV [3], with a consumption of 24 liters per 100 km, the coefficient applicable to diesel is 0.6 kg CO2 / l Km [3].

To calculate the carbon footprint of transporting the biomass, we will consider the path from start to finish, plus the auxiliary elements used in transport - in this case the dampening unit for conservation of biomass in the containers.

3.2.3. Transport's Footprint Score

The different modes of transport employed to mobilize biomass plus the system utilized to conserve moisture will form the elements that generate the emissions which determine the final size of the carbon footprint.

Table 3. *CO_2 transpots emissions.*

		(Km)	Emissions	Total transport emissions
Trucker	Pontevedra-Vigo	25,00 Km	0,60 Kg CO2/ litros Km	15,00 Kg CO_2/ 24,3T
Ship	Vigo-Las Palmas	2234,58 Km	-	802,12 Kg CO_2/ 24,3T
Trucker	Las Palmas-Mogán	80,00 Km	0,60 Kg CO2/litros Km	48,00 Kg CO_2/ 24,3T
Total				823,49 Kg CO_2/ 24,3T

The impact per ton in transport is 34,00 Kg CO_2, where maritime transport accounts for 93 % of emissions and land 7% of the total.

3.3. Schedule Carbon Footprint Production and Transportation

Figures for forest production, processing and transportation from Vigo to Las Palmas of pinewood pellets: emissions are quantified in 193 kg CO_2 / per ton of biomass transported.

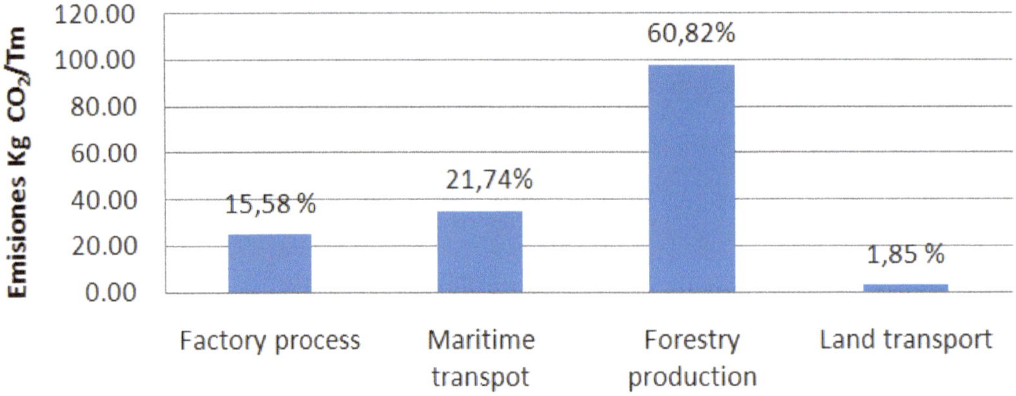

Figure 2. *Distribution of carbon dioxide as activity during production and transportation of pine pellets.*

4. Conclusions

Every ton of biomass consumed by thermal purposes is avoiding issuing 1.350 Kg CO_2 of emission factors as with diesel [2] –where the system of production and transportation of biomass studied represents 13.42% of the emissions that they aim to save, and whose treatment and forestry production represent 8,17 %, with maritime transport 2,92%.

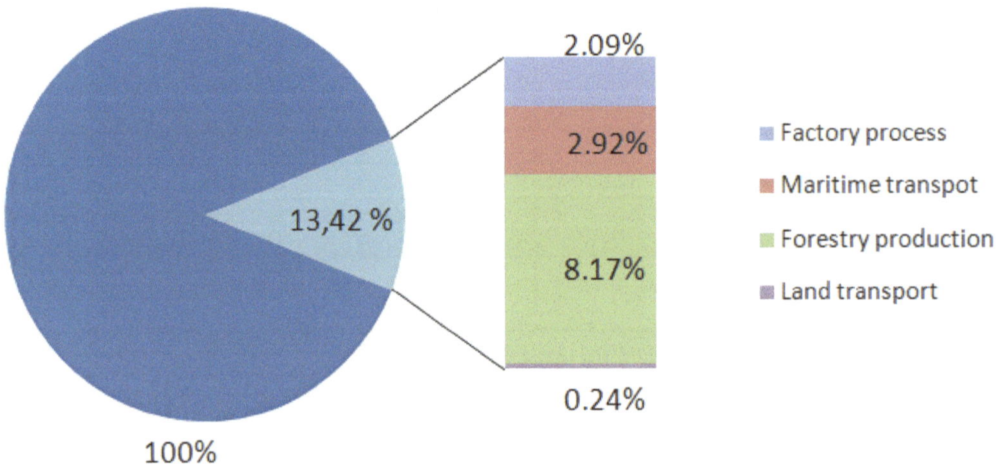

Figure 3. *Distribution of carbon dioxide produced during the production and transport of pellets in total, which aims to save on these emissions per ton.*

In the above, concepts have not been included that would suffer losses to biomass energy through inappropriate transportation conditions and adsorption of moisture during transport [5] in the case of pellets we would be trying to increase.

We propose as an upgrade to the negative impact of excessive emissions during transportation and avoidance of mismatches between the goals of emission reductions and emissions required to benefit from the use of biomass in areas minimal local resources [5]. A policy of energy crop planting of the paulownia type, where the terrain of the Canary Islands is ideal, should be implemented.

The return on investment for crop when used for the production of biomass for thermal purposes is high, especially as the machinery required for the operation is less robust and simpler, requiring less fossil fuels and improving the impact on tasks.

Acknowledgements

This work has been supported by Sedam Management and The Canary Islands Technological Institute. Oliver González Arias, Miguel Pérez Afonso, Gonzalo Pierna Vieja Izquierdo.

References

[1] L.A. Mohamed, M. Kouhila, S. Lahsasni, A. Jamali, A. Idliman, M Rhazi, M. Aghfir, M.Mahrouz. "Equilibrium moisture content and heat of sorption of Gelidiumsesquipedale". Journal of StoredProductsResearch. Vol. 41 N° 2,pp. 199-209. 2005.

[2] Estimación de la huella de carbono de productos forestales"DirecciónNacional de Industrias Ministerio de Industrias, Energía y Minería Consejo Sectorial Forestal-Madera Uruguay 2013.

[3] Guía para el cálculo de la huella de carbono y para la elaboración de un plan de mejora de una mejora de organización, Ministerio del medio ambiente 2014.

[4] La guía para la gestión del combustible, IDAE 2006.

[5] http://www.kn-portal.com/seafreight/environment/calculator.

[6] Díaz R., Contreras G., Guardiola A. "Comparación entre métodos estáticos y dinámicos de cálculo de humedad en biomasa". Dyna energía y sostenibilidad Volumen 3 (1). 2014.

[7] P.MasAlique, F. HerraezGarrido,D. Muñoz Jimenez. "La huella de carbono como ventaja competitiva" Volumen 3 (1). 2014.

[8] Luis Ortíz, Alejandro Tejada, Antonio Vázquez."Aprovechamiento de biomasa forestal". Revista CIS-Madera. 17-32. 2003.

[9] Yolanda Ambosio. APROVECHAMIENTO DE BIOMASA FORESTAL: Maquinaria, sistemas de recogida, rendimientos y costes. Experiencias reales"E.U.I.T Forestal. Universidad Politécnicade Madrid.2009

[10] PAPENDIECK, S. 2010. La Huella de Carbono como Nuevo EstándarAmbiental en el Comercio Internacional de Agroalimentos:Informe Final. ATN/ME-9565-RG BIDFOMIN.

[11] SCHNEIDER, H. y SAMANIEGO, J. 2010. La huella del carbonoen la producción, distribución y consumo de bienes y servicios.Documento de Proyecto. Comisión Económica para América Latinay el Caribe (CEPAL).

8

Physico-Chemical Assessment of Surface and Ground Water for Drinking Purpose in Nawabshah City, Sindh, Pakistan

Abdul Jabbar Kandhro[1], Ali Muhammad Rind[2, *], Abdul Aziz Mastoi[2], Khalida Faryal Almani[2], Sujo Meghwar[3], Muhib Ali Laghari[3], Muhammad Sohaib Rajpout[2]

[1]Department of Community Medicine, Peoples University of Medical & Health Sciences for Women, Nawabshah, Shaheed Benazirabad
[2]Center for Environmental Sciences, University of Sindh, Jamshoro, Sindh, Pakistan
[3]Department of Geography, University of Sindh, Jamshoro, Sindh, Pakistan

Email address:
ali_mohmedr@yahoo.com (A. M. Rind)

Abstract: The work reports the analysis of groundwater and surface water samples, were collected from different towns of Nawabshah City during the end of year 05 Dec, 2013 to 30 Jan, 2014. Sixty five groundwater and sixty surface water samples (water supply scheme) were collected from different parts of the Nawabshah city. Different physico-chemical parameters of water samples were measured at the field and in the laboratory. The conductivity, salinity and Total dissolved solids (TDS) were measured with Orion 115 conductivity meter at the field. pH was recorded with Orion 420A pH meter. Total hardness, chloride and alkalinity were determined by titration with standard EDTA, silver nitrate and hydrochloric acid. Sulfate was determined by turbidity meter as BaSO4 using Hitachi spectrophotometer. The metal ions Na, K, Ca and Mg were determined with Varian Spectr. AA-20 atomic absorption spectrometer with standard burner head and air acetylene flame at conditions recommended by the manufacturer. The results were varied within the ranges; pH 6.64-8.87, EC 240-10170 µS/cm and TDS 158-6050 mg/l, alkalinity 56-1225 mg/l, total hardness 84-1695 mg/l, chloride 32-1852 mg/l, sulfate 25-2170 mg/l. The concentration of essential metals was found in the ranges; Na 34-1725 mg/l, Ca 26-515 mg/l, Mg 13-430 mg/l, K 2-92 mg/l respectively. The analysis revealed that a number of ground water samples (70%) confirm their majority of parameters above the maximum permissible limits prescribed by WHO. Therefore the ground water of Nawabshah city may not be considered as safe to be used for dinking purpose. However, out of 60 samples only four surface water samples may be used for drinking purpose.

Keywords: Canal Water, Water Supply Schemes, Ground Water, Physical, Chemical

1. Introduction

Nowadays Safe drinking water is a major issue mostly in developing countries (1), drinking of polluted water generate serious health issues, may be consumption of contaminated water is a cause of high death rate in developing countries. Most of the cities of world use underground water for their daily use, which is contaminated by sewage. Over all the Word approximately 800 million peoples not getting adequate water supply (2). Water in developed countries used after proper filtration procedure but in developing countries it is inverse, lake of filtration plants leads to use contaminated water to peoples which affected severally. Diarrhea is common in Pakistan due to utilizing contaminated water. Water which safe for drinking purpose called potable water. Water gets polluted by two major sources i.e. point sources and non point sources.

The water in the Indus basin is of variable quality (3-5). It is not saline near sources of recharge, i.e. rivers and major canals but gradually becomes saline with depth as the distance from the recharge source increases (6). Poor microbial quality of drinking-water is the most pressing issue. No urban water supply meets WHO drinking-water quality guidelines (7). The major reasons for this are the intermittent supply through leaking pipes and cross connections with nearby sewer lines (8). People of the district mostly use underground water for drinking purpose, except some towns, where the water supply schemes provide surface water for drinking. The water we drink should meet some national and international standards. It is also predicted that in future Pakistan will face water challenges (9-11). Chemical composition of surface and groundwater is one of the major factors to which the suitability of water for industrial, domestic and agricultural purpose depends (12). The pollution of groundwater results from all the processes and reactions which the water faces from the moment it condensed in the atmosphere to the time it is discharged by a well or hand pump and varies from place to place with depth. Major portion of the rural Inhabitants depend upon the groundwater due to unavailability of water treatment and supply of potable water. 40% of the deaths in Pakistan are caused by water borne diseases directly or indirectly (13).

A number of factors cause pollution of surface as well groundwater including urbanization and industrialization. The quality of groundwater in Drinking Water Quality of Pakistan is deteriorating day by day (14-20). A serious problem appears due to stagnant management of effluent water which becomes penetrating into the soil with the passage of time and can become a part of natural and ground water (6, 21-25).

Lot of work has been carried out on the quality of ground water of different parts of Pakistan (26). Mostly the water of all the sources in Pakistan is polluted (27). The drinking water of different regions of country also contaminated including Peshawar (28, 29), Karachi (30), Lahore (31), Muzaffargarh (32) and Tharparkar (33), but no any reasonable work is reported on the quality of underground water of Nawabshah city.

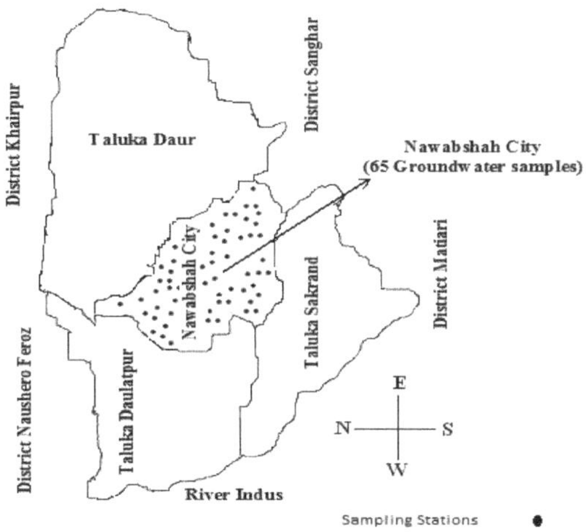

Map 1. District Nawabshah and ground water sampling stations.

The quality of water in domestic use must be tested to know the extent of its pollution and its suitability for human consumption. It is useful to examine the above highlighted problems in an environmental frame work in which the quantity and quality of water resources is a major concern. The object of this study is to assess the drinking water quality in various drinking water sources used in Nawabshah city. In Nawabshah both the underground and surface (Municipal whole water quality of the city and compare it with the standards of World Health Organization. Over all focus of this study was drinking water quality of the city.

2. Materials & Methods

Samples were collected from the municipal water supply and underground water in pre sterilized polyethen bottles, before the collection of samples bottles were washed with sample water twice. All the reagents used were of analytical grade and all the glass ware used was washed properly with double distilled water before use. The metal standard solutions were prepared by dilution from 1000 ppm stock solution of each metal.

2.1. Detection of Physical Parameters

Different physico-chemical parameters of water samples were measured at the field and in the laboratory. The homogenized sample was transferred to a clean 1 L clean plastic bottle. The temperature of air on meter above the surface of water was recorded with mercury thermometer, conductivity, salinity and total dissolved solids (TDS) were measured with Orion 115 conductivity meter at the field. pH was recorded with Orion 420A pH meter.

2.2. Detection of Chemical Parameters

Hardness, chloride and alkalinity were determined by titration with standard EDTA, silver nitrate and hydrochloric acid respectively. Sulfate was determined by turbidity meter as $BaSO_4$ using double beam Hitachi 220 spectrophotometer.

2.3. Detection of Metallic Ions

The essential metal ions (Na, K, Ca and Mg) were determined with Varian Spectra AA-20 atomic absorption spectrometer with standard burner head and air acetylene flame. The analysis was carried out in triplicate with integration time 3 seconds and delay time 3 seconds. The concentration of metal ions (Na, K, Ca and Mg) was determined after appropriate dilution of the sample containing 1ml concentrated nitric acid per 250 ml. The SAR values were calculated using the formula, SAR = Na/ (Ca+Mg) 1/2/2.

3. Results and Discussion

The work reports the analysis of groundwater and some surface water samples from Nawabshah city, located at the center of Sindh province. The results of physical and chemical parameters of the water samples were found in the following ranges. pH 6.64-8.87, TDS 158-6050 mg/L, chloride 32-1852 mg/L, alkalinity-M 56-1225 mg/L, hardness 84-1695 mg/L, SO_4 25-2170 mg/L, The concentration of essential metal ions (Na, Ca, Mg and K) was found in the ranges of 34-1725 mg/L, 26-515 mg/L, 13-430 mg/L and 2-92 mg/L respectively. The results revealed that the majority of ground water samples (40 out of 65) of the study area were not suitable to be used for drinking purpose; on the other hand, 40 water samples based on sodium adsorption ratio (SAR) were found suitable to be uses for irrigation purpose.

In all the natural ecosystems water acts as the primary transport medium for dissolved and particulate matter and determines as well, the rate at which these fluids are added or removed from the system. A complete identification of hydrological characteristics is hence essential for understanding biological, chemical and physical processes that operate within the ecosystem.

The results of physico-chemical parameters of water samples reveal the varying nature of the underground water of the Nawabshah city. The difference in the quality of groundwater may be due to topography of soil; different earth beds and effect of recharge sources (canals etc) on underground water.

3.1. pH

The pH of water samples varied between within the range 6.95-8.87. Sixty three samples confirm pH within the safe guidelines of 6.5-8.5 prescribed by WHO for drinking water, while only two samples were above the limits of WHO. These two samples were collected direct from Gajra (canal) running through Nawabshah city and from a hand pump located near to Gajra (canal). The high values of pH may be due to human activity (domestic and industrial waste) and water logging respectively. The waters with pH below 7.0 are termed as acidic and acidity in water is due to the presence of dissolved carbonic acid. It increases the solubility of different materials including metals ions like Na, K, Ca and Mg. Twenty samples indicated pH below 7.0 but indicated within the permissible limit of 6.5.

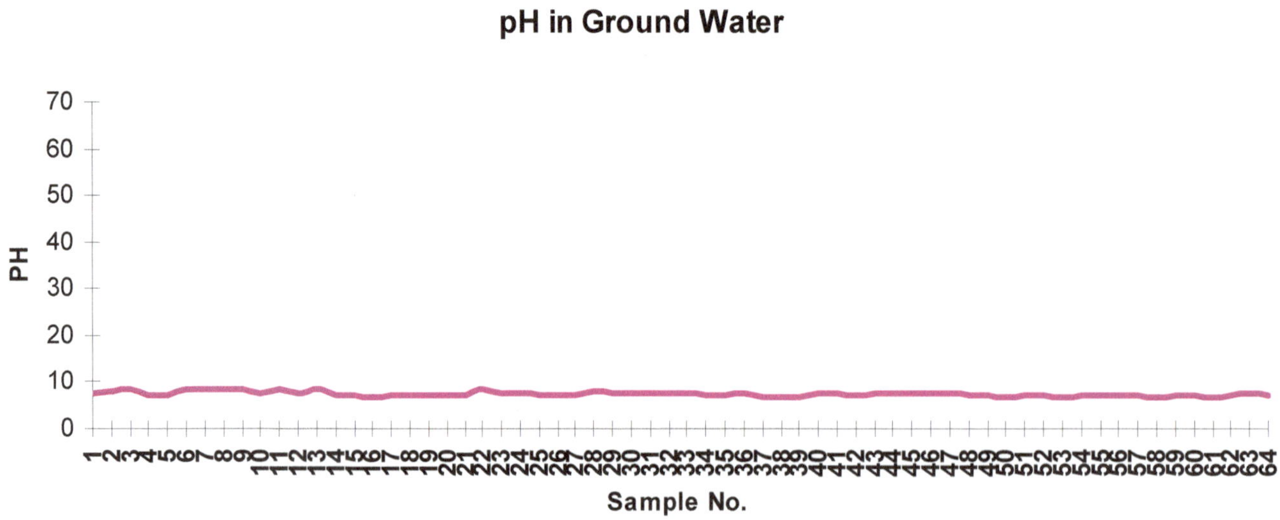

Figure 1. *pH in Ground Water.*

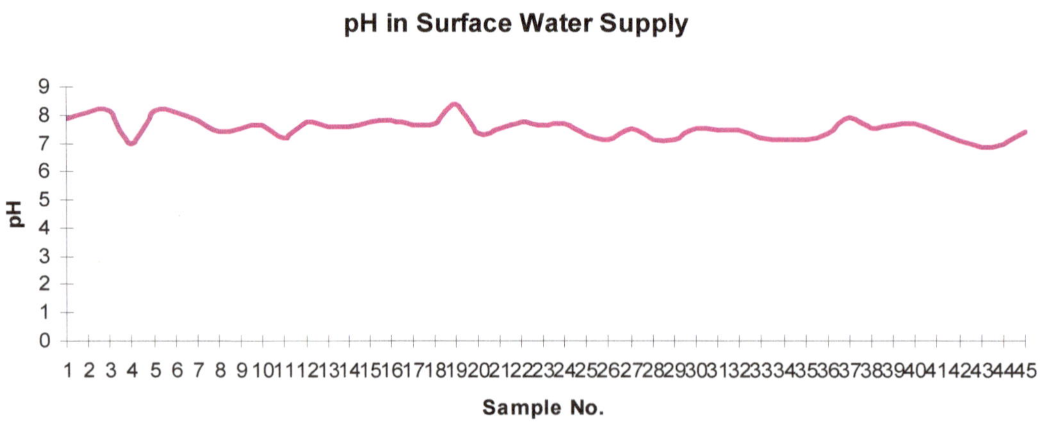

Figure 2. *pH in Surface Water Supply.*

3.2. Total Dissolve Solids (TDS)

The TDS varied between within the range 158-6050 mg/L. A lot of variation was observed in TDS of water samples, only 10 samples were observed within the limits (500 mg/L) of WHO for drinking water. The variation in TDS may be due to different earth beds and recharge sources. Generally the water samples have high values of TDS making them unsuitable for drinking purpose. The water samples with elevated values of TDS may cause several health problems to living organisms and adversely affect the fertility of soil, if used for irrigation purpose.

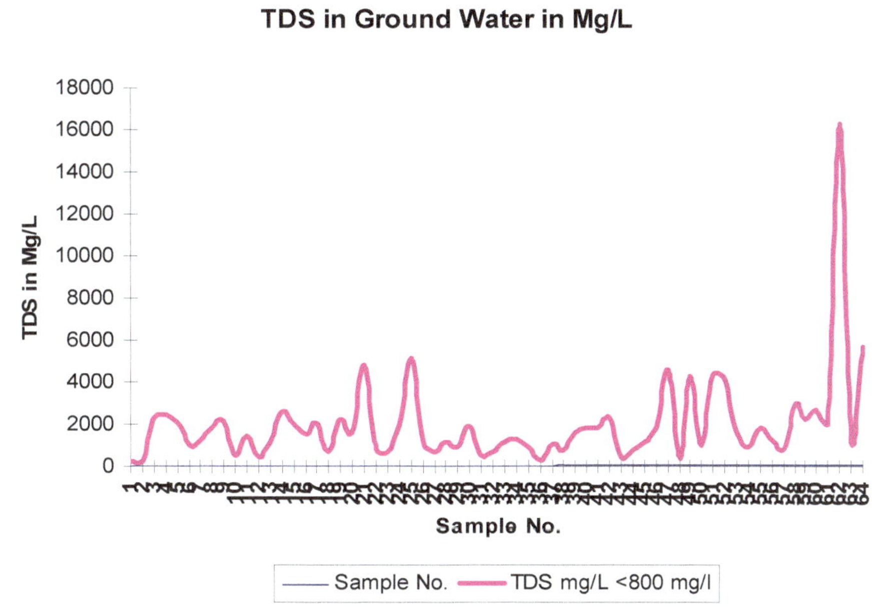

Figure 3. *TDS in Ground Water.*

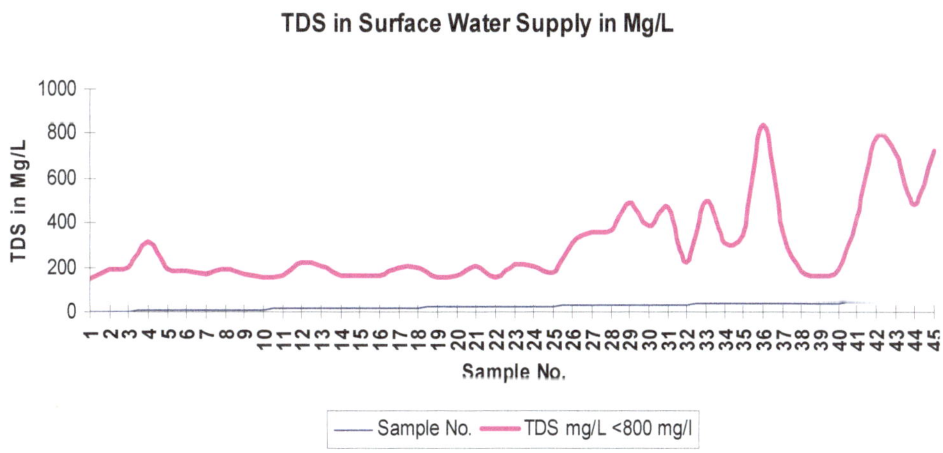

Figure 4. *TDS in Surface Water Supply.*

3.3. Bicarbonates and Hardness

The contents of bicarbonate and hardness fluctuated varied between within the range 56-1225 mg/L and 84-1695 mg/L respectively. A parallel behavior of bicarbonate with hardness was noted. Hardness of 25 samples was within the safe limits prescribed by WHO for drinking water and six samples indicated their hardness above 700 mg/L may be due to geological reasons. The water with elevated hardness than the WHO guidelines may cause gastric problems, dehydration, gas trouble, kidney stone and heart problems.10

3.4. Chloride and Sulfate

The concentration of chloride varied between within the range 32-1752 mg/L. Nearly half of the samples (35 samples) showed their chloride concentration above the regulations (250 mg/L) set by WHO. The chloride in water is present in combination with sodium, calcium and magnesium. Sources of chloride are mostly human waste, mineral rocks, irrigation discharge and industrial effluents like dying and bleaching materials. The water samples with higher concentration of chloride may have toxic effects to health.

The Sources of sulfate in surface and subsurface water are mainly calcium sulfate and sodium sulfate. The sulfates entering in water bodies come from dissolution of minerals containing sulfides and thiosulfates. Sulfate contributes to the permanent hardness to water. The elevated level of sulfate in water causes bad taste of water and also shows corrosive action (34). Sulfate was found between 25- 2170 mg/L. Some of the water samples (20 samples) indicated their sulfate contents above the limits of WHO. These samples were collected from Kazi Ahmed road of Nawabshah city. A parallel trend was found in the concentrations of chloride and sulfate in water samples (35).

The concentration of essential metal ions (Na, Ca, Mg and K) was varied between the ranges of 34-1725 mg/L, 26-515 mg/L, 13-430 mg/L and 2-92 mg/L respectively.

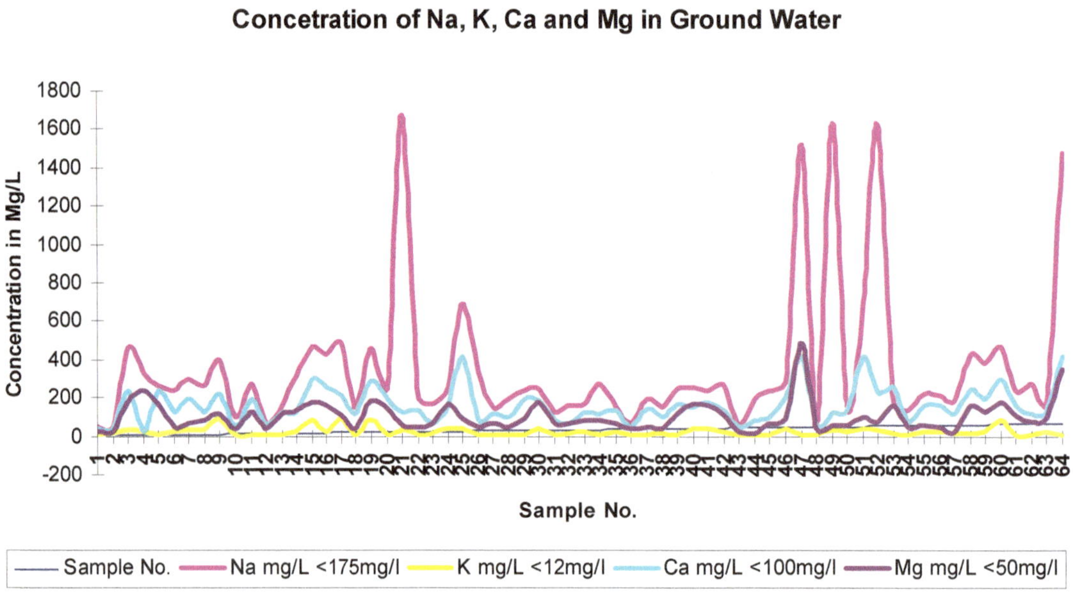

Figure 5. Concentration of Na, K, Ca and Mg in Ground water.

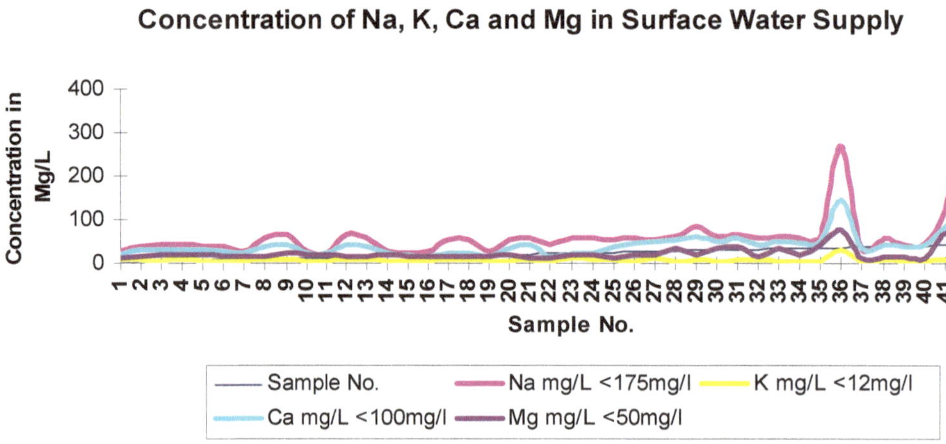

Figure 6. Concentration of Na, K, Ca and Mg in Surface Water Supply.

3.5. Chemistry of Metal Elements

The concentration of major metal ions Na, K, Ca, and Mg varied with high concentration of ground samples. The concentration of major metal ions follows following decreasing order:

$$Na > Ca > Mg > K$$

Sodium is present in all natural waters. The presence of sodium in water depends upon the anions present in that system and the temperature. The high concentration of sodium impart taste to the water and make it unfit for everyday use and leads to cardiovascular diseases and high blood pressure (36). Concentration of Na in water samples of the study area varied as 34-1725 mg/L. The Na ion concentration of the 25 samples was found with in the safe guidelines of 200 mg/L set by WHO for drinking water. All the rest of the samples indicated high values of Na ion concentration than WHO limits.

Potassium plays an important role in the metabolism process of animals and, it is an important micronutrient for living organisms (plants and animals). The WHO threshold of potassium for drinking water is 12 mg/L. The potassium concentration of the water samples studied was varied between 2-92 mg/L, including 40 samples with K ion concentration within the permissible limits.

Calcium and magnesium are abundant in rocks and soil, particularly lime stones and dolomites. They are relatively soluble and dissolve in surface water and then enter into ground water. There are no health concerns associated with calcium and magnesium, but the water containing these metals may contribute towards human dietary needs, however, their high concentration may cause scaling of pipes. The concentrations of Ca and Mg ions in the water samples of the area were found in the ranges of 26-515 mg/L and 13-430 mg/L respectively. Ca in 40 samples and Mg in 45 samples were found to be within the permissible limits of WHO. The rest of the water samples were observed very hard with high concentrations of Ca and Mg.

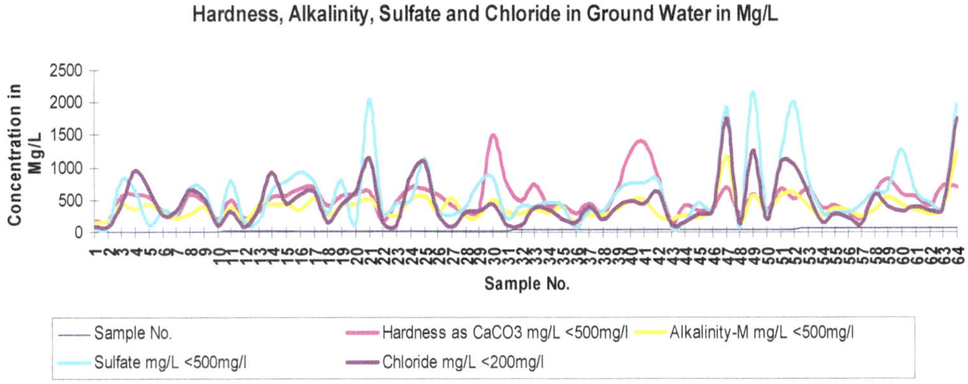

Figure 7. *Hardness, Alkalinity, Sulfate and Chloride in Ground Water in Mg/L.*

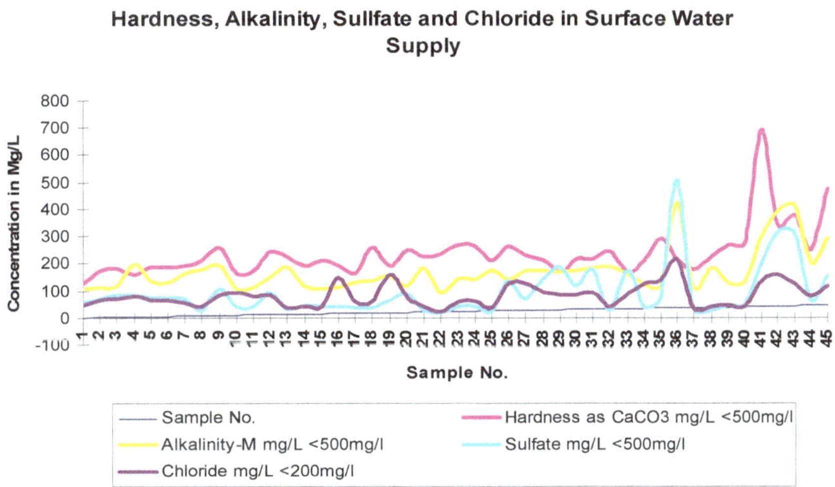

Figure 8. *Hardness, Alkalinity, Sulfate and Chloride in Surface Water Supply Mg/L.*

3.6. Sodium Adsorption Ratio (SAR)

Sodium adsorption ratio (SAR) was calculated to check the suitability of the waters to be used for irrigation purpose. The results revealed that determined 125 samples, only (60 surface water and 40 under groundwater) were suitable for irrigation with SAR value below 6 and remaining 25 samples were unsuitable for irrigation with SAR value above 6, all these 25 samples were under groundwater samples "Figure 9".

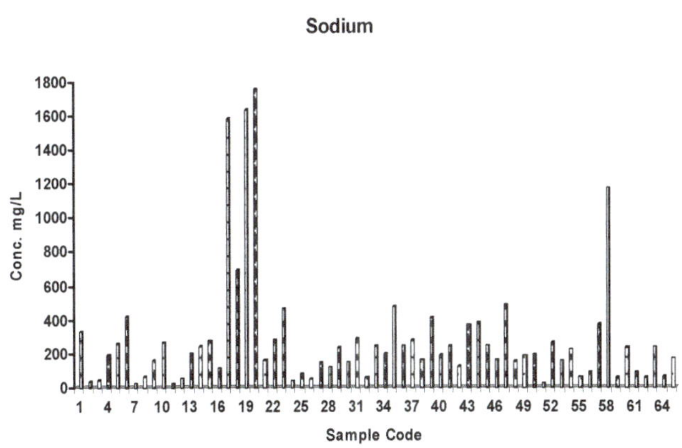

Figure 9. *Graphical representation of Na ion concentration of water samples.*

4. Conclusion

The analysis revealed that a number of ground water samples (70 %) showed their majority of parameters above the maximum permissible limits prescribed by WHO. Majority of the sites (all ground waters) were highly contaminated with toxic metals. The higher the concentrations of metals in ground water may be a concern for human health of Nawabshah city. Therefore the ground water of Nawabshah city may not be considered as safe to be used for dinking purpose. However, out of 60 water supply scheme samples, only 4 water samples may be used for drinking purpose and remaining all the water samples were contaminated.

References

[1] S. Khan, M. Shahnaz, NoorJehan, S. Rehman, M. T. Shah and I. Din. Drinking water quality and human health risk in Charsadda district, Pakistan. Journal of Cleaner Production. Vol. 60, pp. 93–101, December 2013.

[2] WHO, progress on drinking water and sanitation, World Health Organization, Geneva, Switzerland, 2012.

[3] T. Ahmad, P. P. Khanna, G. J. Chakrapani and S. Balakrishnan, Geochemical characteristics of water and sediment of the Indus river, Trans-Himalaya, India: constraints on weathering and erosion. Journal of Asian Earth Sciences. Vol. 16, pp. 333-346 April–June 1998.

[4] A. S. Qureshi, P. G. McCornick, M. Qadir and Z. Aslam, Managing salinity and water logging in the Indus Basin of Pakistan. Agricultural Water Management. Vol. 95, pp. 1-10, January 2008.

[5] S. K. Jain, P. K. Agarwal and V. P. Singh, Indus Basin. Water Science and Technology Library Vol. 57, pp. 473-511, 2007.

[6] M. Y. Khuhawar and S. A. Majidano. An investigation of quality of groundwater of taluka Nawabshah. Pak. J. Chem. Vol. 1, pp. 65-71, 2011.

[7] Guidelines for drinking water quality, 2nd ed. Geneva, WHO, 1995.

[8] Abu-Amr and M. M. Yassin. Microbial contamination of the drinking water distribution system and its impact on human health in Khan Yunis Governarate, Gaza strip. J. Royal Inst. Public Health, 10, pp. 10–16, 2008. PCRWR. National Water Quality Monitoring Program. Water Quality Report 2003–2004.

[9] WWF. Pakistan's waters at risk, Water and health related issues in Pakistan and key recommendations (2007), pp. 1–33 A special report, WWF — Pakistan, Ferozepur Road, Lahore — 54600, Pakistan.

[10] PAK-EPA. State of Environment Report 2005. Pakistan Environmental Protection Agency (Pak-EPA) Ministry of Environment, Govt. of Pakistan, Islamabad, Pakistan 2005.

[11] S. N. H. Mashadi and Muhammad A. Recharge the depleting aquifer of Lahore Metropolis, In Proceedings of regional groundwater management seminar 2000; 209-220 Islamabad.

[12] J. Samina, M. Jaffar and M. H. Shah. Physico-chemical profiling of ground water of

[13] Hazara strip, Pakistan. J. Chem. Soc. Pakistan. 26, pp. 288-292, 2004.

[14] A. Rahman. Groundwater as source of contamination for water supply in rapidly growing megacities of Asia: Case of Karachi, Pakistan. Water Science and Technology. Vol. 34, pp. 285-292, 1996.

[15] K. Khan, Y. Lu, H. Khan, S. Zakir, Ihsanullah, S. Khan, A. A. Khan and L. W. T. Wang. Health risks associated with heavy metals in the drinking water of Swat, northern Pakistan. Journal of Environmental Sciences. Vol. 25, pp. 2003–2013, October 2013.

[16] H. Amjad, I. Hashmi, M. S. Rehman, M. A. Awan, S. Ghaffar and Z. Khan. Cancer and non-cancer risk assessment of trihalomethanes in urban drinking water supplies of Pakistan. Ecotoxicology and Environmental Safety Vol. 91, pp. 25–311, May 2013.

[17] NIH. Survey by the 'Network 'Islamabad: National Institute of Health (NIH), Islamabad-Pakistan. 2004.

[18] UNDP, 2003. Pakistan National Human Development Report, "Water Crisis", Karachi, pp. 7103.

[19] S. Nazeer, M. Z. Hashmi and R. N. Malik. Heavy metals distribution, risk assessment and water quality characterization by water quality index of the River Soan, Pakistan. Ecological Indicators. Vol. 43, pp. 262–270, August 2014.

[20] J. Iqbal, M. H. Shah and G. Akhter. Characterization, source apportionment and health risk assessment of trace metals in freshwater Rawal Lake, Pakistan. Journal of Geochemical Exploration. Vol. 125, pp. 94–101, February 2013.

[21] U. Asghar, F. Parveen, S. K. Alvi, F. A. Khan, I. Siddiqui and T. H. Usmani. Contamination of Arsenic in Public Water supply scheme of Larkana and Mirpur Khas, District of Sindh J. Chem. Soc. Pak. 28, pp. 130-35, 2006.

[22] A. R. Khan, Mumtaz, K. S. Min, G. A. Marwat and M. iaz. Potable water quality characteristics of the urban areas of Peshawar (Pakistan): well water. J. Chem. Soc. Pak. ; 27, pp. 239-245, 2005.

[23] A. R. Khan, A. Haq, A. Waqar, M. Akif, M. Khan and M. Riaz. Quality characteristics of potable water of Mardan city J. Chem. Soc. Pak. 22, pp. 87 -93, 2000.

[24] S. M. Leghari, T. M. Jahangir, M. Y. Khuhawar, A. Laghari and A. D. Rajper, Water quality and biological assessment of ghazi shah and kai springs of district Dadu, Sindh. J. Biol. Sci. Pakistan. 2, pp. 39-42, 2002.

[25] M. Y. Khuhawar and G. M. Mastoi, Studies on some Physico - chemical parameters of Mancher lake Sindh. J. Anal. Environ. Chem. 3, pp. 66-71, 1995.

[26] M. N. Bhutta, M. Ramzan and C. A. Hafeez. Ground water quality and availability in Pakistan. Pakistan Council for Research in Water Resources (PCRWR, Islamabad, Pakistan, 2002.

[27] A. Azizullah, M. N. K. Khattak, P. Richter and D. Hader, Water pollution in Pakistan and its impact on public health — A review. Environment International. Vol. 37, pp. 479–497, February 2011.

[28] Zahoorullah, T. Akhtar and S. Zai. Quality of drinking water in rural Peshawar. Pak J Med Res, 42, pp. 85–89, 2003.

[29] G. Sarwar, J. Khan, R. Iqbal, A. K. Afridi, A. Khan and R. Sarwar. Bacteriological analysis of drinking water from urban and peri-urban areas of Peshawar. Journal of Postgraduate Medical Institute, Khyber Medical College, Peshawar, 18, pp. 64–69, 2004.

[30] M. Hussain, S. A. Rasool, M. T. Khan and A. Wajid. Enterococci vs coliforms as a possible fecal contamination indicator: baseline data for Karachi. Pak J Pharm Sci, 20, pp. 107–111, 2007.

[31] M. S. Anwar, N. A. Chaudhry and M. Tayyib. Qualitative assessment of bacteriological quality and chlorination status of drinking water in Lahore, J Coll Physicians Surg Pak, 14, pp. 157–160, 2004.

[32] R. T. Nickson, J. M. McArthur, B. Shrestha, T. O. Kyaw-Myint and D. Lowry, Arsenic and other drinking water quality issues, Muzaffargarh District, Pakistan Applied Geochemistry. Vol. 20, pp. 55–68, January 2005.

[33] K. D. Brahman, T. G. Kazi, H. I. Afridi, S. Naseem, S. S. Arain and N. Ullah. Evaluation of high levels of fluoride, arsenic species and other physicochemical parameters in underground water of two sub districts of Tharparkar, Pakistan: A multivariate study. Water Research. Vol. 47, pp. 1005–1020, March 2013.

[34] Y. Iqbal, S. Alam, S. Sabir, M. Ishaq and N. Ahmad. Investigation of pollutants in the industrial effluents. J. Chem. Soc. Pak. 22, pp. 239-44, 2000.

[35] Z. H. Lodhi, M. Akif and U. Kalsoom, Evaluation of drinking water from different sources in skardu-Northern area with special reference to heavy metals. J. Chem. Soc. Pak. 25, pp. 110-13, 2003.

[36] J. H. Poverty and R. J. Kopka. Water, Air and Soil Pollution. 57, pp. 399, 1995.

9

Description of Two-Peak Characteristics in Power Engineering

Tetiana Lutchyn

National Technical University of Ukraine "Kyiv Polytechnic Institute", Institute for Energy Saving and Energy Management, Kyiv, Ukraine

Email address:
t.lutchyn@mail.ru

Abstract: Many branches of Power Engineering have the same problems with signal processing. However they can be solved by a general approach. One of these problems is the complexity of signal description with the inertial change (without sharp peaks and dips), but the signal is a complex shape without symmetry and poorly responds to the cycling laws. Also offered, if it is necessary, to use the segmentation of the original signal with further Multiresolutional analysis. As a result, it is possible to make the selection with the most informative wavelet coefficients which can significantly reduce the quantity of the original data set with accuracy within acceptable limits.

Keywords: Multiresolutional Analysis, Wavelet Analysis, Description, Two-Peak Characteristics

1. Introduction

Nowadays there are a lot of methods, but it is very difficult to choose the unique algorithm for the data set automatically description. Especially the solution research is very difficult for asymmetrical signals with aperiodic character and additional conditions. Such type of signals could be with relative insignificant fluctuations, for example, two-peak characteristic and similar to them which can change with some inertness [1].

Description methods for such characteristics are: partial straightening [2], harmonious approach [3], two-level fuzzy models [4] and others. As a rule, signals of the specified character are observed in problems of: water supply [5], magnetic-electroencephalography [6], seismology [7, 8], climate change with taken into account a level of cloudy influence [9], value of precipitation [10], genetics [11], astronomy [12], topology [13] and other directions.

In power engineering, especially often similar signals are detected in tasks of:

- control of level of electricity generation [14];
- choice of electric equipment in distributive networks [15];
- installation of a network configuration [16];
- definition of peak consumption [17];
- calculation of the electricity cost [18];
- design and adjustment of Smart Grid [19];
- classification of consumers by their types [20, 21];
- calculation of losses in the transmission line [22];
- studying of the consumer behaviour [23].

Description approaches are most developed for short-term forecasting of daily graphs. In such tasks, the most various models are:

- according to average differential characteristics [24, 25];
- iterative [26];
- neural network [27–30] and artificial neural network [31];
- ARIMA with the further signals clustering [32];
- wavelet transformation and fuzzy logic [33];
- genetically optimized cascade neural networks with Kohen's modified clustering [34];
- hybrid fuzzy-logical, however these models in some cases provide errors during description of the original data [35, 36].

In the theory of wavelet transformation for the graphs description, the partial signal representation is depicted in [7, 37], and also for air streams [38], seismology [39], medicine [40], meteorology [41], topography [13] and other.

This paper is organized as follows. For creation general algorithm in all next parts new properties and research results are presented. In part 2, main approaches of wavelet transformation and property of the Multiresolutional analysis (MRA) with developed flow scheme of wavelet coefficients are given. Part 3 discusses the description algorithm of the graph similar to two-peak characteristic with features of wavelet coefficients, while part 4 looks into details of data recovering with shifting in wavelet coefficients. In last part, some conclusions are presented.

2. Main Approaches of Wavelet Transformation and Property of the Multiresolutional Analysis

One of the most effective methods for the depiction shape of daily graphs is presented in [42, 43]. However it is noted by authors, this method has hyper sensibility to any changes in a signal. Thus, the variation between calculated and original values can achieve up to the triple divergence on nearly 50 % of all research range.

In this paper, the main accent is directed on the model creation with a possibility of correction of extremely large deviation, in such way the sensitivity decrease to the distribution law of the original data. Minimal quantity of necessary values for recovery signal should be provided. For this purpose the initial signal is transformed MRA. Received wavelet coefficients are more informative, than the original signal without previous processing [42]:

$$k_{m,2n} = \frac{k_{m-1,2n} + k_{m-1,2n+1}}{\sqrt{2}}, \quad k_{m,2n+1} = \frac{k_{m-1,2n} - k_{m-1,2n+1}}{\sqrt{2}}, \tag{1}$$

where $k_{m,2n}$ and $k_{m,2n+1}$ – approximation and detailed wavelet coefficients, respectively; m – level of transformation; n – ordinal number of wavelet coefficient on the mth level of decomposition.

According to the inverse algorithm of MRA [42]

$$k_{m-1,2n} = \frac{k_{m,2n} + k_{m,2n+1}}{\sqrt{2}}, \quad k_{m-1,2n+1} = \frac{k_{m,2n} - k_{m,2n+1}}{\sqrt{2}}, \tag{2}$$

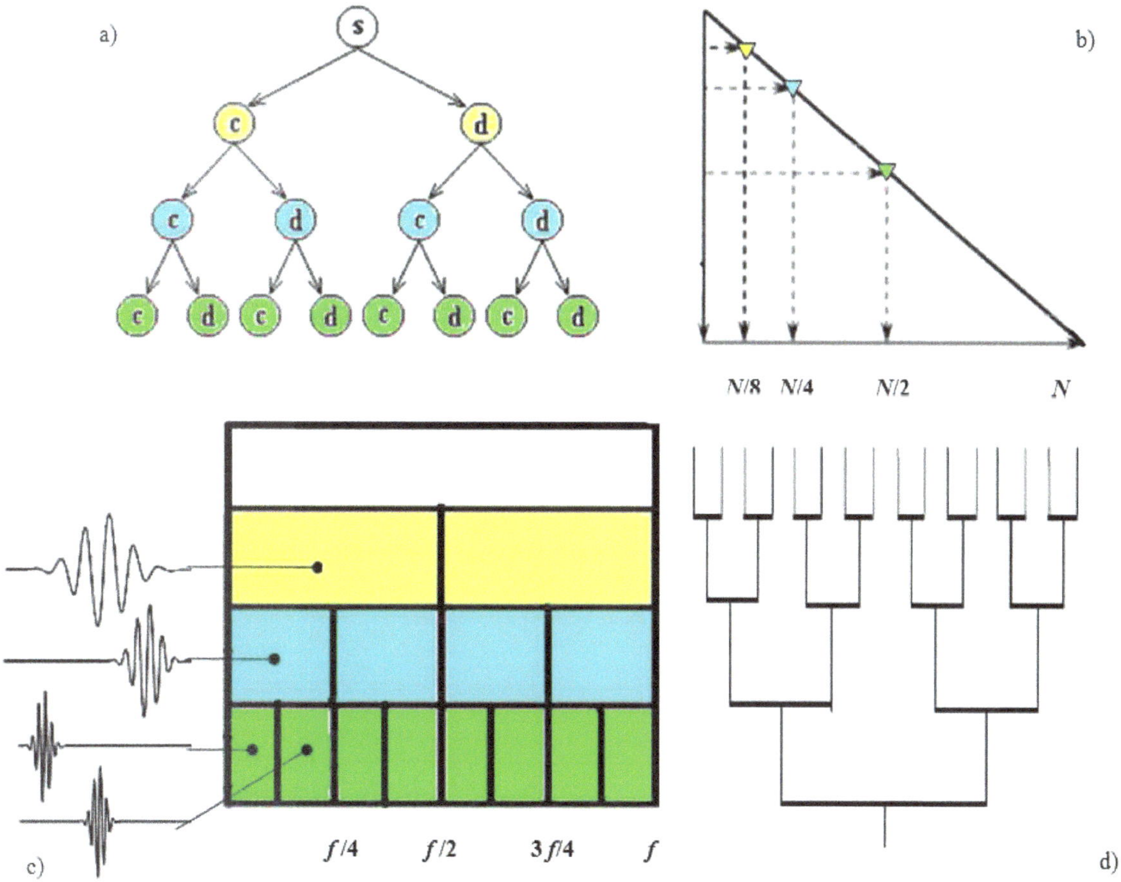

Fig. 1. Three-level decomposition tree of a signal with its main properties.

Let us present schematically a signal after MRA, as an example of three level decomposition tree

In fig. 1, a the most widespread scheme is presented in the wavelet theory. It is the basic principle of signal division according to the MRA main approach. Also for today's time the principles of its construction are well studied. The frequency division of a signal (it is one of the main properties of wavelet transformation) with its change are depicted in details in scale and in local representations (fig. 1, c).

In many papers with methods of data compression, authors

very often miss the description of wavelet coefficients (fig. 1, b). However it is necessary to consider that on each level the calculated coefficients can contain different quantity of values. This problem is very important in the compression tasks for a correct choice of representative parameters for further data recovery. For example, that the original data set contains 16 values. According to the theory of wavelet decomposition, it provides four levels of transformation ($2^m = N$, N – quantity of the initial values). There is a different quantity of wavelet coefficients with the different quantity of values on each level. For example, as a result of compression, it was defined that it was necessary to operate by two wavelet coefficients of the second level, one of the third level and two of the fourth level. It is incorrect to suppose that for full data recovery one can need 2 + 1 + 2 = 5 values (these are wavelet coefficients!!!).

CONCLUSION 1: with increasing level of decomposition the quantity of wavelet coefficients increases in the same turn as the quantity of values decreases in wavelet coefficients.

This statement is confirmed also by fast wavelet decomposition by Mallat algorithm according to which one detailed wavelet coefficient of each level of decomposition and one approximate wavelet coefficient of the last wavelet level apply. For presented example it corresponds to:

$$1 \times 8 + 1 \times 4 + 1 \times 2 + 2 \times 1 = 16.$$

In the classical theory of mathematics it is well known, that if quantity of required values equals to quantity of dissimilar equations, the task is completely defined and solutions are identified unambiguously.

As a research result the operations procedure is presented (fig. 1, d) which needs to be considered for originally non-complete data set ($N \neq 2^m$). The functional properties of wavelet coefficients are considered in detail in paper [1].

3. Description Algorithm of the Graph Similar to Two-Peak Characteristics with Features of Wavelet Coefficients

In practice task it is very often to cut off the peak moments for calculations simplification, because they have small duration. However these values need the most careful control and research.

For example, the electricity tariff for the industrial enterprises depends of the extra consumption. In case of peaks smoothing of the measured signal, the Power Supply Company will have economic losses.

On the basis of the above properties, it was created the general algorithm is applicable for two-peak characteristics with keeping the main properties of a signal.

Let see in detail all stages of below presented algorithm. For better interpretation of the principle of the property definition for the offered algorithm, the approach of the original data depicting in wavelet coefficients with restriction by the quantity will be presented for the third level of wavelet decomposition (fig. 2).

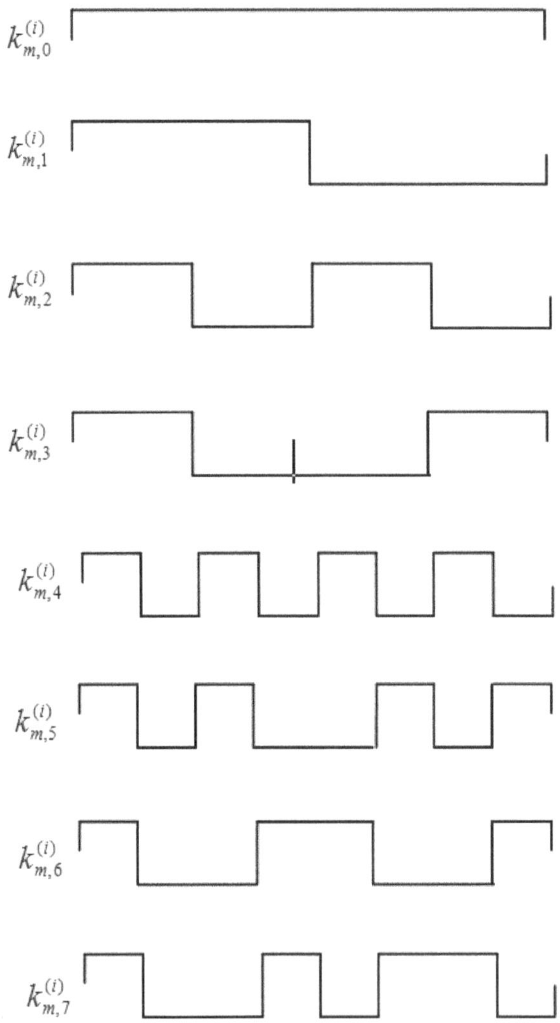

Fig. 2. The principle of the original signal depicting in wavelet coefficients (the quantity of wavelet coefficients in corresponds to the third level).

For daily graphs the segmentation has a large influence. The transition from the $(m-2)$ level of signal decomposition is presented in a matrix form:

$$\begin{pmatrix} k_{m,n} \\ k_{m,n+1} \\ k_{m,n+2} \\ k_{m,n+3} \end{pmatrix} = \begin{vmatrix} + & + & + \\ + & + & - \\ + & - & + \\ + & - & - \end{vmatrix} \quad \begin{matrix} k^{I}_{m-2,n} & k^{II}_{m-2,n} & k^{III}_{m-2,n} \end{matrix} \tag{3}$$

4. Data Recovery with Shifting in Wavelet Coefficients

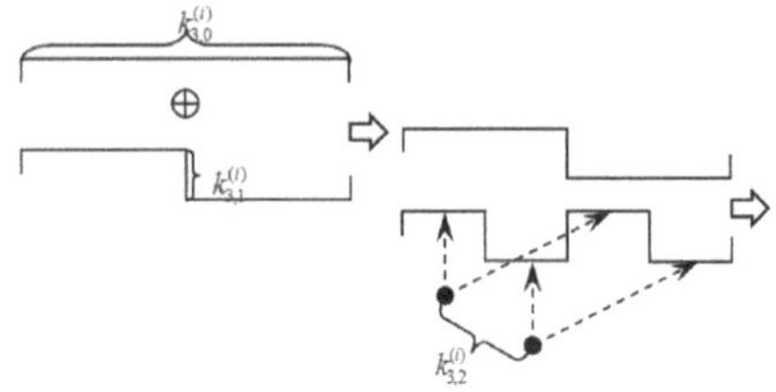

Fig. 3. Information about the original signal in wavelet coefficients.

The developed approach is based on the following property. Wavelet coefficient is responsible for the value shifting to present a difference between half of values of the original signal (fig. 3). Graphical representation (fig. 3) also additional confirms the character of approximated and detailed wavelet coefficients. Especially, it is the property of information accumulation about the main part of a signal and its deviations respectively. Proceeding from fig. 3, the search algorithm can be presented on Table.

Table 1. *Recovery algorithm of signal with functions shifting.*

Proportional value of the total consumption of the graph segment		Domain of the solution identifying			
$k^{(i)}_{(m-2),0}$	$\oplus k^{(i)}_{(m-2),1}$	$\rightarrow$	$\begin{Bmatrix}(0; & N/2];\\(N/2; & N]\end{Bmatrix}$	$\rightarrow$	
	$\rightarrow \oplus k^{(i)}_{(m-2),2}$	$\rightarrow$	$\begin{Bmatrix}(0; & N/4];\\ & \dots \\(3N/4; & N]\end{Bmatrix}$	$\rightarrow$	
	$\rightarrow \oplus k^{(i)}_{(m-2),4}$	$\rightarrow$	$\begin{Bmatrix}(0; & N/8];\\ & \dots \\(7N/8; & N]\end{Bmatrix}$	$\rightarrow$	$k^{(i)}_{(m-2),8}$

On the basis of tabular data presentation and sequence of data recovery, it is possible to make the next conclusion.

CONCLUSION 2: For determination of values in one segment of the original signal, it is necessary to take into account shifting of wavelet coefficients and to use only a half of the transformed values (fig. 2).

Initially, according to the MRA, the calculation of detailed wavelet coefficients is made for a possibility to detect fluctuations in the original signal. However, as a rule, it is widespread to use. Therefore the important properties of other detailed wavelet coefficients are lost which consider the following law of signal distribution (fig. 5). For simplicity, let us to consider the first stage of proposed algorithm (fig. 4 and Table).

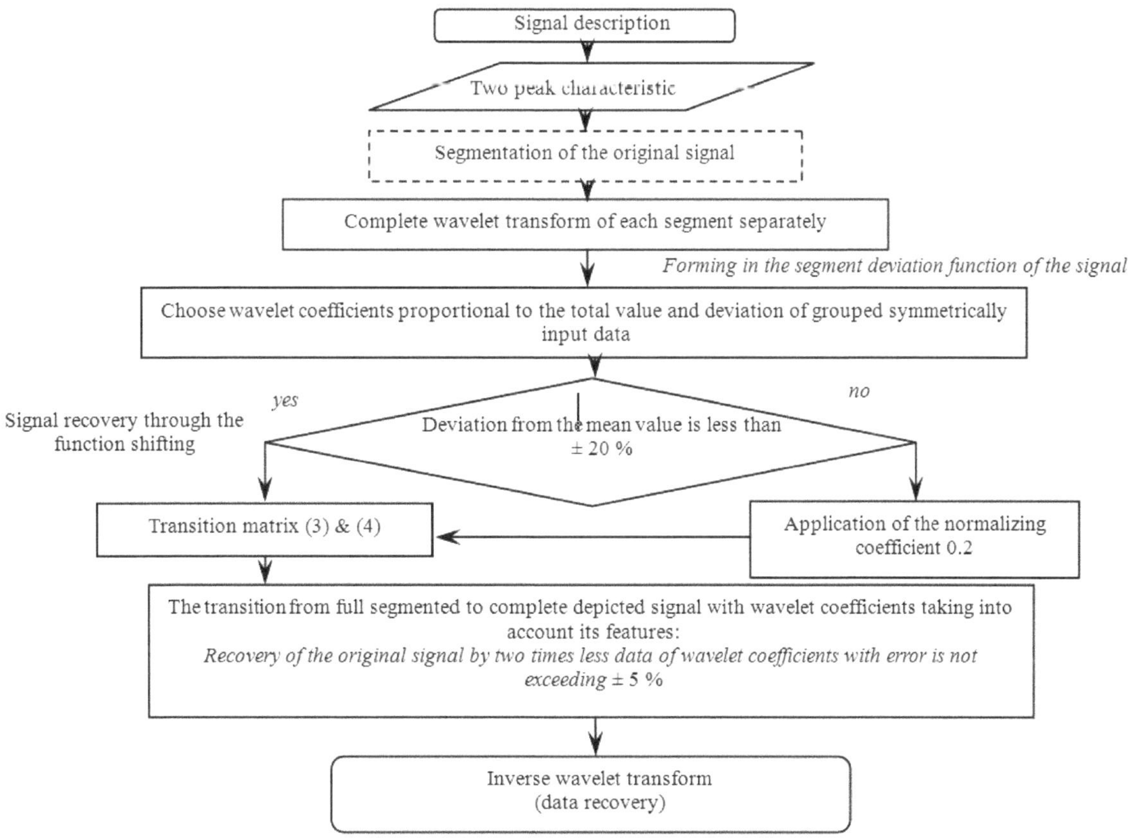

Fig. 4. *Algorithm of the graph description similar to two-peak characteristic with features of wavelet coefficients.*

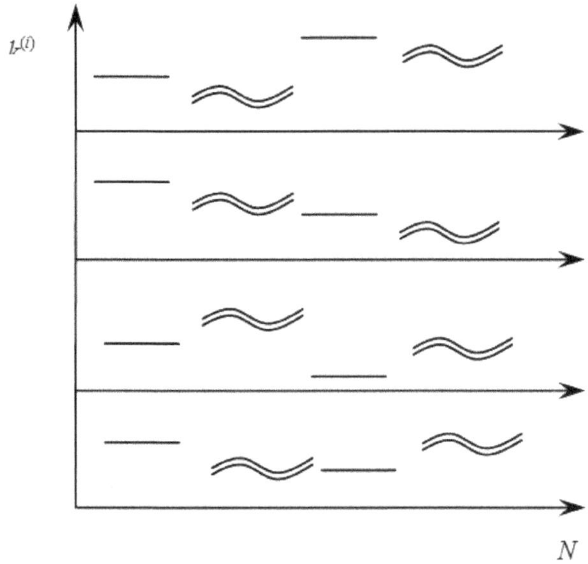

Fig. 5. Possible options of distribution of restored values with functions shifting.

Pair recovering of the summarized data of the original signal at present level of wavelet decomposition is carried out according to the equation:

$$\begin{cases} k_{(m-2),2(1)} = \dfrac{k_{(m-2),1(1)} + k_{(m-2),2}}{2}; & k_{(m-2),2(2)} = \dfrac{k_{(m-2),1(1)} - k_{(m-2),2}}{2} \\ k_{(m-2),2(3)} = \dfrac{k_{(m-2),1(2)} + k_{(m-2),2}}{2}; & k_{(m-2),2(4)} = \dfrac{k_{(m-2),1(2)} - k_{(m-2),2}}{2} \end{cases} \quad (4)$$

NOTE. Stage-by-stage transition between levels is provided with well-known equations (1) and (2).

CONCLUSIONS. As a research result, the description algorithm of nonlinear graphs is developed and offered which allows reducing twice the quantity of original data. However the error does not exceed normally acceptable limits which can be used in Power Engineering. Also in the paper, the scheme of properties of the Multiresolutional analysis is generalized that allows to present and to apply more effective detailed wavelet coefficients for creation new models and methods of signal processing.

References

[1] A. Voloshko, T. Lutchyn, "Principle of Determining Informative Wavelet Transformed Values in a Partial Restoration of Original Data Sample", Registration, Storing and Data Process, 14:4 (2012), 33-40. – (in Rus.).

[2] W. Lian, A. Mao, L. Zhang, "A Method for Cascading Failure Simulation based on Static Security Analysis", 10.1109/APPEEC.2009.4918053.

[3] A. Vahidnia, A. Dastfan, M. Banejad, "Determination of Harmonic Load Characteristics in Distribution Networks of Cities" Power Engineering, Energy and Electrical Drives, 2009. POWERENG '09. Intern. Conf., (2009), 442-446.

[4] Z. Zakaria, K. L. Lo, "Two-stage Fuzzy Clustering Approach for Load Profiling", Universities Power Engineering Conference (UPEC), 2009 Proc. of the 44thIntern., (2009), 1-5.

[5] X. Ge, L. Zhang, Ya. Yang, "Multi-objective Hydro Optimal Scheduling with Flow Time", 10.1109/PowerCon.2012.6401259.

[6] L. Kipinski, R. Konig, C. Sieluzycki, "Application of modern tests for stationarity to singletrial MEG data. Transferring powerful statistical tools from econometrics to neuroscience", Biological Cybernetics, 105 (2011), 183–195.

[7] Yu. Yagi, Yu. Fukahata, "Introduction of uncertainty of Green's function into waveform inversion for seismic source processes", Geophysical Journal International, 186 (2011), 711–720.

[8] A. C. Iliopoulos, G. P. Pavlos, "Global Low Dimensional Seismic Chaos in the Hellenic Region", International Journal of Bifurcation and Chaos, 20:7 (2010), 2071-2095.

[9] C. L. Bain, J. de Paz, J. Kramer, "Detecting the ITCZ in Instantaneous Satellite Data using Spatiotemporal Statistical Modeling: ITCZ Climatology in the East Pacific", American Meteorological Society, 24 (2011), 216-230.

[10] T. Kozu, K. Masuzawa, T. Shimomai, "Estimation of No* for the Two-Scale Gamma Raindrop Size Distribution Model and Its Statistical Properties at Several Locations in Asia", Journal of Applied Metrology and Climatology, 49 (2010), 676–686.

[11] F. Qi, X. Liu, Yi. Ma, "Synthesis of neural tree models by improved breeder genetic programming", Neural Computing & Applications, 21 (2012), 515–521.

[12] M. M. Katsova, I. M. Livshits, J. Sykora, "The Rotation of the Sun as a Star from the Green-Line Emission of the Entire Corona", Astronomy Reports 86:4 (2009), 343-354.

[13] Th. Candela, F. ois Renard, M. Bouchon, "Characterization of Fault Roughness at Various Scales: Implications of Three-Dimensional High Resolution Topography Measurements", Pure & Applied Geophysic, 166 (2009), 1817–1851.

[14] R. Liu, M. Chen, H.-X. Wang, "Study on Grouping Control Strategy of AGC Units", 10.1109/ISGT-Asia.2012.6303348.

[15] J. A. Jardini, C. M. V. Tahan, E. L. Ferrari, S. U. Ahn, "Selection of Distribution Transformer Based on Economic Criteria", Electricity Distribution. Part 1: Contributions. CIRED. 14th Intern. Conf. and Exhibition, 6 (1997), 7.

[16] M. H. Shariatkhah, M. R. Haghifam, "Determining of Annual Distribution Feeder Configuration Using Load Curves Clustering", Electrical Power Distribution Networks (EPDC), 2011 16th Conf., (2011)

[17] Ch. Preet, G. Yadwinder, B. Kanwardeep, "Incentive Based Demand Response Program: an Effective Way to Tackle Peaking Electricity Crisis", 10.1109/CCECE.2012.6334813.

[18] G. I. Rhodes, "A Method of Studying Power Costs with Reference to the Load Curve and Overload Economies" American Institute of Electrical Engineers, (1912), 81-100.

[19] S. Kahrobaee, R. A. Rajabzadeh, L-K. Soh, "A Multiagent Modeling and Investigation of Smart Homes With Power Generation, Storage, and Trading Features", Smart Grid, IEEE Transactions, 4:2 (2012), 659-668.

[20] S. V. Verdu, M. O. Garcia, F. J. G. Franco, "Characterization and Identification of Electrical Customers Through the Use of Self-Organizing Maps and Daily Load Parameters", Power Systems Conf. and Exposition, 2004. IEEE PES, 2 (2004), 899-906.

[21] J. A. Jardini, C. M. V. Tahan, M. R. Gouvea, "Daily Load Profiles for Residential, Commercial and Industrial Low Voltage Consumers", Power Delivery, IEEE Transactions, 15:1 (2000), 375-380.

[22] A. L. Shenkman, "Energy Loss Computation by Using Statistical Techniques", IEEE Transactions on Power Delivery, 5:1 (1990), 254-258.

[23] P. Kadar, "Understanding customer behavior", 10.1109/TDC-LA.2008.4641800.

[24] Sh. Yang, Yi. Zhang, "Short-term Load Forecast Based on Decomposition of Daily Load Curve", 2010 International Conference on Computer, Mechatronics, Control and Electronic Engineering (CMCE) 3 (2010), 65-68.

[25] Yu. Sabri, N. Hariyanto, F. Fitriana, "Spatial Short-Term Load Forecasting using Grey Dynamic Model Specific in Tropical Area", 10.1109/ICEEI.2011.6021776.

[26] Q. Ding, J. Lu, H. Liao, "A Practical Super Short Term Load Forecast Method and Its Implementations", Power Systems Conference and Exposition, 2004. IEEE PES, 1 (2004), 483-486.

[27] M. B. Tasre, V. N. Ghate, P. P. Bedekar, "Comparative Analysis of Hourly Load Forecast for a Small Load Area", 2012 International Conference on Computing, Electronics and Electrical Technologies (ICCEET) (2012), 80-85.

[28] F. X. Xie, M. Huang, W. G. Zhang, "Research on Electric Vehicle Charging Station Load Forecasting", The International Conference on Advanced Power System Automation and Protection 2011 IEEE (2011), 2055-2060.

[29] R. Afkhami, F. M. Yazdi, "Application of Neural Networks for Short-Term Load Forecasting", 10.1109/POWERI.2006.1632536.

[30] M. Lopez, S. Valero, C. Senabre, "A SOM Neural Network Approach to Load Forecasting. Meteorological and Time Frame Influence", 10.1109/PowerEng.2011.6036553.

[31] J. Nazarko, A. Jurczuk, W. Zalewski, "ARIMA Models in Load Modelling with Clustering Approach", 10.1109/PTC.2005.4524719.

[32] P. T. T. Binh, N. T. Hung, P. Q. Dung, "Load Forecasting Based on Wavelet Transform and Fuzzy Logic", 10.1109/PowerCon.2012.6401281.

[33] I. Erkmen, A. Ozdogan, "Short Term Load Forecasting Using Genetically Optimized Neural network Cascaded with a Modified Kohonen Clustering Process", Proceedings of the 12th IEEE International Symposium on Intelligent Control (1997), 107-112.

[34] D. Srinivasan, A.C. Liew, C.S. Chang, "Forecasting daily load curves using a hybrid fuzzyneutral approach", IEE Processing-Generation, Transmission, Distribution, 141:6 (1994), 561-567.

[35] D. M. FalcZio, H.O. Henriques, "Load Estimation in Radar Distribution Systems Using Networks and Fuzzy Set Techniques", Power Engineering Society Summer Meeting, 2 (2001), 1002-1006.

[36] M. Elsayed, "An overview of wavelet analysis and its application to ocean wind waves", Journal of Coastal Research, 26:3 (2010), 535–540.

[37] G. M. Menanno, A. Mazzotti, "Deconvolution of multicomponent seismic data by means of quaternions: Theory and preliminary results", Geophysical Prospecting, 60 (2012), 217–238.

[38] M. R. Homaeinezhad, M. Aghaee, H. N. Toosi, "Application of the Discrete Wavelet Transform for the Robust Detection of the Impulsive Incidences: Application to Arterial Blood Pressure Characteristic Events Detection-Delineation", International Journal of Wavelets, Multiresolution and Information Processing, 9:5 (2011), 813–842.

[39] G. Hall, S. Woodborne, M. Pienaar, "Rainfall control of the δ13C ratios of Mimusops caffra from KwaZulu-Natal, South Africa", The Holocene, 19:2 (2009), 251–260.

[40] Kh. Mistry, R. Roy, "Enhancement of Voltage Stability Index of Distribution System by Network Reconfiguration Including Static Load Model and Daily Load Curve", Innovative Smart Grid Technologies - India (ISGT India), 2011 IEEE PES21 (2011), 65-68.

[41] D. Raisz, A. M. Dan, "Ripple Control as a Possible Tool for Daily Load Balancing in an Open Electricity Market Environment", 10.1109/PTC.2005.4524451.

[42] I. M. Dremin, O. V. Ivanov, V. A. Nechytailo, "Wavelets and its Usage", SFS, 171:5 (2001), 465–501. – (in Rus.).

[43] A. Voloshko, T. Lutchyn, "Analysis of the Efficiency of Measures Compress of Schedules of Electrical Loads Аналіз ефективності міри стиснення графіків електричних навантажень", Vinnitsa NTU, 16:4 (2011), 209-214. – (in Ukr.).

10

Assessment of Rn and U Concentrations in the Soil of Qadafery, Kalar and Zarayan Located in Sulaimani Governorate of Kurdistan Region- Iraq

Kamal O. Abdullah, Salar Z. Muhammed, Adil M. Hussein*

Physic Dep., Faculty of Science and Science Education, University of Sulaimani, Kurdistan region, Iraq

Email address:
kamaloa13@gmail.com (K. O. Abdullah), adiljaff@gmail.com (A. M. Hussein)

Abstract: The profile of radon (222-Rn) and uranium (238-U) concentrations in15 surface soil samples collected from the three agricultural areas of town Qadafery, Kalar and Zarayan which are located at the east of Sulaimani governorate in Kurdistan Region – Iraq has been determined using solid- state nuclear track detector (CR-39), where Qadafery located in the margin of Suren Mountain, also located in the north of the famous plain of the Middle East region called Sharazoor plain and Zarayan located in the west north of the Sharazoor plain, also its surrounds in another direction by Qaradagh and Barznja mountains. They (Qadafery, Kalar and Zarayan) have been shown that the maximum and minimum values of radon concentration of the air inside the tube were 2242.263±143.152 Bq/m^3 and 154.099±2.579 Bq/m^3, respectively, with average value 741.709±27.234 Bq/m^3 while the maximum and minimum values of uranium concentration are 18.225±12.905 ppm and 1.253±0.233 pmm, respectively, with average value 6.029±2.455 pmm. It appears that higher concentrations of radon and uranium was in soil sample Qadafery_5 and the minimum value was in soil sample Kalar_4 and by comparison with the world values there are some positions in Qadafery and Zarayan with higher level concentrations that is due to the geology information of these regions.

Keywords: Soil, SSNTD, CR-39, Radon, Uranium

1. Introductions

The natural radioactive nuclides like radon is the main source of radiation exposure for human when they are emitted from soil, building materials and natural energy sources which contain ^{238}U traces, and the ^{222}Rn with half-life 3.8 d can be considered to be one of the most dangerous radioactive elements [1,2], when uranium is high in the soil it can cause to ^{222}Rn and its airborne daughter is a significant internal health hazard [3]. Uranium as a heavy metal is a natural constituent of the earth's crust and the presences of it in soil are natural and anthropogenic when the main sources of anthropogenic exist in various industrial sources and agricultural activities such using fertilizer while the result of these activities are causing to soil pollution by these types of heavy metals. The maximum permissible levels of metal concentrations in soil were complying with the idea of preventing toxic exposure of plants, animals and humans [4]. Generally, the distribution of such heavy metals is influenced by kind of main source, climate and relative mobility depending on soil parameters such as mineralogy, texture and category of soil and PH [5].

Noble radon gas (^{222}Rn) originates from radioactive transformation of ^{226}Ra in the ^{238}U decay chain in the earth's crust. Only a fraction of the radon atoms so created is able to emanate from the mineral grains and enter the void space, then radon moves further by correspondence to: A. Gregory diffusion and advection dissolved either in water or in carrier gases. Eventually, it exhales into the atmosphere. The radon emanation depends mainly on ^{226}Ra content and mineral grain size, its transport in the earth governed by geophysical and geochemical parameters, while exhalation is controlled by hydro meteorological conditions [6].

Radon concentrations in soil gas within a few meters of the surface of the ground are clearly important in determining radon rates of entry into pore spaces and subsequently into the atmosphere and it depend on the radium concentration in the bedrock and on the permeability of the soil [7].

In the present study, measurements of radon and uranium concentrations of 15 soil samples collected in three regions, the eastern of Sulaimani governorate, were carried out using Solid State Nuclear Track Detector (SSNTD) technique, in order to assess whether the soil and the general public is at any risk due to this exposure to excessive levels of radon because these regions were used as a farm mostly for agriculture in this area. The results of the present study shall help in determining the positions which have the highest activity of the natural radiation background (Uranium) directly or indirectly affects the human health. The results have been compared to the limits of worldly values.

2. Experimental Method

2.1. Sampling

Fifteen samples were collected from surface soil (until depth 20cm) at each agriculture farm in the studied area in Sulaimani (Kurdistan Region-Iraq). The places Zarayan and Qadafary were situated at the east of Sulaimani, their are geographical coordinates are 35° 31^{l} 88^{ll} North, 45° 67^{l} 77^{ll} East and 35° 33^{l} 39^{ll} North, 45° 94^{l} 03^{ll} East, respectively, but Kalar was situated at the South of Sulaimani with geographical coordinates of 34° 37' 33" North, 45° 18' 31" East, as shown from the map of the studied area in Fig.(1).

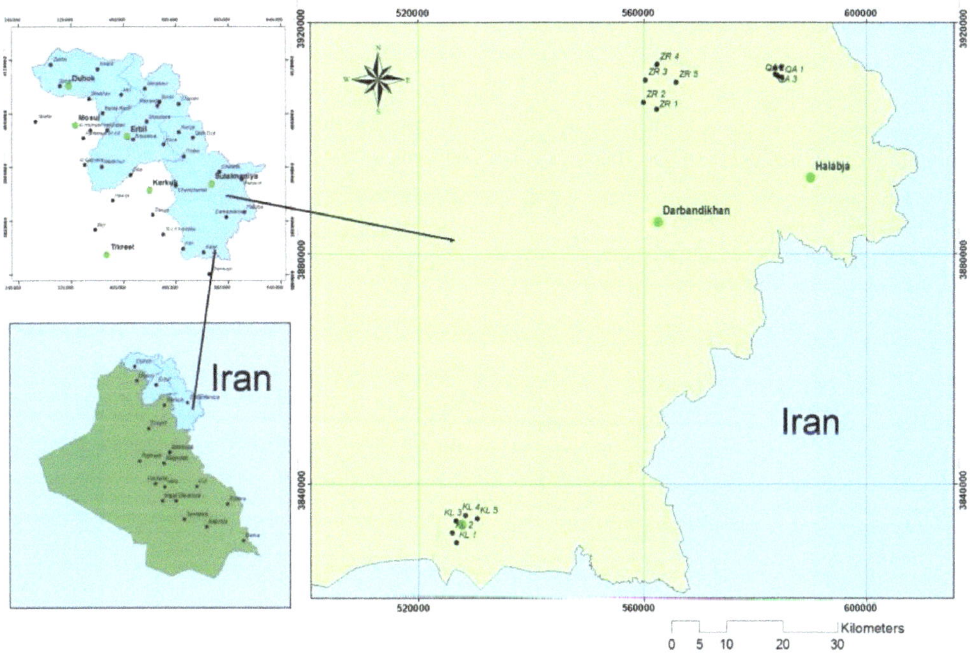

Fig. (1). Studied sample map in Kalar, Qadafery and Zarayan.

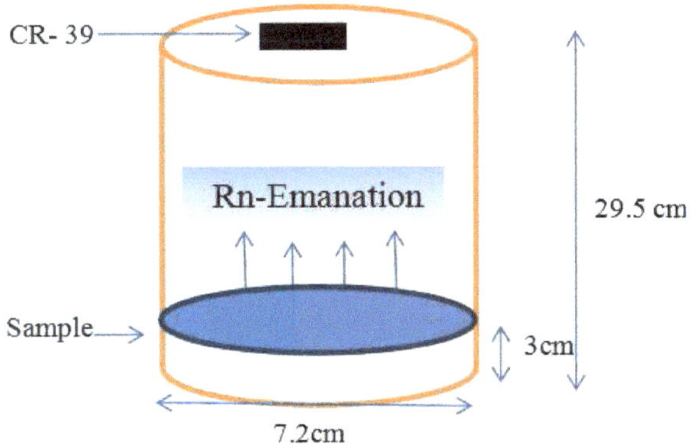

Fig. (2). Schematic Diagram of Long-Tube for SSNTD Technique.

The sample weight were 150 gm prepared after drying and placed on the bottom of a sealed polyethylene bottle (cylindrical plastic) and the (CR-39) detector with pieces of area 1cm x 1.5cm on the other side of cylindrical plastic suspended as shown in Fig. (2), schematic diagram of long-tube, then stored for two months to allow radioactive equilibrium between ^{222}Rn and ^{226}Ra. Exposed detector was collected and etched chemically using 6.25 Molarity NaOH at 70 C° for 6 hours after washing these detectors in distilled water, the track density on CR-39 (tr. /cm^2) was counted using an optical microscope with magnification 400x [8].

2.2. Radon Concentration Measurement

The concentrations of ^{222}Rn in the air in the tube, that contain soil samples by using CR-39, can be calculated from relation [9,10]:

$$C_a = \rho / \eta \, T \tag{1}$$

where ρ - is the track density on the exposed detector CR-39 (Tr/cm^2).

T - is the exposure time of the samples (60 d)

η - is the detection efficiency (0.05774 Tr/cm^2/d recorded per Bq/m^3)

Then (η) is determined from this relation using the dimensions of the tube [8]:

$$\eta = \frac{1}{4} r \left(2\cos\theta_t - \frac{r}{R_\alpha}\right) \tag{2}$$

where r - is the tube radius for the diffusion volume (3.6 cm).
θ_t - is the threshold angle of the CR-39 detector (35°) [11]
R_α - is the range of α- particle in air from Rn
R_α can be calculated from relation [12]:

$$R_\alpha = (0.005\ E_\alpha + 0.285)\ E_\alpha^{3/2} \quad (3)$$

$$= 4.019 \text{ cm (for energy } E_\alpha = 5.49 \text{ MeV)}$$

The value of the diffusion constant (η) according to the dimensions of the present system equals to (0.057744)Tr.cm^{-2}. d^{-1} / Bq . m^{-3}, where Tr- is a trace , d^{-1} -is day period of time.

To calculate ^{222}Rn concentration in the samples, the following relation can be used [13]:

$$C_s = \frac{\lambda_{Rn}\, C_a\, H\, T}{L} \quad (4)$$

where C_s - is the ^{222}Rn concentration in the samples (Bq\m^3)
C_a – is the ^{222}Rn concentration in air space inside the tube (Bq\m^3)
λ_{Rn} – is the Decay constant of Rn^{222} (0.1814 day^{-1})
H – is the Height of air space in the tube (29.5 cm)
L – is the Thickness of the sample in the tube (3 cm)
T - is the Time of irradiation (60 days)

To calculate the uranium concentrations by ppm must be calculated the number of radon atoms uing relation [12]:

$$A_{Rn} = \lambda_{Rn}\, N_{Rn} \quad (5)$$

From N_{Rn} one can determine the number of atoms of uranium in the samples, using the equation of secular equilibrium [14,15]:

$$\lambda_U\, N_U = \lambda_{Rn} N_{Rn} \quad (6)$$

where λ_U is the decay constant of uranium (4.883x10^{-18} sec^{-1}), the weight of uranium in the samples can be calculated from [16]:

$$W_U = \frac{N_U\, M_u}{N_A} \quad (7)$$

where Mu - is the mass number of uranium ^{238}U
N_A - is the Avogadro number (6.02x10^{23} mol^{-1})

Thus, the concentration of uranium in ppm can be obtained through:

$$C_U\,(ppm) = \frac{W_U}{M_S} * 10^6 \quad (8)$$

where Ms – is the mass of the studied samples.

3. Result and Discussion

We summarized in Table (1) the number of track density ρ (Tr/cm²), the value of radon concentration for air inside the tube C_a (Bq/m) and for soil samples Cs (Bq/m³) with the value of uranium concentration C_U (ppm). The analysis of the measured values shows that there are two positions of Qadqfary's soil samples (Qadafery_2 and Qdafery_5) with high values of radon and uranium concentrations of (145523.227 ± 699.319 and 239980.470 ± 1480.952) Bq/m^3 for radon and (11.052 ± 6.094 and 18.225 ± 12.905) pmm for uranium, also the high values appeared in the soil samples in (Kalar_1, Zarayan_3 and Zarayan_4) with values (85373.627 ± 314.240) Bq/m^3, (102836.414 ± 415.429) Bq/m^{3}and (73731.769 ± 252.208) Bq/m^3 for radon and (6.484±2.738) ppm, (7.810±3.62) pmm and (5.599±2.198) pmm for uranium as shown in the figures (3) and (4), this is due to these regions were agriculture farms and farmers used a fertilizer every years is caused to high concentrations of uranium [17] and according to geology information the soil on type clay and limestone gravel with kometan formation which causes also to high concentrations of uranium [18] while the color of soil in Qadafery (high value) positions was black when the color on the other positions was brown which effect on concentrations [19].

Table (1). *Radon and Uranium concentrations in the soil samples.*

Sample	ρ (Tr/cm²)	C_a (Bq/m)	Cs(Bq/m³)	C_U (ppm)
Kalar1	2763.2	797.691±30.375	85373.627±314.240	6.484±2.738
Kalar2	1444.4	416.975±11.48	44627.123±118.761	3.389±1.035
Kalar3	1507.2	435.104±12.236	46567.433±126.590	3.537±1.103
Kalar4	533.8	154.099±2.579	16492.632±26.681	1.253±0.233
Kalar5	1664.2	480.427±14.197	51418.207±146.876	3.905±1.28
Qadafary1	2323.6	670.785±23.423	71791.459±242.318	5.452±2.112
Qadafary2	4710	1359.700±67.598	145523.227±699.319	11.052±6.094
Qadafary3	2449.2	707.044±25.348	75672.078±262.229	5.747±2.285
Qadafary4	2072.4	598.268±19.729	64030.220±204.105	4.863±1.779
Qadafary5	7767.2	2242.263±143.152	239980.470±1480.952	18.225±12.905
Zarayan1	1884	543.880±17.101	58209.291±176.915	4.421±1.542
Zarayan2	1695.6	489.492±14.601	52388.362±151.053	3.979±1.316
Zarayan3	3328.4	960.855±40.156	102836.414±415.429	7.810±3.62
Zarayan4	2386.4	688.915±24.379	73731.769±252.208	5.599±2.198
Zarayan5	2009.6	580.139±18.839	62089.910±194.898	4.715±1.698
Ave.		741.709±27.234	79382.148±281.748	6.029±2.455
Max.		2242.263±143.152	239980.470±1480.952	18.225±12.905
Min.		154.099±2.579	16492.632±26.681	1.253±0.233

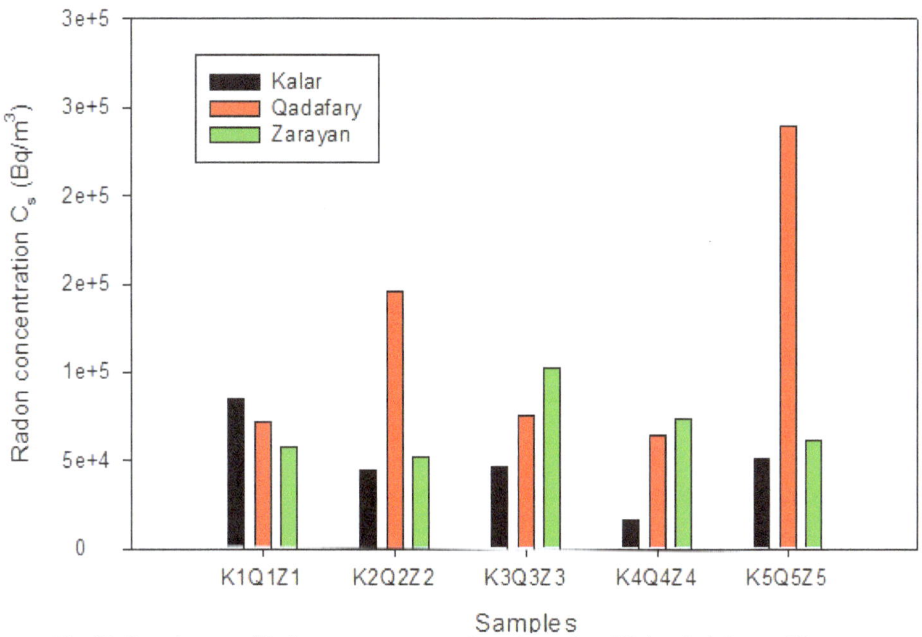

Fig. (3). *Distributions of Radon concentration in the soil samples of Kalar, Qadafary and Zarayan.*

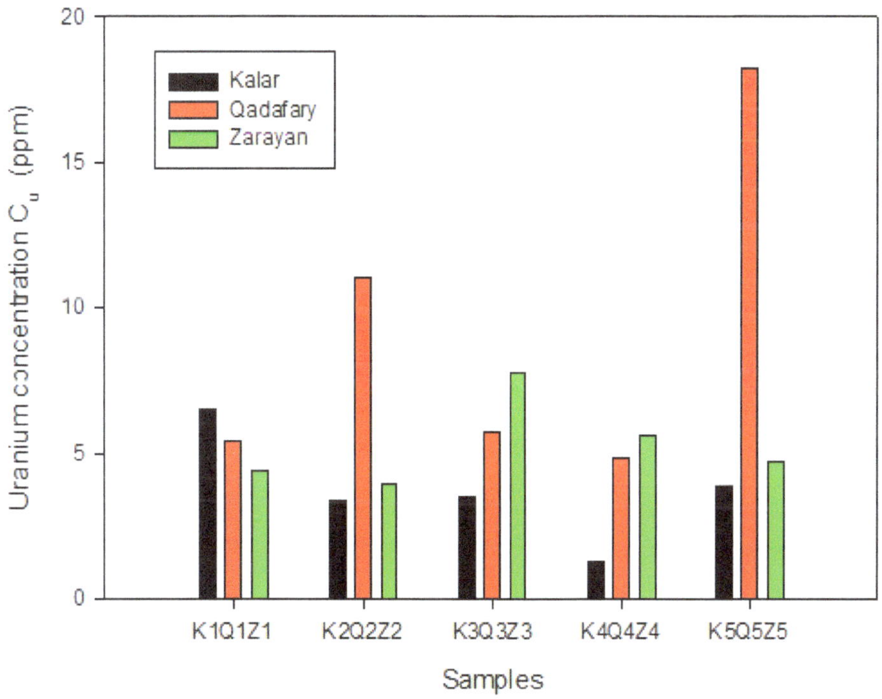

Fig. (4). *Distributions of Uranium concentration in the soil samples of Kalar, Qadafary and Zarayan.*

4. Conclusion

The concentrations of radon and uranium were measured by using a SSNTD technique for 15 soil samples which collected from three regions in Sulaimani area (Qadafery, Kalar and Zarayan). It is concluded that the high concentrations of radon and uranium was in the soil sample Qadafery region more than the others and the reason of higher level concentrations of radon and uranium due to the geology information of these regions.

References

[1] Banman A. et al., Natural Radiation Environment. Willy R. Ltd, New Delhi 1982.

[2] Mansoor H.," Measurement of indoor radon levels in Arbil by using SSNTD", Radiation Measurement. Vol.40,pp 544, 2005.

[3] Lubin J.," Studies of Radon and Lung Cancer in North America and China" Radiation Protection Dosimetry", Vol.104 (4), pp315, 2003.

[4] Shapiro JB et al., "A preliminary screening technique for selected metals at waste sites", J Radio analytical Nuclear Chemistry, Vol. 192 (2), pp275, 1995.

[5] [Al-Khashman O., "Heavy metal distribution in dust, street dust and soil from the work place in Karak industrial Estate, Jordan", Atmosphere Environment, vol.38, pp6803, 2004.

[6] Etiope, G. and Martinelli, G.," Migration of carrier and trace gases in the Geosphere: an overview", Phys. Earth Planet. In., vol. 129, pp185, 2002.

[7] Annex B., Sources and effects of ionizing radiation. United Nations Scientific Committee on the Effect of Atomic Radiation, UNSCEAR, United Nations, New York, 2000.

[8] Omer N., Electrochemical Etching, Msc. Dissertation, University of Mousl, 1999.

[9] Azam, A. et al.,"Radium Concentration and Radon Exhalation Detectors", Nuclear Geophysics, vol.9 (6), pp653, 1995.

[10] Ahmad M, "Employment SSNTD (CR-39) to Detect Radioactive Pollution of DU in a Particular Region from Salahadeen Governorate" Proceeding in Ist Scientific Conference for College of Science University of Tikrit, Iraq, pp295, 2011.

[11] Barillon R. et al.," Comparison of Effectiveness of three Radon Detectors (LR-115, CR-39 and Si. Diode pin) Placed in Cylindrical Device- Theory an Experimental Techniques, Nuclear Track" Radiation Measurement, vol.22 (1-4), pp281, 1993.

[12] Michael F. Annunziata," Radioactivity Introduction and History", Ist edition USA, 2007.

[13] Ahmad M., "Employment SSNTD (CR-39) to Detect Radioactive Pollution of DU in a Particular Region from Salahadeen Governorate" Proceeding in Ist Scientific Conference for College of Science University of Tikrit. Iraq, pp295, 2011.

[14] Podgorsak E.B., "Basic Radiation Physic", Radiation oncology physics, IAEA in Austria, Ch.1, pp1, 2005.

[15] Sroor A. et al.," Natural Radioactivity and Radon Exhalation rate of Soil in Southern Egypt" Applied Radiation and Isotopes, vol.55, pp873, 2001.

[16] Samuel S.M.," Introductory Nuclear Physic", 2nd Edition, University of Toronto, 2004.

[17] T. EL-Zakla et al." Natural Radioactivity of Some Local Fertilizers", Rom. J. Physics, vol.52, pp731, 2007.

[18] www.geology-kurdistan.net (2013)

[19] Kamal O. and Rasheed M.,"Natural Radioactivity Measurements of soil and water in Sulaimani Governoraye", Ph.D. Dissertation, University of Sulaimani, 2013.

A sustainable planning approach to resolving transportation hub problems in Egyptian cities "proposed measurement matrix"

Waleed Hussein Ali Hussein

Department of Architectural Engineering, Faculty of Engineering, Fayoum University, Fayoum, Egypt

Email address:
Waleed.a.hussien@gmail.com

Abstract: The research study discusses the environmental aspects of automated and pedestrian transport that need to be addressed when devising general and detailed urban plans, particularly for new urban conglomerations. It describes the adverse effects resulting from motor vehicles dominating roads and highways, including environmental hazards such as air and noise pollution, and identifies environmental concerns to be taken into account in the planning of automated and pedestrian traffic routes and parking spaces and underlines the need for upgrading transport and traffic policies in new urban conglomerates. The study focuses on environmental issues that can be resolved through the application of laws, policies and implementation mechanisms. It also identifies environmental considerations that should be included in all general urbanization plans, pointing out that the review of these issues by environment experts (before and during plan implementation) is a key developmental approach to resolving transportation hub problems in Egypt. In addition, the study presents an account of the components and basic principles of transportation hub design and the international criteria used in assessing the sustainability of transport routes. Local and international examples of sustainable transportation hubs are used as a reference to help in formulating the principal findings and conclusions that constitute the developmental approach to resolving this problem.

Keywords: Sustainable Planning, Road Networks, Pedestrian Traffic Hubs, Environmental Pollution, Traffic Hubs, Urban Development, Sustainable Development

1. Research Problem

The research problem concerns the obvious shortcomings in the development and resolution of transportation hub problems in Egypt, which necessitate the intensification of efforts and the use of analytical methods aimed at developing and upgrading this important component of urban planning for Egyptian cities.

2. Research Objective

The aim of this study is to formulate a developmental approach designed to reduce the adverse environmental effects (air and noise pollution) of transportation hubs, to identify the environmental aspects of transportation hub design and to benefit from international examples already in place, without losing sight of the aesthetic dimension.

3. Introduction Urban Transportation

Transportation and road networks today constitute a large part of any city's area and vital facilities and they are the lifeline that provides it with the means for living. Additionally, the style and form of a city's road network contribute to the shaping of its morphological identity.

The various components of a city's transport sector account for approximately 32% of its total area. In Canada, the transport sector constitutes 37% of the country's total inhabited area and, according to researcher Murphy's estimation; the transport sector in the United States comprises 34% of the total area of the country's cities1.

Pederson found that, on average, highways take up 15% to 20% of a city's area and other transport facilities, including

1 Dr. Zein El Abedeen Ali, *Mabade' takhteet al naql al 7adary*, p.30.

ports and airports, account for a further 5%2.

Figure 1. *High way Roads*

4. Transport and Traffic Sector Problems in Egypt

4.1. Road Accidents

Accidents are among the traffic problems that affect human lives the most. The accident incidence rate increases with the growth of urbanization generally. In addition to environmental factors such as rain, wind and storms, road accidents are often the result of a disregard for traffic rules and regulations. Preventing road accidents completely is impossible, but sensible planning can certainly go a long way towards reducing their number and alleviating their gravity.

4.2. Congestion

Figure 2. *Traffic congestion in Eastern Cities.*

Traffic congestion is the most widespread urban transport problem in Arab cities, especially during the peak hours of the day when people are going to work or heading home. In most cases, congestion is caused by the low absorptive capacity of road networks and by inadequate transportation systems that do not meet the needs of the size of traffic in the cities. During peak hours, travel speed often falls to 4 – 5 km / hour, leading to an extensive waste of time and energy. Long hours spent on congested roads also affect people emotionally, creating mental pressure that affects how they function at work and at home.

4.3. Respect for Traffic Rules

Traffic violations are a principal cause of the problems discussed in this paper. In addition to their social, psychological, educational and cultural connotations, they are one of the gauges that measure the behavior of motorists driving private and public vehicles.

4.4. Noise Pollution

Traffic sounds (automobile horns, braking, wheel friction etc) are a source of disturbance to building residents, people using public facilities and pedestrians on the roads, putting them under considerable mental strain.

4.5. Air Pollution

Traffic-related air pollution is one of the greatest environmental challenges of transportation in Arab cities. Air pollution increases with the increase of fuel (all varieties) combustion in all vehicle types and is very high because of the congested traffic conditions and low speeds mentioned above. This is the case particularly in densely populated urban areas, where large numbers of people are affected by high levels of traffic-related air pollution within a short time period - a very serious threat to public health and the environment.

"In his book on the relation between urbanization and the development of transport services, Dr. HaydarKamouna identifies some of the most important measures that should be taken to minimize transport and traffic problems in Arab cities"

Further research and additional field studies on the transport sector to obtain the information on actual urban developments in cities, (changes in population density and road accident statistics etc) required for planning3.

5. Factors and Variables Affecting Urban Transportation

Research on urban transportation and traffic planning reveals that many urban elements affect the generation of traffic mobility, and the transportation sector as a whole. In addition to factors related to land use in urban areas, these include urban development factors (horizontal extension of urbanization and the increased urban density in some areas or cities resulting from vertical construction expansion). Other elements that affect the transportation sector are related to the type of traffic (whether traffic is internal or external, locally generated or passing through), and its relationship to urban road network patterns, their absorptive capacity, and how effectively they meet existing traffic needs. The existence or absence of sufficient parking space in urban districts,

2 Pederson, E. O., *Transportation in Cities*, Pergamon Press, 1980, p. 14.

3 Dr. Haydar Kamouna, Al 3elaqa bayn al ta7adur watatawor al khadamat – A study published on the Magalet al 3ooloom al 'egtema3eya Forum

particularly in central areas, is another important factor.

All the above factors, however, are influenced by certain important variables that have the most impact on daily traffic mobility generation.

5.1. Population Growth

Population size is an important variable in the statistical analysis of transportation and traffic, and is one of the factors that influence total daily traffic mobility. The population factor is more dynamic than all the other variables, because there is a direct correlation between the number and density of an area's inhabitants and the number of road trips per day in that area. Furthermore, urban transport studies show a clear correlation between increases in the number of family members and increases in the number of road trips, finding that daily trips increase at the rate of 0.8 daily trips per additional family member. Studies also reveal that the number of road trips increases with the increase of the number of employed members in a single family4.

5.2. Family's Average Monthly Income

A relation exists between the socio-economic traits of a population and traffic movement in general. Most transport-related urban planning studies underline the correlation between a family's average monthly income and the number of road trips per day of its members. This is because the higher a family's income is, the greater its purchasing power and the more varied its needs; a greater of number of trips are required to satisfy family needs.

5.3. Vehicle Ownership

There is a marked increase in the use of private vehicles today. The ownership of private vehicles is one of the important factors affecting the generation of daily trips. It contributes to traffic congestion on urban roads and is the cause of many road accidents. In most countries, the per capita share of private vehicle ownership has increased. In Iraq, for example, statistics show that private automobile ownership rose from five vehicles for every 1000 persons in 1970 to 82 vehicles for every 1000 persons in 1995 (35 vehicles per 1000 persons in Baghdad in 1980). In Cairo, the ratio was 25 privately owned vehicles for every 1000 persons in 1980, a proportion that has almost doubled today. Private vehicle ownership is one of the indicators used in estimating future urban transport demand5.

5.4. Energy and Transportation

Energy consumption is a necessity of socio-economic development. Transport (the sector that consumes the most energy) constitutes one of the major challenges facing sustainable development because it is a source of pollution that is harmful to both the environment (at global, regional and local levels) and to human health. This dilemma has long baffled countries across the globe. The two principal challenges that face the transport and energy sectors are to insure a more widespread use of cleaner fuels and to convert existing modes of transport into cleaner and more efficient ones.

At the World Summit on Sustainable Development, the nations of the world agreed on the need for increased access to affordable energy resources. They decided to work towards enhancing energy efficiency, promoting the use of renewable energy resources, implementing advanced, cleaner, energy technologies, applying transport strategies that support sustainable development, limiting environmentally harmful energy subsidization policies and encouraging the use of cleaner fuels6.

Urban planning techniques, or methods, are among the most important factors affecting the urban transport system and the volume of its energy consumption. European urban plans are based on transport systems that depend largely on railway networks and urban developments concentrated around city centers, whereas urban plans in the U.S. depend on wide highways and horizontal urbanization.The planning of most U.S. cities makes automobiles indispensable to the life of their inhabitants, and these cities grow and expand horizontally for tens of miles.

A recent study by Dr. Edward Klisser of Harvard University found that only 22% of the inhabitants of the largest U.S. cities live within a three-mile radius from the city center (the principal commercial and industrial area), and almost 80% at distances of up to 30 miles outside the city borders. Obviously, this leads to a high rate of energy consumption.

In Europe, on the other hand, one can finish a day's work in Geneva, Switzerland and arrive in central Paris, France less than two hours later, getting off the train before the doorstep of the place of destination (hotel or company etc). This is because most European cities depend on a model rail network that connects all European city centers.

Some Arab countries, such as the U.A.E., Bahrain, Kuwait and Oman, have adopted the British urban planning system of narrow, winding roads but disregarded the most important factor; railway networks. As a result, these countries suffer from the narrow roads and vehicle parking shortages of the European urban planning system and are now moving towards the broad highways of the U.S. urban planning system. This is a noticeable tendency in the Arab world today. Note Dubai's initiative of building the first metro line in the Arab Gulf, which came into service in September 2009, and the adoption of the 2030 maritime transport plan in Abu Dhabi, which features the simultaneous construction of metro, tramway and railway lines. The above reveals the close relation between resolving energy problems, the control of a city's growth rate and public transport solutions (such as railway networks connecting large cities and ports)7.

4 Dr. Zein El Abedeen Ali, Mabade' takhteet al naql al 7adary, pp 82 - 84

5 Ibid.

6 World Summit on Sustainable Development (WSSD) publications – Johannesburg, South Africa, 26 August to 4 September 2002

7 Dr. Abdallah Ibrahim Al Fayez, *Takhteet al modon wa moshkelat al e3temad 3ala al taqasharl abodminho; Hansen, C.J. (2011) The role of policy-making and planning cultures for sustainable transport?. European Transport, 47 (2011), 89-108.*

Figure 3. *Train station in eastern cities*

6. Environmental Aspects of Traffic Mobility in the Thinking of the Pioneers of Urban Planning

Whereas the eighteenth century witnessed rapid progress and successful achievements in the manufacturing and construction sectors, the nineteenth century is characterized by the development and upgrading of scientific processes, sensible planning, economic reconstruction and the increase of population migration from rural to urban areas. This led to:

- Poor technical and architectural building specifications
- Inadequate infrastructure network planning
- Air, water and soil pollution

These factors led to the rise of a new generation of thinkers, writers, architects and other leading scholars whose objective was to re-plan and re-shape the modern city as follows:

Table 1. *Environmental Aspects of Traffic Mobility in the Thinking of the Pioneers of Urban Planning.*

Points of Comparison:	**Frank Lloyd Wright** **The Decentralized, Organic City**	**Le Corbusier1887 – 1965** **The Raised City**	**Toni Garnier1909** **The Industrial City8**	**Ebenezer Howard1889** **The Garden City9**
Main Site:	A decentralized city integrated into the natural (organic) environment, where buildings are separated by large spaces and automobiles are needed to connect them. Decentralized services are spread across the surrounding countryside.	A city of high, geometrically-shaped buildings, surrounded by vast open spaces	Natural topography has a strong influence on the main site. City orientation depends on the direction of the wind - separation of industrial and residential areas	A ring road (beyond which is a greenbelt to cut back future urban spread) surrounds the city. Factories and warehouses are located beyond the ring road and farmland and pastures surround the city.
Roads:	Automobiles are used to connect the various parts of the city. There is no city center; public services are spread across the city.	A network of freeways (longitudinal design) ensures travel speed	Road orientation allows for the penetration of sun and air into homes	A railway line, bordered by gardens, surrounds the city. To reduce pollution, a large square separates residential areas from industrial zones.
City Center:	There is no city center. The city is not designed on a scale and land use is not planned. The size of public utility buildings is kept to a minimum to facilitate their distribution across the city, and long distances separate different elements of the city, and automobiles are the principal means of transportations.	City center of medium density –Road traffic speed promotes distances between public utility buildings	Topographic factors are used to separate the city center from industrial zones. Climate conditions affect the orientation of city elements.	Public utility buildings are located in the city center, surrounded by green spaces – a compact, mixed-use city
Mobility & Transportation:	Dependence on automobiles to link buildings and various services leads to long travel distances, resulting in a waste of time and energy on in-city transport.	Freeway network based on a longitudinal geometric pattern for automobile travel – Dependence on automobiles creates numerous environmental problems, from the use of green spaces to build roads, to air pollution, waste of energy resources, traffic congestion and a high incidence of road accidents	The plan depends chiefly on the railway line as a means of transport and travel. The railway passes between the industrial zone and the city.	Because the 'Garden City' plan was drawn up before the spread of automobiles as a means of transport and travel, the main plan depends on the railway as an ideal solution for pre-automobile industrial cities.
Energy Conservation:	Dependence on automobiles to connect different elements of the city led to wasting energy resources on roads and	Dependence on automobiles as a means of transport led to a considerable waste of energy resources.	The use of railway lines conserves energy resources.	Dependence on railway lines reduces energy consumption.

8 Catanese, Anthony J. & James C. Snyder, Introduction to Urban Planning, McGraw - Hill Book Company - New York, 1986

9 Burtenshaw, D. and Batenman, M. and Ashworth, G. J., *The City in The West Europe* - John Wiley & sons, UK, 1981

Points of Comparison:	Frank Lloyd Wright The Decentralized, Organic City	Le Corbusier1887 – 1965 The Raised City	Toni Garnier1909 The Industrial City8	Ebenezer Howard1889 The Garden City9
	transportation.			
Reducing Pollution:	Dependence on automobiles contributed to raising air pollution levels in cities.	Abundant use of automobiles raised the level of air pollution in cities.	Isolation of the industrial zone in the south at a lower land level limits the spread of air pollution in cities. Environmental home designs provide a healthy environment for inhabitants.	Positioning the industrial zone in a region separate from the residential area, and surrounding it with a greenbelt helped to minimize air pollution levels in cities; the urban greenbelt is considered the city's lungs.

Source: Catanese, Anthony J. & James C. Snyder, Introduction to Urban Planning: 1986, McGraw- Hill Book Company - New York

7. Environmental Considerations in the Planning of Motor Vehicle Travel Routes

- Choice of road type and network integration
- The role of laws and regulations in upgrading road environment

7.1. Choice of Road Type and Network Integration

The following considerations should be taken into consideration when selecting a model road network system as part of the overall plan:

- Prior to selecting the appropriate road network pattern or model, the primary function of the traffic hub and the secondary road functions must be identified, and the planners must have a correct estimation of the anticipated road traffic density of traffic. This will lead to the selection of the appropriate design and avoid future traffic congestion problems.
- Various types of road networks should complement one another. For example, automobile road networks do not do away with the need for pedestrian and cycling routes; indeed, neglecting to design pedestrian networks leads to increased pressure on automobile networks, multiplying the chances of road accidents and raising pollution levels.
- Sensible planning of secondary road functions, such as temporary parking space, pedestrian crossings and car parks. These should be the object of separate plans that are revised periodically.
- Public transport routes (railway lines, public busses etc) must be provided. Service stations and passenger waiting areas should be delineated in conjunction with the planning of traffic hubs or roads.

Table 2. *Types of Roads and Traffic Hubs and their Environmental Effects.*

Type of Road	Advantages	Environmental Problems	Solutions (if available)
Radial Roads	Link city centers to suburbs and constitute a major population attraction factor	The need for ring road and highway junctions; the insufficient number of these junctions leads to traffic congestion and higher pollution levels.	Building free ring roads to spread traffic flow around cities or city centers
Ring Roads	Considered an integral part of the radial road network – Despite being the longest road type, they reduce travel time, since most trips are outside of crowded cities and need not pass through city centers, thus avoiding highly congested roads	There are no environmental problems, except that the length of the road may necessitate additional lighting after dark as well as telephone and other services, while not encroaching upon ecological areas of a special nature.	
Gridiron Roads	Promote the homogeneousness of city and suburb traffic travelling over the network as a whole – The possibility of providing one-way lanes with a parallel lane in the opposite direction –Bridges and / or tunnels can be used as crossings	Gridiron road systems can result in very long roads that may lead to excessive speed and result in road accidents	Effective traffic management and speed control policies
Linear Pattern	Economic considerations may necessitate the use if this pattern in certain geographical locations	Traffic flows into a major linear road used for both inner city and outer city travel, resulting in a city that is divided in two and a high rate of traffic congestion	Building roads parallel to the principal linear road to spread traffic among the different roads

Source: Stones, Tabor R.: *Beyond the Automobile* - Prentice Hall, Inc., N. J., 1992; Commonwealth of Australia. (2011).Our Cities, Our Future: A national urban policy for a productive, sustainable and livable future. Canberra: Department of Infrastructure and Transport.

7.2. The Role of Laws and Regulations in Upgrading Road Environment

Laws and regulations play a leading role in influencing environmental considerations during the planning of traffic routes:

- They shape the environmental policies in cities and minimize negative environmental impact. By controlling speed limits, traffic laws and regulation promote safety and lower the incidence of road accidents, while laws prohibiting excessive use of automobile horns reduce noise pollution.
- Road management regulations establish stipulations, such prohibiting lorries inside cities and restricting their passage to perimeter roads outside city limits, or closing off certain roads to traffic and designating them for pedestrian use.
- To ensure that the road safety conditions are met, traffic laws set specific speed limits for vehicles travelling on bridges, through tunnels or in residential areas.

8. Environmental Considerations in the Planning of Automobile Parking Spaces

The planning of parking space is a key element of the transportation system, particularly in light of the huge increase in the number of automobiles in a city and the numerous environmental consequences, such as:

- Obstruction of automobile traffic flow on the roads
- Obstruction of pedestrian passage in front of building entrances
- The takeover of public squares, playgrounds and green spaces for use as parking space
- Increased air pollution from exhaust emitted when starting up the engines of parked vehicles
- Distortion of cities' aesthetic value (visual pollution)

Negative Environmental Impact of the High Number of Automobile Parking Spaces in New Cities:

- Loss of public spaces, squares and social meeting places that have been converted into parking space (a single automobile takes up $20m^2$ to $30\ m^2$ of parking space)
- Air and noise pollution: fuel combustion when starting up engines emits exhaust gas and is also noisy
- Blocking the entrances and disfiguring the facades of buildings: lines of parked cars in front of commercial outlet, public and private building entrances mar the city's appearance and increase traffic density in road lanes, resulting in crowded roads that do not perform their function.
- Impeding the passage of vehicles: sometimes, lines of parked vehicles in narrow side streets impede the passage of automobiles, which causes traffic problems and slows down mobility on the main roads into which these side streets lead
- Preoccupation with automobile-related problems led to the neglect of public transport development programs, resulting in increased dependence on automobiles and the environmental problems caused by them, at the expense of developing public transport needs, such as special public transport routes and adequate numbers of passenger waiting spots.

Table 3. *Methods Used to Provide Automobile Parking Space10.*

Method	Advantages	Environmental Problems
Grouped garages:	Accommodate large numbers of cars - free road space - creation of green spaces - city beautification	Crowding of garage entrances and roads when vehicles enter or leave grouped garages during rush hours
Parking space located in suburb & outlying areas:	This method entails leaving automobiles in parking spaces located on the outer perimeters of congested conglomerates and riding public transport from the stops surrounding the parking area. This is a partial solution to the problem of parking in crowded cities. It also promotes the use of public transport as a means of circulation.	None
Organizing on-road parking: Parking on narrow roads:		
One-way road: Two-way road:	Road must be at least 4 meters wide Road must be at least 5.75 meters wide	Double parking may bring traffic flow on the road to a standstill
Building, garden and other entrances:	Parking in front of public and private service buildings is prohibited	Blocking the entrances and exits of public and private service buildings creates social problems
Pedestrian crossings:	Cars may not park on pedestrian crossings (which should be at least 8 meters wide)	Non-observance of traffic rules causes panic and confusion among pedestrians and endangers their lives
Bridges and tunnels:		Bridges and tunnels were designed to resolve traffic problems. Parking is only allowed in cases of emergency, such as sudden breakdowns or collisions
Special locations (hospital emergency centers, fire stations, bus stops etc)	Parking in front of bus stops, fire stations and hospital emergency centers is forbidden.	Inhibits the functioning of public services

Source: O'flanerty, C. A.: Highway Traffic Planning and Engineering, Volume 1; Commonwealth of Australia. (2011).Our Cities, Our Future: A national urban policy for a productive, sustainable and livable future.

10 O'flanerty, C. A.: *Highway Traffic Planning and Engineering*, Volume 1 - University of Leeds, UK

Canberra: Department of Infrastructure and Transport

Figure 4. *Parking types and spaces*

9. Environmental Considerations in the Planning of Pedestrian Traffic

The pedestrian traffic planning stage is a key element of a city's road and transport system. To identify the environmental problems related to this issue, we must first shed light on the issue of pedestrian traffic.

9.1. Pedestrian Paths

This is the place reserved for pedestrians on motorway sidewalks; paths reserved only for pedestrians where there are no trees, overhangs, pillars or dangerous obstructions. They should be fit for use by pedestrians as well as by the physically impaired. Pedestrian paths (sidewalks) must be at least 1.525 meters (60 inches) wide, to allow two people to walk side by side. Depending on population density, sidewalks in commercial sectors and similar places may be wider11.

Figure 5. *Pedstrian Paths.*

11 Daleel tasmeem al toroq – ketab tasmeem al toroq, 1-2, 2-2 (Ministry of Transport);Low, N. (2011).Australian Cities, Transport Integration and the Global Dilemma of Mobility. Retrieved April 2, 2012, from State of Australian Cities 2011: http://soac2011.com.au/full-papers-list.php

9.2. Pedestrian Problem

Beginning in the 1960's, and due to the enormous increase in the number of automobiles, conditions in cities (particularly for pedestrians and public transport passengers) began to deteriorate. This period witnessed an absence of:

Policies aimed at protecting pedestrians and cyclists

Construction of bridges and tunnels for pedestrian crossing of motorways

Investments in the development and upgrading of public transport systems

Policies aiming at encouraging the use of public transport over automobiles in congested areas

Figure 6. *Bike Lane.*

Some Proposed Solutions to the Pedestrian Problem:

- Converting one of the streets in the city center into a pedestrian street
- Using thick tree borders to separate motorways from pedestrian paths (sidewalks)
- Exploring possibilities for connecting the newly-converted pedestrian street to a public transport stop, and providing automobile parking space
- Creating special lanes for motorcycles (a non-polluting means of transport)
- The use of public transport and electric powered busses to reduce environmental pollution and congestion caused by automobiles.

10. The Aesthetic Dimension of Roads

10.1. Neglect of Aesthetic Standards in Cities (Visual Pollution)

The great effort expended on the construction of various types of roads and bridges in cities led to the neglect of many of the standards related to aesthetics and tranquility that characterized cities before the age of the automobile. Some important examples of this phenomenon are:

- Increasing the width of highways and city center streets at the expense of pedestrian paths, squares and meeting places

- Road and traffic hub construction has engulfed green spaces, which has had a huge negative impact on the natural and urban environment.

10.2. Aesthetic Elements of Pedestrian Passages and Transportation

The temptation to beautify pedestrian and automobile passages is important in the planning and design of pedestrian lanes. Great care must be taken in the positioning of plants and trees on roads; they must be planted in the right spots to perform the right function. Trees and bushes may be planted on passages that are largely reserved for pedestrian use. However, planting trees on the sides of highways or in the median strips separating two-way streets is not recommended. In fact, trees in the middle of highways can be the cause of some of the worst types of road accidents when automobiles crash into them12.

Figure 7. *internal pedestrian walk way.*

11. Standards and Parameters for the Design of Sustainable Transportation Hubs

11.1. Road Design Categories

This refers to the grading, for architectural design purposes, of a number of principal roads according to the level of traffic service they provide to users. Each of the four design urban road categories provides a particular level of traffic and other services to the area through which it passes. The architectural specifications and characteristics for each category are designed to match its particular conditions13.

11.1.1. Road Classification Factors

A number of factors and technical specifications serve as guidelines for the functional classification of road networks. These include anticipated size and quality of road traffic, average travel duration on a given road and the position of the road in relation to other roads on the road network.

11.1.2. Size and Quality of Road Traffic

Roads that accommodate heavy traffic naturally require appropriate architectural specifications and standards to provide an acceptable level of service that ensures the smooth flow of traffic while minimizing travel time and cost, thus ensuring suitable travel speeds and safeguarding security. Furthermore, traffic patterns (in addition to local traffic, traffic between far-flung destinations can be either passing through or long haul) directly affect the choice of a suitable category. In general, and depending on traffic size and quality, road networks can be classified into categories according to their needs, as follows:

Outer-City Roads:

- Highways and freeways: principally used by heavy traffic, they connect distant transport destinations and usually carry both uninterrupted and interrupted traffic flows. Highways are equipped with ramp intersections, while the intersections on freeways are all free.
- Side roads: these make up the road network that links highways or freeways to local roads. Medium category design elements, suitable to the size of traffic and allowing for the appropriate speed levels, are used for side roads.
- Local roads: these are roads that lead directly to nearby services. They must allow for acceptable traffic speeds that suit these services and the anticipated interference by other services

Urban Streets

In accordance with the above functional classification concepts for outer-city roads and in keeping with international and local classification concepts, urban roads are divided into the following categories:

- Highways and freeways: principally used by heavy traffic, they connect distant transport destinations and usually carry both uninterrupted and interrupted traffic flows. Highways are equipped with ramp intersections, while the intersections on freeways are all free.

Side roads:

Connect highways and freeways to local roads

11.1.3. Average Road Travel Distance

The design elements used for roads designated for long distance travel must allow for traffic speeds that are appropriate to long distances.

11.1.4. Position of the Road on the Network

The overall hierarchical categorization of road networks has a direct effect on the choice of road functions that are suitable to the road's position on the network. The function assigned to a road must be in keeping with the grading (categorization) sequence of roads on the network.

11.1.5. Cycling Lanes

The design of cycling lanes on urban roads depends chiefly on the factors of comfort and security. Cycling lanes must be provided on urban roads where bicycle traffic size exceeds 500 bicycles a day.

---

12 Department of Transport and Main Roads (2011). Connecting SEQ 2031 – An Integrated Regional Transport Plan for South East Queensland. Brisbane: Queensland Government.

13 Ibid.

11.1.6. Space Requirements

A bicycle lane must have an empty space width of 1.10 meters (0.60 meters), the standard width of a bicycle, plus 0.25 meters of free space on either side to allow for the free flow of movement (other bicycles, pedestrians or vehicles).

11.2. Safety

The following points ensure safety on the road:

- Appropriate design factors (protective barriers – pedestrian paths – cycling lanes – lateral slopes – the number of lanes etc)
- Coordination between vertical and horizontal planning to ensure that drivers have a safe sight distance all along the road
- Making use of existing safety studies and expertise concerning road network components (vertical and horizontal planning – intersections – pedestrian and cycling lanes – railway crossings etc)
- Compatibility of road equipment (lighting – traffic lights and traffic signs – ground markings – traffic control equipment etc)

11.3. Environmental Issues

Transport and road network projects affect the environment, so it is crucial to identify and adopt transport design elements and factors that contribute to reducing negative environmental effects. For instance, careful note must be taken concerning a road's environmental impact regarding:

- Air pollution from exhaust fuel emitted by vehicles using the road, particularly in urban areas, and its effect on the public health of living creatures and plant life
- Noise emitted by vehicles on the road, lorries and large trucks in particular

11.4. Economic Factors

The study of economic considerations is one of the important elements that influence the choice of the foundations, design components and selected routes for road projects. Economics studies for roads comprise the following components:

- Road construction costs
- Road maintenance costs
- Vehicle running costs

11.4.1. Road Construction Costs

These include the cost of design studies and of expropriating any private property that may be located on the chosen route. Costs of similar road projects in the region may be used as a reference.

11.4.2. Road Maintenance Costs

Calculated according to the project's annual maintenance costs (includes the cost of routine and periodical maintenance)

11.4.3. Vehicle Running Costs

This is the cost of running vehicles on a road and is calculated per kilometer vehicle. Vehicle running costs depend on the model and average speed of each vehicle and it is also affected by the structural features of the road, such as horizontal curves, longitudinal slopes and the type and condition of road paving. Vehicle running costs include fuel, oil and tire consumption costs, servicing costs and vehicle depreciation resulting from mileage covered. The cost of time consumed on the road is also taken into account14.

12. Description and Analysis of International and Local Examples

12.1. London

Driving super personal electric transport pods to be installed near London

Milton Keynes, a small city north of London, has just announced plans to install one hundred automated "pod cars" to run between the city's central station and the city center. These ULTRA Personal Electric Transportation Pods will travel along the platforms, at speeds of up to 12 miles per hour. The project, which costs 104 million USD, is part of a five-year pilot plan expected to be completed in 201715.

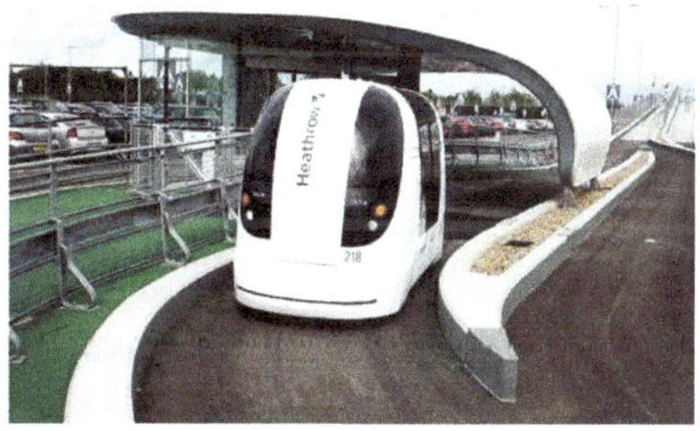

Figure 8. *super personal electric transport pods.*

Figure 9. *super personal electric transport pods.*

14 Oglesby, Clarkson, H., *Highway Engineering*, (Arabic translation of third edition)

15 Newman, P., & Kenworthy, J. (2011, May 26). Peak car use: urban planning needs to change say Newman and Kenworthy. Retrieved April 10, 2012, from The Fifth Estate :http://www.thefifthestate.com.au/archives/24214

ULTRA Personal Electric Transportation Pods

Pedestrians will be able to get a taste of future centuries through smart mobility applications. This application was first used at London's Heathrow Airport in 2011, but the British government is now looking into applying it to public transport systems, in order to free the roads of pollution emitted by buses.

The pods are computer-driven vehicles and each can transport up to four passengers, along with their luggage. They are battery-operated, have rubber tires and can navigate complex routes with lightweight infrastructure. When running, the ULTRA Personal Electric Transportation Pods are silent, produce no noise, zero emissions and almost no external vibrations16.

Lessons learnt from the example

- Safety
- Use of electric energy
- Privacy
- Protecting the environment against pollution

12.2. New York City

Figure 10. 300 MPH Maglev Train

Figure 11. 300 MPH Maglev Train.

300 MPH Maglev Train Could Carry Passengers from Washington D.C. to New York in 60 Minutes

Revised plans have recently been revealed concerning the 300 miles per hour Maglev bullet train that could transport passengers from Washington, D.C. to New York City in just 60 minutes. Driven by powerful electro-magnates, the train would be elevated, which would reduce friction and enable it to travel at high speeds with rapid acceleration. Growing interest in applying high-speed technologies to public transport (such as the Hyperloop) is behind the reinstatement of this project17.

Lessons Learnt from the Example

- Travel speed and resulting time savings
- Use of electro-magnetic energy

12.3. United Arab Emirates

The United Arab Emirates is a leading country in the field of sustainable transport. Examples of main streets where sustainable design standards have been applied are presented and examined below. The best urban streets are those that can easily accommodate a balanced mix of pedestrians, public transport passengers and drivers of motorcycles and other vehicles. Efficient and safe road networks necessarily depend on an integrated design process that takes into account different types of transport and the various users of a transport system. A number of variables must be considered before embarking on the process of urban road design. In the case of Abu Dhabi, it is very important for urban road design to consider climatic conditions and local context (in addition to standard design priorities and factors)18.

Example 1: Abu Dhabi: Al Moroor Street – Side Lane Added to a City Avenue

Al Moroor Street, located in the business center of the city of Abu Dhabi, is chosen as an example illustrating the methodology used in the three-step process of updating urban streets. The street's conditions are representative of the prevailing pattern existing in many of the city's streets:

- Chaotic advertisement billboards and signs
- Scarcity of seats and shade canopies at bus stops and in areas reserved for pedestrian use
- Narrow side barriers and passenger waiting areas at bus stops (2 meters)
- Wide (5 meters) one-way side barrier
- No safe pedestrian crossing at bus stops
- Perpendicular parking spaces, at a 5.6 meter angle
- Narrow, restricted pedestrian passages (2.65 meters wide)

Step One: Cleaning

Removal of unnecessary billboards and sign posts

Affixing signs unto multi-functional posts

Replacing damaged trash bins and fixing the new bins unto sign posts

Step Two: Upgrading Pedestrian Areas

Installing elevated crossings leading to bus stops

Repaving pedestrian passages using quality materials

Adding short pillars on the elevated crossings

16 Schiller, P., Bruun, E., & Kenworthy, J. (2010). An Introduction to Sustainable Transportation: Policy, Planning and Implementation. London: Earth scan.

17 Schiller, P., Bruun, E., &Kenworthy, J. (2010). An Introduction to Sustainable Transportation: Policy, Planning and Implementation. London: Earth scan.

18 Daleel al shaware3 al7adareya – Abu Dhabi,2012

Providing new fixtures (such as seats) in pedestrian facility areas and sun protection canopies at bus stops, and providing a 2-meter wide free space around bus stops

Pavement extensions at elevated crossings, to reduce the length of the crossing

Providing additional space for street fixtures, then reducing the width of the side barrier to a distance of between five and three meters from the pavement

Figure 12. Parking location, pedestrian crossing old statues.

Figure 13. Parking location, pedestrian crossing new.

Step Three: Street Reconstruction

- Consulting the Department of Transport about replacing the perpendicular parking spaces on the right side of the road with parallel parking spaces
- Narrowing the side barrier to a width of four meters in order to accommodate perpendicular parking , and to a width of three meters at pavements extensions to slow down vehicle speed
- Adding tiles all along passages in front of building facades to slow down traffic speed
- Planting new trees or installing sun canopies
- After consulting with the Department of Transport, widening the side barrier by narrowing perpendicular parking to 5.3 meters (on the left side of the street)

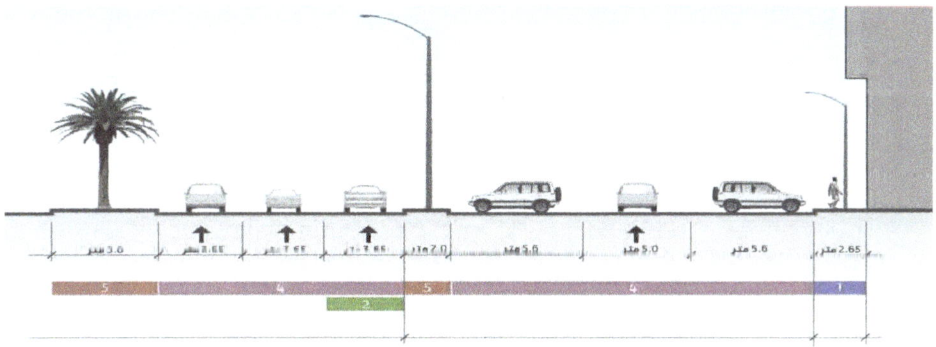

Figure 14. proposed road cross section.

Below are some examples of solutions applied to various streets in the U.A.E.:

The following figures show the differences between transport practices in Abu Dhabi before and after the application of urban street design standards and guidelines related to intersection design.

The elevated pedestrian crossings will encourage vehicles to give right-of-way to pedestrians.

(Before)

(After)

Figure 15. *Pedestrian crossing treatments.*

The broad median divider islands provide safe and comfortable waiting areas for pedestrians.

(Before)

(After)

Figure 16. *Pedestrian crossing treatments.*

Reducing traffic lane width contributes to slowing down vehicle speed at pedestrian crossings and decreases the distance pedestrians must walk to cross the street.

12.4. Applied Project in Shebeen El Kom, Egypt

The aim of the project is to promote the use of non-polluting, non-automotive means transportation (such as bicycle and pedestrian transportation) in certain traffic hubs of Shebeen El Kom. It also aims to provide bicycle parking space and to encourage the university and a number of companies to sell bicycles to students and employees at easy payment terms. Moreover, the project aims at encouraging university graduates to set up bicycle repair shops and at implementing some of the activities included in the National Campaign to Encourage Bicycle and Pedestrian Transport in Al Menoufeya Governorate, specifically in the city of Shebeen El Kom19.

Figure 17. *Shebeen El Kom, Egypt.*

Shebeen El Kom Project Implementation Objectives:

- To contribute to the amelioration of air quality and the reduction of noise levels
- Energy conservation
- To reduce greenhouse gas emissions, thus contributing to global efforts to combat climate change
- To keep pace with the global trend towards promoting bicycle transport, because of the significant health, economic, environmental and other benefits it provides

Shebeen El Kom Project Components:

Creating safe hubs for the use of bicycles in some of the city's transportation hubs

Creating safe bicycle parking space in selected locations

Creating a market selling bicycles at easy payment terms

Preliminary application of the National Campaign to Encourage Bicycle Use

First Component:

Creating Safe Hubs for the Use of Bicycles in some of the City's Transportation hubs

Facilitating bicycle circulation in the following selected transport hubs in Shebeen El Kom city: Gamal Abdel Nasser Street, Al Masna3 Street, a section of SabryAbouAlam Street and Tala3t Harb Street (dividers, road marks and signs will be

19 Ministry of Transport, Report on the Sustainable Transport in Shebeen El KomProject,2012

installed to mark the course of cycling lanes).

Second Component:

Creating Safe Bicycle Parking Space in Selected Locations

In coordination with the Governorate, bicycle parking space will be provided in certain central locations, such as the vicinity of various university faculties, a number of preparatory and secondary schools, government agencies and the factory, among others. Provision of special parking space for bicycles will encourage their use, reassure people about the safety of their bicycles and decrease the incidence of bicycle theft.

Third Component:

Creating a Market Selling Bicycles at Easy Payment Terms

- Preparing an implementation program for selling bicycles at easy payment terms to students and employees (reduced prices or long-term installment payment plans)
- Working with the Environmental Protection Fund (EPF) of the Egyptian Environmental Affairs Agency (EEAA) and the Social Fund for Development (SFD) to assist young people to set up bicycle rental and maintenance projects in Shebeen El Kom, providing maintenance and rental services to bicycles users in the city20

This project is a *preliminary application of the* National Campaign to Encourage Bicycle Use. After evaluation, the project will be implemented in other medium-size cities.

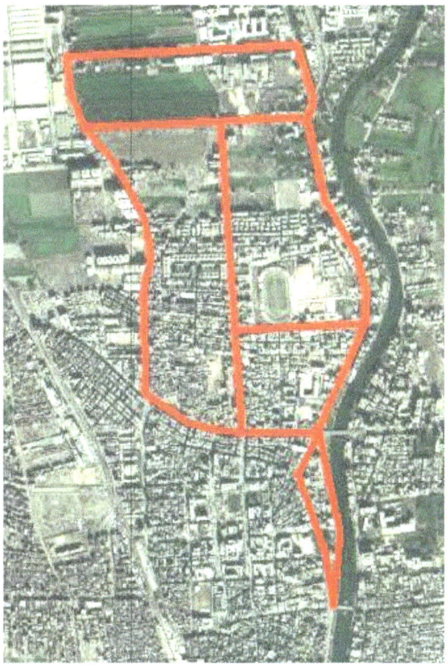

Figure18. *a preliminary application of the National Campaign to Encourage Bicycle Use.*

Figure 19. *Using a Bike Lane with the road design.*

Figure 20. *Bike lane as a majore transport facility.*

Based on the above presentation and analysis, proposed design approach components and the relevant inventory of tests can be formulated as follows:

"To facilitate the process of drawing conclusions, care has been taken to ensure that the measurements used in the inventory of tests are applicable, testable and falsifiable."

Conclusions and Recommandations

- Length of travel time that contributes to increasing fuel consumption and higher carbon monoxide emissions, calling for devising sensible environmental plans to reduce travel times.
- Below average fuel combustion rates in engines are caused by the quality of fuel and the manner of driving.
- Size and density of traffic require appropriate transport hub planning
- Pollution levels vary according to the type of vehicle (automobile, railway, lorry)
- Road topography (mountain roads, desert roads, coastal roads, roads passing through plains): flat roads consume less energy than mountainous or rough roads
- Trees and vegetation on the sides of the road contribute significantly to reducing the negative effects of car exhaust on urban structures and the environment generally
- By using local construction methods, combined with a little technology, it is possible to implement sustainable urban transport plans in Egyptian cities at reasonable cost
- It is important to benefit from Arab and international sustainable transport planning examples while preserving national and economic features that are in line with both the

20 Ibid.

Egyptian architectural environment and modern technical and architectural techniques, particular green architecture

- Sustainable transport is an integral part of green architecture (a comprehensive concept whose components must be applied together and not separately). Green architecture is in harmony with the environment and seeks to limit adverse environmental effects, promote efficient energy use, to encourage the use of renewable energy resources and the efficient use and recycling of materials and resources. It respects and interacts with the site, while adapting to climatic conditions and ensuring the comfort of users.
- Upgrading the present standards for the design and planning of automated traffic routes and pedestrian areas in urban contexts will contribute significantly to reducing the incidence of road accidents.

Table 4.*proposed measurement matrix for sustainable roads design.*

NO.	ITEMS DESCRIPTION	CHECK				remarks
		Yes		No	N/A	
		Full	Partial			
1.0	Suitability of roads to site topography and minimization of digging and backfilling					
1.1	Appropriateness of road network design to contour lines					
1.2	Digging and excavation works do not exceed the volume allowed by the owner					
1.3	The excavation methods selected, and the use of environmentally friendly material in backfilling					
2.0	Regulation of traffic hub movement and ease and safety of movement					
2.1	The proposed design facilitates transportation to and from the various site sections					
2.2	The design ensures prevention of present and future traffic congestion					
2.3	The design ensures that pedestrian and automated traffic are separate and that there is no discordance between them					
2.4	The design ensures that pedestrian networks, crossings and traffic are clearly marked, particularly for the disabled					
3.0	Promoting pollution reduction in traffic hub design					
3.1	The design includes natural vegetation elements that promote pollution reduction					
3.2	The design designates specific garbage collection areas according to modern environmental specifications					
3.3	The design includes notification methods to minimize environmental, noise and visual pollution					
4.0	Incorporation of road gradation and integrated networks					
4.1	Existence of a clear hierarchy of traffic hub networks					
4.2	The design ensures the complementarity of traffic hub networks and traffic flow from and to all service sectors					
4.3	The design ensures that infrastructure and electricity networks are interconnected					
5.0	Provision of optimal methods and systems for automobile parking space					
5.1	The design provides automobile parking space that is appropriate to site conditions					
5.2	The design provides for the required number of automobile parking spaces					
5.3	The design provides appropriate solutions for indicating parking space locations and facilitating pedestrian movement from and to parking spaces					
6	Taking the aesthetic dimension into account in traffic hub planning and design					
6.1	The aesthetic dimension is considered in the layout of the area surrounding roads					
6.2	The design encourages walking and cycling					
7.0	Observing traffic hub sustainable design standards and components					
8.0	Using suitable technologies to monitor and improve traffic hub performance					
9.0	Provision of mechanisms to ensure the continuity of periodic maintenance and development works necessary for preserving traffic hub conditions					

Recommendations

Advising state institutions and agencies of the need to update laws related to transport design by adding stipulations concerning environmental protection and the prevention of environmental infringements.

Likewise, advising educational and academic institutions of the need to raise awareness concerning the importance of maintaining favorable road conditions and abstaining from behavior that could adversely affect them in order to resolve the problems that threaten society. It is therefore necessary to develop educational curriculums that explain environmental conservation theories and their relation to green transport and to organize public awareness conferences, seminars and lectures concerning the environmental effects of personal decisions and actions.

In addition, the study recommends taking immediate steps to identify special design standards for pedestrian areas and to apply the principles of road network planning and design that aim to guarantee pedestrian comfort in places where pedestrian and automated traffic meet. Above surface solutions should be used first, followed by tunneling systems, while the safety and comfort of users should be a priority at all times. The next step should be to build bridges with gradient ascents and descents, rather than steps. Pedestrian traffic is one of the essential components of a city's structure, and priority must be given to planning pedestrian crossings of automated traffic roads. The above recommendations should be the subject of clear plans that are integrated into overall city plans. These plans should include the creation of pedestrian passages, since it is necessary to provide streets for automated traffic and separate places for pedestrians that are safe, functional and aesthetically attractive. The study focuses on the importance of separating pedestrian and automated traffic by creating underground paths, pedestrian bridges etc to keep the movement of city inhabitants away from road traffic. It further recommends investigating possibilities for installing underground or aboveground trains to encourage the use of public transport, similarly to modern cities all over the world.

The importance of improving air quality and reducing noise levels is stressed. In keeping with the global trend towards encouraging cycling as a mode of transport, the study highlights the example of the executive program facilitating the sale of bicycles to students and employees at reduced prices or in easy installments, due to the health, economic, environmental and other benefits of this mode of transport. Finally, the study recommends that utmost importance be accorded to sustainable transport planning in Egypt.

References

[1] Schiller, P., Bruun, E., & Kenworthy, J. (2010). An Introduction to Sustainable Transportation: Policy, Planning and Implementation. London: Earth scan.

[2] Commonwealth of Australia. (2011).Our Cities, Our Future: A national urban policy for a productive, sustainable and livable future.Canberra: Department of Infrastructure and Transport.

[3] Department of Transport and Main Roads (2011). Connecting SEQ 2031 – An Integrated Regional Transport Plan for South East Queensland. Brisbane: Queensland Government.

[4] Hansen, C.J. (2011) The role of policy-making and planning cultures for sustainable transport?. European Transport, 47 (2011), 89-108.

[5] Low, N. (2011). Australian Cities, Transport Integration and the Global Dilemma of Mobility. Retrieved April 2, 2012, from State of Australian Cities 2011: http://soac2011.com.au/full-papers-list.php

[6] Newman, P., & Kenworthy, J. (2011, May 26).Peak car use: urban planning needs to change say Newman and Kenworthy. Retrieved April 10, 2012, from The Fifth Estate :http://www.thefifthestate.com.au/archives/24214

[7] Hay, William, W. An Introduction to Transportation Engineering (translated by Drs. Saeed Abdel Rahman Al Qadi and Anis Abdallah Al Taneer, King Saud University Publications, 1999

[8] Dr. Mohamed Tewfiq Salem, Handasset al naqlwa al moroor – Dar al rateb al game3eya, 1985

[9] Oglesby, Clarkson, H., Highway Engineering (Translated by Dr. Ali Suleiman Huzayn and others, Dr. Tarek Youssef Al Ridi, Dr. Mohamed Salah Al Din Al Howary) - John Wiley & Sons, New York

[10] Burtenshaw, D. and Batenman, M. and Ashworth, G. J., The City in The West Europe - John Wiley & sons, UK, 1981

[11] Catanese, Anthony J. & James C. Snyder, Introduction to Urban Planning: 1986, McGraw- Hill Book Company - New York

[12] Dr. Ali Mohamed Abdel Moneim Hassan, Handassat al naqlwa al moroor (mabade' takhteet al naqlwa al moroor dakhel al modon) - Dar al rateb al game3eya, 1994

[13] Dr. Khaleel Ahmed Abou Ahmed, Al tasmeem wa al takhteet al handasee leltoroq al 7adareya wa al khalaweya– Dar al rateb al game3eya

[14] Dr. Zein El Abedeen Ali, Mabade' takhteet al naql al 7adary – Dar safa'alelnashrwa al tawzee3, Amman, Jordan, 2000

[15] Dr. Sabah Mahmoud Mohamed, Al madkhal fi takhteet al naql al 7adari – Mo'assaset al waraqlelnashrwa al tawzee3, Amman, Jordan, 2002

[16] Al e3tebarat al bee'eya fi takhteet ma7awer al 7araka belmodon al gadeeda fi misr

[17] Mohamed Abdel Ba'I Mohamed Ibrahim, Daleel tasmeem al toroq, Ketab tasmeem al toroq 1-2, 2-2 (Ministry of Transport, Egypt)

[18] Al tasmeem al handassi leltoroq (General Authority for Roads and Bridges - Ministry of Transport, Egypt)

[19] Oglesby, Clarkson, H., Highway Engineering, (Arabic translation of third edition)

[20] Report on the Sustainable Transport in Shebeen El Kom Project (Ministry of Transport, Egypt)

[21] "Tatawor wasa'el al naql wa almowassalat", published on Sciences Way website (http://www.sciencesway.com/vb/showthread.php?t=5839)

[22] Fahd Ben Khaled Al Fawzan, "Qeta3 al naql wa dawroh fi al tanmeya al eqtessadeya" (http://www.suhuf.net.sa/2001jazhd/jun/26/ec.htm)

[23] Haydar Kamouna, "Alhameyat qeta3 al naqlwa al morror fi al tanmeya al eqtesadeyawa al egtema3eya" (http://www.almadapaper.com/sub/06-405/p19.htm)

[24] World Summit on Sustainable Development (WSSD) publications – Johannesburg, South Africa, 26 August to 4 September 2002 (http://www.un.org/arabic/esa/desa/aboutus/keyissues.html)

[25] Engineer Mohamed Hassan Sayed, "Ba3d semat 7arakat al moroor wa athareha", Arab Urban Network website (http://www.araburban.net/author/muhammed.html)

[26] Adel Hamoud, "Al naql/ tareekhoh, wasaeloh, tataworoh (http://www.alsabah.com/paper.php?source=akbar&mlf=copy &sid=37221

[27] Dr. Abdallah Ibrahim Al Fayez, "Takhteet al modon wa moshkelat al e3temad 3ala altaqa sharlabodminho" (http://www.aleqt.com/2207/01/28/article_7726.html)

[28] "Al mawsoo3a al goghrapheya / derassat al naqlwa al moroor medaneyan" (http://www.4geography.com/vb/showthread.php?p=2782)

[29] Dr. Haydar Kamouna, "Al 3elaqa bayn al ta7adur wa tatawor al khadamat", Magalet al 3ooloom al 'egtema3eya Forum (http://swmsa.net/forum/archive/index.php/t-10193.html)

[30] Pederson, E.O., Transportation in Cities, Pergamon Press Offices, New York, USA, 1980

[31] Litman, Todd, "Introduction to Multi-Modal Transportation Planning Principles and Practices", 18 November 2008, Victoria Transport Policy Institute

[32] Federal Highway Administration, Federal Transit Administration, "The Transportation Planning Process: Key Issues – Part-1", updated September 2007, Publication Number: FHWA-HEP-07-039

[33] "Guiding Principles for Sustainable Transportation" – The Global Development Research Center – Conference organized by the Organization of Economic and Social Development (OECD), Vancouver, Canada, 24 – 27 March 1996

[34] Bradley, Bradley M.J. & Associates - Comparison of Energy Use & CO2 Emissions From Different Transportation Modes - Submitted to American Bus Association, Washington, D.C., May 2007

[35] Green Transportation - powered by Feed Burner_files-O'flanerty, C. A.: Highway Traffic

[36] Planning and Engineering, Volume 1 - University of Leeds, UK—Stones

[37] Tabor R. Stone, Beyond the Automobile; reshaping the transportation environment, Prentice Hall, Inc., N. J., 1992

[38] AASHTO, A Policy on Geometric Design of Highways and Streets, 1994

[39] http://en.wikipedia.org/wiki/Sustainable_transport

12

Use of information technology in emergency and disaster management

Biswajit Mukhopadhyay, Buddhadev Bhattacherjee

Faculty: Civil Engineering; Institute: Kolkata Metropolitan Water & Sanitation Authority, City: Kolkata, India

Email address:
biswajitmukherjee23@rediffmail.com (B. Mukhopadhyay), bbudda1234@gmail.com (B. Bhattacherjee)

Abstract: Disaster management being a key part of emergency management encompasses wide range activities. Emergency management programme are developed and implemented through the analysis and information. The majority of information is spatial and can be mapped. GIS allows emergency management to identify onset of any disaster. For this data has to be gathered organized and displayed logically to determine the size and scope of emergency management so that appropriate steps can be taken in case any disaster.

Keywords: Manmade, GIS, Preparedness, Mitigation

1. Introduction

Disaster management activities in case of a disaster are focused on three primary objects like protecting life, property and environment. Disaster is of two types –

1) Natural and 2) Man- made.

Earthquakes, flood, storm, Hurricane, Tornado, Cyclone, Forest fires etc. are the natural disaster where collapses of high rise buildings, plane crush, fire in building or mal functioning of nuclear facility are the man- made disaster which throw a challenge to the public emergency services for mitigating such disaster efficiently and co-ordinate manner. The optimal provision of information concerning the situation is an essential pre-requisite. Since co-ordination requires current information both from the occurring place and the management groups within a reasonable time which require an integrated communication and information system for disaster management that provides efficient, reliable, secure exchange and processing relevant information.

Disaster management is standing on three major phases like preparedness, prevention and mitigation. Disaster preparedness highlights on warnings and forecasts of impending disasters and often entails processes which are quite dynamic and results in a "rapid onset" disaster. Disaster prevention is a long term activity where satellite monitoring of relevant factors such as changing of land use, population growth is the main criteria. Communication of information about a disaster to a population at risk and appropriate actions to mitigate that hazard is an important part of information technology. The hazard or disaster sometimes cannot be avoided but the negative impacts on the people's property can be mitigated through public warning which is primary goal of the information technology and to ensure to the greatest extend so that hazard may not be a disaster. The success of information technology in case of disaster management is hidden in the seed of communication technology such as evacuation of the people from the hazardous site, to reduce damage to the property and thus minimizing human suffering. It is very important to note that in case of disaster, information technology is not a singular technology to act on but a system constituting the identification, detection, risk assessment of the hazard and accurate identification of the vulnerability of the population at risk and finally communication of information to the vulnerable population about the wide spread activity of the hazards and its rescue measures a sufficient time before so that people can take actions to abort negative consequences to a great extent

2. Key Players in Disaster Warning

The following components are considered as the key parties that play major roles in disaster management process-

1. Communities- The communities residing at the bank of the sea or in the hilly area or in the earthquake zone or other disaster prone areas are to be alerted through radio or television messages of on setting any hazard or

disaster. Community can also be made aware of potential negative impacts to which they are exposed and specific actions to be taken to minimize the threat of loss or damage.

2. Local Government- Local Government should have considerable knowledge of the hazards to which their community is exposed. They must be involved in design and maintenance of early warning systems and able to advice or instruct local population for their safety and reduces the potential loss of resources.
3. National Governments- National Governments are responsible for policies and framework to facilitate the early warnings in addition to the technical system for the preparation and issuance of timely and effective warning to the respective areas of the country. They should ensure the warnings and related responses are directed towards to the most vulnerable population through the design of holistic disaster response and early warnings framework that helps specific needs of related micro and macro level population. Moreover NGO (Non Governmental Organization), private sectors, media and scientific community have a critical and efficient role for mitigation of any natural hazards by supporting scientific and systematic monitoring and warning services to the people's at risk.

3. Components of Information Technology Used in Disaster Management

1. Awareness program: To combat disaster, awareness of the risk population about the possible disaster and its combat –strategies is utmost important. It can be done in various ways. If target population are mostly common illiterate people, audio visual aid like Video conferencing, tale- communication even with the experts who are not available at field level can play a great role in educating people to mitigate disaster like situation.
2. Radio & television – The most traditional electronic media used for disaster warning and a widespread effect to the people. The effectiveness of this media in the developing countries and rural environments where the teledensity is relatively low and there it can be used to spread the warning quickly to a broad population. A study on this media revealed that the effectiveness of radio and television with a easy understandable language of warning can reduce the potential death toll of catastrophic, cyclonic and tidal bore. But only drawback of this system is that at night this media are generally switched off.
3. Telephone/Mobile- The landlines and Mobile phones has an important role in warning the communities at stake for an incoming danger of a disaster. A phone call with warning saved many lives in South Asian Countries in 2004 Tsunami mainly at coastal region. However there are two drawbacks to warn the people through telephone are –
   i. Installation of telephone in the rural areas in all India still is not satisfactory even if there is an exponential use of mobile phone in recent past has mainly bounded in urban areas – particularly in rural and coastal areas the communities are mainly fisherman communities still think that mobile phone is a luxury.
   ii. The other drawback is congestion of phone lines before and during the disaster resulting incomplete phone calls at vital period.
4. Short Message Service.(SMS) – the SMS is allowable in most of the mobile phone and permits sending of short messages amount the mobile phone an even land line. In case of failure of network the SMS can work on a different band and can be sent or received even when phone lines are congested. During 2005 Hurricane Katrina disaster in USA, many residence of affected coastal areas were unable to contact relatives and friends through telephone but they could be able to communicate to each other through SMS.

Other communication technologies are used like Cell broadcasting, satellite radio, internet or e-mail, amateur and community radio, etc. to warn the people at stake in case of a impending disaster.

5. Auto- ALERT through Siren /Emergency bell: People in villages who ate not exposed to mobile phones etc can be brought under the auto alert systems of sirens or emergency bells which starts on its own whenever any alert message is detected. They can be installed at offices of local administrations or police stations in the locality.

4. GIS and Remote Sensing in Disaster Management

GIS (Geographical Information System) can be defined as a system of hardware and software used for storage, retrieval, mapping and analysis of geographic data. GIS can be used for scientific investigation, resource management and development planning.

Remote Sensing is a measurement of acquisition of information about a phenomenon by a recording device and gathered information about the environment without physical or intimate contact with the object or phenomena.

As disaster management work usually involves different agencies working in different areas. The need for detailed geographical information in order to make critical decision is hike. By utilizing GIS agencies involved in response can share information through data bases on computer generation maps in one location. For disaster management, workers have to access a number of department managers with their unique data and unique maps which in case of a disasters do not allow time to gather these resources. GIS provides the mechanism to centralize and visual display critical information during an emergency. GIS based spaced technology solutions have become an integral part of disaster management in developed as well as developing countries.

However incase of failure of GIS system in affected area by thunderstorm or flood like natural disaster, Disaster Management group has to take on physically though data of GIS in pre-failure state can be used to a great extent by the group depending on their expertise and post-failure scenario of GIS. The use of GIS in different phases can be illustrated as follows-

1. Planning
2. Mitigation
3. Preparedness
4. Response

1. *Planning*- Emergency management in case of a disaster begins with locating and identifying potential emergency problems. The use of GIS can become the backbone of emergency management as it provides a mechanism to centralize and visually display critical information during an emergency. Using a GIS official's can be triggered the bull's eye of the hazards and evaluation of the consequences of potential emergencies or disasters can easily be made. When a disaster like earthquake faults, fire hazards, flood zones, etc are viewed with other map data such as buildings, residential areas, rivers and waterways, streets, pipelines, power lines, etc an official can formulate mitigation, preparedness and possible recovery needs with the help of the information derived from remote sensing and satellite imagery play an importance in disaster management and crisis period. Before an effective emergency management program implementation thorough analysis and planning must be done. GIS facilitates the process by allowing planners to view appropriate combinations of different data to computer generated maps.

2. *Mitigation*- Mitigation is the activity that actually eliminates or reduces the probability of a disaster by taking preventive measures (e.g. legislation that require stringent building code in earthquake prone area or in case of a flood zones). Mitigation may include the implementation of strict legislation considering the magnitude of an earthquake, characteristics of soil and other geological data, landslide data, vegetation, topography, weather, etc in different areas and take proper action to avoid or to reduce the effects of unavoidable disaster. A GIS can identify certain soil types in an adjacent to an earthquake impact zone whereas bridges and overpass are at risk and GIS can identify the path of a flood based on topographic features or spread of a coastal oil spill based on current and winds. The habited property and human life at risk can easily be identified and targeted for protective actions. Thus based on GIS data safe Zones of habitation can be mapped and accordingly people can be guided properly to choose the safe places.

3. *Preparedness*- In the preparedness face the Governments developed plans to save lives and minimize the disaster damage. GIS can accurately support better response planning in areas such as determining evacuation routes or locating vulnerable infrastructures and vital lifelines, etc. it also supports logistical planning to be able to provide relief supplies by displaying previously available information on roads, bridges, airports, railways, etc. apart from these activities such as evacuee camp planning can also be done by GIS. GIS can display real time monitoring for emergency early warning. Remote weather stations can provide current weather indexes based on locations and surrounding areas, wind direction, temperature and relative humidity, etc. Wind information is vital in predicting an early report the movement of chemical cloud release or anticipating the direction of wild fire spread upon.

4. *Response* – Response are the activities followed by a disaster. These activities are designed to provide emergency assistance for victims (e.g. search and rescue emergency shelter, medical care, mass feeding, etc). GIS can provide one of the primary components for Computer aided dispatch (CAD) system. Emergency response units based at fixed locations can be selected and routed for emergency response. The quickest response units can be selected, routed and dispatched to the emergency zone. Depending on the emergency a GIS can provide a detailed information before the first unit arrived. For example during a commercial; building fire the location of the closest hydrants, electrical panels, floor planning of the building can be identified in case of emergency through GIS.

5. Purification and Treatment of Water During Disaster

Water is essential requirement for the victims of disasters before during and after their disastrous condition. Storage of water in an hygienic condition is very important. Water management is considered as prerequisite at all stages of disaster management: Preparedness, reduction and mitigation. Several conventional water storage techniques and know-how have been successfully implemented for combat of the disaster situation for the benefit of people. The disaster management groups use the fresh water to the affected community in both pre and post disaster situation that include (1) storage of water, (2) purification and treatment of water.

(1) Storage of water: In the outbreak of any disaster like Cyclone, Earthquake or any other disaster, the affected community might not have the access to water for days or even for weeks also. Information Technology can play an important role to educate the people of disaster prone area by letting them know technical know-how through public address system like radio, Television, mobile phones to keep the safe storage water in purified condition for a week so that requirement of water for their different house hold purposes during the time of disaster may be adopted.

Containers for water storage: of the many type of containers available for water storage, the glass/ fiber glass, plastic or metal containers are used widely.

Disinfection of water containers: water containers should

be disinfected before filling them with purified water. In case of disaster the purified and disinfected water are filled up mainly in plastic packets and are distributed to the affected population. In case of house hold disinfection the water can be boiled for 10 minutes for disinfection of water.

Chemical disinfection like sodium hypo-chloride or bleaching powder may be used for disinfection of water in case of outbreak of any disaster.

(2) Purification and treatment of water: In the disaster situation mainly at the time of cyclone, floods or earthquake the water lines and sewerage lines are damaged and they mixed with each other. Then outbreak of many water borne diseases like Cholera, Hepatitis 'A', Dysentery, Typhoid etc can occur due to contamination of water may be treated in various methods like heating, chemical disinfection, filtration or an appropriate combination of these methods. The one method is not sufficient for purification but a series of combination of methods may give full proof disinfected water. Heating or boiling or chlorination are the process which can kill the microbes but other contaminants like dissolve solids, heavy metals. So, for purification purpose, starting from sedimentation for removing suspended particles to filtration through Rapid gravity or slow sand filtration is done for removing dissolved solids. Many advanced method of water purification have been developed and marketed recently which could have potential application in case of disaster situation. In rural area when any outbreak of disaster occur people used two or three earthen jar placed one on top of the other having different diameters of holes starting from 0.5 mm to 2.0 mm through which water is passing away and from the bottommost jar the water is taken out and one drop of hypo-chloride solution in 1 liter of water is used for disinfection. Information technology can play a role in mapping of disaster affected areas and help to divide it in several zones based on the information received regarding intensity of disaster in each affected zone and class of people live in that Zone. Accordingly Sectorial action plan is taken – which suggests what kind of water treatment processes to be adopted for each of the affected Zones. Information technology can further be used in quality control of treatment process by two way communications with the people at field to the experts who are physically miles apart but virtually present.

6. Conclusion

In the time of disaster the people get puzzle and during outbreak of any disaster like cyclone, flood or earthquake or any other natural calamities the preparedness and response are very important part of mitigation management. Radio, television, mobile phone are the communication media help us in preparedness and arrangements for mitigation of the disaster situation by storing sufficient food stuff, purified water bottle or plastic pouch are very much essential. Thus the use of Information Technology plays a very important role in case of disaster or emergency management.

References

[1] Barrett, E.C., Beaumont, M.J. and Herschy, R.W. (1990). "Satellite Remote Sensing for Operational Hydrology: Present Needs and Future Opportunities." Remote Sensing Review, 4(2), 351-466.

[2] Borneman, R. (1988). "Satellite Rainfall Estimating Program of NOAA/NESDIS Satellite Analysis branch."National Weather Digest, 13(2), 7-15.

[3] Chavez, E., Ide, R., Kirste, T. (1999). Interactive applications of personal situation aware assistants. Computers & Graphics, 23(6):903-915, 1999.

[4] Guttman, E. (2001). Auto configuration for IP Networking: Enabling Local Communication, IEEE Internet Computing, May/June 2001.

[5] Mobile Broadband for Emergency and Safety Applications: MESA Project, *http://www.projectmesa.org/*

[6] Barber, D.G., Hochheim, K.P., Dixon, R., Mossocrop, D.R. and McMullan, M.J. (1996). " The role of Earth Observation Technologies in Flood mapping: A Manitoba Case Study". Canadian Journal of Remote Sensing, 22(1), 137-143.

[7] Bruynzeel, L.A. (1990). Hydrology of moist tropical forests and effects of conservation: a state of knowledge review. UNESCO, 224.

[8] Farley, J.A., Hecht, L.G. (1999). Open computing technologies as infrastructures for disaster management. In GDIN conference, Mexico city, May 1999.

13

Community based sewage treatment through hybrid constructed wetlands system for improved heath & hygiene and for enhanced agriculture productivity / livelihood generation in rural water scarce environments-Pakistan

Col Islam-ul-Haque[1,*], Asim Saleem[2]

[1]Chairman Eco-Logical Sustainability Through Environment Protection Services, Inc, (Eco-STEPS), Islamabad, Pakistan
[2]Country WASH advisor, Plan International Pakistan, Islamabad, Pakistan

Email address
islamhaq3@yahoo.com (C. Islam-ul-Haque)

Abstract: World over, fresh water scarcity problems are growing in scope & complexity to multiple factors, like population growth, climatic impacts, excessive water uses pattern in agriculture. A substantial quantum of treated wastewater , can be made available through constructed wetlands, which can supplement the fresh water scarcity. Almost, 80 % of the total generated raw sewage is being discharged into water bodies or being used for agriculture purposes, specially for vegetable farming, which contributes towards fresh water sources and food chain contamination. Centralized mechanical types of wastewater treatment plants are in use which is capital and recurring costs intensive. Resultantly, due to lack of sufficient recurring finances and energy shortages, these sewage treatment plants become non-operational and un-manageable. Moreover, huge centralized sewage treatment plants are difficult to manage due to longer chain of wastewater collection system and inflexibility of reuse of treated water in the areas of its generation. Therefore, keeping in view these obvious impediments, it was envisaged to plan & implement community based constructed wetlands system for sewage treatment, based on bio and phyto-remediation processes at village Chattal, district Chakwal (an arid area) which is functioning well and being sustained from selling the treated water to farmers and harvesting the various aquatic plants. Temporal analysis was carried out for verification of treatment system efficiency, whereby, monthly wastewater samples were collected at Inlet point (treated water reservoir) and got analyzed for Chemical Oxygen Demand (COD) and Biological Oxygen Demand (BOD) from PCRWR-Islamabad. Analysis report showed significant reduction of COD & BOD i.e. 86.1% & 91.4% respectively which are under permissible limits of Pak-EPA. Aesthetic value of treated water is remarkably improved after treatment as turbidity reduced from 15 NTU to 2 NTU. Detailed Health survey was carried out at village Chattal, to ascertain, the intervention efficiency. The result showed significant reduction in diarrheal & Hepatitis diseases i.e. 92.5% & 40% respectively and substantial improvements was observed in socio-economic, biodiversity and health & environment conditions of the rural population of village Chattal. Pakistan- the study area. This model will prove to be equally good in the rural environments of developing countries, especially in South Asia and Africa.

Keywords: Constructed Wetlands, Contamination, Community Based, Capital Cost, Investment, Reuse, Wastewater, Sewage Treatment

1. Introduction

1.1. Background

The population of world is increasing by alarming rate, mostly in urban center of Asia. The United Nations center for human settlements specifies that the urban population in the Asia and pacific region between 1991 and 2020 is expected to increase from 991 million to 2.44 billion. This means an additional 1.4 billion people will be living in urban centers of

Asian cities. This quantum increase in population will undermine the provisioning of basic services, like water and sanitation to the people. The increasing rate of rural population to urban migration is the main cause of population growth in urban areas. The year wise water availability in Pakistan is shown in figure-1, below;

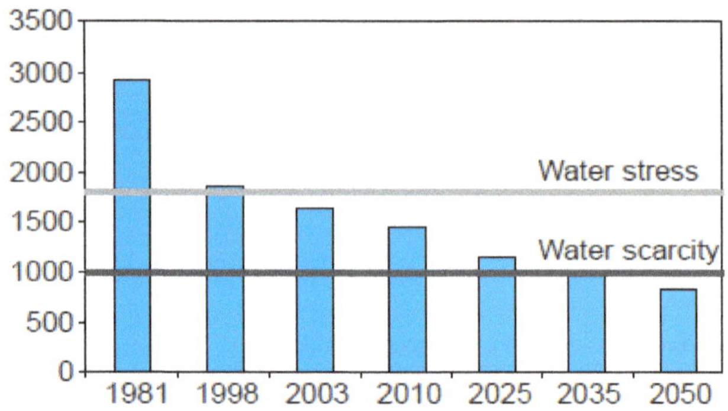

Figure 1. *The yearwise water availability in Pakistan*

In Pakistan, to meet this demand of increased urban population, mushroom growth of housing societies have been noticed which are unable to fulfill the basic services of water and sanitation. The wastewater from these localities is directly being discharged into natural water bodies like lakes, streams and rivers which is causing massive ground and surface water contamination. The same scenarios are also in place rural areas where wastewater is normally collected in a pond, located in the center of the village, thus causing health & hygiene problems. Despite drafting national strategy for sustainable environment, provision of safe drinking water and sanitation, the country is still confronted with major challenges in providing sufficient access to drinking water and sanitation, particularly establishing wastewater treatment system. Some of the public health indicators, due to mismanagement of wastewater, in Pakistan are as under;

1.2. Water Quality and Health Statistics - Pakistan

The overall water quality and health indicators in Pakistan are appended below;

- 87% of water in Pakistan is unsafe for human consumption (PCRWR, 2007)
- More than 1.6 million DALYs (Disability Adjusted Life Years) are lost annually as a result of death and disease due to diarrhea, and almost 900,000 as a result of typhoid (World Bank report 2006)
- The total health cost are estimated at Rs. 428 billion, or approximately 5.81 percent of GDP (Pakistan Economic Survey 2011-12)
- Alarmingly, over 2 million people are drinking unsafe water, some with high arsenic concentrations (WWF Report, 2007)
- An estimated over 250,000 child deaths occur each year in Pakistan due to water borne diseases (USAID Report).

The water quality situation both at source and at consumer's end in the project area, district (Chakwal) (PCRWR - report on technical assessment of water supply scheme, 2011); is given in figure-2, below;

Quality of Water at Source in District Chakwal

Tehsil	Functional Water Supply Schemes	Samples Collected	Safe for Drinking		Unsafe for Drinking		Major Causes of Contamination
			No	%	No	%	
Chakwal	50	56	4	7	52	93	biological, turbidity, sodium
Talagang	36	41	14	34	27	66	biological, nitrate, sodium, turbidity, iron
Choa Saiden Shah	22	24	3	13	21	88	biological, turbidity, fluoride, hardness
Kaller Kahar	18	19	-	-	19	100	biological, turbidity, fluoride
District Total	126	140	21	15	119	85	

Quality of Water at Consumer's end in District Chakwal

Tehsil	Functional Water Supply Schemes	Samples Collected	Safe for Drinking		Unsafe for Drinking		Major Causes of Contamination
			No	%	No	%	
Chakwal	50	100	9	9	91	91	biological, turbidity, sodium, nitrate, chloride, TDS
Talagang	36	72	12	17	60	83	biological, nitrate, sodium, turbidity, sulfate, TDS
Choa Saiden Shah	22	44	7	16	37	84	biological, turbidity, TDS, nitrate, sodium
Kaller Kahar	18	30	-	-	30	100	biological, turbidity, fluoride
District Total	126	246	28	11	218	89	

Figure 2. *Water quality at source and at consumer's end district Chakwal*

Hence, keeping in mind the adverse impacts of un-treated wastewater and over all fresh water scarcity situations in Pakistan, Hybrid constructed wetlands sewage treatment, embedded with value addition interventions, a sustainable , cost-effective and environment friendly pilot project was planed and implemented in village Chattal, district Chakwal - Pakistan . The salient project objectives are as under;

2. Project Objectives 2

- To protect fresh / drinking water sources from contamination, being caused by un-managed wastewaters in rural environments.
- To create improved health and hygiene environments by reducing mortality and morbidity rates.
- To increase the water resources potentials for irrigation by recycling the treated domestic wastewater through hybrid constructed wetlands system.
- To enhance food security and income generation opportunities at local / village and community levels, specially to marginalized communities / families.
- To create awareness and advocacy among the community for adaptation of improved health and hygiene practices and replication of this model elsewhere.

3. Study Area Overview

3.1. Geographical Location of the Project

Chattal is located in Chakwal district of the Punjab province of Pakistan which is in the south-east of federal capital Islamabad at a distance of 90 km and village Chattal is 12 km from Chakwal city headquarters and can be accessed from both GT road and Motorway -M-2. The aerial view of village Chattal is shown in figure-3, below;

Figure 3. Aerial View of Village Chattal

3.2. Pre-Project Condition and Need Assessment

Village Chattal comprises 400 households with approximately a population of 2400 nos. There was no proper mechanism for waste water collection and disposal. 120000 US Gallons/Day (approximately) was being discharged through open drains into an open pond, located within the village premises. The stagnant waste water pond was a big nuisance to local residents and was causing drinking water contamination. The villagers were suffering from water borne diseases due to ground water contamination, the only drinking water source in the area.

The location of existing wastewater ponds, located near water supply scheme is shown in figure-3, below;

Figure 4. *Location of existing foul smelling wastewater collection pond /arrangements- Village Chattal- district Chakwal, Pakistan*

4. Literature Review

In Nepal, low-cost, technology processes were introduced, by which wetlands are constructed and wastewater treated in them. (Krishna Lal Shrestha, 2008). The Environmental Public Health Organisation (ENPHO) introduced constructed wetlands system in Dulikhel (Nepal) Hospital in 1997. Since then this has been followed by 13 such systems at various institutions (e.g. hospitals, schools, university, and monastery) and individual households. The system is found to be highly effective in removing pollutants such as suspended particles, ammonia-nitrogen, BOD, COD and pathogens (Bhushan Tuladhar et al). Organic contaminants removal efficiency is high in all types of constructed wetlands. Domestic sewage normally contains organic matters and, therefore, constructed wetlands technology for sewage treatment is appropriate. (Jan Vymazal, 2010)

5. Method and Material

Plan international Pakistan, an International non-governmental organization working on child health care & well being, funded this pilot project as a role model for government to ensure sustainable low cost sanitation solutions with reuse of treated wastewater for irrigation & livelihood generation. Primary objective of this project was to establish model solution at village level, based on low cost technology, sustainable and easy for replication. Therefore a decentralized wastewater treatment system was needed to provide improved health and hygiene environments. The following methodology was adopted/chalked out;

5.1. Community Mobilization

Community development facilitator from Plan International Pakistan, carried out community mobilization and their participation for the project ownership to ensure post project sustainability. The district local government representatives were also engaged for requisite government buying.

5.2. Hybrid Constructed Wetlands Project Design

Following two major project components are as under;

5.2.1. Wastewater Collection through Covered Surface Drains

Plan International Pakistan engaged National Rural Support Program (NRSP) for technical support to community for development of drainage system in village for which community also provided financial contribution, while major funding for construction of covered drains was made available by NRSP through their own sources. The construction of drains was managed/ monitored by the community.

5.2.2. Hybrid Constructed Wetlands

Community based Hybrid constructed wetlands system, embedded with value addition interventions was designed due to poor socio-economic conditions of the people. The harvesting of aquatic plants, fish breeding and selling of treated water will generate enough finances to maintain the facility and sufficient livelihood for the work force. Being the newly adopted natural bio-technology in Pakistan, the construction provides for an applied science perspective on this "low cost" wastewater treatment process which uses a combination of aerobic and anaerobic natural treatment systems when combined, called hybrid wetland bioremediation and phyto-remediation waste water treatment technology. The application of this technology demonstrates a sanitation system which protects natural resources, promotes bio-diversity and sustainability. The salient features of the project shown in table-1, below;

Table 1. *Project overview , showing various about planning and implementation methodologies*

Type of Project	Small Scale, rural development, low cost sanitation development demonstration
Project Completion Period	Start of project: July 2011 Completion of Project: May 2012
Design Value	400 households, population 2400 discharge 38400 gallons /day
Capital Cost	Rs. 3.3 Million or US 32000 $
Planning, Designing, Construction, Execution and Development	NBS and Eco-STEPS Inc (Team Members)
Funding Agency	Plan International, Pakistan
O&M Cost	-Village Committee manage the site, as there is no electric & mechanical appliances and the entire in-fluent & effluent flow design is based on gravity flow regime. -Gravity flow based system assures the project sustainability and makes O&M easy. - Only minimal aquatic plants maintenance will be required , for which finances will be mustered from value addition interventions

5.2.3. Technical Description of the Project

Bioremediation is a naturally occurring phenomenon in which different plant species accumulate heavy metals, inorganic, toxic and hazardous substances from water and soil and reclaim the water for reuse. Similarly there are many kinds of microbial strains including aerobic and anaerobic, which degrade the organic pollutants into non-toxic smaller fractions. These microbes also accumulate certain heavy metals including chromium, mercury, lead, zinc etc. certain microbes produce such extra cellular enzymes which chelates the heavy metals and due to sedimentation of chelates the wastewater become safe for irrigation. This entire phenomenon is known as biodegradation. The application of biodegradation for waste water treatment and soil reclamation is called bioremediation.

NBS and Eco-STEPS Inc. (Eco-logical Sustainability Through Environment Protection Services) employs combination of four technologies i.e. Microbial Degradation, Phyto-remediation, Constructed Wetlands and Up flow Anaerobic Baffled Reactor (ABR) simultaneously for making process more appropriate and efficient degradation of pollutants from wastewater .

The unit operations and processes employed for domestic waste water treatment at Village Chattal are given in table - 2, below:

Table 2. *Unit Operations/Processes, Their Functions and Units Used for Domestic Wastewater Treatment*

Unit Operations/Processes	Functions	Treatment Devices
Screening	Removal of large floating, suspended and settle able solids	Screens of various description
Grit Removal	Removal of inorganic suspended solids solids	Screening Unit
Primary Sedimentation	Removal of organic/inorganic settle able solids	Primary sedimentation tank
Anaerobic biological growth Processes	Conversion of organic matter into CH4 & CO2 and relatively stable organic residue	Anaerobic Baffled Reactor (ABR)
Anaerobic Stabilization of Organic Sludges	same as above	Same as Above
Biological Treatment and Filtration	Accumulation, conversion of toxic substances and removal pathogenic strains	Sub surface water flow constructed wetland
Aerobic Biological Suspended Growth Process	Conversion of colloidal, dissolved and residual suspended organic matter into settle able biofloc and stable in-organics	Free Water Surface Flow Constructed Wetlands
Aerobic Biological Attached Growth Process	same as above	Rotating biological Sand Filter

5.2.4. Structural Design Information

The raw sewage must be treated before it is discharged into the natural system. The extent of treatment required to be given depends upon, not only on the characteristics and quality of the sewage but also upon the source of disposal, its quality and capacity to tolerate the impurities present in the sewage effluents without itself getting potentially polluted. The schematic view of the project site is shown in figure-5

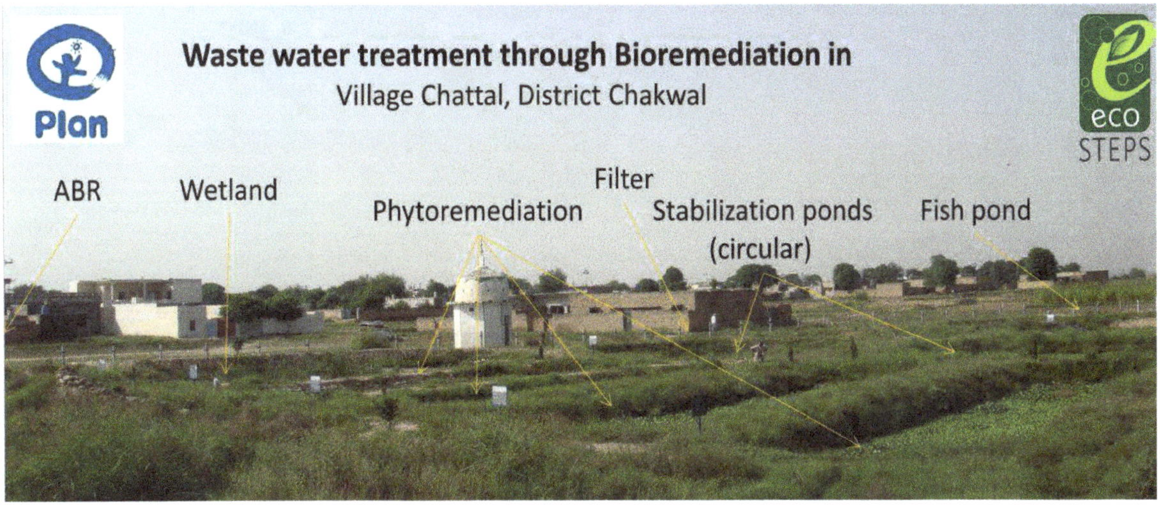

Figure 5. *Flow Diagram of Constructed Wetland at Village Chattal*

At Village Chattal combination of different biological process were employed for treatment which include anaerobic process achieved through Anaerobic Baffled Reactor (ABR) provide up to 70% reduction in both Chemical oxygen demand (COD) & Biological oxygen demand (BOD). ABR developed as secondary treatment system after primary screening and grit removal. Capacities of various components are shown in table-3, below;

Table 3. *Capacities of Different Components employed at Village Chattal*

S. No.	Components/Processes Employed	Capacity (US Gallons)
1	Inlet Drain	3,291
2	Screening Unit	6,676
3	Anaerobic Baffled Reactor	32,201
4	Subsurface Horizontal Flow Constructed Wetland	69,115
5	Free Water Surface Flow Constructed Wetland-I	87,965
6	Free Water Surface Flow Constructed Wetland-II	87,965
7	Free Water Surface Flow Constructed Wetland-III	87,965
8	Free Water Surface Flow Constructed Wetland-IV	87,965
9	Free Water Surface Flow Constructed Wetland-V	128,898
10	Free Water Surface Flow Constructed Wetland-VI	128,898
11	Biological Filter	87,965
12	Treated Water Reservoir & Rain Water Harvesting Pond	763,723
	Total	1,572,626

For tertiary treatment series of constructed wetlands were established which includes subsurface horizontal flow constructed wetland, free water surface flow constructed wetland which serves as facultative anaerobic, facultative aerobic and aerobic stabilization ponds, where aquatic plants fixed atmospheric oxygen in treated water. Treated water reservoir was established for water storage pond for further reuse in agriculture and fish breeding. Prior to treated water reservoir sand filter was established for removal of leftover microbial flora from waste water. Periodic microbial consortia application is shown in figure-6, below

Figure 6. *Application of Microbial Consortia & Seeding of Aquatic Plants in Constructed Wetland*

5.2.5. Water Reservoir as Fish Pond and Composting Pits

Treated water after anaerobic and aerobic treatment requires storage for harvesting for irrigation purposes so water reservoir was built after constructed wetland and this water reservoir is being used as a fish pond. Fish is a biological indicator for treated water and fish species can only survive with 4-5 mg/L dissolved oxygen (DO).

If there is less DO in treated waste water, then fish can't survive which shows that treated water is not of the desired quality. Fish production also maximize the income generation from the project. Similarly different aquatic plants with higher protein, nitrogen, phosphorus & potash contents used in bioremediation facility will be harvested regularly and further process for composting. For this purposes composting pits were constructed at project site.

5.3. Operations and Maintenance

Bioremediation technology is also very cost effective in its capital and reoccurring costs as compared to mechanical treatment system. It does not involve any kind of chemical and expensive electrical aeration process in meeting water treatment process. Due to these very reasons, local community was mobilized for taking physical ownership of the project. Operation & maintenance only involves, removal of sludge from bio-digester and grit chamber and removal of debris material from screens. Some services are also required regarding harvesting of aquatic plants on regular basis and overall maintenance of treatment site to be used as recreational site. Requisite capacity building sessions of the project committee was carried out for up-keeping the facility.

6. Results and Conclusion

The Hybrid constructed wetlands, embedded with value addition interventions for post project sustainability, was established in the target village (Chattal-Pakistan). Waste water treatment through this facility is fully functional with promising results. Various water quality analysis have been carried out and the results are according to national waste water standards. The physical appearance of treated water can be compared, as shown in figure-7, below

Figure 7. *Visual Condition of Water at Inlet & Outlet of Constructed Wetland Project at Village Chattal*

Waste water collected at Inlet point before entering into screening unit and treated water collected at outlet, exit from treated water reservoir was analyzed for Chemical Oxygen Demand (COD) and Biological Oxygen Demand (BOD)

from PCRWR-Islamabad. Analysis report showed significant reduction of COD & BOD i.e. 86.1% & 91.4% respectively which are well under permissible limits of Pak-EPA.

Aesthetic value of treated water is remarkably improved after treatment as turbidity reduced from 15 NTU to 2 NTU as shown in Fig. 5. There is no foul smell at constructed wetland site and whole site look like a water park. The previous waste water pond land has been reclaimed as shown in figure-8, below;

Figure 8. *Impact of CW project upon village surrounding areas, with acceptable aesthetic values*

6.1. Measured and Observed Impacts

6.1.1. Reduction in Water Related Diseases at Village Chattal, District Chakwal

The district health representatives carried out detailed survey of village Chattal, District Chakwal, where village sewage is being treated through Hybrid constructed wetlands, to ascertain the efficiency of the intervention. The summary of the survey showed significant reduction of water born diseases in village Chattal after establishment of Constructed wetland project i.e. 92.66% reduction in diarrheal disease & 40% reduction in Hepatitis as shown in table-4 and Fig 9, below

Table 4. *Reduction in water related before and after constructed wetlands interventions*

S#1	Ratio of Diseases like diarrhea Before the Vet land project	Ratio of Diseases like diarrhea after the Vet land project
Analysis	130-150 patients per month	10-11 patients per month
S#2	Ratio of Diseases like Hepatitis Before the Vet land project	Ratio of Diseases like Hepatitis after the Vet land project
Analysis	15 Patients from Village	09 Patients from Village out of 15

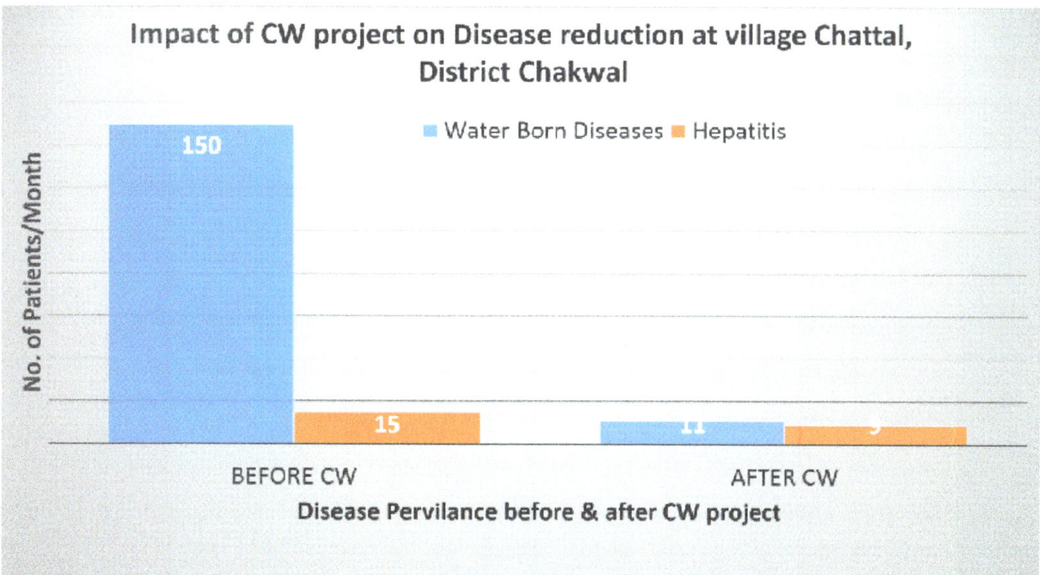

Figure 9. *Result of Health Survey carried out District Health representative before and after CW project*

Appreciable reduction in COD & BOD was noted which indicates the Hybrid constructed wetlands efficiency and effectiveness, as shown in figure-10, below;

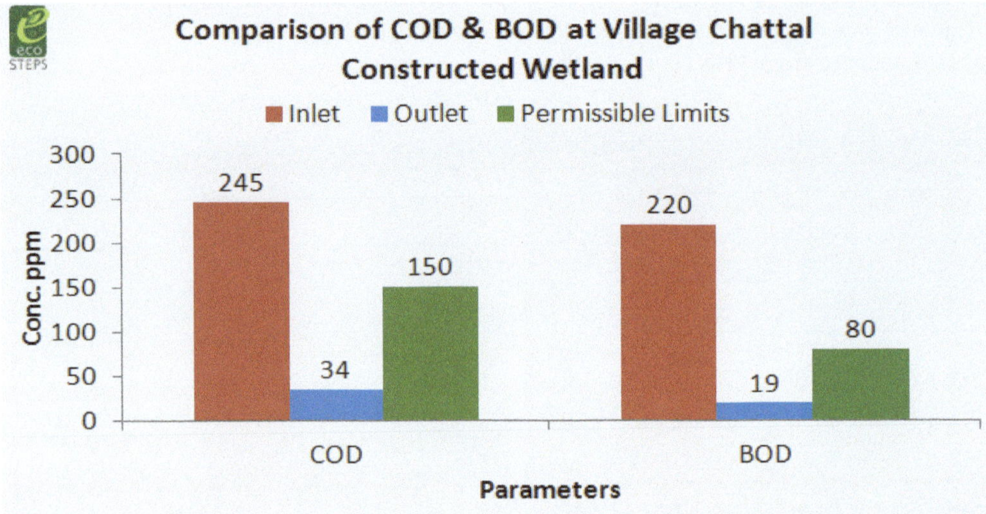

Figure 10. Comparison of COD and BOD before and after CW project

6.1.2. Revenue Generation- from Vale Addition Interventions

Sale of treated water - used for agriculture purposes: The area is arid in nature, therefore, availability of treated waste water to be used in agriculture has motivated the local participation manifolds and in return contributed towards revenue generation.

Fish Breeding in Treated Water Pond: Treated waste water reservoir was further augmented with rain water harvesting also which is being used for fish breeding. Since fish breeding is somewhat a technical field, therefore, wetlands site management committee has outsourced fish breeding activity and now earning approximately Rs 2000 per month.

Harvesting of Aquatic Plants: Regular harvesting of various aquatic plants is being carried out for production of poultry feed, animal fodder and manure, which also generate sizeable revenue for operation & maintenance of the wetlands system.

Enhanced Bio-Diversity: Due to enabling environment, various flora & fauna species have been seen in and around wetlands, specially turtles', reptiles and migratory birds etc, as shown in the figure-11 , below

Figure 11. Improved biodiversity conditions -various flora and fauna species

Revenue generation from fruit orchard: Fruit orchard, comprising of various fruit plants species, was established and sizeable revenue is being generated which is being utilized for the further development of the facility. High value fruit species are now being planned, with technical/expert support of national agriculture research centre (NARC) for more revenue earnings.

Herbal garden: Various varieties of herbs have been planted, in consultation with renounced pharmaceutical companies which will train the village management committee for efficient sowing of required herbs. Market linkages will be established for profitable disposal of these herbs. Beside this, local community will also be using these products. If properly managed and maintained, this intervention can fetch lot of revenue.

Integrated Farming: This is another main revenue generation intervention which consists of poultry, live stock and vegetable farming, as aquatic plants provide sufficient quantity of feed and compost for these activities.

The wetlands site has further been developed / landscaped which is now being used as a recreational place. Various species of fruit trees have been planted, specially the lemons, which will further contribute to maximize the revenue generation. The lush green constructed wetlands site provides a picnic spot to the inhabitants, as shown in figure-12, below;

Figure 12. *Pictorial View of Site after Completion- a process of bio & phyto-remediation*

7. Conclusion

Community based Hybrid constructed wetlands system at village Chattal was made partially operational in December 2012 and became fully functional in May 2013. The waste water (influent and effluents) analysis was promising and according to national standards and guidelines which contributed towards reduction in ground /drinking water contamination. There was tremendous reduction in water related diseases and improvement in health indicators. Since, the project design was embedded with value addition interventions/ livelihood generation, therefore, the local community did not face any financial problem to maintain the facility.

This small / decentralized community based sewage treatment facility is cost effective, environment friendly, financially & economically viable and sustainable. This model is suitable for the rural communities and can be replicated anywhere in the world.

In addition, various stakeholders (policymakers, water management institutes, health institutes, international development agencies) can use this facility / site as a model and resource centre for demonstration and promotional platform for mainstreaming the decentralized Hybrid constructed wetlands wastewater treatment system, best suited for small and medium sized communities in rural environments. Most of the developing & under developed countries are confronted with such problems, therefore, this model can be replicated in rural environments of these countries.

Reference

[1] National Water Quality Monitoring Programme (Fifth Monitoring Report 2005-06) June 2007, Pakistan Council of Research in Water Resources (PCRWR), E - mail: pcrwr@isb.comsats.net.pk Website: www.pcrwr.govt.pk.

[2] Pakistan Strategic Country Environmental Assessment, South Asia Environment and Social Development Unit, Document of the World Bank October, 2006.

[3] Pakistan Economic Survey 2011-12, Ministry of Finance, Government of Pakistan Islamabad.

[4] Pakistan's Water at Risk, Water and Health related issues in Pakistan & Key Recommendations, Freshwater & Toxics Programme, WWF – Pakistan, Published 2007 by WWF – Pakistan.

[5] PuR Water Purification Reaches Pakistan, United States Agency for International Development USAID Report, http://www.usaid.gov/.

[6] (Krishna Lal Shrestha, 2008). Decentralised waste water management using constructed wetlands in Nepal.

[7] Bhushan Tuladhar, environment public health organization, Decentralised wastewater management using constructed wetlands.

[8] Jan Vymazal, *Water* 2010, *2*, 530-549; doi:10.3390/w2030530- Review, Constructed Wetlands for Wastewater Treatment.

14

Spatial and temporal dynamics and parameters of growth of toxic cyanobacteria in Nuwara wewa and Nachchaduwa wewa in Sri Lanka

Mohamed Cassim Mohamed Zakeel[1], Prathiba Aruni Weerasinghe[1, *], Banumu Arachchige Dilhani Ganga Kumari[2], Hetti Arachchige Mangalika Wickremasinghe[3]

[1]Department of Plant Sciences, Faculty of Agriculture, Rajarata University of Sri Lanka, Puliyankulama, Anuradhapura, Sri Lanka
[2]Postgraduate Institute of Agriculture, University of Peradeniya, Peradeniya, Sri Lanka
[3]Department of Agricultural Biology, Faculty of Agriculture, University of Peradeniya, Peradeniya, Sri Lanka

Email address:
zakeelag48@yahoo.com (M. C. M. Zakeel), aruni500@yahoo.com (P. A. Weerasinghe), dilkumari@yahoo.com (B. A. D. G. Kumari), mangalikaw@pdn.ac.lk (H. A. M. Wickremasinghe)

Abstract: Cyanobacterial toxins available in drinking water pose a major health problem to humans and animals. Nuwara wewa and Nachchaduwa wewa are two important reservoirs supplying water for domestic purposes in Anuradhapura, Sri Lanka. As cyanobacterial toxins are resistant to deactivation by heat, it is important to control the proliferation of cyanobacteria in these reservoirs. Therefore, making an inventory of cyanobacterial variations with respect to time and space is imperative. Distinct temporal variations of temperature, pH, turbidity, dissolved oxygen and cyanobacteria density were observed in both reservoirs. Strong positive correlations existed between the above physico-chemical parameters of water and the cyanobacterial density in both reservoirs. These parameters can therefore be used as good indicators to assess the status these reservoirs with respect to cyanobacterial density and distribution.

Keywords: Cyanobacterial Toxins, Chronic Renal Failure, Microcystin, Cyanobacterial Bloom, Nuwara Wewa, Nachchaduwa Wewa

1. Introduction

Chronic renal failure of unknown etiology is a major health issue in Anuradhapura district which is one of the hotspots of chronic renal failure of unknown etiology in Sri Lanka. Although many factors such as excessive cadmium in drinking water and food stuff especially rice, high level of fluoride in drinking water, use of low quality aluminium utensils for cooking, presence of arsenic in hard water with higher levels of minerals and toxins generated by cyanobacteria are suspected to be contributing to this disease [1], a recent study in Sri Lanka has proved that one of the major causal agents for renal failure is cyanobacterial toxins which cannot be easily removed by boiling [2]. As stated in [3], toxic cyanobacterial blooms in fresh waters pose a health risk to people due to the production of toxic compounds such as microcystins and nodularin. Acute exposure to microcystins, which is the most widespread cyanobacterial toxin present in fresh waters, can cause liver failure and death [4] whereas exposure to low levels of microcystins causes chronic effects in humans [5, 6, 7, 8, 9].

Ionization and cupric treatments are the only promising measures to get rid of the cyanobacterial toxins [10]. Cupric treatment kills the toxic cyanobacteria and it seems to be highly potential. More than 0.05 mg/l of Copper can be toxic to algae. Copper is an essential nutrient for animals at low concentration while toxic at higher concentration [11]. However, frequent use of this treatment is not advisable as other types of organisms may also be executed. Nuwara wewa and Nachchaduwa wewa are two important reservoirs serving as major sources of water for domestic purposes. People take water either directly from these reservoirs or irrigations channels from these reservoirs or even as tap-borne water supplied by the National Water Supply and Drainage Board in Sri Lanka.

Therefore, it is important to make an inventory of toxic cyanobacteria that populate Nuwara wewa and Nachchaduwa wewa and their seasonal dynamics with the variation of physical and chemical parameters of the water in the lakes. This will help in applying measures to prevent the growth of toxic cyanobacteria and the accumulation of cyanobacterial toxins in drinking water. This study therefore investigated the space-time dynamics of toxic cyanobacteria that populated the Nuwara wewa and Nachchaduwa wewa in Anuradhapura district during January – December, 2012.

2. Materials and Methods

2.1. Study Site

Two man-made reservoirs namely Nuwara wewa (8^0 20'08.95" North (N) and 80^0 25'34.19" East (E) with an elevation of 292 feet) and Nachchaduwa wewa (8^0 14' 51.07" N and 80^0 29' 05.25" E with an elevation of 342 feet) in Anuradhapura were selected for sampling in this study (Figure 1).

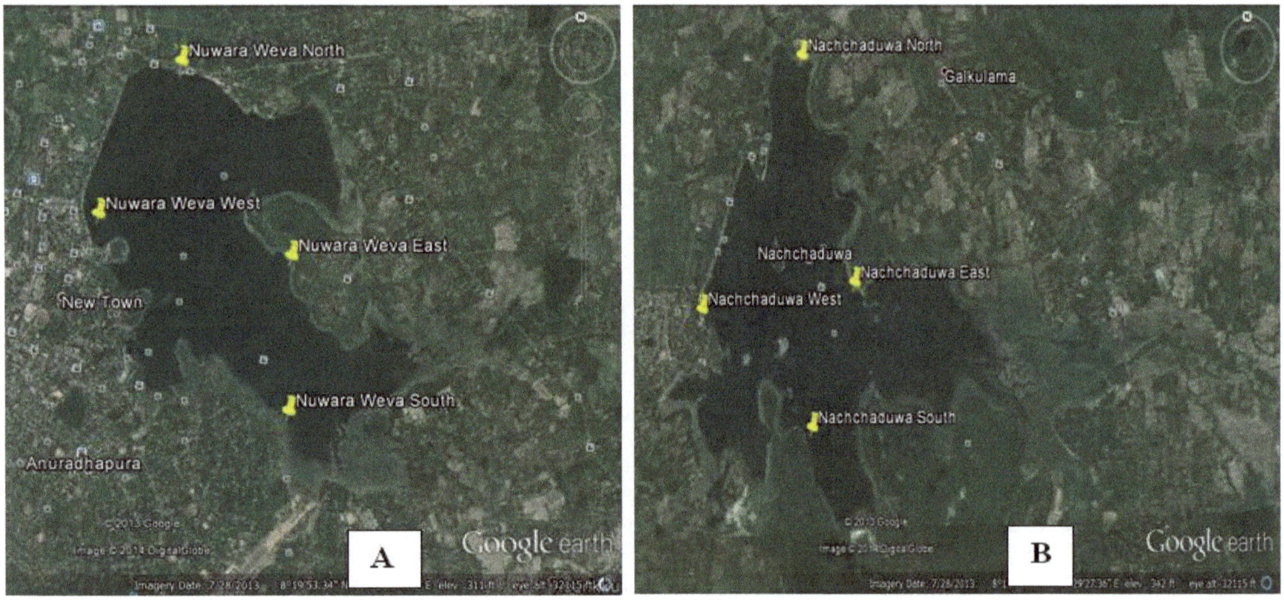

Figure 1. Sampling sites in Nuwara wewa (A) and Nachchaduwa wewa (B)

2.2. Sample Collection

Water sampling was carried out at four sites by main four directions; North, South, East and West. The coordinates of North, South, East and West sampling sites in Nuwara wewa were 8^0 21'12.94" N & 80^0 25' 36.44" E & 306 feet, 8^0 19'06.77" N & 80^0 26' 10.21" E & 289 feet, 8^0 20'06.16" N & 80^0 26' 11.55" E & 297 feet and 8^0 20'25.11" N & 80^0 25' 00.23" E & 291 feet, respectively (Figure 1). North, South, East and West sampling sites of Nachchaduwa wewa had the coordinates of 8^0 16' 29.87" N & 80^0 28' 55.37" E & 346 feet, 8^0 14'03.17" N & 80^0 28' 56.26" E & 361 feet, 8^0 15'13.73" N & 80^0 29' 13.06" E & 345 feet and 8^0 15'07.60" N & 80^0 28' 13.50" E & 335 feet, respectively (Figure 1). The samples were collected from two different layers; surface layer and bottom layer (one meter below the surface), at a rate of three samples per each layer at a sampling site with a total of 24 samples per reservoir per month.

Water samples were collected in sterile brown glass containers every month starting from January 2012 to December 2012. The samples were immediately tested for physico-chemical parameters within a few hours in addition to primary microscopic observations. Formalin solutions of 4% and 6% were added into aliquots of samples to prevent the proliferation of cyanobacteria with time in order to preserve the water samples for short and long terms, respectively [12]. For the proliferation of microorganisms, aliquots of samples without formalin were kept under room temperature.

2.3. Physico-chemical Parameters of Water

Temperature, pH, dissolved oxygen in water and turbidity were measured as physico-chemical parameters of the collected water samples. The temperature was measured using a standard Celsius thermometer. Turbidity meter (WalkLab) and pH meter (PCE 228) were used to measure the turbidity and the pH of water samples, respectively. All these parameters were measured at the time of sampling. The turbidity of water was measured in the unit of Nephelometric Turbidity Units (NTU).

2.4. Determination of Cyanobacterial Density

Enumeration of the toxic cyanobacteria was performed under light microscope. A homogenized sample of 0.1 ml was spread out between a slide and a cover slip. The smear was observed according to horizontal course over the entire length of the cover slip [12]. As stated in [12], this operation was repeated five times while shifting clearly over the width of the plate, of approximately a field of microscope to avoid overlapping observations. At the time of these five horizontal courses, all cyanobacteria present were enumerated with the magnification of 10 X 40 [12]. The average monthly density was calculated using the following formula.

Average Monthly Density (AMD) = $X \cdot 10^4 / 5*4*6$

X = Total number of individuals

5 = the number of times the slide prepared with one sample

4*6 = six samples taken from each of 4 sites

2.5. Generic Identification of Cyanobacteria

Collected water samples were concentrated by centrifugation at 3500 rpm for ten minutes. The resultant pellet was serially diluted and inoculated onto BG11 medium to isolate cyanobacteria [13]. The generic identification of cyanobacteria was carried out based on morphological criteria according to keys of identification [14]. Colour, size, shape of colonies and trichome, and the presence of akinete, gas vacuoles, heterocyst and gelatinous sheaths were taken into account [12].

2.6. Maintenance of Cyanobacterial Cultures

BG11 medium (Table 1) was prepared to maintain the cyanobacterial cultures [15]. After inoculation, the cultures were incubated at room temperature with 12 hours light-dark cycling until the culture grew well. Sub sampling was done to obtain a pure sample of cyanobacteria for further investigations.

Table 1. Composition of BG 11 Medium

Ingredients	Quantity/liter
$NaNO_3$	1.500 g
$K_2HPO_4.3H_2O$	0.040 g
$MgSO_4.7H_2O$	0.075 g
$CaCl_2.2H_2O$	0.036 g
Citric acid	0.006 g
Ferric ammonium citrate	0.006 g
EDTA	0.001 g
Na_2CO_3	0.020 g
Trace metal mix A5+Co*	1.000 mL
Deionized water	1.000 L

*Trace metal mix should be consisted of H_3BO_3, 2.86 gl^{-1}; $MnCl_2.4H_2O$, 1.81 gl^{-1}; $ZnSO_4.7H_2O$, 0.222 gl^{-1}; $Na_2MoO_4.2H_2O$, 0.390 gl^{-1}; $CuSO_4$.$5H_2O$, 0.079 gl^{-1}; $Co(NO_3)_2$.$6H_2O$, 0-0494 gl^{-1}

2.7. Statistical Analysis

Descriptive statistics and Pearson's correlations between cyanobacterial density and different physico-chemical parameters of water were found using SPSS (version 17.0) statistical software.

3. Results and Discussion

3.1. Temporal Variations of Physico-chemical Parameters of Water

Analysis of monthly mean pH, temperature, turbidity and dissolved oxygen in water of Nuwara wewa and Nachchaduwa wewa were carried out to compare the changes over time. According to the Department of Meteorology [16], Anuradhapura receives rainfall during the North East monsoon from October to February. However, a different pattern of rainfall was observed in 2012. The rain started in November and continued till May. Therefore, the water capacity of the study reservoirs measured in height increased during these months (Figures 2 and 3).

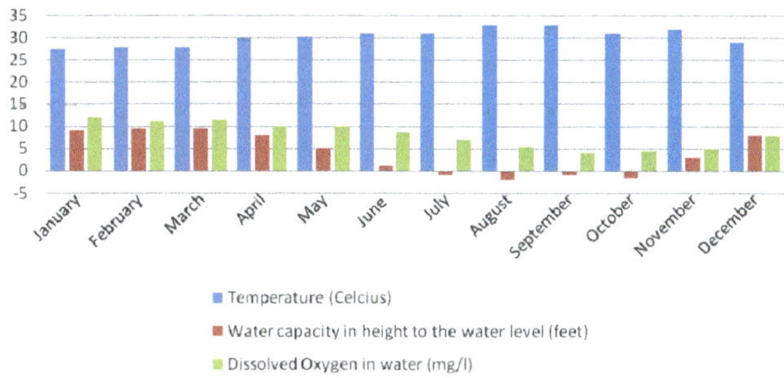

Figure 2. Temporal dynamics of physico-chemical parameters of water in Nuwara wewa

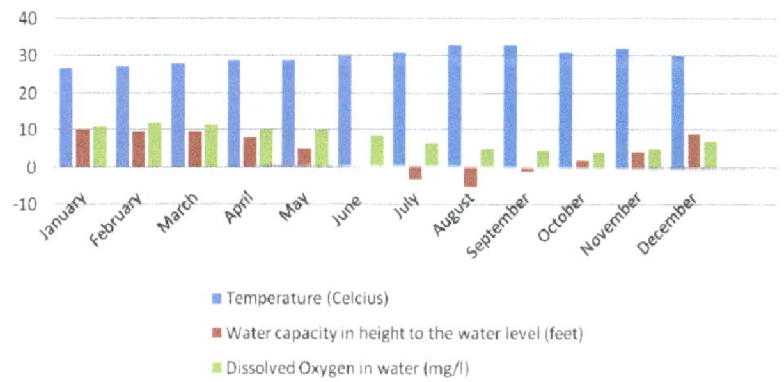

Figure 3. Temporal dynamics of physico-chemical parameters of water in Nachchaduwa wewa

During rainy season, the temperature was observed to be lower compared to dry spell (Figures 2 and 3). The highest temperature recorded was 30 °C in August and September in both Nuwara wewa and Nachchaduwa wewa (Tables 2 and 3). However, there was no much temperature fluctuation observed across the year. Dissolved oxygen levels in waters were low during the months which recorded lower water capacity (Figures 2 and 3). The highest pH values were recorded in the month of September in both Nuwara wewa and Nachchaduwa wewa and they were 8.78±0.09 and 8.89±0.08, respectively (Tables 2 and 3). The lowest pH was observed in the month of December in both Nuwara wewa and Nachchaduwa wewa with a value of 6.71±0.04 and 6.44±0.16, respectively (Tables 2 and 3).

Table 2. Temporal dynamics of physico-chemical parameters of water and average density of cyanobacteria in Nuwara wewa

Month	Avg. Density of Cyanobacteria (x10^3)	Mean pH	Temperature (^{0}C)	Water level (feet)	Turbidity (NTU)	Dissolved O_2 (mg/l)
January	0.17	6.96±0.07	27.5	9	3	12
February	0.33	7.21±0.09	28	9.5	2	11
March	0.08	7.34±0.02	28	9.5	2	11.5
April	0.25	7.78±0.09	30	8	2.5	10
May	0.08	7.09±0.18	30.5	5	5	10
June	0.17	8.36±0.09	31	1	6	8.5
July	0.92	8.35±0.11	31	-1	7	7
August	2.08	8.35±0.12	33	-2	10	5.5
September	1.25	8.78±0.09	33	-1	12	4
October	0.42	7.10±0.23	31	-1.5	8	4.5
November	0.08	7.22±0.27	32	3	8.5	5
December	0.08	6.71±0.04	29	8	5	8

Table 3. Temporal dynamics of physico-chemical parameters of water and average density of cyanobacteria in Nachchaduwa wewa

Month	Avg. Density of Cyanobacteria(x10^3)	Mean pH	Temperature (^{0}C)	Water level (feet)	Turbidity (NTU)	Dissolved O_2 (mg/l)
January	0.25	6.88±0.12	27	10	5	11
February	0.25	7.10±0.09	27.4	9.5	3	12
March	0.17	7.20±0.09	28	9.5	2	11.5
April	0.08	8.04±0.16	29	8	7	10
May	0.17	7.03±0.13	29	5	8	10
June	0.42	8.65±0.11	30	1	10	8.5
July	1.50	8.41±0.20	31	-3	12	6.5
August	2.92	8.36±0.13	33	-5	18	5
September	2.08	8.89±0.08	33	-1	12	4.5
October	0.83	7.59±0.30	31	2	10	4
November	0.17	7.01±0.13	32	4	9	5
December	0.08	6.44±0.16	30	9	6	7

It was revealed that with higher water levels in the reservoirs, the dissolved oxygen in water was comparatively higher in both reservoirs (Tables 2 and 3). A very strong positive correlation was observed between the water level in Nuwara wewa and the dissolved oxygen in water of the reservoir (Table 4). These two parameters exhibited a strong positive correlation in Nachchaduwa wewa (Table 5) and it was significant at the α value of 0.01. These results indicate that dissolved oxygen levels might have elevated due to mixing of water with atmospheric oxygen during rain fall as well as by runoff water from catchment areas.

Table 4. Pearson's correlation coefficients between different parameters of water in Nuwara wewa

	Cyanobacterial Density	pH	Temperature	Water Level	Turbidity	Dissolved Oxygen
Cyanobacterial Density	1.00	0.69*	0.65*	-0.67*	0.67*	-0.54
pH	0.69*	1.00	0.67*	-0.66*	0.57	-0.43
Temperature	0.65*	0.67*	1.00	-0.88**	0.92**	-0.87**
Water Level	-0.66*	-0.66*	-0.88**	1.00	-0.88**	0.85**
Turbidity	0.67*	0.57	0.92**	-0.88**	1.00	-0.93**
Dissolved Oxygen	-0.54	-0.43	-0.87**	0.85**	-0.93**	1.00

*. Correlation is significant at the 0.05 level (2-tailed).

**. Correlation is significant at the 0.01 level (2-tailed).

Both reservoirs showed that the turbidity of water declined with the increase in water levels. The highest turbidity in Nuwara wewa being 12 NTU was recorded in September which received comparatively lower rain fall (Table 2). The Pearson's correlation analysis revealed that a very strong negative correlation (r = -0.88) existed between the water level in Nuwara wewa and the turbidity (Table 4). Similar pattern was observed in Nachchaduwa wewa as well. A correlation coefficient of -0.77 was obtained between the water level and turbidity in Nachchaduwa wewa (Table 5). In rainy seasons especially in the North East monsoon, the turbidity was lower due to higher water capacity and flushing of solids during water discharge. According to the World Health Organization, suitable standard turbidity value for consumption should be less than 1 NTU. Therefore, the water from both of the reservoirs is not suitable for direct consumption without treatment.

Table 5. Pearson's correlation coefficients between different parameters of water in Nachchaduwa wewa

	Cyanobacterial Density	pH	Temperature	Water Level	Turbidity	Dissolved Oxygen
Cyanobacterial Density	1.00	0.68*	0.74**	-0.85**	0.85**	-0.61*
pH	0.68*	1.00	0.56	-0.77**	0.67*	-0.40
Temperature	0.74**	0.56	1.00	-0.84**	0.86**	-0.93**
Water Level	-0.85**	-0.77**	-0.84**	1.00	-0.94**	0.74**
Turbidity	0.85**	0.67*	0.86**	-0.94**	1.00	-0.77**
Dissolved Oxygen	-0.61*	-0.37	-0.93**	0.74**	-0.77**	1.00

*. Correlation is significant at the 0.05 level (2-tailed).
**. Correlation is significant at the 0.01 level (2-tailed).

Pearson's correlation analysis between turbidity and dissolved oxygen in both reservoirs disclosed a negative correlation with the coefficients of - 0.93 and -0.77 in Nuwara wewa and Nachchaduwa wewa, respectively (Tables 4 and 5). The elevated turbidity might be due to mixing and concentrating of organic materials which in turn might have undergone decomposition reducing the total dissolved oxygen in waters.

A very strong negative correlation between the temperature and the dissolved oxygen was observed (Tables 4 and 5). This may be due to two reasons: (i) dissolved oxygen in water gets evolved from water with elevated temperature and (ii) at temperature ranging from 30-33 °C, cyanobacteria proliferate making the reservoir eutrophic, prevent penetration of sunlight into the reservoir and decrease the physical contact of atmosphere with water, all contributing to the decline in dissolved oxygen [12].

3.2. Space-Time Dynamics of Cyanobacteria

Bottom and the water column of each reservoir had been always changing with the changes of the environment and weather parameters. The correlation analysis revealed that a strong positive correlation existed between the pH of the water in the study reservoirs and the cyanobacterial density (Tables 4 and 5). The highest cyanobacterial density was observed at a pH range of 8.00 – 8.50 (Table 2 and 3). This result is in line with the findings by Soumati *et al.* [12] in Cheffia dam in the North-East of Algeria.

Temperature also had a strong positive correlation with the cyanobacterial density within a particular range (Tables 4 and 5). Soumati *et al.* [12] stated that the presence of toxic cyanobacteria correlates with a moderate temperature (15 – 30 °C) [17]. However, the cyanobacteria density and the dissolved oxygen in water exhibited a moderate negative correlation (Tables 4 and 5). This may be owing to the proliferation of cyanobacteria. Cyanobacterial bloom makes a mat of algae on the surface of waters preventing the physical contact with atmosphere and decreasing the penetration of sunlight into water [18]. Poor penetration of sunlight into water lowers the photosynthesis by submerged autotrophic flora in the water. This leads to net decline in dissolved oxygen in water. Further, respiration during nighttime can deplete dissolved oxygen concentration to a critical level. Oxidative decomposition of organic compounds derived from dead planktons and wastes of fish and aquatic animal would have also decreased the level of dissolved oxygen. These processes could have contributed to the depletion of the available dissolved oxygen in the water. Therefore, these parameters (pH, temperature and dissolved oxygen) can be used a good indicators to evaluate the status of a water body in terms of cyanobacterial density and distribution. In both reservoirs, the cyanobacterial density declined with the increase in water level (Figure 4). This may be due to low density of the cyanobacterial population because of higher volume of water as well as the unfavorable condition of water in the reservoirs for cyanobacterial proliferation. Chorus and Bartram [8] stated that when temperature, light and nutrient status are conducive, surface water may host augmented growth of cyanobacteria. The period from July to September was a dry spell and the temperature was also slightly high favoring the proliferation of cyanobacteria. Toxic cyanobacteria produce cyanobacterial toxins like microcystin during high growth or high metabolism [3]. Therefore, measures should be taken to reduce the cyanobacterial population or completely eliminate them during this period.

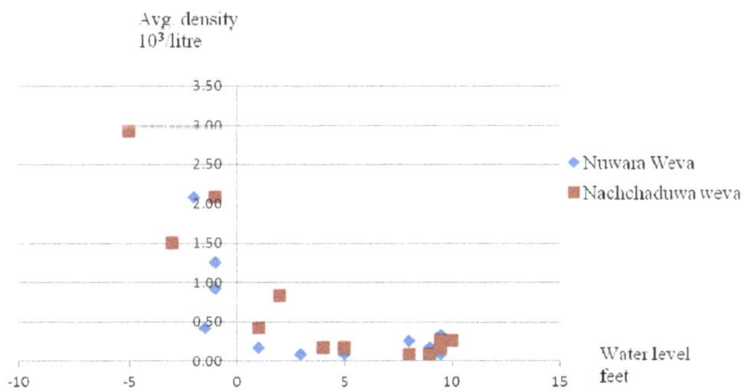

Figure 4. *Variation of average monthly density of cyanobacteria with water level*

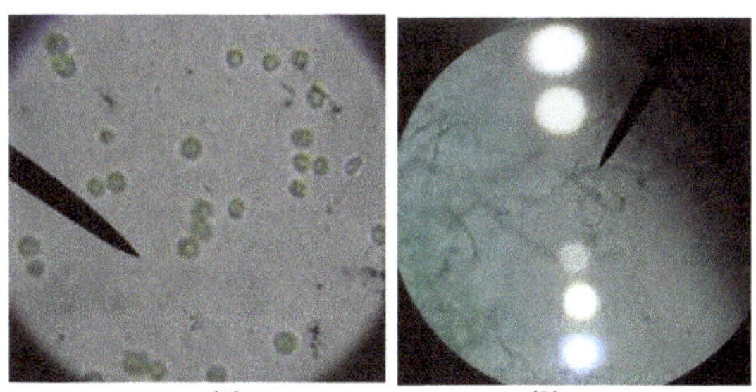

(Magnification x 1000)

Figure 5. *Morphology of Chroococcidiopsis (1) and Cylindrospermopsis (2) through light microscope*

In most of the eutrophic and hypertrophic lakes, Microcystis, Planktothrix and Anabaena are the major microcystin producing cyanobacterial genera [3, 8, 19, 20, 21]. According to the standard identification keys, many of the isolates recovered in this study were identified as *Chroococcidiopsis* and *Cylindrospermopsis* isolates through microscopic observations (Figure 5). Magana-Arachchi and Wanigatunge [22] also disclosed the prevalence of the genus *Chroococcidiopsis* in dry zone waters in Sri Lanka.

4. Conclusion

Strong positive correlations exist between physico-chemical parameters (pH, temperature and dissolved oxygen) of water in both Nuwara wewa and Nachchaduwa wewa and the cyanobacterial density. These parameters can be used as good indicators to evaluate the status these reservoirs in terms of cyanobacterial density and distribution.

Acknowledgements

We would like to express our sincere gratitude to the National Research Council, Sri Lanka for the financial support through NRC Grants 2011 (NRC Grant No. 11-07).

References

[1] M.C.M. Zakeel, (2001), Chronic Renal Failure and Its Etiology (Feature article). YSF Newsletter, vol. 1, Young Scientists Forum, National Science and Technology Commission, Sri Lanka.

[2] D.M. Dissanayake, J.M. Jayasekera, P. Ratnayake, W. Wickramasinghe and Y.A. Radella, (2011), The short term effect of cyanobacterial toxin extracts on mice kidney. Proceedings of the Peradeniya University Research Sessions: University of Peradeniya (LK), p. 13.

[3] W. Ye, X. Liu, J. Tan, D. Li and H. Yang, (2009), Diversity and dynamics of microcystin-producing cyanobacteria in China's third largest lake, Lake Taihu. Harmful Algae, vol. 8, pp. 637-644.

[4] W.W. Carmichael, S.M.F.O. Azevedo, J.S. An, J.R. Molica, E.M. Jochimsen, S. Lau, K.L. Rinehart, G.R. Shaw and G.K. Eaglesham, (2001), Human Fatalities from Cyanobacteria: Chemical and Biological Evidence for Cyanotoxins. Envirn. Health Persp., vol. 109, pp. 663-668.

[5] I.R. Falconer, J.V. Smith, A.R.B. Jackson, A. Jones and M.T.C. Runnegar, (1988), Oral toxicity of abloom of the cyanobacterium *Microcystis aeruginosa* administered to mice over periods up to 1 year. J. Toxicol. Environ. Health, vol. 24, pp. 291-305.

[6] P. Cohen, (1989), The structure and regulation of protein phosphatases. Annu. Rev. Biochem, vol. 58, pp. 453-508.

[7] R.B. Fitzgeorge, S.A. Clark and C.W. Keevil, (1994), Routes of intoxication. In: International Symposium on Detection Methods for Cyanobacterial (Blue-Green Algal) Toxins. pp. 69–74. G.A. Codd, T.M. Jeffries, C.W. Keevil and E. Potter (eds.). Royal Society of Chemistry, Cambridge, UK.

[8] I. Chorus and J. Bartram, (1999), Toxic Cyanobacteria in Water: A Guide to their Public Health Consequences, Monitoring and Management. Spoon, London, pp. 41– 111.

[9] K. Sivonen and G. Jones, (1999), Toxic cyanobacteria in water: a guide to public health significance, monitoring and management. In: Chorus I., Bertram J. Cyanobacterial toxins, E & FN Spon, London, United Kingdom, pp. 41-111.

[10] T.W. Lambert, C.F.B. Holmes and S.E. Hrudey, (1996), Absorption of microcystins – LR by activated carbon and removal in full scalewater treatment. Wat. Res., vol. 30, pp. 1411-1422.

[11] C.E. Boyd, (2005), Production Sustainable Aquaculture Practices; Copper Treatments Control Phytoplankton. Global Aquaculture Advocate, p. 69.

[12] B. Soumati, H. Nasri, A. Meddour, S. Kadri and N. Loucif, (2005), Space-time dynamics and parameters of growth of toxic cyanobacteria in freshwaters in cheffia dam (North-East of Algeria). Ninth International Water Technology Conference, Egypt, pp. 755-764.

[13] R.Y. Stanier, R. Kunisawa, M. Mandel and G. Cohen-Bazire, (1971), Purification and properties of unicellular blue-green algae (order Chroccoccales). Bacteriol. Rev., vol. 35, pp. 171-205.

[14] L. Lawton, B. Marsalek, J. Padisak, I. Chorus and J. Bartram, (1999), Determination of Cyanobacteria in the Laboratory (Chapter 12). Toxic Cyanobacteria in Water: A guide to their public health consequences, monitoring and management. WHO, ISBN 0-419-23930-8.

[15] R. Rippka, J. Deruelles, J.B. Waterbury, M. Herdman and R.Y. Stanier, (1979), Generic Assignments, Strain Histories and Properties of Pure Cultures of Cyanobacteria, J. Gen. Microbiol., vol. 111, pp. 1-61.

[16] Department of Meteorology, Sri Lanka. Available at: www.meteo.gov.lk. (Accessed: 16 July 2014).

[17] Prentice and M. James, (2008), Temporal and spatial variations of cyanobacteria in Karori Reservoir, Wellington. MSc Thesis, The University of Wellington. Available at: http://waikato.researchgateway.ac.nz/ (Accessed: 16 July 2014).

[18] K. Kunlasak, C. Chitmanat, N. Whangchai, J. Promya and L. Label, (2013), Relationship of dissolved oxygen with chlorophyll-a and phytoplankton composition in Tilapia ponds, Int. J. Geosci., vol. 4, pp. 46-53.

[19] W.W. Carmichael and I.R. Falconer, (1993). Diseases related to freshwater blue green algal toxins, and control measures. In: Algal Toxins in Seafood and Drinking Water, pp. 187–209. I.R. Falconer (ed.)., Academic Press, New York.

[20] W.W. Carmichael, (1994), The toxins of cyanobacteria. Sci. Am., vol. 270, pp. 78-86.

[21] L.E. Fleming and W. Stephan, (2001). Blue green algae, their toxins and public health issues. The Florida Harmful Algal Bloom Taskforce, Florida St. Petersburg, Florida.

[22] D.N. Magana-Arachchi and R.P. Wanigatunge, (2013), First report of genus *Chroococcidiopsis* (cyanobacteria) from Sri Lanka: a potential threat to human health. J. Natn. Sci. Foundation Sri Lanka, vol. 41, pp. 65-68.

Above- and Below-Ground Reserved Carbon in Danaba Community Forest of Oromia Region, Ethiopia: Implications for CO_2 Emission Balance

Muluken Nega Bazezew[1, *], Teshome Soromessa[2], Eyale Bayable[2]

[1]College of Agriculture and Natural Resources, Dilla University, Dilla, Ethiopia
[2]Center for Environmental Science, College of Natural Science, Addis Ababa University, Addis Ababa, Ethiopia

Email address:
fullday.gc2002@gmail.com (Bazezew M. N.)

Abstract: Forests can capture and retain enormous amount of carbon over long periods of time. Their role in carbon emission balance is also well documented. However, especially in developing country wide spread deforestation and forest degradation is continuing unknowingly and deliberately. This study was conducted to estimate CO_2 mitigation capacity of the dry Afromontane forest of Danaba found in Oromia Regional State of Ethiopia. A systematic sampling method through Global Positioning System (GPS) was used to identify each sampling point. Results revealed that the total mean carbon density of the CF was 507.29 (1861.75 CO_2 equivalents) $t \cdot ha^{-1}$ whereas trees share 319.43 (1172.31 CO_2 equivalents) $t \cdot ha^{-1}$, undergrowth shrubs 0.40 (1.47 CO_2 equivalents) $t \cdot ha^{-1}$, litter, herbs and grasses (LHGs) 1.06 (3.89 CO_2 equivalents) $t \cdot ha^{-1}$ and soil organic carbon (SOC) 186.40 (684.09 CO_2 equivalents) $t \cdot ha^{-1}$ (up to 30 cm depth). The ultimate result implies that Danaba CF is a reservoir of high atmospheric CO_2. To enhance sustainability of the forest potentiality, the carbon sequestration should be integrated with Reduced Emission from Deforestation and Degradation (REDD$^+$) and Clean Development Mechanism (CDM) carbon trading system of the Kyoto Protocol to get monetary benefit of CO_2 mitigation.

Keywords: Carbon Sequestration, Climate Change, Community Forest, Geo-Position, Mitigation

1. Introduction

Forests are known to play an important role in regulating the global climate. The response of forests to the rising of atmospheric CO_2 concentrations is crucial for the global carbon cycle as they have huge potential in sequestering and storing more carbon than any terrestrial ecosystem [1]. Though the role of forests in climate change mitigation is widely recognized, the recent assessment shows carbon stocks in forest biomass decreased by an estimated 0.5 gigatonne annually during the period 2005–2010 because of a reduction in the global forest area [2]. Loss of forest biomass through deforestation and forest degradation makes up 12 to 20% of annual greenhouse gas emission, which is more than all forms of transportation combined [3]. Especially, in Africa forest degradation is very high compared to the rest of the world.

Ethiopia is facing rapid deforestation and degradation of forest resources and experiencing the effects of climate change such as an increase in average temperature, and rainfall pattern variability, and is one of most vulnerable countries to climate change [4]. As Ethiopia is dependent on natural resources and agriculture, it is less able to cope with the shocks of climate change-induced droughts, floods, soil erosion and other natural disasters. People will find it hard to escape poverty if vulnerability to climate change persists. The Government of the Federal Democratic Republic of Ethiopia has therefore implemented National REDD$^+$ working document in 2008 and Climate Resilience Green Economy (CRGE) Framework in 2011 by means of protecting and re-establishing forests for their economic, ecosystem services and carbon storage. Even if the strategic frameworks focus on carbon emission management, Ethiopia does not have carbon accumulation records and databank to monitor and enhance carbon sequestration potential of different forests.

Working in CFs would have high importance in achieving carbon sequestration and conservation of biodiversity on the one hand, and empowering communities to take part and improve their living condition on the other hand since state owned forests are unsuccessful in their sustainability in the past decades. Since many Ethiopian people are living close to forests, the relationship of these people to forests has to be taken as an opportunity for spreading CFs to improve carbon sequestration.

An integrated forest management approach has been initiated in 2000 and named Forest Dwellers Association in Danaba community forest (CF). Danaba CF is heavily exploited remnant coniferous forests found in West-Arsi Zone of Oromia Regional State of Ethiopia. Ongoing threats of observed human activities such as agricultural expansion, livestock grazing, illegal charcoal production and harvesting for firewood and construction which will likely diminish all carbon pools unless effective measures are enforced.

Therefore, the study was designed to estimate the reserved biomass carbon in trees, shrubs, LHGs and soil of Danaba CF which would have high important as an information basis that can create the environment to attract climate change mitigation finances and so to expand and conserve CFs in Ethiopia.

2. Material and Methods

2.1. Study Area

Danaba CF is a 5437 ha forest that belongs to Adaba-Dodola CF priority areas under the administrative of Community Forest User Groups (CFUGs). The area is located in West Arsi Zone of Oromia National Regional State located 5–11 km South-East of Dodola town and 320 km South-East of Addis Ababa, Ethiopia (Figure 1). It lies between 06°54'20"N and 6°54'3"N latitude and between 39°8'19"E and39°13'50"E longitude with an elevation ranging between 2490–3218 m a.s.l. According to Ethiopian National Meteorology agency weather data from 1995–2013, the mean minimum and maximum temperature of the study area is 3.6° C and 24.3° C, respectively. The mean annual rainfall is 964 mm, of which 70-80% received in main wet season of June to early September and 20-30% from remaining less pronounced wet periods. Vegetation of Danaba CF falls under dry-evergreen montane forest with strongly dominated by *Juniperus procera* and *Podocarpus falcatus* species. The soil in most part of the study area is Luvisols type with Cherozem occurring in some place at lower altitudes [5].

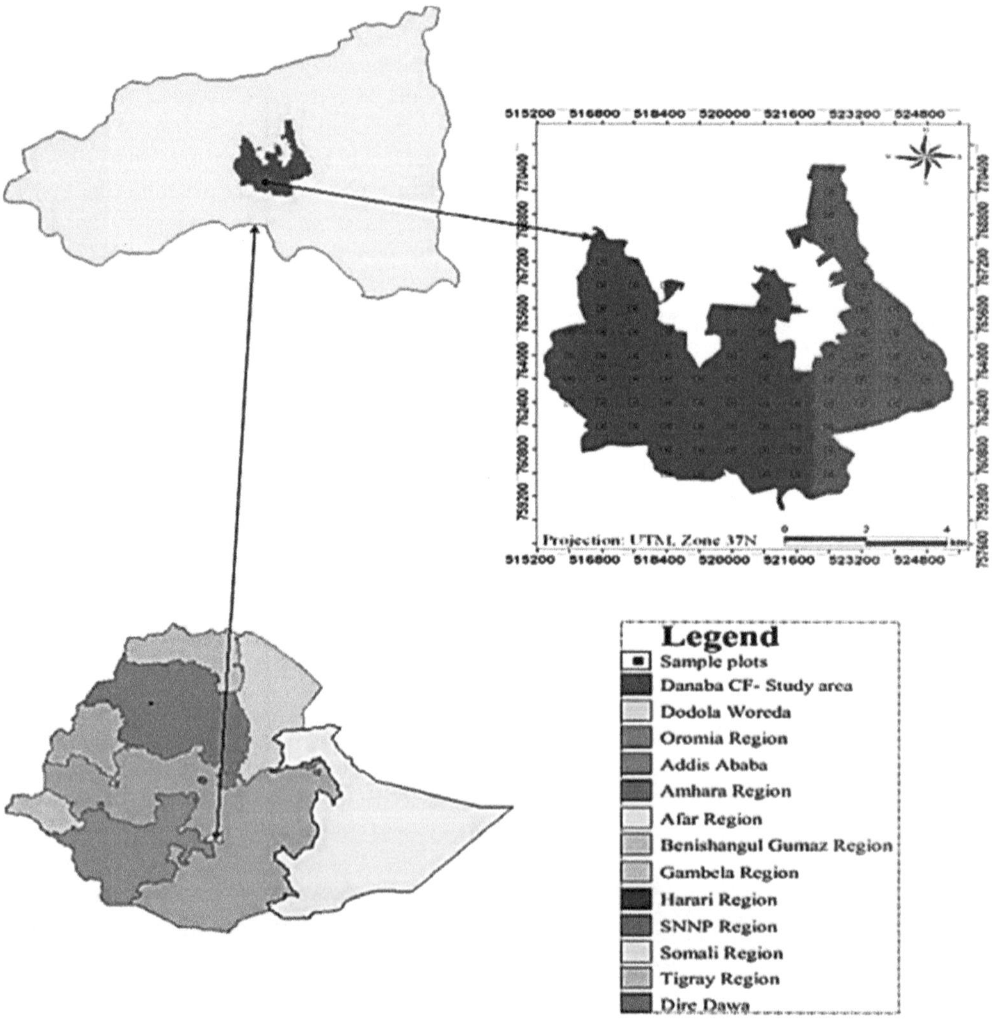

Figure 1. *Map of Ethiopia showing Regional States, Danaba CF and location of acquired sample plots.*

2.2. Sampling Design and Measurements

The field work for forest inventory was conducted from September 2013 to March 2014. A systematic sampling method was used for identification of sampling points distant 800 m from each other resulting in a total of 83 intersection points (Figure 1). In each intersection, 20 m × 20 m (400 m^2 equivalent to 0.04 ha) of plots were established for biomass inventory and identified using GPS in the field.

In each biomass plot, all tree species were identified and had their Diameter at Breast Height (DBH ≥ 2.5 cm) and height measured using diameter tape and Suunto Hypsometer, respectively. Above-ground biomass calculation for trees used a two-way method: For tress ≥ 5 cm DBH, [6] model was used while trees having between ≥ 2.5 cm and < 5cm DBH (1), an allometric model of biomass and volume tables with species description for community forest management developed by [7] was applied to calculate biomass (2).

$$Y = \text{Exp.} \{-2.187 + 0.916 \ln (D^2 \times H \times S)\} \quad (1)$$

Where: Y– Above-ground biomass (kg), H– Height of tree (m), D– Diameter (cm) at breast height (1.3 m), and S– Wood density ($t.m^{-3}$) for specific species [8-10].

$$\ln (AGSB) = a + b \ln (D) \quad (2)$$

Where: AGSB– Above-ground sapling biomass (Kg), a and b– Species specific constants [11, 7], and D – Diameter (cm) at breast height (1.3 m).

Below-ground biomass of tree species was calculated considering 15% of the aboveground biomass [12]. The biomass of stock density was converted to carbon stock density by multiplying 0.47 fraction of the [13] default value.

Additionally, at the center of each main plot a 5 m × 5 m sub-plots were used for shrub species sampling. The species were counted and their samples were uprooted and divided into above- and below-ground through identifying the collar region and fresh weights recorded. A procedure adapted by [14] of the loss on ignition (LOI) method was used to estimate percentage of carbon in shrub species. In this method, initially taken fresh weight of samples was dried at 65°C in the oven for 48 hours to take dry weight. Oven dried grind samples were taken (3.00 g) in pre-weighted crucibles, and then put in the furnace at 550°C for one hour to ignite. The crucibles were cooled slowly inside the furnace. After cooling, the crucibles with ash were weighed and percentage of organic carbon was calculated according to [15]. Equation 3 showed below:

$$A_{sh} = (W_3 - W_1) / (W_2 - W_1) \times 100 \quad (3)$$

C (%) = (100 – % A_{sh}) × 0.58 (Considering 58% carbon in ash-free litter material)

Where: C – Biomass carbon stock, W_1 – Weight of crucible, W_2 – Weight of the oven-dried grind sample and crucible, and W_3 – Weight of ash and crucible.

For sampling of LHGs, a 1 m × 1 m sub-plots at all corner and middle positions of each main plot were used. LHGs within five 1 m^2 quadrats of each main plot were collected and weighed on the field, and 100 g of evenly mixed sub-samples were brought to the laboratory to determine dry biomass and percentage of carbon. To estimate the biomass carbon stock, the sub-samples taken in the field were used to determine an oven-dry-to-wet mass ratio that was used to convert the total wet mass to oven dry mass according to [16]. The amount of biomass per unit area was calculated according to equation 4:

$$LHGs = \frac{W_{field}}{A} * \frac{W_{sub-sample,dry}}{W_{sub-sample,wet}} * \frac{1}{10,000} \quad (4)$$

Where: LHGs – Biomass of leaf litter, herbs, and grasses ($t.ha^{-1}$), W_{field} – Weight of the fresh field sample of leaf litter, herbs, and grasses- destructively sampled within an area of size A (g), A – Size of the area in which leaf litter, herbs, and grasses were collected (ha), $W_{sub\text{-}sample,\ dry}$ – Weight of the oven-dry sub-sample of leaf litter, herbs, and grasses taken to the laboratory to determine moisture content (g), and $W_{sub\text{-}sample,\ wet}$ – Weight of the fresh sub-sample of leaf litter, herbs, and grasses taken to the laboratory to determine moisture content (g). To determine percent of carbon in LHGs, the loss on ignition (LOI) method of [15] was applied. The carbon density of LHGs was then calculated by multiplying biomass of LHGs per unit area with the percentage of carbon determined for each sample.

For SOC determination, soil samples were collected within five 1 m^2 quadrats in which LHGs samples were taken. Soil samples were collected up to 30 cm in depth (between 0–10, 10–20 and 20–30 cm depths) using a calibrated soil auger [13]. A composite sample was obtained by mixing soil from three layers taken from five sub-plots of each main plot in order to determine bulk density and organic carbon concentration. About 150 g of composite samples were collected from each main plot. To determine SOC, field moist soil were dried in an oven at 105°C for 12 hours in laboratory to determine moisture content and dry bulk density. To estimate the percentage of organic carbon, samples were analysed by the wet oxidation method [17]. The carbon stock density of soil organic carbon was calculated as recommended by [16] from the percentage of carbon and bulk density of soil at predetermined depth of the samples were taken.

$$SOC = \%\, C \times \rho \times d \quad (5)$$

Where:

SOC – Soil organic carbon stock per unit area ($t.ha^{-1}$), %C – Carbon concentration (%), d – Soil depth (cm), and ρ – Bulk density ($g.cm^{-3}$).

The carbon stock is then converted to tons of CO_2 equivalent by multiplying it by 44/12 or 3.67 of Molecular weight ratio of CO_2 to O_2 in order to understand CO_2 mitigation potential of the study area. Carbon density in trees, shrubs, LHGs and organic soil were analysed using SPSS statistical software package.

3. Results and Discussion

3.1. Accumulated Carbon in Tree Species of Danaba CF

The assessment of [18] and [19] on biome-average tropical forest biomass carbon stock estimates, the average carbon stock of Sub-Saharan Africa, Tropical Asia and Brazilian Amazon forests are 143, 151, 186 t.ha^{-1}, respectively. On the other hand the mean biomass carbon stocks of trees in the Natural Forest of Bangladesh is 110.94 t·ha^{-1} [14] and Community Forest of Mid Hill Region of Nepal is 71.36 t·ha^{-1} [20]. The current study showed that out of the sixteen major tree species recorded in the study area, *Juniperus procera* and *Podocarpus falcatus* stored enormous density of carbon with 179.17 (657.55 CO_2 equivalents) (56.1%) and 105.73 (388.03 CO_2 equivalents) (33.1%)t·ha^{-1}, respectively; that amount accounts for approximately 90% of the Danaba CF carbon stock. The lowest carbon was recorded for *Myrsine africana* with 0.03 and 0.004 t·ha^{-1} of above- and below-ground carbon stock, respectively (Table 1). Hence, the present study was exceedingly higher than those continental and countries study. Carbon stored in trees above- and below-ground was comparable to the previous Ethiopian studies of Egdu Forest [21] and Tara Gedam Forest [22]. The variation might come from variation of age of the trees, existing species and management of the forests. The use of an allometric model for biomass estimation might also help explaining for the difference in estimated value as explained that reliance on allometric equations could be one of the limitations resulting in large variations in such estimates [23].

Table 1. *Biomass carbon stock of tree species (t.ha^{-1}).*

Scientific name	Family	TAGB	TBGB	TB	TAGC	TBGC	TC	CO_2 eq.
Juniperus proceraHochst. Ex Endl.	Cupressaceae	331.46	49.72	381.18	155.79	23.36	179.17	657.55
Podocarpus falcatus (Thunb.) R. B. ex Mirb.	Podocarpaceae	195.63	29.34	224.97	91.95	13.79	105.73	388.03
Cupressus lusitanica Miller	Cupressaceae	23.01	3.45	26.46	10.81	1.62	12.44	45.65
Maytenus arbutifolia (A. Rich.) Wilczek	Celasteraceae	18.16	2.72	20.88	8.54	1.28	9.82	36.04
Hagenia abyssinica (Bruce) J.F. Gmel.	Rosaceae	7.44	1.12	8.56	3.50	0.52	4.02	14.75
Eucalyptus globulus Labill.	Myrtaceae	5.63	0.84	6.47	2.65	0.40	3.04	11.16
Myrsine melanophloeos (L) R. Br.	Myrsinaceae	5.10	0.77	5.87	2.40	0.36	2.76	10.13
Ilex mitis (L) Radlk	Aquifoliaceae	1.72	0.26	1.98	0.81	0.12	0.93	3.41
Eucalyptus camaldulensisDehnh.	Myrtaceae	0.95	0.14	1.09	0.45	0.07	0.51	1.87
Osyris quadripartita Salzm. exDecne.	Santalceae	0.69	0.10	0.79	0.32	0.05	0.37	1.36
Oncoba spinosa Forssk.	Flacourtiaceae	0.39	0.06	0.45	0.18	0.03	0.21	0.77
Olea europaea L.	Olacaceae	0.36	0.05	0.41	0.17	0.03	0.19	0.70
Galiniera saxifraga (Hochst.) Bridson	Rubiaceae	0.21	0.03	0.24	0.10	0.01	0.11	0.40
Pittosporum viridiflorum Sims	Pittosporaceae	0.12	0.02	0.14	0.06	0.01	0.06	0.22
Hypericum revolutum Vahl	Hypericaceae	0.07	0.01	0.08	0.03	0.01	0.04	0.15
Myrsine africana L.	Myrsinaceae	0.06	0.01	0.07	0.03	0.004	0.03	0.11

TAGB, TBGB– total above- and below-ground biomass respectively; TB– total biomass; TAGC, TBGC– total above- and below-ground carbon respectively; TC– total carbon; eq– equivalents

3.2. Carbon Store Share within DBH and Height Classes of Tree Species

Within eight category of DBH classes, 5–20 cm DBH class had the highest density of trees with 401 trees ha^{-1} (41.8%) while trees with DBH greater than 120 cm were the least dominant in the study area and consisting of 4 trees ha^{-1} (0.5%). Irrespective of the highest density of DBH class, the highest corresponding carbon reserves were found in DBH class of >80–100 (25.3%), >60–80 (20.1%) and >100–120 (15.8%) cm with 80.74, 64.20 and 50.60 t·ha^{-1} of corresponding carbon density, respectively. DBH class of 2.5–<5 cm was the reservoir of least carbon stock in the CF with 3.65 (0.5%) t·ha^{-1} of the total stock density (Figure 2). The height of tree species were categorized into eight classes, of which height class of >10–15 m had the highest density of 355 trees ha^{-1} (34.5%) while least density of trees were found within the uppermost canopy of trees with >35 m of height class by accounting 3 trees ha^{-1} (0.3%). From the mean total mean carbon stock of the study area stored in above- and below-ground biomass of tree species of the study area, the highest carbon reserves were found in height class of >25–30 (25.4%), >20–25 (23.1%) and >15–20 (18.8%) m with corresponding stock density of 81.27, 73.79 and 59.97 t·ha^{-1}, respectively (Figure 3).

More than 50% of the trees were found in <20 cm DBH class. Hence, the study showed the forest is dominated by young trees after the implementation of community forest management through plantation and natural regenerations. The ultimate inference indicates that, there is high potential of increasing biomass carbon stock in the future if appropriate management of the forest is implemented.

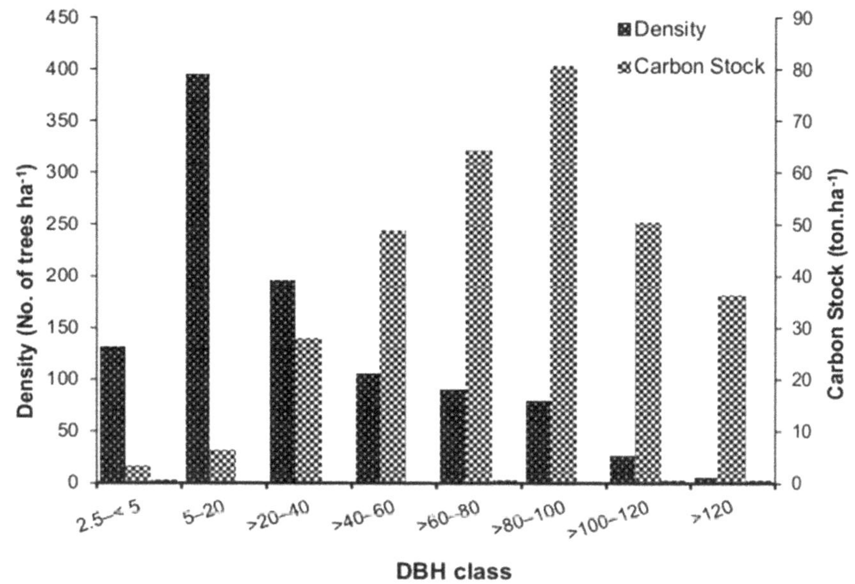

Figure 2. Carbon stock distribution within DBH Classes.

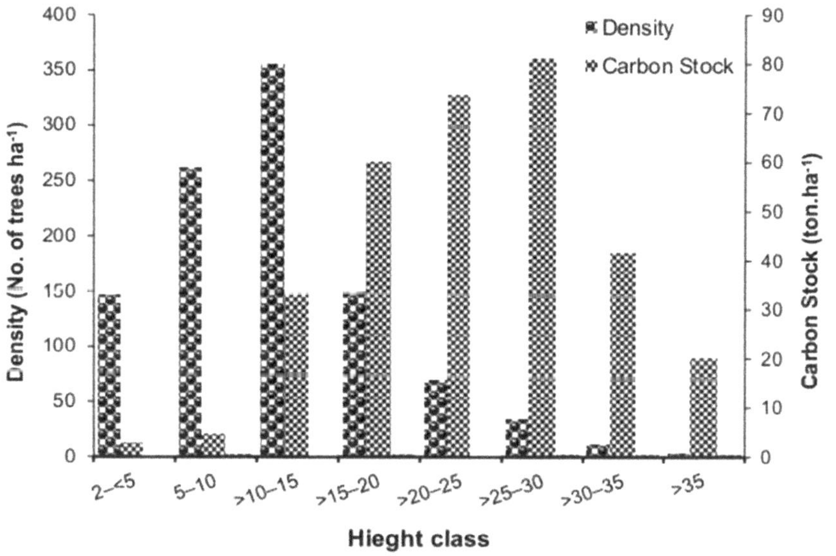

Figure 3. Carbon stock distribution within Height Classes.

3.3. Accumulated Carbon in Shrub Species of Danaba CF

Among six frequently occurring shrub species of the study area, mean carbon density of 0.40 ± 0.16 (1.47 CO_2 equivalents) t.ha^{-1} was recorded. *Conyza hypoleuca* and *Carissa spinarum* were the highest and least store of carbon with 0.18 (46.3%) and 0.03 (7.3%) t·ha^{-1}, respectively (Table 2). Shrub species of Danaba CF contributed small biomass carbon. Due to huge canopies and observed seasonal plantation of tree species makes unsuitable for shrub species regeneration.

Table 2. Biomass carbon stock of shrub species (t.ha^{-1}).

Scientific name	Family	TAGB	TBGB	TB	TAGC	TBGC	TC	CO_2 eq.
Conyza hypoleucaA.Rich.	Asteraceae	0.30	0.15	0.45	0.13	0.05	0.18	0.66
Maytenus undata (Thunb.) Blakelock	Celastraceae	0.18	0.05	0.23	0.03	0.02	0.06	0.22
Rosa abyssinica Lindley	Rosaceae	0.09	0.02	0.11	0.04	0.01	0.05	0.18
Dovyalis abyssinica (A.Rich) Warb.	Flacourtiaceae	0.08	0.02	0.10	0.03	0.01	0.04	0.15
Erica arborea L.	Ericaceae	0.07	0.01	0.08	0.03	0.01	0.04	0.15
Carissa spinarum L.	Apocynaceae	0.08	0.03	0.11	0.02	0.01	0.03	0.11

TAGB, TBGB– total above- and below-ground biomass respectively; TB– total biomass; TAGC, TBGC– total above- and below-ground carbon respectively; TC– total carbon; eq– equivalents

3.4. Accumulated Carbon in LHGs and Organic Soil

LHGs biomass carbon shared small amount of carbon in the CF. The assessment on mean litter carbon of tropical forests varies between 2.6–3.8 t·ha^{-1} as reported by [24] and 2–16 t.ha^{-1} by [18]. In current inventory of Danaba CF, mean value of 1.06 ± 0.31 t·ha^{-1} carbon density with highest store seems to be in grasses. Hence, 3.89 t·ha^{-1} of CO_2 equivalents were stored in LHGs biomass. The result was lower than those ranges. The mean stock density was also lower than most previous studied of Ethiopian forest. The reason for the small carbon stock of LHGs is due to huge closed canopies of *J. procera* and *P. falcatus* up to the near ground making the growth of herbs and grasses unsuitable. This dominance with evergreen tree species of Danaba CF has also contributed for the existence of small litter falls. As the study area had mountainous manifestation, litter run off occurred and might cause for small carbon account in this pool.

The average bulk density of soil in the CF was estimated to be 0.937 ± 0.0535 g.cm^{-3}. The percentages of carbon content of the soil in the study area ranges from 2.27–15.85% with mean value of 6.38 ± 2.6764 %. Thus, the current average soil organic carbon investigated in the study area was found to be 186.40 ± 76.5465 t·ha^{-1}. Accordingly, the study area could possibly store 684.088 t·ha^{-1} of CO_2 equivalents within organic soil. SOC estimates of Afromontane rain forests varies between 252 and 581 t·ha^{-1} [25]. The result of present study was lower than this range. Besides, the value was also lower than that of dry Afromontane Tara Gedam [23] and Egdu Forests [21] of Ethiopia. Rainfall and temperature variation of the studies might have contribution for this variation. Mountainous manifestation of the study area might cause early run off litter, herbs and grasses which in turn contributed for soil organic matter in decomposition.

The SOC share was varied at different soil depths. Table 3 showed the variation of SOC among different soil profile. The average bulk density of the study area increased with depth increment. The mean values of bulk density from top, middle and deep soil profile were 0.82, 0.96 and 0.99 g.cm^{-3}, respectively; however, SOC decreased with depth increment. The pattern indicates that soil carbon decreased significantly with soil depth which revealed major trends in carbon accumulation which shows that it is found in the upper soil layers (Table 3). This may be due to the accumulation and rapid decomposition of forest litter in the top soil. [14, 26, 27] found that more SOC was stocked at the soil depth of 0–14 cm. Therefore, the result was in high conformity with those findings.

Table 3. *Soil organic carbon stock at different soil depths.*

Depth of soil (cm)	Bulk density (g.cm^{-3})	Organic carbon (%)	SOC (t.ha^{-1}.depth^{-1})
0–10	0.82 ± 0.101	11.70 ± 3.243	95.01 ± 15.550
10–20	0.96 ± 0.162	6.35 ± 3.182	59.34 ± 12.441
20–30	0.99 ± 0.151	5.42 ± 2.549	52.11 ± 13.893
F	6.61	19.05	11.87
P	0.0032**	0.0000**	0.0001**

** Values significant at α=0.05 (95%); SOC- Soil Organic Carbon

Thus, this study showed that the carbon density of trees, shrubs, LHGs and organic soil were found to be 319.43 (1172.31 CO_2 eq.), 0.40 (1.47 CO_2 eq.), 1.06 (3.89 CO_2 eq.) and 186.40 (684.09 CO_2 eq.) t·ha^{-1}, respectively. Hence, in the current study the total carbon stock in Danaba CF was 507.29 (1861.75 CO_2 eq.) t·ha^{-1} (Table 4). Accordingly, the maximum quantity of carbon stock was found in tree species with reservoir of 63% of the total carbon. The forest soil organic carbon ranked the second reservoir of carbon which has accumulated 36.7% of the total carbon in the study area. Shrubs and LHGs' biomass contributes small amount of carbon; stored only 0.1 and 0.2 % of the total carbon, respectively (Figure 4).

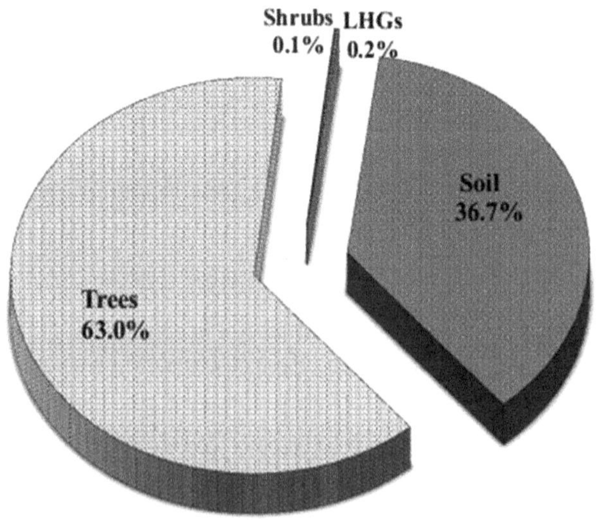

Figure 4. *Carbon stock percentage in different forest strata.*

Table 4. *Total reserved carbon (t.ha^{-1}) in Danaba CF of Oromia Regional State, Ethiopia in 2014.*

Carbon pools	TAGB	TBGB	TB	TAGC	TBGC	C(LHGs)	SOC	TC
Trees	591.00	88.65	679.65	277.78	41.65			
Shrubs	0.80	0.28	1.08	0.29	0.11			
LHGs						1.06		
Soil							186.40	
Total	591.80	88.93	680.73	278.03	41.76	1.06	186.40	507.29

TAGB, TBGB– total above- and below-ground biomass respectively; TB– total biomass; TAGC, TBGC– total above- and below-ground carbon respectively; C(LHGs) – Litter, herbs and grasses biomass carbon; SOC – soil organic carbon; TC– total carbon

4. Conclusion

As a conclusion, we observed that tree species sequestered the highest carbon of all carbon pools and *J. procera* reserved the highest biomass carbon. Undergrowth shrubs, LHGs and organic soil were also important pools that contributed for carbon sink in the CF. Danaba CF was the reservoir of potentially high amount of carbon compared to similar areas in the tropics particularly in tropical Africa, Asia and Latin America. Currently, the CF had the capacity to store 507.29 $t \cdot ha^{-1}$ carbon; helping in sequestering 1861.75 $t \cdot ha^{-1}$ of CO_2 equivalents which implies that remarkable carbon finance benefit has to be demanded. The carbon sequestration should be integrated with Reduced Emission from Deforestation and Degradation (REDD+) and Clean Development Mechanism (CDM) carbon trading system of the Kyoto Protocol to get monetary benefit of carbon dioxide mitigation which can be helpful for the sustainability of the forest.

Acknowledgements

The authors acknowledge Addis Ababa University, Dilla University, Wondo Genet College of Forestry and Natural Resources, Oromia Forest and Wildlife Enterprise and Ethiopian National Meteorology Agency, for the assistance during research period. Fund was obtained from the Thematic Research of Addis Ababa University; "Floral and Fungal Diversity, Ethnobotany and Carbon Sequestration Potential of Western and Southwestern Forests of Ethiopia".

References

[1] E. Sundquist, B. Robert, F. Stephen, G. Robert, H. Jennifer, K. Yousif, T. Larry, and W. Mark, "Carbon Sequestration to Mitigate Climate Change", USA, New York: U.S. Geological Survey, pp. 1–4, 2008.

[2] FAO, "Global Forest Resource Assessment", Rome Italy: FAO, Main report, Forest paper 103, 2010.

[3] Saatchi, S.M., Harris, S., Brown, M., Lefsky, E., Mitchard, W., Salas, B., Zutta, W., Buermann, S., Lewis, S., Hagen, S., Petrova, L., White, M., and Silman, A., "Benchmark map of carbon forest carbon stocks in tropical regions across three continents", Proceedings of the Natural of Academy Science, 108(24), 9899–9904, 2011.

[4] World Bank, "Convenient solutions to an inconvenient truth: Ecosystem-based approaches to climate change", Washington D.C.:World Bank, 19 pp, 2009.

[5] Digital soil and terrain data base of East Africa, "Soil and Terrain Database for Northeastern Africa",Rome, Italy: Food and Agricultural Organization of the United Nations, 1997.

[6] Chave, J., Andalo, C., Brown, S., Cairns, M., Chambers, J., Eamus, D., Folster, H., Fromard, F., Higuchi, N., and Kira T., 2005,"Tree allometry and improved estimation of carbon stocks and balance in tropical forests", Oecologia, 145, 87–99.

[7] P.R. Tamrakar, "Biomass and Volume Tables with Species Description for Community Forest Management", Kathmandu, Nepal: Ministry of Forest and Soil Conservation, 2000.

[8] Morales, J.B., "Wood specific gravity from two tropical forests in Mexico",IAWA Bulletin New Series, 8(2), 143–148, 1987.

[9] G. Reyes, S. Brown, J. Chapman, and A.E., "Wood Density of Tropical Tree Species", New Orleans, Louisiana, Winrock International, Southern Research Station: USDA Forest Service Publication, pp. 10–30, 1992.

[10] IPCC, "Good Practice Guidance for Land Use, Land Use Change and Forestry, J. Penman, M. Gytarsky, T. Hiraishi, T. Krug, D. Kruger, R. Pipatti, L. Buendia, K. Miwa, T. Ngara, K. Tanabe, and F. Wagner", Eds. Hayama, Japan: IPCC National Greenhouse Gas Inventories Programme, Institute for Global Environmental Strategies, 2003.

[11] E.R. Sharma, and J. Pukkala, "Volume Tables for Forest Trees of Nepal", Kathmandu, Nepal: Ministry of Forest and Soil Conservation, 84 pp, 1990.

[12] K.G. MacDicken, "A Guide to Monitoring Carbon Storage in Forestry and Agroforestry Projects", Arlington, Virginia: Forest Carbon Monitoring Program Winrock International Institute for Agricultural Development, 84 pp, 1997.

[13] IPCC, "Guidelines for National Greenhouse Gas Inventories",Hayama, Japan: National Greenhouse Gas Inventories Program, Institute for Global Environmental Strategies Publishing, No. 4, 2006.

[14] Ullah, M.R., and Al-Amin M., "Above- and below-ground carbon stock estimation in a natural forest of Bangladesh", Journal of Forest Science, 58(8), 372–379, 2012.

[15] S.E. Allen, H.M. Grimshaw, and A.P. Rowland, "Chemical analysis, In: Methods in Plant Ecology, P.D. Moore, and S.B. Chapman", Eds. Boston, USA: Blackwell Scientific Publications, pp. 285–300, 1986.

[16] T. Pearson, S. Walker, and S. Brown, "Sourcebook for Land-use, Land-use Change and Forestry Projects", Arlington, USA: Winrock International and the Bio-carbon Fund of the World Bank, pp. 18–35, 2005.

[17] S.M.I. Huq, and M.D. Alam, "A Handbook on Analyses of Soil, Plant and Water", Bangladesh: BACER-DU, University of Dhaka, 246 pp., 2005.

[18] S. Brown, "Estimating Biomass and Biomass Change of Tropical Forests", Rome, Italy: A primer, pp. 20–55, 1997.

[19] Achard, F., Eva, H.D., Mayaux, P., Stibig, H.J., and Belward, A.,"Improved estimates of net carbon emissions from land cover change in the tropics for the 1990s Glob",Biogeochem. Cycles, 18, 1029–42, 2004.

[20] Anup, K.C., Bhandari, G., Joshi, G.R., and Aryal, S.,"Climate change mitigation potential from carbon sequestration of community forest in mid hill region of Nepal", International Journal of Environmental Protection, 3(7), 33–40, 2013.

[21] Adugna, F., Teshome, S., and Mekuria, A., "Forest carbon stocks and variations along altitudinal gradients in Egdu Forest: Implications of managing forests for climate change mitigation", Science, Technology and Arts Research Journal, 2(4), 40–46, 2013.

[22] Mohammed, G., Teshome, S., and Satishkumar, B., "Forest carbon stocks in woody plants of Tara Gedam forest: Implication for climate change mitigation", Science, Technology and Arts Research Journal, 3(1), 101–107, 2014.

[23] R.d.Lasco, F.B. Pulhin, R.G. Visco, D.A. Racelis, I.Q. Guillermo, and R.F. Sales, "Carbon stocks assessment of Philippine forest ecosystems", Philips, Bogor, pp. 28–29, 2000.

[24] Brown, S., and Lugo, A.E., "The storage and production of organic matter in tropical forests and their role in the global carbon cycle",Biotropical, 14, 161–187, 1982.

[25] Munishi, P.K., and Shear, T., "Carbon storage of two afromontane rain forests in the eastern arc mountains of Tanzania", Journal of Tropical Forest Science, 16(1), 78–93, 2014.

[26] Mendoza-Vega, J., Karltun, E., and Olsson M., "Estimations of amounts of soil organic carbon and fine root carbon in land use and land cover classes, and soil types of Chiapas highlands, Mexico", Forest Ecology Management, 177, 191–206, 2003.

[27] Chowdhury, M.S.H., Biswas, S., Halim, M.A., Haque, S.M.S., Muhammed, N., and Koike M., "Comparative analysis of some selected macronutrients of soil in orange orchard and degraded forests in Chittagong hill tracts, Bangladesh", Journal of Forest Science, 1, 27–30, 2007.

Physico-chemical analysis of drinking water quality at Jigjiga City, Ethiopia

Adhena Ayaliew Werkneh*, Belay Zimbelachew Medhanit, Angaw Kelemework Abay, Jemal Yimer Damte

Department of Chemistry, College of Natural Science, Jigjiga University, Jigjiga, Ethiopia

Email address:
adhena1988@gmail.com (A. A. Werkneh)

Abstract: Water is one of the most important of all natural resources known on the earth. The safety of drinking water is important for the health. The safety of drinking water is affected by various contaminants which included chemical and microbiological. Such contaminants cause serious health problems. The physico-chemical analyses of drinking water quality at Jigjiga city, Somali Region, Ethiopia were studied. Temperature, pH, electrical conductivity, total dissolved solids, Total hardness have been determined along various water quality profiles. The experimental procedures were set according to the international drinking water standards set by WHO (1999). Average temperature, pH, electrical conductivity, total dissolved solids and total hardness values are 13.6 °C, 7.4, 1.143 S/m, 571.67 mg/l and 362.67 mg/l in $CaCO_3$ respectively. The result shows that, except the total hardness and electrical conductivity all the parameters fulfill the minimum and maximum permissible limit for drinking water guidelines.

Keywords: Physico-Chemical Parameters, Groundwater, Water Quality

1. Introduction

Water quality is a critical factor affecting human health and welfare. Studies showed that approximately 3.1% of deaths (1.7 million) and 3.7% of disability-adjusted-life-years [1] (54.2 million) worldwide are attributable to unsafe water, poor sanitation and hygiene [1]. The problem is the backward socio-economic development resulting in one of the lowest standard of living, poor environmental conditions and low level of social services [1], [2].

The functioning of an aquatic ecosystem and its stability to support life forms depend to a great extent, on the physico-chemical characteristics of its water. Physico-chemical parameters are highly important with respect to the occurrence and abundance of species. Ground water is by far more abundant than surface water and its quality is as important quantity. Water meant for drinking must therefore meet quality standards. Water quality is essentially determined by its physical and chemical characteristics [1]. Naturally, ground water contains mineral ions. These ions slowly dissolves from soil particles, sediments, and rocks as the water travels along mineral surfaces in the pores or fractures of the unsaturated zone and aquifer.

Good quality of water resources depends on a large number of physico-chemical parameters and biological characteristics. To asses that monitoring of these parameters is essential to identify magnitude and source of any pollution load [3].

The only source of water for drinking and agricultural purpose throughout the Ethiopian Somali regional state is the underground water. In nature, the hydrochemistry of the water sources were affected by a rich of metal ions and other physical factors that leads the water more polluted.

In this work, the physico-chemical analysis of drinking water quality was studied at Jigjiga city, Ethiopia. The main aim of this study was to carried out different physco-chemical parameters of water samples collected from different sites of Jigjiga City and to recommend the whether it is potable or not. The major water quality parameters considered for the examination in this study are like pH, temperature, total dissolved solids (TDS), dissolved oxygen (DO), total hardness and alkalinity.

2. Materials and Methods

2.1. Description of Sampling Locations

Jigjiga is a city in eastern Ethiopia and the capital of the Somali Region of the country. The city is located in the Jigjiga Zone approximately 106 km east of Harar and 628 km from Addis Ababa. This city has an elevation of 1,609 meters above sea level. Based on figures from the central statistical agencyin 2005, Jigjiga has an estimated total population of 98,076 of whom 50,355 are men and 47,721 are women. The climate of Jigjiga is a subtropical highland climate with the influence of mountain climate, with hot and dry summers and cold winters. This is attributed to the fact that Jigjiga is located on a plain surrounded by mountains and to its distance to the sea and its effects. The average temperature and rainfall range are between 25 to 31 °C and 11 to 712 mm respectively. The % humidity was in the range of 45 to 70% [Wikipedia).

2.2. Sampling and Preservation

Tap water samples were collected from three activesites across the Jigjiga city in April 8, 2014. All the samples collected were colorless and odorless. Samples were drawn with the aid of plastic drawer into three same types of polyethylene bottles i.e.1.5 L for physic -chemical parameters in the two sites. The plastics bottles were previous washed and soaked overnight with 5% HNO_3 solution [4]. To avoid any kind of contamination during sampling extra care was taken and the bottles were rinsed several times with the water being collected or filled. However, on-site analyses was comprised for temperature; electrical conductivity (EC) and pH were urgently determined when receive the sample in the laboratory because of their unstable nature. A 0.75 L of the sample was acidified with 1.5 ml concentrated HNO_3 (Analytical grade). Samples were then transported to laboratory and kept at 4^oC prior the time of analyses.

2.3. Materials and Chemicals

Apparatus and instruments: Polyethylene bottle, pH meter, oven, balance, glass bottles, Thermometer, desiccators, Conductivity Meter, Measuring Cylinder, Filter paper, Conical flask, Burette, Pipette, Beaker, Measuring flask.

Reagents and Chemicals: 0.1 MEDTA solution, EBT indicator, Basic buffer solution (NH_4OH and NH_4Cl), Standard hard water, given water sample.

2.4. Experimental Analysis

2.4.1. Temperature, Conductivity and pH Measurements

Parameter like temperature was determined in the field due to their unstable nature. A mercury filled centigrade thermometer calibrated from 0^oc to 100^oC was used for temperature measurements. The pH of the water sample was measured with a pH meter. The electrical conductivity was measured using digital conductivity meter. All analyses were carried out at a standardized laboratory using international regulatory methods. The evaluation of water quality was in accordance with regulatory standard. The approach ensures that the samples collected were tested in accordance with agreed requirements using competent personnel as well as appropriate equipment and materials. For physco-chemical analysis all the chemicals used were analytical grade. Temperature, pH and EC were measured by universal water analysis kit and manual method. Total hardness of water was estimated by complexometric titration with EDTA.

2.4.2. Determinations of Total Dissolved Solids

A clear dry glass beaker was taken (which was kept at 103^0Cin an oven for 1 hour) of 150ml capacity and put appropriate identification mark on it. Weight of the beaker was noted. Take a 100 ml. of sample and filter it through a double layered filter paper and collect the filtrate in a beaker. Place the beaker in an oven maintained at 103^0C for 24hours. After 24 hours, cool the beaker and weight. Find out the weight of solids in the beaker by subtracting the weight of the clean beaker determined in the first step. Total dissolved solids (TDS) can be determined as follows:

Calculating total dissolved solids concentration:

$$\text{Mg dissolved solids/l} = \frac{(A-B)}{mLofsample} \text{ x } 1000$$

A = weight of dried residue + dish, mg
B = weight of dish, mg

2.4.3. Total Hardness Determination of Water Samples Standardization of EDTA Solution with Standard Hard Water

20 ml of standard hard water was pipetted out in a washed conical flask. 5ml of basic buffer solution and 2-3 drops of Eriochrome Black-T indicator were added, the colour of the solution turns to wine red. This solution was titrated against EDTA solution which was taken in the burette until the color changes from wine red to clear blue at the end. The final reading of the burette was noted and the titration was repeated to get a concordant value

Estimation of Total Hardness of Given Water Sample:

20 ml of given hard water was pipette out in a washed conical flask. 5ml basic buffer solution and 2-3 drops of Eriochrome Black-T indicator were added, the colour of the solution turns wine red. This solution was titrated against EDTA solution taken in the burette until the colour changes from wine red to clear blue at the end. The final reading of the burette was noted and the titration was repeated to get concordant value. Finally using the analytical calculation, total hardness of given water sample was determined in terms of ppm of $CaCO_3$.

3. Results and Discussion

The mean values of different selected physico-chemical parameters have been tabulated below in the table.

Table 1. *Physco-chemical parameters for tap water from ground water sources at Jigjiga town*

No	Parameters	Values obtained			WHO value
		JJU tap water	04 kebele tap water	05 kebele water sample	
1	Temperature	13.5°c	13.6°c	13.8°c	Not exceed 15°c
2	pH	7.2	7.6	7.4	6.5 to 8.5
3	Total dissolved Solids	570 mg/l	580 mg/l	565 mg/l	500/1000 mg/l
4	Total hardness	350 mg/l of $CaCO_3$	378 mg/l$CaCO_3$	360 mg/l $CaCO_3$	500 mg/l
5	Electrical conductivity	1.12S/m	1.23S/m	1.08 S/m	3S/m

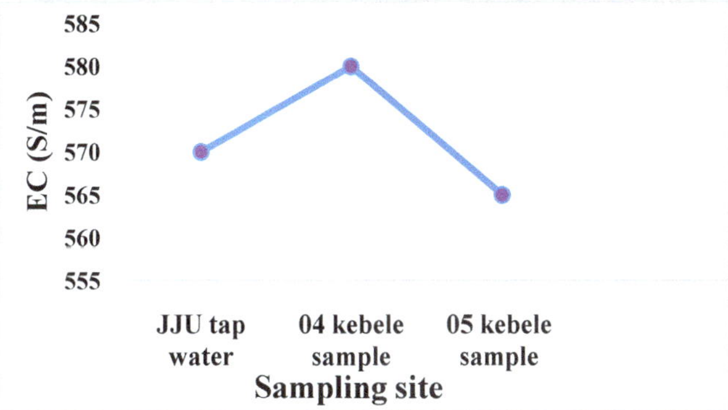

Fig 3.1. *Total dissolved solids (mg/l) values of various water samples*

Temperature: As shown in fig 3.2 below, the temperatures of the samples were noted at the sampling point itself. As indicated in graph below, the temperature was 13.5, 13.6 and 13.8°C in the three sites. During the present investigation, there was no great difference between the temperatures of the tap water from different sources of ground water and it is related to the WHO standards i.e. 15 °c.

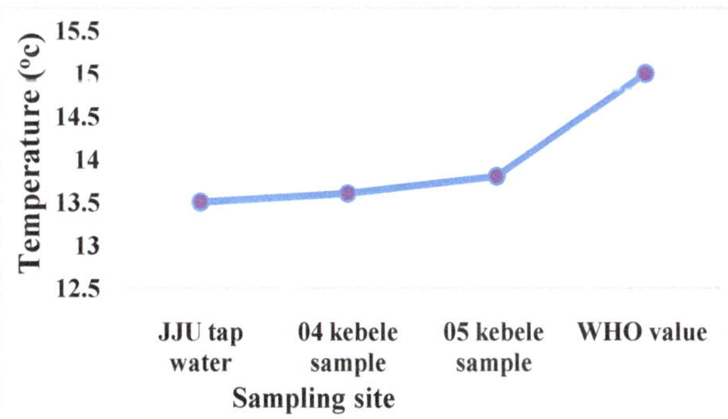

Fig 3.2. *Temperature (°c) values of various water samples*

Hydrogen Ion concentration (pH): is an important parameter which is important in evaluating the acid-base balance of water. Also it is the indicator of acidic or alkaline condition of water status. WHO has recommended maximum permissible limit of pH from 6.5 to 8.5.As shown in fig 3.3 the current investigation were 7.2, 7.6, and 7.4 which are in the range of WHO standards. The overall result indicates that the water sources are within the desirable and suitable range. Basically, the pH is determined by the amount of dissolved carbon dioxide [CO_2], which forms carbonic acid in water. According to [6], pH of ground water can also be lowered by organic acids from decaying vegetation, or the dissolution of sulfide minerals. The slight basic nature of the bore well water may be mainly due to the limestone basin of the all the locations.

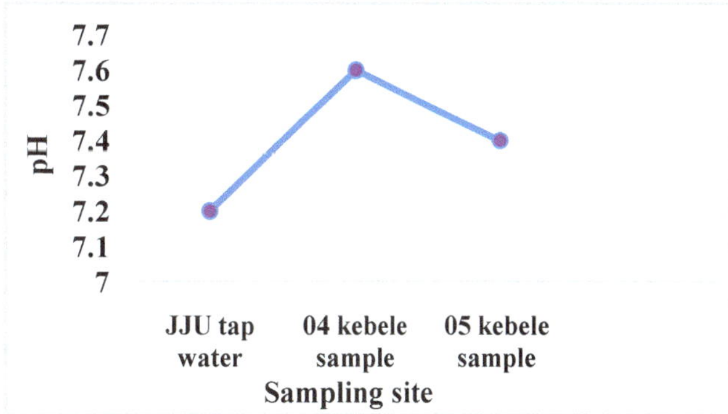

Fig 3.3. *pH values of various water samples*

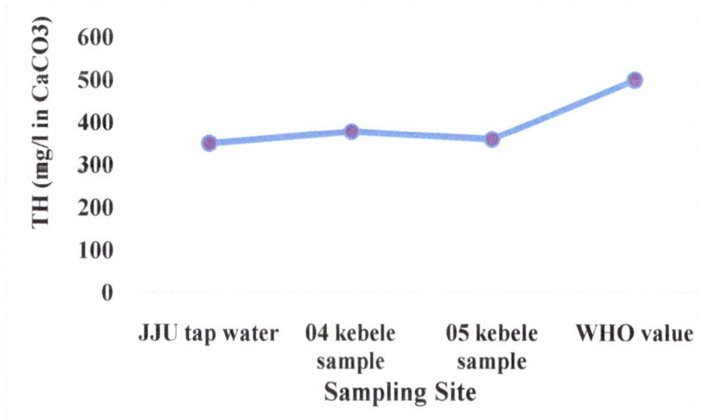

Fig 3.4. *Total hardness (mg/l in $CaCO_3$) values of various water samples*

Electrical Conductivity (EC): The result was shown in fig 3.5 below. EC measurement is an excellent indicator of TDS, which is a measure of salinity that affects the taste of potable water. The conductivity of water is a measure of capacity of a solution to conduct electrical current through it and depends on the concentration of ions and load of nutrients. Electrical conductivity is used to indicate the total ionized constituent of water. It is directly related to sum of the cations and anions. As most of the salts in water are present in ionic forms, they make water capable for conducting current. The conductivity, thus serves as a good and rapid measure of the total dissolved solids in water. EC measurement is an excellent indicator of TDS, which is a measure of salinity that affects the taste of potable water.

As shown in fig 3.5, the mean electrical conductivity (EC) of the tap water samples are 1.12, 1.23 and 1.08 respectively;

which is above the standard limit of 0.8 S/m.

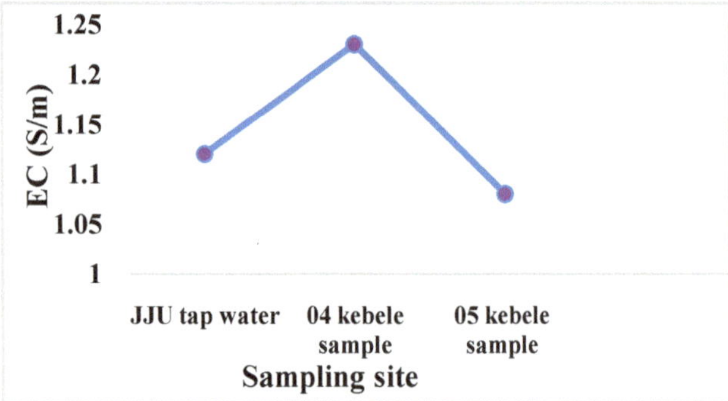

Fig 3.5. Electircal conductivity (S/m) values of various water samples

Thus the water has very high electrical conductivity, implying the presence of reduced level of ionic species.

Total Dissolved Solids (TDS): The result was shown in figure 3.1 above. This is the important parameter for the use of water. The water with high TDS value indicates that water is highly mineralized. Desirable limit for TDS is 500 mg/l and maximum limit is 1000 mg/l prescribed for drinking purpose. The concentration of TDS in present study is observed in the range of 647 and 537 mg/l. The mean total dissolved solids concentration in Jigjiga town tap water was found to be 592 mg/L and it is within the limit. High values of TDS in ground water are generally not harmful to human beings but high concentration of these may affect persons, who are suffering from kidney and heart diseases. Water containing high solid may cause laxative or constipation effects. According to [12], potable water should not contain more than 1000 mg/l of total dissolved solids (TDS).

Total Hardness: Hardness is a very important parameter in decreasing the toxic effect of poisonous element. As shown in fig 3.4, the hardness was found to be in the range of 350, 378 and 360 mg/L in $CaCO_3$ and with average value of 373.5 mg/l in CaCO3 in all the sampling sites. This value is within the hard level. None of the samples cross the maximum permissible limits of 500 of WHO and standards [13].

4. Conclusions

The result obtained during study was compared with ISI standards. Potable water is water safe enough to be consumed by humans or used with low risk of immediate or long term harm. The study assessed the evolution of water quality in tap water from a groundwater source of Jigjiga town. A comparative study of both type of bore well water and surface water was carried out by taking certain important parameters like temperature, pH, total dissolved solid, conductivity and total hardness. In this present investigation it was found that the maximum parameters were not at a level of pollution.

Acknowledgements

I am grateful to department of chemistry Jigjiga University for supporting necessary materials during the analysis.

References

[1] WHO, (2004): Guidelines for drinking water quality. 3rd edn. World Health Organisation, Geneva.

[2] Reza, R. and Singh. G. (2009). Physico- Chemical Analysis of Ground Water in Angul-Talcher Region of Orissa, India, J. Am. Sci. 5(5):53-58.

[3] F. A. Ushie & P. A Amadi, Chemical characteristics of ground water from parts of the basement complex of Oban massif and Obudu Plateau, South Eastern Nigeria.

[4] Rao, S. M. and Mamatha, P. (2004) Water quality in sustainable water management. Cur. sci., 87 (7): 942-947.

[5] Venkateswara Rao. B, (2011). Physicochemical analysis of selected ground water samples of Vijayawada rural and urban in Krishna district, Andhra Pradesh, India, International Journal of environmental sciences Volume 2, No 2, pp 710-714

[6] Narain, P. Singh and Pal P. Quality of underground water in the semi-arid zone, Ann. Of arid zone 1076, 15: 16.

[7] Manivaskam N, (2005). Physicochemical examination of water sewage and industrial effluent, 5th Ed. Pragati Prakashan Meerut.

[8] Sushma Jain and Monika Agarwal, (2012). Study on physico-chemical characteristics of ground water of various villages around Raisar, Journal of Chemical, Biological and Physical Sciences, Vol.2. No. 3, pp 1551-1555.

[9] Ramakrishnan and L. Leena Hebsibai, (2011). Studies on Physicochemical Analysis of Ground Water in Amaravathi River Basin at Karur (Tamil Nadu), India, Water Research & Development Vol. 1 No.1 pp 36-39.

[10] Shittu, O.B., Olaitan, J.O. and Amusa, T. S. Jkk. (2008). Physico-Chemical and Bacteriological Analyses of Water Used for Drinking and Swimming Purposes in Abeokuta, Nigeria. African Journal of Biomedical Research, Vol. 11, 285 -290

[11] M. Soylak, F. Armagan Aydin, S. Saracoglu, L. Elci, M. Dogan Xfghj. (2002). Chemical Analysis of Drinking Water Samples from Yozgat, Turkey. Polish Journal of Environmental Studies Vol. 11, No. 2, 151-156

[12] S Sasikaran, K Sritharan, S Balakumar, V Arasaratnam. (2012). Physical, chemical and microbial analysis of bottled drinking water. Ceylon Medical Journal. Vol. 57, No. 3, 111-116.

[13] Kawther F, Alwakeel S. Mineral and microbial contents of bottled and tap water in Riyadh, Saudi Arabia. Middle-East Journal of Scientific Research 2007; 3: 151-6.

Preconcentration and Determination of Traces of Heavy Metals with Polymer Chelating Sorbents in the Analysis of Natural and Waste Water

Abdunnaser Mohamed Etorki[1, *], Ibrahim Salem Shaban[2]

[1]Department of Chemistry, University of Tripoli, Tripoli, Libya
[2]Department of Environmental Science and Engineering, Libyan Academy, Tripoli, Libya

Email address:
a.etorki@ch.uot.edu.ly (A. M. Etorki)

Abstract: Monitoring the concentrations of heavy metals in natural and waste water at and below the level of their maximum permissible concentrations is an urgent environmental problem. Hence, new procedures for the preconcentration of heavy metals with their subsequent determination by different methods are required. Along with other sorbents, significant attention is attracted to polymer chelating sorbents, which provide individual or group extraction of trace elements, eliminate matrix effects, and provide high concentration factors. The effect of the various parameters such as electrochemically and chemically synthesis methods, physical oxidation state of the polymer, polymer thickness, solution pH and metal ion concentration on the adsorption, kinetics and efficiency were investigated. The results showed a vary broad concentration range of the heavy metals from (0.05 to 10 mg/L) can be adsorbed on different kinds of polymers at different pH values and different efficiently. The adsorption capacity of the polymer to different concentrations of heavy metals was evaluated as the milligram of metal ions by one gram of various forms of the polymer. The DC conductivity measurements were also employed on the solid polymer before and after adsorption of metal ions. The experimental adsorption date was fitted to different mathematical isotherms to estimate the binding constant of heavy metals with the polymer in both single and mixed ion solutions. The method provides the extraction of analytes from natural water of complex composition containing high concentrations of alkali, alkaline-earth and other elements and is characterized by rapidly, selectivity, low detection limits, and a high reproducibility of the results. The relative standard deviation is 2-4%. The technique was test with real waste water samples.

Keywords: Polymers Sorbents, Heavy Metals, Preconcentration, Adsorption Isotherms, Mixed Ion Solutions, Selectivity

1. Introduction

The selective adsorption and removal of metal ions in aqueous environmental samples prior to determination have received increasing importance in environmental evaluation and protection in recent years. Adsorption is the concentration of a substance at the surface or interface [1]. The adsorption at a surface or interface is largely as a result of binding forces between atom, molecules, and ions of the adsorbate on the surface [2]. The commercial adsorbents used today for the removal of heavy metals from solutions include a variety of clays, activated carbon, gels, alumina, silica, zeolites, and other resinous material. Spectrometry techniques, including atomic absorption spectrometry (AAS) and inductively coupled plasma optical emission spectrometry (ICP-OES) as well as inductively coupled plasma mass spectrometry (ICP-MS) are the most commonly used analytical methods for trace metals determination. In practice for complex matrices, however, the accurate determination of trace metals is frequently very difficult due to the presence of interfering effects, comprising spectroscopic and non-spectroscopic interferences. Moreover, in some cases, analyte concentration might be to low to be analysed directly. For the purpose of separation of metal ions either for purification or enrichment, various kinds of different sorbents have been developed. This has been the subject of numerous papers and review articles related to synthesis,

Characterization and metal ion uptake studies.

Intrinsic conducting polymers with conjugated double bonds have been attracted much attention as advances materials due to their multifunctional properties and potential applications. They can be easily synthesized in both aqueous and organic solvents, chemically and electrochemically. Recently, the use of aniline, pyrrole and phenyl-based polymers as sorbents in solid phase extraction technique has been demonstrated [3–5]. Polypyrrole was applied for extraction of phenol and chlorophenols [6], while applicability of polyaniline was investigated for the preconcentration and speciation of mercury [7] as well as for determination of Cd, Cu, Pb and Sb in potassium iodide medium in biological matrices [8]. The microparticles of sulfodiphenylamine and diaminonaphtalene copolymers were tested for adsorption of Ag(I) [9], while poly(m-phenyldiamine) microparticles were used for recovery of lead ions [10].

Due to be rapidly applications of polymer chelating ligands in environmental analytical chemistry, the present study is undertaken with the following specific objectives:

1. to examine the performance and effectiveness of electrochemically and chemically synthesized poly-(2-isopropyl aniline) in removal of Cd(II), Pb(II), and Zn(II) ions by adsorption from waste water.
2. to determine the effect of initial metal concentration, pH and temperature on the adsorption capacity of the electrochemically and chemically synthesized poly-(2-isopropyl aniline).
3. to study the applicability of the Langmuir, Freundlich and Frumkin isotherms;
4. to provide additional information on the adsorption of the relevant heavy metal cations by the electrochemically and chemically synthesize poly-(2-isopropyl aniline); and
5. to understand the mechanism of adsorption of the relevant metal cations by the electrochemically and chemically synthesize poly-(2-isopropyl aniline) based on parameters such as adsorbent capacity, free energy change, enthalpy change and entropy change.

2. Experimental

2.1. Material and Methods

Deionized doubly distilled water was used to prepare all solutions. Stock solutions of metal ions were prepared from their nitrate salts $Cd(NO_3)_2$ $.4H_2O$, $Zn(NO_3)_2$ $.6H_2O$ and $Pb(NO_3)_2$ (Fluka). All metal working standards were prepared by dissolving appropriate amount of metal salt into deionized water. Nitric acid and sodium hydroxide were both analytical reagents grade (Merck). 2-isopropyl aniline was used without further purification and supplied from (Aldrich).The solution of the polymer was prepared in deionized doubly distilled water.

2.2. Preparation of the Polymer (Adsorbent)

poly-(2-isopropyl aniline) with different molecular weights: 90,000, 120,000 and 180,000, were synthesised by the oxidative polymerization of 2-isopropyl aniline in aqueous HCl solution using potassium dichromate as the oxidant, 11.5 mmol of monomer and 2 mmol $K_2Cr_2O_7$ were dissolved in 10 ml of 1.5M HCl solution, separately. The $K_2Cr_2O_7$ solution was added drop wise to the stirred monomer solution at room temperature. The black precipitate was filtered, after a 1h reaction period, and then washed with 1.0M HCl solution until the filtrate was clear. The polymer was dried under vacuum at 80°C for 48h.

2.3. Apparatus

An atomic emission spectrometer (ICP-OES, Perkin Elemer, Optima 5300, USA) with a wipro Dell computer was used. A WTW digital pH/mV meter (model pH-340-A/SET2) with a combination glass electrode (WTW D82362 Weiheim) was used for all pH measurements. A Damon model centrifuge equipped to timer and speed control was employed for centrifuging of samples. The DC conductivity measurements were carried out on compressed pellets by using a tow probe technique at room temperature.

2.4. Sampling

The tannery wastewater samples were collected in pre-washed (with detergent, deionized doubly distilled water, dilute HNO_3 and deionized doubly distilled water respectively) polyethylene bottles from Abe Kammash Industrial Region (Zwara)-Libya. The samples were filtered through Millipore cellulose nitrate membrane of pore size 0.45 μM. The samples were stored in 1 litre polyethylene bottles and acidified to 1% with nitric acid and were subsequently stored at 4°C in a refrigerator.

2.5. Adsorption and Analytical Procedure

The adsorption of heavy metals on poly-(2-isopropyl aniline) was carried out using the batch method. Batch adsorption experiments were conducted using 100 mg of adsorbent with 20 mL of solutions containing heavy metal ions of desired concentrations at different temperature 25 °C and pH values (3.0, 5.0, 6.0 and 7.0) in 50 mL glass stoppered flasks. The pH of solutions was adjusted by using dilute NaOH or HCl solutions before adsorption analysis. The flasks were shaken in a temperature-controlled water bath for 30 min with the mixing rate of 350 rpm and the solutions containing heavy metals were filtered through Whatman filter paper (No. 42) after the contact time of 30 min. The exact concentration of metal ions and filterable metal concentrations were determined by (ICP-OES).

The maximum loading capacity of the polymer was determined by applying the above procedure with higher concentration of analyte. To obtain a completely metal-free polymer, the procedure was repeated four times. To regenerate the polymer, i.e., to obtain a metal-free polymer again in order to use further in other experiments, after the full procedure was completed, a 3 ml of 4 mol L^{-1} HNO_3 was added on to the polymer and analyte elements which remained from a previous procedures. The polymer was used in other experiments again.

2.6. Constants

The percent adsorption (%) and distribution ratio (K_d) were calculated using the equations

$$\% \text{ Adsorption} = [(C_i - C_f)/C_i]*100 \quad (1)$$

where C_i and C_f are the concentrations of the metal ion in initial and final solutions, respectively, and

$$K_d = (\text{amount of metal ion in adsorbent})/(\text{amount of metal ion in solution}) \times V/m \text{ (mL/g)}, \quad (2)$$

where V is the volume of the solution (mL) and m is the weight of the adsorbent (g) [11].

3. Results and Discussion

3.1. Effect of pH on Recoveries of Analytes

The effect of pH on the recoveries of the lead, zinc and cadmium is shown in Figure 1. If the pH of the sample and polymer mixture was higher than 6.0, cadmium, zinc and lead were quantitatively recovered with high precision. Since quantitative recoveries were reproducibly obtained. It can be concluded that cadmium ions were completely bound to the polymer and there was no need to wait during steps of the procedures. The effect of initial cadmium concentration in the aqueous solution on the removal of Cd(II) by the polymer is shown in Figure 2. As expected the removal capacity of the polymer increased with increasing initial concentration. The removal of Cd(II) increased from 10.0 % to 99.2 % .

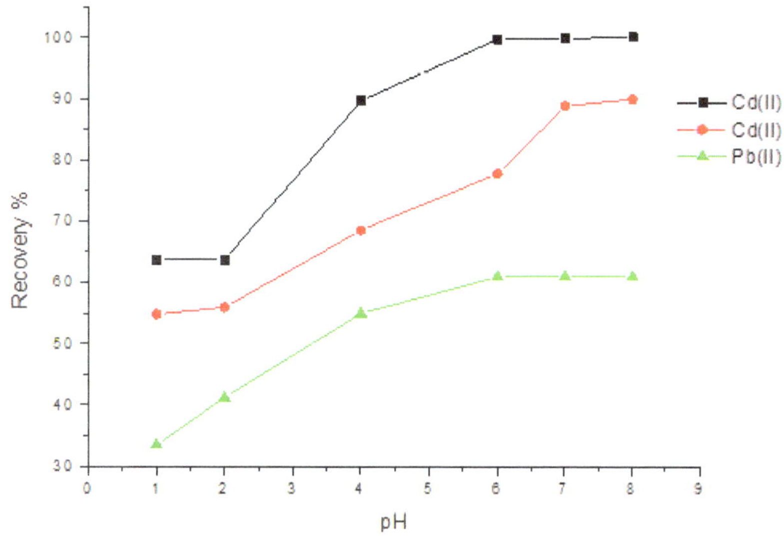

Figure 1. *Effect of pH on removal of cadmium.*

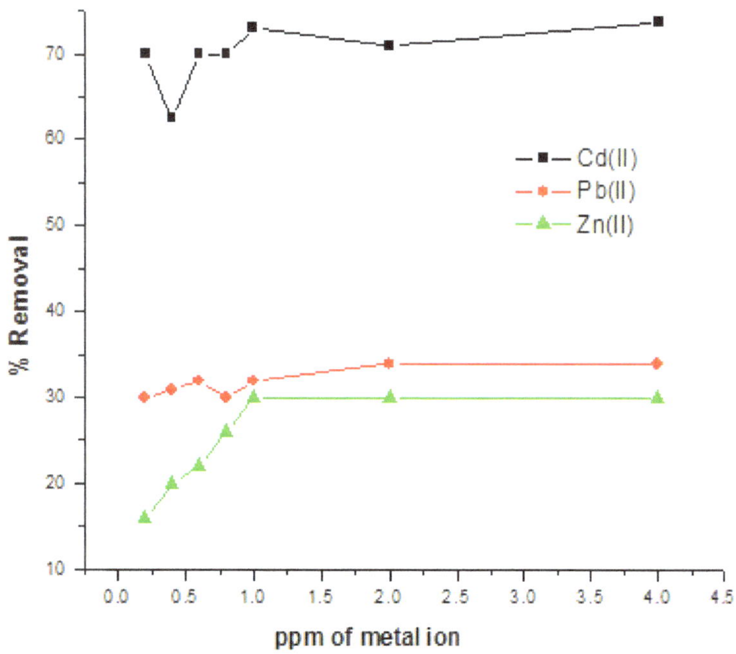

Figure 2. *Effect of initial concentration on removal of cadmium.*

3.2. Adsorption Isotherm

The experimental adsorption isotherms were fitted to mathematical Langmuir and Freundlich isotherms models. For the Langmuir model, the following equation was used:

$$q_{eq} = q_m \, C_{eq}(K_d + C_{eq}) \qquad (1)$$

where q_{eq} and C_{eq} correspond to the milligrams of Hg adsorbed per one gram of the polymer and the residual Hg concentration in the solution when in equilibrium and K_d and q_m are Langmuir constant and maximum capacity of adsorption for this model. Freundlich isotherm constants were calculated with the equation:

$$\ln q_{eq} = n \ln C_{eq} + \ln K_F \qquad (2)$$

where K_F and n are the Freundlich constants characteristic of the polymer system. The Langmuir model is valid for modelling monolayer adsorption onto a homogenous surface with constant adsorption energy; the Freundlich equation posits a heterogeneous surface and considers that molecules attached to a surface site will have an effect on the neighbouring sites. The adsorption data from the different initial concentrations of Cd(II) , Pb(II) and Zn(II) were analyzed in terms of both Langmuir and Freundlich equations at pH= 6.4 and 25°C, and the values obtained for the respective constants are shown in Table 1.

Table 1. *Langmuir and Freundlich constants for Cd(II) Pb(II) and Zn(II) / poly-(2-isopropyl aniline).*

Metals	q_m(mg g^{-1})	K_d (L mg^{-1})	1/n (L g^{-1})	K_F (mg g^{-1})
Cd(II)	77.2	0.412	0.243	10.04
Pb(II)	13.3	0.180	0.122	4.22
Zn(II)	11.7	0.164	0.101	3.58

The Langmuir isotherm equation ($r^2 = 0.9973$) seemed to describe better the adsorption process of Cd(II) by poly-(2-isopropyl aniline) than the Freundlich isotherm ($r^2=0.9910$). the determination coefficients (r^2), used to describe the fitness between the experimental and theoretical models, were high for both equations (Figure 3), and the applicability of both Langmuir and Freundlich isotherms implied that both monolayer adsorption and heterogeneous surface conditions exist under the experimental conditions used here, with the latter process prevailing. According to reference [9], if 0.1< 1/n <1.0, the adsorption of an adsorbent on the polymer is favourable. In this study 1/n was 0.33, indicating that the polymer can be used effectively for removal of Cd(II) from aqueous solutions.

3.3. DC Conductivity Measurements

The DC conductivity measurements were also employed on the solid polymer before and after adsorption of cadmium ions. The results are shown in table 2. The DC conductivity of the poly(2-isopropyl aniline) alone was 0.9 x10^{-6} ohm^{-1} cm^{-1} The results shown that the DC conductivity increases with increase concentration of cadmium ions in solution after the polymer adsorbed cadmium ions from single metal ion solutions.

Table 2. *DC conductivities for poly(2-isopropyl aniline) after adsorbed cadmium ions from solution.*

Ppm of Cd(II)	DC Conductivity x$10^{-6}$$Ohm^{-1}$ cm^{-1}
0.2	19.2
0.4	19.4
0.6	40.0
0.8	37.5
1.2	57.5
1.5	69.6
2.0	79.5
2.5	87.3
3.0	87.4
3.5	99.3
4.0	97.4
4.5	97.2

3.4. Removal of Cd(II) Ions from Wastewater Samples

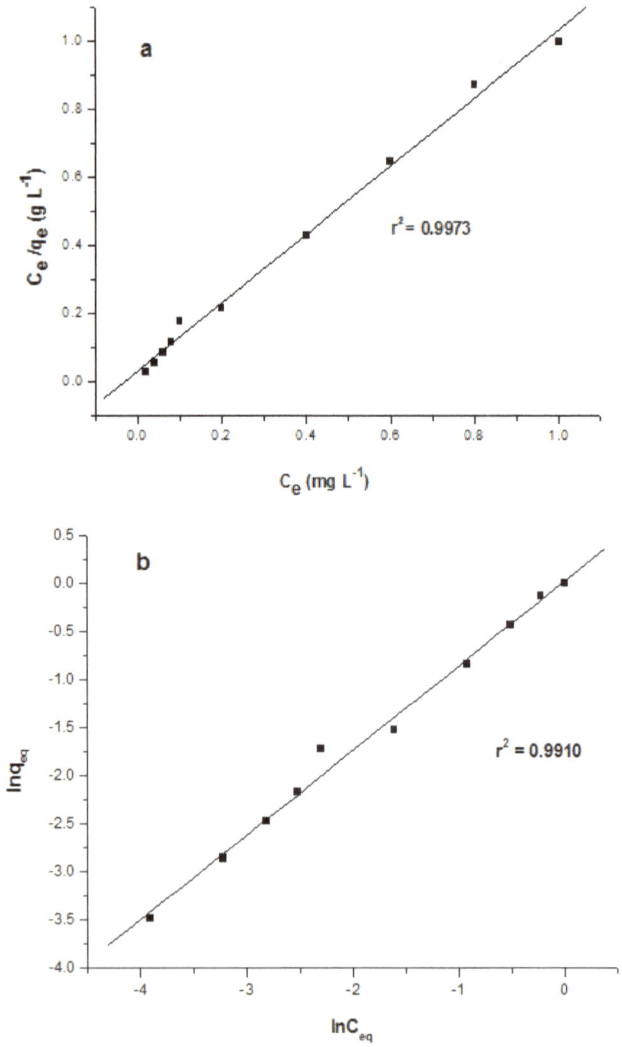

Figure 3. *Cd(II) adsorption isotherm based on the (a) Langmuir and (b) Freundlich models.*

The wastewater samples were firstly characterised by ASTM and ISO standard methods. The wastewater samples were found to contain 0.1-5.0 and 0.05-2.0 mg l^{-1} of cadmium (II), respectively. Six samples were digested with nitric acid, followed by pH adjustment and treatment with the poly-(2-isopropyl aniline) under the optimized conditions described above .Assay of mercury in the final effluents indicates 97.2-99.5% removal of the original concentrations. The mean standard deviation is ±1.3%. The adsorption capacities were 70.5 mg/g for for Cd(II) , 15.3 mg/g for Pb(II) and 12.4 mg/g for Zn(II).

4. Conclusions

Poly-(2-isopropyl aniline) could be successfully applied for the preconcentration and separation of cadmium, lead and zinc ions from aqueous solutions prior to their analysis by ICP-OES. The suggested method has been applied to different wastewater samples with the satisfactory precision and accuracy.

References

[1] C. Kadirvelu, C. Faur-Brasquet, P. Le Cloirec, "Removal of Cu(II), Pb(II) and Ni(II) by Adsorption onto activated carbon cloths" Langmuir 16, pp. 8404–8409 (2000)

[2] D. Sedlak, J.T. Phinney, W.W. Bedsworth, "Strongly complexed Cu and Ni in wastewater effluents and surface runoff," Environ. Sci. Technol, vol. 31, pp. . 3010– 3016, 1997.

[3] R.A. Issac, L. Gil, A.N. Cooperman, K. Hulme, B. Eddy, M. Ruiz, K. Jacobson, C. Larson, O.C. Pancorbo, "Corrosion in drinking water distribution systems: A major contribution of copper and lead to wastewaters and effluents," Environ. Sci. Technol, vol.31, pp. 3198–3203, 1997.

[4] M.A. Schneegurt, J.C. Jain, J.A. Menicucci Jr., S.A. Brown, K.M. Kemner, D.F. Garofalo, M.R. Qualick, C.R. Neal, C.F. Kulpa Jr., "Biomass by-products for The remediation of wastewaters contaminated with toxic metals," Environ. Sci.Technol,. vol 35, pp.3786–3791, 2001.

[5] C.F.Poole, New trends in solid-phase extraction," Trends in Analytical Chemistry," vol.22, pp.362-373, 2003.

[6] R. Rao, T.R.Kala and S.Danial, "Metal ion-imprinted polymers novel materials for selective recognition of inorganics," Analytical Chim Acta, vol 578, pp.105- 116, 2006.

[7] A. Muck and A. Svatos.,"Chemical modification of polymeric microchip devices," Talanta, vol 74, pp. 333-341, 2007.

[8] A.C.Sahayamg,"Determination of Cd, Cu, Pb, and Sb in environmental samples by ICP-AES using polyanline for separation,"Fresenius J Anal Chem, vol 362, pp.285-288,1998.

[9] J.R.Fischer, D.Pang and T.S.Beatty, "Silica-polyamine composite materials for heavy metal ion removal, recovery and recycling. II. Metal ion separations from mine wastewater and soft metal ion extraction efficiency," Separation Science and Technology, vol 34, pp.3125-3137, 1999.

[10] W.A.Ibrahim, L.I.AbdoAli, A.Sulaiman, M.M.Sanagi, H.Y.Abdoul-Enein, "Application of solid phase extraction for trace elements in environmental and biological samples," Critical Reviews in analytical Chemistry, vol 44, pp. 233-254, 2014.

[11] M.A.Oskooie,M.M.H.Heravi, "Preparation and application of poly(2-aminothiophenol)MWCNTs nanocomposites for adsorption and separation of cadmium and lead ions via solid phase extraction", J. Hazard. Mater, vol 203-204, pp.93-100, 2012.

Pollution of the Chari River in the Slaughterhouse Farcha (N'djamena) in Chad

Tchadanaye New Mahamat[1], Noumi Guy Bertrand[2,*], Paloumi Yabe[1,2]

[1]Department of Chemistry, Faculty of Exact and Applied Sciences, University of N'djamena, N'djamena, Chad
[2]Department of Chemistry, Faculty of Science, University of Ngaoundere, Ngaoundere, Cameroon

Email address:
gnoumi@yahoo.fr (N. G. Bertrand)

Abstract: This study was carried out to evaluate the influence of the discharge of effluents Farcha slaughterhouses pollution of the Chari River. To overcome this, water and sediment samples were taken upstream to the outlet and downstream of Farcha slaughterhouse. These samples were subjected to some physicochemical analysis. Slaughterhouse effluent showed a Chemical Oxygen Demand (COD) ranging from 96 $mg.L^{-1}$ to 100 $mg.L^{-1}$; Biochemical Oxygen Demand (BOD_5) 54 $mg.L^{-1}$ to 70 $mg.L^{-1}$ and the suspended solids (TSS) 64 $mg.L^{-1}$ to 108 $mg.L^{-1}$. The physicochemical characterization of water samples Chari gave a change in hydrogen potential (pH) between 6 to 7; a temperature ranging from 28.35°C to 28.70°C with an electric conductivity ranging from 50 $\mu S.cm^{-1}$ to 129 $\mu S.cm^{-1}$. The dissolved oxygen levels in water shows an appreciable oxygenation (5.02 - 7.37 mg $O_2.L^{-1}$). As the mineral substances, the chloride ions (Cl^-) vary from 9.05 $mg.L^{-1}$ to 29.25 $mg.L^{-1}$; ferrous ions (Fe^{2+}) of 0.48 $mg.L^{-1}$ to 11.10 $mg.L^{-1}$; sulphate ions (SO_4^{2-}) of 21.50 $mg.L^{-1}$ to 38 $mg.L^{-1}$; phosphate ions (PO_4^{3-}) of 20.45 mg/L to 84.35 mg/L; nitrate ions (NO_3^-) of 15 $mg.L^{-1}$ to 39.50 $mg.L^{-1}$. In opposite, the concentrations of nitrite (NO_2^-) increased from 7.45 $mg.L^{-1}$ to 13 $mg.L^{-1}$; that of ammonium ions (NH_4^+) 1.49 $mg.L^{-1}$ to 4.95 $mg.L^{-1}$. The concentrations of dissolved ions showed higher values in July than in September. We observed significant variations in these parameters, some of which values were in excess of the WHO standard. However; the rate of COD/BOD_5 = 1.67 show that these wastewater Farcha slaughterhouses are biodegradable.

Keywords: Chari River, N'djamena, Slaughterhouse, Effluent, Physicochemical Parameters

1. Introduction

Generally, the aquatic environment quality is significantly affected by several natural and anthropic activities which provide many contaminants [1-4].

The Chari River (1200 km) undergoes physical, chemical and biological constraints through several cities in Chad. This river is the main tributary of Lake Chad, which is experiencing a significant crowding in recent years. The lake area who was 25,000 km^2 in 1960 decreased to 1,500 km^2 in 2000 [5]. At the city of N'djamena, wastewater is discharged directly into the river or in the gutters drain rainwater draining the Chari River. These domestic water from hotels, hospitals, households and other domestic purposes. Also, diseases such as gastroenteritis and typhoid were detected among residents because of the downstream water consumption from the river and slaughterhouses Farcha [6].

Slaughterhouses Farcha is a typical example of industry where the water is used for washing the by-products (offal) and disposal of waste (faeces, debris paunch and blood). Built in 1958, the slaughterhouse Farcha (La Société Moderne des Abattoirs SMA/AFF) has a capacity of 110 tons of meat per day and daily production is about 200 cattle, 200 to 300 small ruminants and 30-80 camels [7]. Located on the river, the slaughterhouse Farcha is connected to the Chari River through its outflow channel. The effluents of these slaughterhouses are then discharged into the Chari River without prior treatment. Thus the water system is subject to important risks of pollution that threatens water quality. This study thus proposes to contribute to the assessment of pollution in the river Chari by the slaughterhouse Farcha (N'djamena) in Chad.

2. Materials and Methods

2.1. Overview of the Study Area

The Chari River (1200 km) is our environmental space and

study more precisely the area from the 27 camp on the side of the right bank of the river alongside the zootechnical laboratory Farcha, slaughterhouse outfall Farcha up Nursery Milezi. This area is included in the first district of the city of N'djamena. The Chari River, very heterogeneous, with an average flow of 537 $m^3.s^{-1}$ is weak and unstable because of natural hazards that describe climate variability (evaporation, climate change, inter annual variation in rainfall distribution), and intensification of market garden [8]. All industrial effluents, sewage and household waste are discharged into the river without proper treatment. *La Société Moderne des Abattoirs* of Farcha is one of the food industry do not have wastewater treatment plant that discharges its effluents directly into the liquid Chari River.

2.2. Sampling

The selection of sample points was made taking into account the location of the wastewater discharge point slaughterhouses Farcha and various identified activities (domestic sewage, industrial, agricultural, leaching) in the area. Thus, three points (AMO, EXU and AVA) upstream to the outlet and downstream slaughterhouses were selected for this study. These points have been designated as follows: P1 (AMO): Sampling point upstream slaughterhouses, 100 m from the outfall; P2 (EXU): Outlet slaughterhouses Farcha and P3 (AVA): Downstream sampling point slaughterhouses, located 100 m from the outfall.

Figure 1 shows the map of the study area and the location of sampling points.

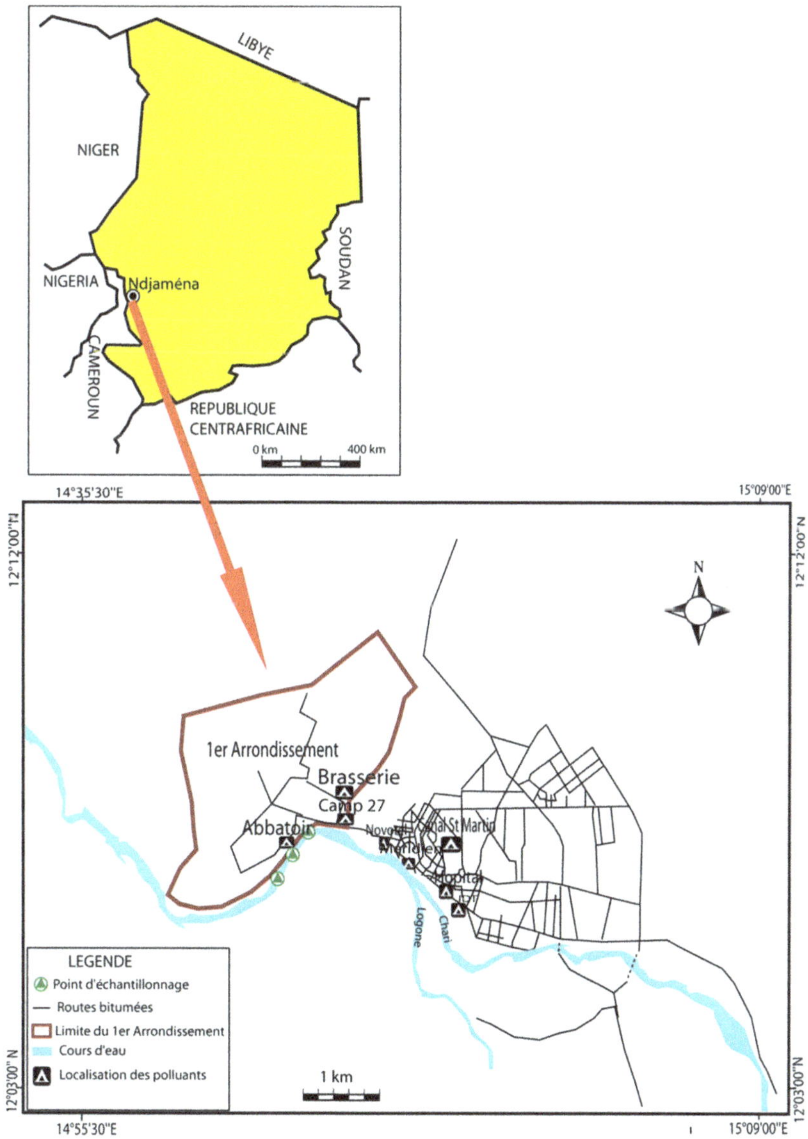

Figure 1. Study area and location of sampling points.

Two sets of samples were taken during the study, one in low water period (July 2015) and the other in winter period (September 2015). The water samples were collected in polyethylene bottles of 250 mL, previously washed and rinsed with distilled water. All samples were stored at low temperature (4°C) and transported to the laboratory. These water samples were filtered using a Whatman filter paper (1.2 µm, GF/C). 50 mL of each sample was acidified with concentrated hydrochloric acid (0.5 mL) and stored in vials until analysis [9]. About 100 g sediment were collected in the same water sampling points and stored in polyethylene plastic and were then dried 24 hours in the laboratory. The dry sediment was ground, sieved through a sieve with 1 mm mesh. A mixture of 10 g of sediment and 50 mL of ammonium acetate was subjected to stirring for 1 hour, then filtered through a filter paper (Whatman, GF/C, 47 mm) in the 100 mL flask. For each sample (water, sediment) a triplicate of samples was taken.

2.3. Analytical Methods

The physicochemical study of water was done on sixteen parameters: temperature, hydrogen potential (pH), conductivity (EC), turbidity and dissolved oxygen have been measured in the field. The suspended solids (TSS), nitrogen forms (NO_3^-, NO_2^-, NH_4^+), phosphate (PO_4^{3-}), sulphates (SO_4^{2-}), chloride (Cl^-), iron (Fe^{2+}) and organic pollution (BOD_5, COD) was carried out in Laboratory of Water and Environment of the Faculty of Exact Sciences and Applied of the University of N'djamena following the technics and methods described by Rodier [9]. The pH, conductivity and temperature of the water was measured using a pH meter Cyberscan-pH110 brand fitted with a thermometer, and a conductivity meter. Dissolved oxygen was measured using a HACH Oximeter. Phosphates, ammonium ions, nitrates, nitrites, sulphates, chlorides, iron were measured by the photometer-7100 (Palintest). The chemical oxygen demand, suspended solids were measured by spectrophotometer HACH brand. The turbidity was read by the photometer-7100. The biochemical oxygen demand was measured by a measuring BOD in a thermoregulatory AQUALYTIC under temperature of 20°C.

3. Statistical Analysis

The results of this work have been analysed by the software Microsoft Office Excel 2007. The principal components analysis (PCA) was performed using the Statgraphics software and XLSTAT 2007 version 8.04 software.

4. Results and Discussion

4.1. Water Temperature

It's important to know the water temperature with good accuracy because it is an important ecological factor. Temperatures range between 28.35°C and 28.64°C in July and between 28.45°C and 28.70°C for the month of September (Figure 2). The high value is EXUj, this can be explained by the release of water from the slaughterhouse in contact with the waters of the river Chari. In September the maximum is AMOs and this could be reflected in releases of the upstream brewery slaughter or time of sampling. These results are close to the results of previous work: 27°C to 29,14°C [8]. However, the values obtained are less than 30°C (WHO Standard) [10].

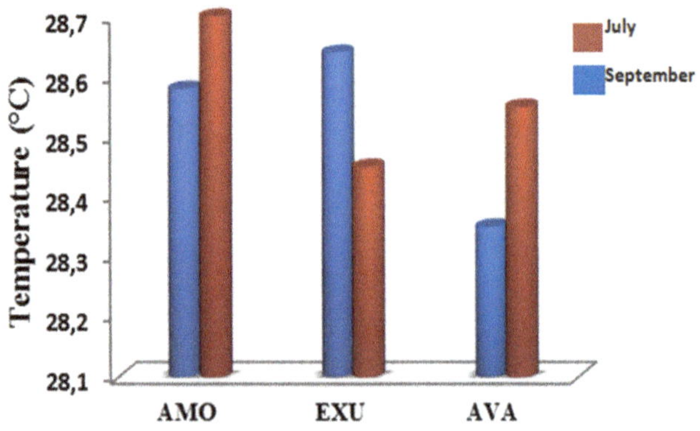

Figure 2. *Temperature change.*

4.2. Hydrogen Potential

This parameter determined a large number of physical and chemical equilibria, and depends on many factors, including temperature and the source of the water, it represents an important indication as regards the aggressiveness of the water (Suitability dissolve the limestone) [11, 12]. The pH values (Figure 3) ranges from 7.10 to 7.28 in July and 6.96 to 7.15 in September. Regarding the series of July, the pH decreases from upstream to downstream in AMOj (7.28). The same decrease is observed in September with a small point AVAs sudden jump (7.15). These variations may be due to the leaching activity in the zone, using basic products and the other hand to the photosynthetic activity which releases CO_2 resulting in a pH increase. The results obtained at different sampling points seem to be in agreement with the water agency and the WHO limit (pH < 8.5). These results are comparable to those of work [8] on the Chari River (7.03 to 8.14).

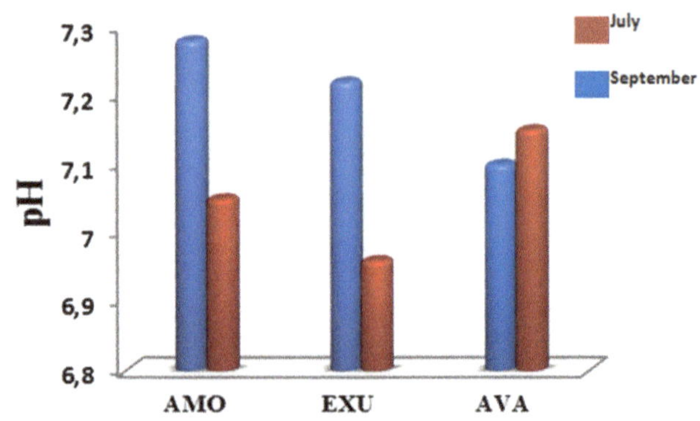

Figure 3. *Change in pH.*

4.3. The Electrical Conductivity

The conductivity measurement is a good assessment of the degree of mineralization of water which is each ion by its concentration and specific conductivity. The recorded average values (Figure 4) fluctuate between 125.25 µS/cm (EXUj) and 125.15 µS/cm (AVAj) for the series in July. There is a drop in conductivity in September, while maintaining a growth upstream respectively downstream 50.90 µS/cm (AMOs); 53.80 µS/cm (EXUs); and 69.70 µS/cm (AVAs). The higher values of conductivity measured in July may be explained by the increase in temperature causes an increase in dissolved minerals from the IDT (Chad printing press) and zootechnical laboratory by water evaporation phenomenon. On the one hand, the relatively low values in September can be justified by the decrease in temperature. Furthermore the dilution effect during the flood period, waters Chari would have led to a decrease in the concentrations of dissolved ions. The values obtained are below the WHO/Chad Directive (2007) set to 1000 µS/cm for surface waters and confirm previous work [8] conducted on the Chari River. However, higher values (360 µS/cm) were obtained from slaughterhouse waste of Kenitra [13].

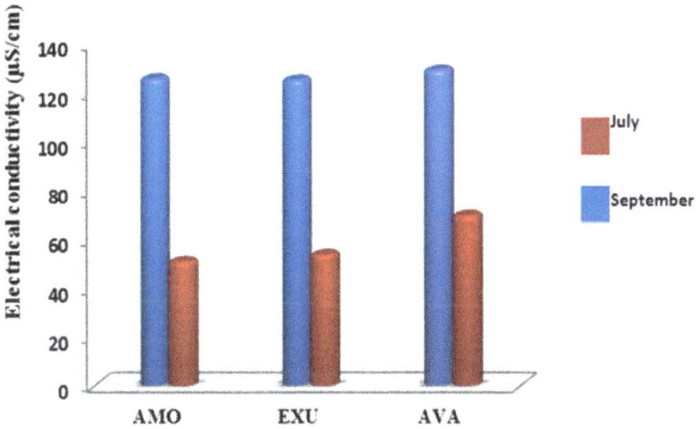

Figure 4. *Change in electrical conductivity.*

4.4. Turbidity

Turbidity is a sensory parameter and an expression of the optical properties of a water absorbing and/or diffusing light is due to the presence of finely-divided suspended solids (clays, silts) [14]. Analysis of variance of the turbidity values of water samples summarized in figure 5 shows that there is a significant difference ($p < 0.05$) between the two sample sets. High values (Figure 5) of the turbidity of the different sampling points are observed at the outlet in July (415 NTU) until September (127.50 NTU). This turbidity is a function of seasons and sampling periods. Indeed, high turbidity were observed water withdrawn in July which marks the beginning of the Chari flood. The increased turbidity results from rainwater run-off phenomenon and that of re-suspension of sediments previously deposited along the Chari River. They are much more marked at the outlet of slaughterhouses due to suspended solids discharges from the slaughterhouses.

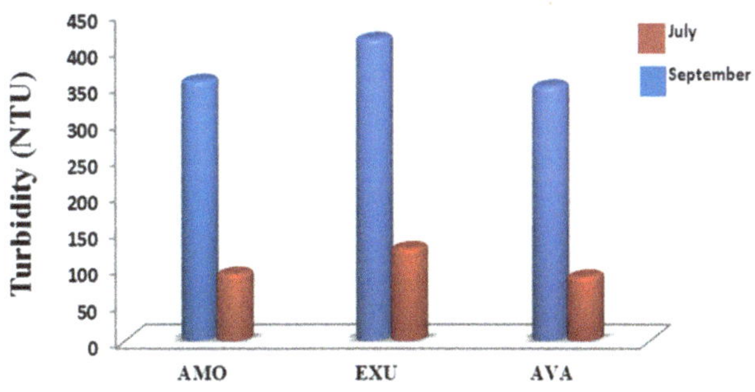

Figure 5. *Change in turbidity.*

4.5. The Suspended Solids (TSS)

The amount of suspended solids depends on the nature of the area to be crossed, season, rainfall, flow water regime, the nature of releases, etc. The corresponding averages (Figure 6) in each series are respectively 82 mg.L^{-1} and 84 mg.L^{-1} for the month of July and September and show that there is no significant difference ($p < 0.05$) the two sample sets. Changes in levels of suspended solids (TSS) Chari River water around the Farcha slaughterhouses are almost similar for both periods: the maximum values are recorded at the outlet of the slaughterhouse. These high levels may be the result of a brutal hydrological event (flood), the burden of TSS can be attributed to intense erosion and runoff and gutters of the city, following rainstorms that cause an increase suspended solids. This is likely due to effluents from slaughterhouses and liquid waste generated by coastal centers (IDT, Chad Brewery) as well as solid waste randomly deposited on the banks and gutters installed upstream slaughterhouses.

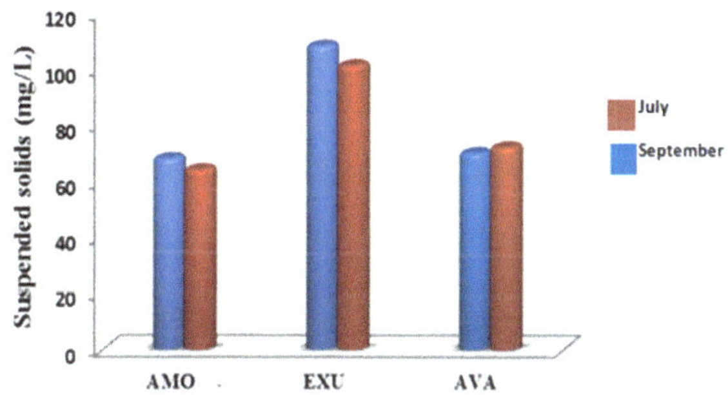

Figure 6. *Change in suspended solids.*

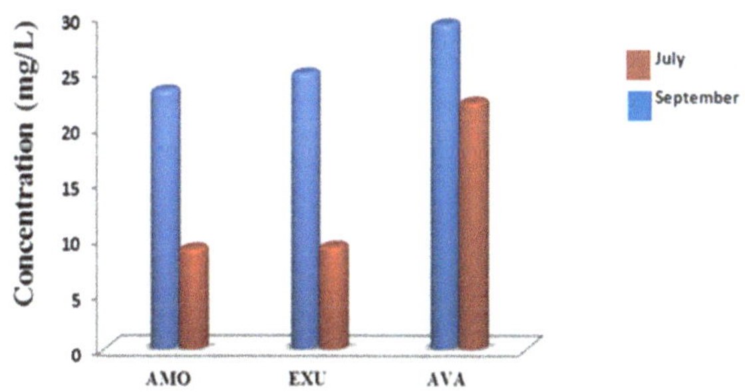

Figure 7. *Concentration of chloride ions.*

4.6. Chloride Ions (Cl^-)

The chlorides are often used as an indicator of pollution. The concentrations of chloride ions into the waters identified (Figure 7) show the upstream downstream of 23.30 mg.L^{-1} to 29.25 mg.L^{-1} in July. This variation is also observed in September with the high content of 22.24 mg.L^{-1} developed AVAs. High concentrations are recorded in the downstream part of slaughterhouses could indicate the contribution of slaughterhouses both from urban and industrial. For both study periods, chlorides record levels that do not exceed WHO standards set at 250 mg.L^{-1}. These levels are close to the work of Ngaram *et al.* (2011) [8] and Belghyti *et al.* (2009) [13] performed respectively on the Chari River and wastewater slaughterhouse Kenitra.

4.7. The Iron Ions (Fe^{2+})

Analysis variance of water samples iron contents (Figure 8) shows that there is a significant difference ($p < 0.05$) between the two sampling periods. Indeed, the concentrations obtained (Figure 8) shows that concentrations ranged from (7.25 ± 2.19) mg.L^{-1} and (11.10 ± 2.68) mg.L^{-1} in July when they are between (0.48 ± 0.60) mg.L^{-1} and (1.70 ± 0.56) mg.L^{-1} in September. The low values obtained in September may be justified by the dilution effect. The September month actually marks the Chari flood waters. The concentration is always high at the outlet compared to other points which attests that intakes of iron by slaughterhouses Farcha are not negligible. The iron contents of the Chari river water exceed the WHO limit (0.30 mg.L^{-1}) for consumer waters. These results are lower than those of the work Haoua *et al.* (2011) [15] on the Niger River (39.19 mg.L^{-1}).

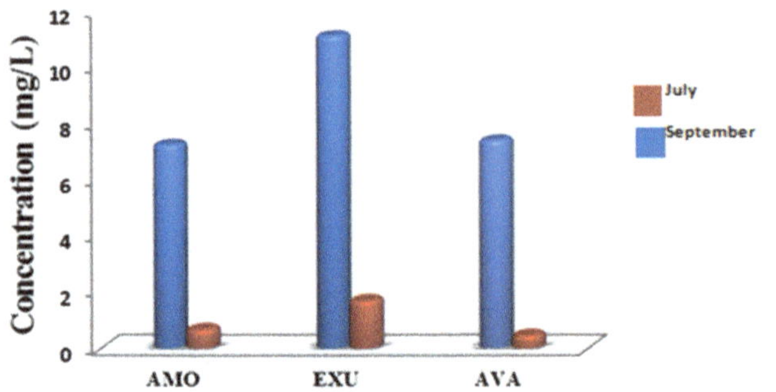

Figure 8. *Concentration of Iron ion.*

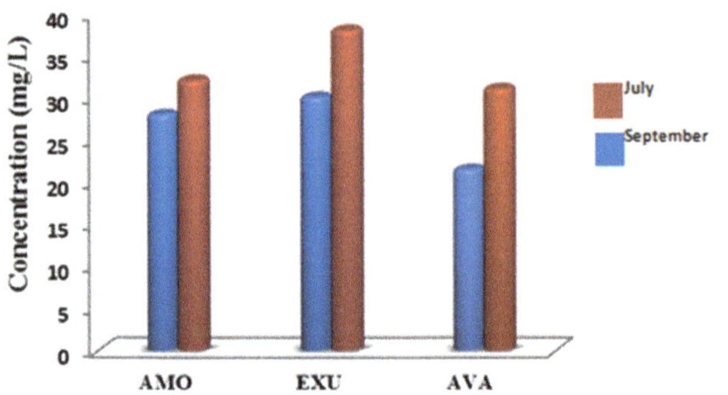

Figure 9. *Concentration of sulphate ions.*

4.8. The Sulphate Ions (SO_4^{2-})

The concentrations of sulphate ions (Figure 9) show that they are higher in September compared to July and showed a significant difference ($p < 0.05$) between the two periods. The average values range from (21.50 ± 2.10) mg.L^{-1} to (30.00 ± 0.52) mg.L^{-1}. In September they vary from (31.00 ± 6.70) mg.L^{-1} to (38.00 ± 3.50 mg.L^{-1}). The highest values observed at the outlet of slaughterhouses, may result in the intake of sulphate ions from effluents slaughterhouses. The high levels in September suggest also consider other source discharges from the brewery. We find that the average concentrations of sulphate decreased from upstream downstream slaughterhouses. This may be due to settling or to any complexion sulphates along the path of water flow. The average levels of sulphate are much lower than the limit value set by the WHO as regards the quality of drinking water (500 mg.L^{-1}).

4.9. The Phosphate Ions (PO_4^{3-})

The phosphate concentrations in the waters of the Chari River in the vicinity of the discharge of slaughterhouses Farcha range from (20.45 ± 5.02) mg.L^{-1} to (32.60 ± 6.92) mg.L^{-1} and (42.85 ± 17.32) mg.L^{-1} to (85.35 ± 9.40) mg.L^{-1}, respectively, for the month of July and September (Figure 10). First, these high levels in September can be justified by the use of phosphate fertilizers in the area at that time. On the other hand, this strong availability of phosphates can be explained by urban waste of the city in addition to contributions slaughterhouses. These results are much higher than those of previous work (Ngaram *et al.*, 2011) [8].

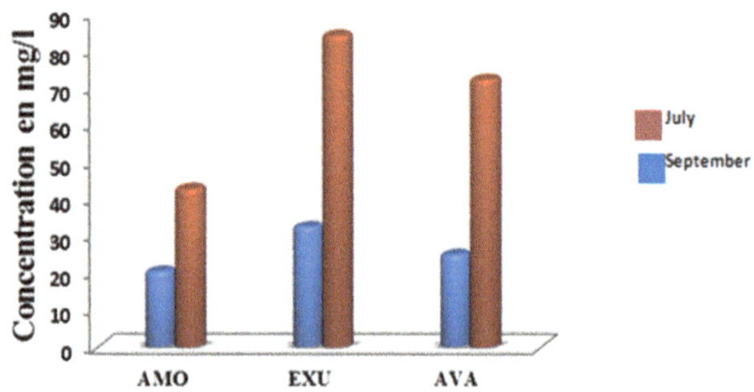

Figure 10. *Concentration of phosphate ions.*

4.10. Nitrogen Forms

4.10.1. Nitrate Ions (NO_3^-)

The growth is observed in the concentration of nitrates upstream downstream with a statistically significant difference ($p < 0.05$) during the two sampling periods (Figure 11a). Average grades range from (15.00 ± 7.07) mg.L^{-1} to (27.00 ± 0.00) mg.L^{-1} points in July. They range from (20.00 ± 1.41) mg.L^{-1} to (39.50 ± 1.76 mg.L^{-1}) in September. The high values in both periods could be attributed to releases from bellies, manure, urine, faeces of animals contained in effluent from slaughterhouses. This can also be justified by discharges of wastewater from hotels, gutters, hospital (HGRN), much to the leaching of fertilizers used in agricultural soils situated on the outskirts of the Chari which have been no treatment. These results are in agreement with those of Ngaram *et al.* (2011) [8] on the Chari River (21.60 mg.L^{-1}). Nitrate levels recorded in Chari surface water at the outlet of Farcha slaughterhouses are well above the standard suggested by WHO (1-10 mg.L^{-1}) [10].

4.10.2. Nitrite Ions (NO_2^-)

Analysis of variance of the concentrations of nitrite ions in water shows that there is not a statistically significant difference ($p < 0.05$) for the two series (Figure 11b). Indeed, concentrations of nitrites analysed waters vary between 7.45 ± 8.83 mg.L^{-1} to 10.80 ± 1.13 mg.L^{-1} and of 9.30 ± 42 mg.L^{-1} to 13.21 ± 20.50 mg.L^{-1}, respectively, during the months of July and September. These values are beyond the WHO threshold (3 mg.L^{-1}). Nitrites are generally from the either of an incomplete degradation of ammonia or a nitrate reduction, they represent only an intermediate stage and readily oxidized to nitrate [16]. The high concentrations of nitrites encountered, could be explained by the fact that they come from effluent from slaughterhouses and nutrient rich agricultural area NPK (vegetable crops along the river).

4.10.3. Ammonium ion (NH_4^+)

Analysis of variance of ammonium ion concentrations show a statistically significant difference ($p < 0.05$) between two sampling series (Figure 11c). Average grades vary between 2.75 ± 0.63 mg.L^{-1} to 4.95 ± 0.63 mg.L^{-1} and 1.49 ± 1.12 mg.L^{-1} to 1.89 ± 1.26 mg.L^{-1}. The ammonium ion high values in July translate the incomplete degradation process of organic matter. The grades obtained in September are lower than in the period July especially downstream of the study area. Which reflects the effect of dilution and shows oxygenation of water thus causing the oxidation of nitrogen. The limit set by the WHO standard is 10 mg.L^{-1} for the protection of contamination of aquatic organisms.

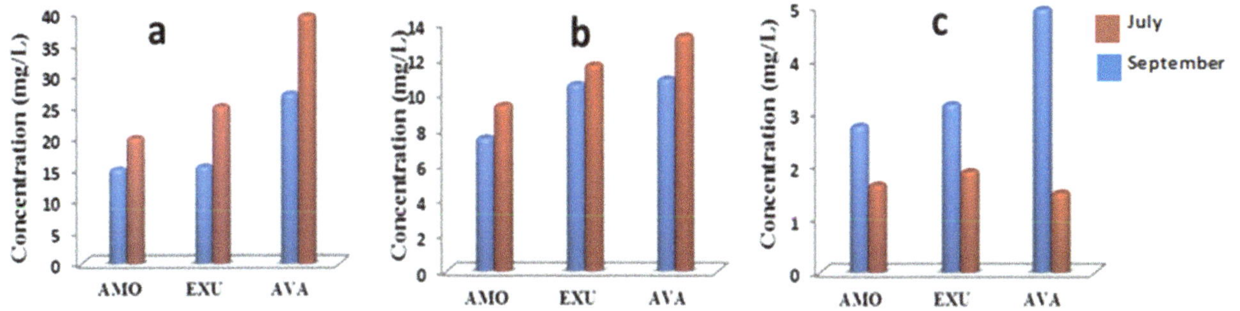

Figure 11. Ion concentration a) nitrate: b) nitrite; c) ammonium.

4.11. Evaluation of Organic Waste Water Pollution

4.11.1. The Dissolved Oxygen Content

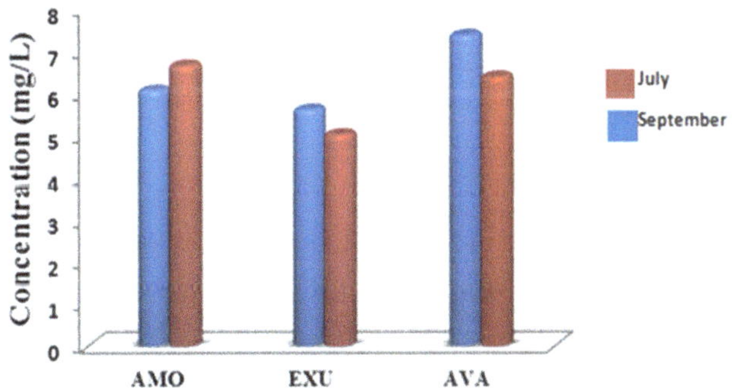

Figure 12. Variation of the dissolved oxygen concentration of water samples.

Analysis of variance of dissolved oxygen concentrations (Figure 12) shows that there is not a statistically significant difference ($p < 0.05$) in the two sampling periods. The average contents of dissolved oxygen in various samples range from: 5.61 ± 0.42 mg.L^{-1} to 7.37 ± 0.70 mg.L^{-1} in July and 5.02 ± 0.41 mg.L^{-1} to 6.65 ± 0.15 mg.L^{-1} in September. These results show regular and close values of previous work 5.97 mg.L^{-1} to 7.97 mg.L^{-1} conducted on the waters of the Chari River in the city of N'djamena [8]. This may be mainly due to the decrease of the water temperature; as a cold water has a higher amount of dissolved oxygen than warm water [17]. The recorded average levels are above 5 mg.L^{-1}, which is acceptable to the WHO/Chad standard. However, the data obtained show that river waters are polluted by slaughterhouses and wastewater of industrial origin, urban upstream slaughterhouses.

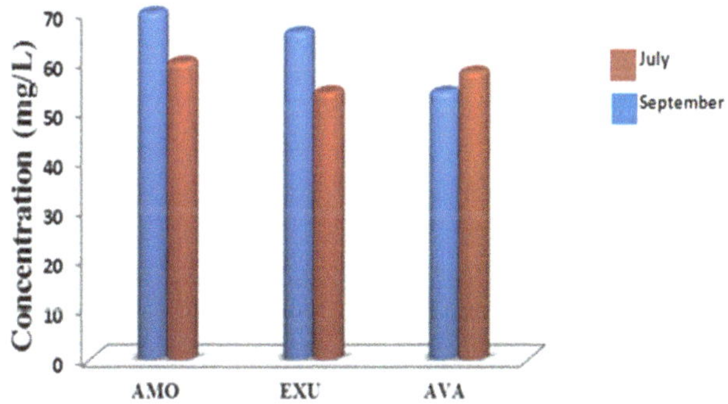

Figure 13. Variation of BOD5 content of water samples.

4.11.2. The Biochemical Oxygen Demand in Five Days (BOD_5)

BOD_5 of waters around the Chari slaughterhouses Farcha decreases from upstream to downstream with a statistically significant difference ($p < 0.05$) for both sampling series (Figure 13). Indeed, the values vary between BOD_5 70.00 ± 0.42 mg.L^{-1} to 54.00 ± 3.00 mg.L^{-1} and 60.00 ± 2.44 mg.L^{-1} to 54.00 ± 3.20 mg.L^{-1} respectively for the months of July and September. Increased levels of BOD_5 in waters around the Chari slaughterhouse is due to significant releases of organic matter in these slaughterhouses Farcha. This can also be due to runoff result in numerous organic substances into the river. For the upstream portion of the study area receives raw sewage rich in organic matter and nutrients from urban facilities, causing a considerable increase in the organic load in the waters of the Chari River.

4.11.3. Chemical Oxygen Demand (COD)

Analysis of variance of the COD (Figure 14) shows that there is not a statistically significant difference ($p < 0.05$) between the two series. Average grades obtained COD at the water samples are from 97.80 ± 0.50 $mg.L^{-1}$ and 99.10 ± 1.30 $mg.L^{-1}$ in July and between 96.30 ± 0.56 $mg.L^{-1}$ to 100.70 ± 0.70 $mg.L^{-1}$ in September. The increase of COD can be explained by the additional input of organic matter slaughterhouses and runoff materials during the dry season and entrained by rainwater. Note that the results are superior to those obtained previously by Ngaram *et al.* (2011) [8] on the Chari (30.50-30.75 $mg.L^{-1}$).

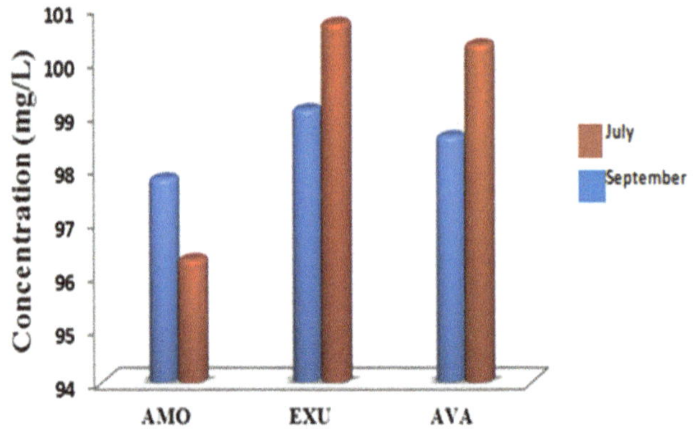

Figure 14. *Variation of the COD content of the water samples.*

4.11.4. Ratio COD / BOD_5

For a better understanding of the origin of the waste water, the calculation of COD/BOD_5 has very important interests. The ratio COD/BOD_5 can estimate the biodegradability of a given effluent. It is generally agreed the following limits: 2 > COD/BOD5: the effluent is easily biodegradable; 2 < COD/BOD_5 < 3: effluent is biodegradable with selected strains and 3 < COD/BOD_5: the effluent is not biodegradable [18].

The wastewater from the slaughterhouse Farcha have ratios equal to 1.47; 1.64 and 1.55 respectively upstream, at the outlet and downstream of the slaughterhouse. These reports from different points are lower than 2 which showed the wastewater Farcha slaughterhouses are readily biodegradable. However, a biological treatment is absolutely necessary.

5. Characterization of Sediments

The concentrations of different parameters are relatively low for the July season (Figure 15).

These levels are much observed in September. This may be due to a transport of these nutrients by rain runoff from the market gardens of the intense agriculture made by the residents and the massive use of detergents by money along the river through the city N'djamena. These concentrations ranged from 0.021 $mg.L^{-1}$ (AMOj) to 0.16 $mg.L^{-1}$ (AVAs) for nitrite. They range from 0.23 $mg.L^{-1}$ (AVAj) to 4.20 $mg.L^{-1}$ (AVAs) for nitrates. Phosphates vary from 1.90 $mg.L^{-1}$ (AMOj) to 5.20 $mg.L^{-1}$ (AVAs). Sulphates grow 1.20 mg/L (AMOj) to 4.26 $mg.L^{-1}$ (AVAs) and chlorides range from 1.80 $mg.L^{-1}$ (AMOj) to 3.50 $mg.L^{-1}$ (AVAs). The ammoniacal nitrogen 3 $mg.L^{-1}$ (AVAj) is reduced to 0.27 $mg.L^{-1}$ (AVAs). While low concentrations of iron 0.01 $mg.L^{-1}$ (AMOj) increased to 9.20 $mg.L^{-1}$ (AVAs).

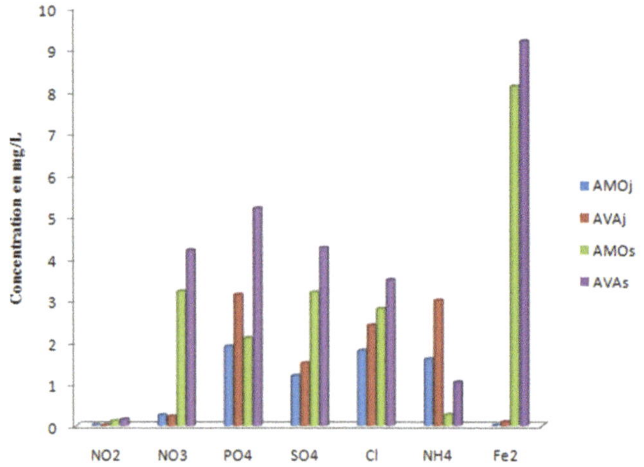

Figure 15. *Ion concentration in sediments.*

6. Principal Component Analysis (PCA)

The purpose of this review is to summarize the data structure described by quantitative variables, while obtaining uncorrelated factors between them. The principal component analysis was performed on both samples studied.

The 2 factors F1 and F2 explain 66.65% of the variances of water samples (Figure 16). It is observed that: the pH, Conductivity, Turbidity, BOD_5, the PO_4^{3-}, NH_4^+, Cl^-, SO_4^{2-}, on the Fe^{2+} are the parameters shown on the axis F1 (Table 1) for a variance of 45.246%.

By their homogeneous distribution along this axis and the presence of parameters such as Conductivity The Turbidity, BOD_5, pH, so there is mineralization (NH_4^+, Cl^-, the Fe^{2+} mainly) in the area, we can observed that this axis represents parameters from human activities mainly agriculture, atmospheric deposition and discharges from the city. TSS, the NO_3^- are represented positively F2 and O_2 negatively for a variation of 21.314%. The strong contribution of TSS, O_2 and NO_3^- on this axis suggests a source of effluent from slaughterhouses Farcha.

For samples of the sediment, both F1 and F2 factors represent 88.476% of the total variance (Figure 17). Table 2 show that the factor F1 represents a variance of 69.49% includes the NO_3^-, Fe_2^+, the PO_4^{3-}, Cl^- having a high contribution. NH_4^+ and NO_2^- (negatively) have a significant contribution. All these nutrients are original effluent from the slaughterhouse Farcha. They would be much more related to agriculture in the area N'Guéli rice mill and the heavy traffic of releases of facilities (Novotel, Meridien HGNR) and atmospheric deposition from air traffic from the international airport of N'Djaména. The factor F2 is marked by the strong contribution of original SO_4^{2-} would much effluent from slaughterhouses Farcha because the levels are higher SO_4^{2-} EXUj and EXUs.

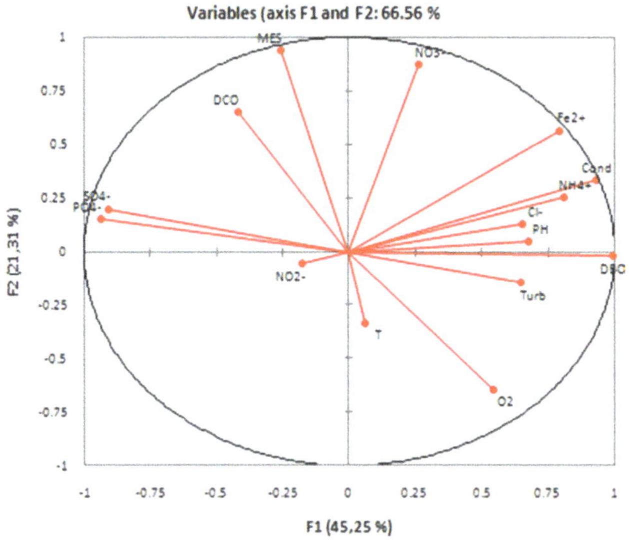

Figure 16. Correlation circle water samples (F1 vs. F2).

Table 1. Variable contributions for F1 and F2 factors in the waters of the Chari River.

	F1	F2
T °C	0.059	-0.331
pH	0.678	0.048
Conductivity	0.932	0.335
Turbidity	0.648	-0.146
DCO	-0.416	0.653
DBO_5	0.994	-0.020
MES	-0.257	0.940
O_2	0.547	-0.640
NO_2^-	-0.174	-0.052
NO_3^-	0.260	0.875
PO_4^{3-}	-0.938	0.150
NH_4^+	0.808	0.252
Cl^-	0.654	0.127
SO_4^{2-}	-0.905	0.199
Fe^{2+}	0.795	0.568

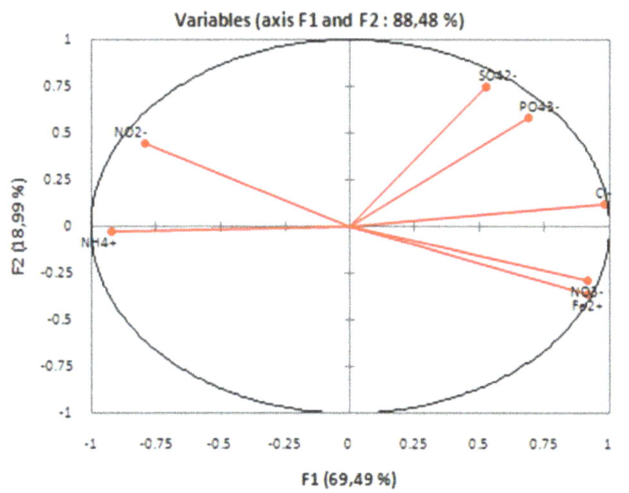

Figure 17. Correlation circle sediment samples (F1 vs F2).

Table 2. Variable contributions for F1 and F2 factors in sediments of the Chari River.

	F1	F2
NO_2^-	-0.792	0.446
NO_3^-	0.917	-0.284
PO_4^{3-}	0.692	0.581
SO_4^{2-}	0.526	0,755
NH_4^+	-0.921	-0.025
Fe^{2+}	0.914	-0.356
Cl-	0.979	0.120

7. Conclusion

Assessing the degree of Chari River pollution by waste water from slaughterhouses SMA/AFF Farcha was conducted during this study. The results of characterization of the three selected points of sampling generally, for the majority of the analysed parameters, pollution of river water is evident. The values obtained in TSS, BOD_5 and COD at the outlet of Farcha slaughterhouses are high and are representative of a possible organic pollution. This pollution is representative of possible eutrophisation of the aquatic environment. The overall results show that the water of the Chari River at the outlet of slaughterhouses Farcha shows signs of degradation. The contents of nitrate, nitrite, iron, ammonia nitrogen in water and sediment are largely beyond the threshold set by WHO for water consumption and surfaces. High values in these nutrients are found at the outlet of the slaughterhouse demonstrating that slaughterhouses Farcha contribute to the pollution of the Chari River.

References

[1] Noumi G. B., Fabane F., Siéliéchi J. M., Speciation of phosphorus in the sediments sampled of lake Dang, Ngaoundere – Cameroon, *Environmental Science and Pollution Research*, 2015, 22 (4), 3098-3106.

[2] Siéliéchi J. M., Noumi G. B., Fadimatou M., Ali A., Kapseu C., Speciation of heavy metals in sediments sampled from different pollution sources of lake Dang, Ngaoundere – Cameroon, *International Journal of Environment Protection*, 2013, 2 (3), 10–19.

[3] Tchoroun M, Noumi G. B., Tchadanaye N. M., Dangwang D. J. M., Heavy Metals Pollution Level in Water, Fish and Sediments from the Logone River Within Moundou City (Chad), *International Journal of Environmental Monitoring and Analysis*, 2015; 3 (5): 275-28.

[4] Siéliéchi J. M., Dangwang Dikdim J. M., Noumi G.B., Speciation phosphorus in the sediments sampled of lake Bini, Ngaoundere – Cameroon, *Environment Technology*, 2014, 35 (14), 1831–1839.

[5] Abderahmane M. A., Profil National du Tchad sur la gestion des produits chimiques, Ministère de l'Environnement et de l'eau, République du Tchad, 2001, 101p.

[6] Moctar Ismaël, Etude de la chaine de valeur de la filière cuir au Tchad, 2010.

[7] Athman M., Ettriqui A., Hassane A., Étude sur les abattoirs d'animaux de boucherie en Afrique centrale (Cameroun-Congo-Gabon-Tchad), Organisation des nations unies pour l'alimentation et l'agriculture, 2013, 70p.

[8] Ngaram N., Tchadanaye N. M., Merle A., Bauteri P., Baskali-Bouregaa N, Caractérisation physicochimique des eaux du fleuve Chari au niveau de la ville de N'djaména (Tchad), *Anales de l'Université de Ndjamena*-Série C –N°05, 2011, 102p.

[9] Rodier J, *L'Analyse de l'Eau*, 9ème édition, Dunod, Paris, 2009, 1579p.

[10] WHO, Global pollution and health results of related environmental monitoring, Global Environnement Monitoring system, WHO, 1987.

[11] Gomez E., Durillon C., Rofes G., Picot B., Phosphate adsorption and release from sediments of brackish lagoons: pH, O_2 and loading influence, *Water Res.*, 1999, 33, 2437–2447.

[12] Sorme L., Lagerkvist R., Sources of heavy metals in urban wastewater in Stockholm, *Sci. Total Environ.*, 2002, 298, 131–145.

[13] Belghyti D., Caractérisation physico-chimique des eaux usées d'abattoir en vue de la mise en œuvre d'un traitement adéquat: cas de Kénitra au Maroc, *Afrique* Science *05 (2), 2009, 199–216.*

[14] Obaidy M. J. A. H., Adel H. T., Shahad R. Z., Environmental Assessment of heavy metal distribution in sediments of Tigris River within Baghdad, *International Journal of Advanced Research*, 2014, 2 (8), 947–952.

[15] Haoua A., Mahaman S.L., Abdou S.M, Évaluation des rejets d'eaux usées de la ville de Niamey dans le fleuve Niger, *Afrique Science,* 2011, 07 (2), 43–55.

[16] Thomas. O, (1995). Métrologie des eaux résiduaire. *Ed. Cebedoc* / Tec et Doc. 11, Liège - 75384, Paris.

[17] Hebert S., Legre S., Suivi de la qualité de l'eau des rivières et petits cours d'eau, Direction du suivi de l'état de l'environnement, Ministère de l'Environnement Gouvernement du Québec, 2000, 5p.

[18] Bliefert C., Perraud R., *Chimie de l'environnement: air, eau, sols, déchets*, 2ème édition, Edition De Boeck Université, 2009, 478p.

Valorization of Grape by-Products

Cristina Damian*, Ana Olteanu, Mircea Oroian, Ana Leahu, Sorina Ropciuc

Stefan cel Mare University of Suceava, Faculty of Food Engineering, 13 Universitatii, Suceava, Romania

Email address:
cristinadamian@fia.usv.ro (C. Damian)

Abstract: The by-products of processed grape (grape seeds and grape peels) can be recovered for further food applications. Extracts of grape by-products were obtained with solvents of different polarity (aqueous acetone, methanol and ethanol) and assayed for their total phenolic content and antioxidant activity. The high amount of total phenols was found in grape seeds. Higher yields of phenolic antioxidants were recovered with acetone. The recovery of phenolic antioxidants from the peels and seeds of processed grape could be a valuable alternative to traditional disposal routes (including landfill), in particular for cooking grape varieties. The recycling process could enhance the growth of traditional culinary markets thanks to the new business opportunities for the peel-derived and seed-derived materials.

Keywords: Waste Valorisation, Peel and Seeds Polyphenols, Antioxidant Value

1. Introduction

The agricultural sector produces a large amount of by-products and wastes, representing an increasing interest as industrial crops both due to the economic reasons and from the environmental concerns.

Grapes (Vitis vinifera) belong to Vitaceae family. There are many reports on the benefits of eating grapes as they are known to be packed with nutrients such as magnesium, vitamins (A, B1, B2, B6 and C) and possess antioxidants properties. Grapes are effective as anti-ageing agents through the effects of resveratrol, a molecule in the skin pulp [AHMAD, W., 2014].

The grape is one of the major fruit crops worldwide and its harvest is about 60 millions tones per year [LAFKA, T.I., 2007]. About 80% of the harvest is utilized for winemaking and the grape waste is about 20% of the weight of processed grapes [MAZZA, G., 1993]. Grape pomace consists in the skin, stems and seeds of grapes that remain after processing in the wine and juice industry [MUŇOZ-GONZÁLES, C., 2013]. Grape processing wastes can be an important economical problem to producers besides the environmental impact caused by the large amount of these types of residues generated during the harvest season. The majority of this pomace is discarded as natural waste or distilled to produce alcohol and other distilled beverages. Therefore, the wineries will have new economic difficulties with winery waste management.

Some traditional applications of grape pomace are for animal feed formulations or compost production, without any pre-treatment. Recently, the scientific works carried out on the characterization of the chemical components of grape waste by-products have allowed looking for different applications in trying to obtain high added value ingredients. Some of these applications are the production of grape-seed oil [FIORI, L., 2007] or biodesel from it [FERNÁNDEZ, C.M., 2010], obtaining dietary fibre [IGARTUBURU, J.M., 1998] and the extraction of polyphenols with antioxidant properties [PINELO, M., 2005].

Marcs, stems and dregs (sludgy residual deposits at the bottom of fermentation vats) represent sources of antioxidants that have been relatively unexploited to date, but are of increasing industrial interests. If stalks are stripped from grape prior to crushing, winery marc consists of approximately 30% seeds and 70% skin and pulp. Studies on grape seeds are rather limited, despite their richness in polyphenolic substances, mainly monomeric and oligomericflavanols [TOUNSI, M.S., 2009]. Phenolic compounds have received considerable attention due to their pharmacological effects, including antimicrobial and antioxidant activities [DELIORMANORHAN, D., 2009]. Alternatively, polyphenols from grape pomaces are suggested to be a valuable crop for the production of adhesives [MENDES, J.A.S., 2013]. There are different efficient detection methods [RZEPECKA-STOJKO, A., 2010].

In recent years, antioxidants have gained more importance because of their positive involvement as health promoters in conditions such as cardiovascular problems, atherosclerosis, treatment of many forms of cancer, and the ageing process. Many antioxidants compounds, naturally occurring in plant sources have been identified as free radical scavengers [DUH, P.D., 1998]. Nowadays, interest has considerably increased in finding naturally occurring antioxidants for use in food or medicinal materials to replace synthetic antioxidants, which are being restricted due to their side effects such as carcinogenity [ZHENG, W., 2001, HABEEBULLAH, S.F.K., 2010]. Natural antioxidants are required at higher level than artificial products, so the importance of identifying active components and optimizing usage is emphasized [MOHAMED, H.M.A.,].

The antioxidant activity may be determined using different free radicals. One of the methods use DPPH or 1,1-diphenyl-2-picrylhydrazyl, which as radical is violet and in reduced form is yellow. The reduction is made with antioxidants and the color variation is used to evaluate the antioxidant concentration needed to reduce a certain amount of radicals. Spectrophotometric measurements can determine the color variation [SÁNCHEZ-MORENO, C., 2002].

$$\underset{Violet}{DPPH} \xrightarrow{antioxidant} \underset{Yellow}{DPPH-H}$$

***Figure1.** DPPH method scheme.*

The utilization of by-products from winemaking is an urged problem in Europe within context of recently posed regulations presuming measures against burying of wine by-products affecting the soil erosion/compaction and the quality of groundwater [CCE, 2006].

In the present investigation, phenolic content of seed and peel methanolic, ethanolic and aqueous acetone extracts of Vitisvinifera (Babeasca, Fraguta, and Tamaioasa) were investigated by spectrophotometry. Antioxidants activity against DPPH radical of these extracts were assayed too.

2. Materials and Methodology

2.1. Chemicals

DPPH (1,1-diphenyl-2-picrylhydrazyl) and Folin-Ciocalteau reagent were purchased from Aldrich. These solutions were wrapped in aluminium foil and stored at 40C. All other chemicals used were of analytical grade.

2.2. Plant Material

Our study was carried out on rape seeds and peels of grape of three varieties (Vitis vinifera L.) Babeasca (red grape), Fraguta (rosé grape) and Tamaioasa (white grape). 1 kg of samples has been harvested, dating from September 2013, from the local market. Seeds and peels were manually separated and dried at ambient temperature in dark until used.

2.3. Polyphenol Extraction

The air-dried seeds and peels were ground. 1 g of this ground material was extracted by stirring with 10 mL of pure methanol, ethanol and aqueous acetone for 30 min. The extracts were then kept for 24 h at 40C, filtered through a Whatman No4 filter paper, and evaporated under vacuum to dryness and stored at 4^0C until analyzed [MAU, J.L., 2001].

2.4. Total Phenolic Content

Total phenolic contents were assayed using the Folin-Ciocalteau reagent, following Singleton's method slightly modified by Dewanto et al. [DEWANTO, V., 2002]. An aliquot (0.125 mL) of a suitable diluted methanolic, ethanolic and aqueous acetone seed and peel extract (0.25 mg.mL-1) was added to 0.5 mL of deionized water and 0.125 mL of the Folin-Ciocalteau reagent. The mixture was shaken and allowed to stand for 6 min, before adding 1.25 mL of 7% Na_2CO_3 solution. The solution was then adjusted with deionized water to a final volume of 3 mL and mixed thoroughly. After incubation for 90 min at 230C, the absorbance versus prepared blank was read at 760 nm. The phenolic content of seeds and peels (three replicates per treatment) was expressed as mg gallic acid equivalents (GAE) per gram of dry weight through the calibration curve with gallic acid. The calibration curve range was 50-400 $mg.mL^{-1}$ ($R^2 = 0.99$).

2.5. DPPH Assay

The electron donation ability of the obtained methanol, ethanol and aqueous acetone extracts was measured by bleaching of the purple-colored solution of 1,1-diphenyl-2-picrylhydrazyl radical (DPPH) according to the method of Hanato et al. [HANATO, T., 1988]. 2 mL of methanolic grape seed and peel extracts were added to a 0.5 mL of a 0.2 mmol.L-1 DPPHmethanolic solution. The mixture was shaken vigorously and left standing at room temperature for 30 min. The absorbance of the resulting solution was then measured at 517 nm after 30 min. The antiradical activity (three replicates per treatment) was expressed as IC50 (mg.mL-1), the concentration required to cause a 50% DPPH inhibition. A lower IC50 value corresponds to a higher antioxidant activity of seed and peel extract. The ability to scavenge the DPPH radical was calculated using the following equation:

$$DPPHscavengingeffect(\%) = \left[\frac{(A_0 - A_1)}{A_0}\right] \times 100$$

where A_0 is the absorbance of the control at 30 min, and A_1 is the absorbance of the sample at 30 min. Samples were analyzed in triplicate.

2.6. Statistics

Samples were assayed in triplicate and results are given as averages ± SD. Student's *t* test was used for the statistical evaluation and $p<0.05$ was considered statistically significant.

3. Result and Discussion

3.1. Total Phenolics Contents

Based on the absorbance values of methanolic extract solutions reacted with the Folin-Ciocalteau reagent and compared with the standard solutions of gallic acid equivalents, the total phenolic content is given in Figure 2. The phenolic content of seed extracts varied between varieties and ranged from 134.88 to 413.88 mg.GAE-1. Babeasca seeds have a higher phenolic amount (413.88 mg.GAE/g) than Fraguta ones (267.53 mg.GAE/g). Tamaioasa seeds presented only 134.88 mg.GAE/g.

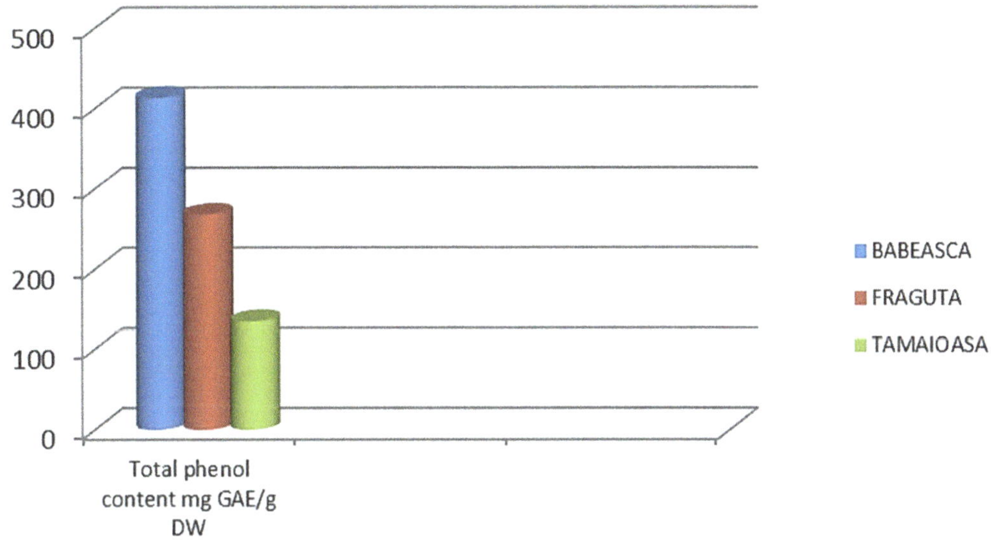

Figure 2. *Total phenolic content in methanolic extracts from Babeasca, Fraguta and Tamaioasa seeds (mg gallic acid equivalents (GAE) per gram of dry weight).*

Our results were different to those of Berrin et al. (2008) who mentioned the presence of moderate phenol amounts (88.11-105.7 mg.GAE/g DW) of four international varieties from Turkish Vitis vinifera seeds. However, Nilgün et al. (2004) noted that total phenolic content in seeds of Vitis vinifera L. species can reach 667.98 mg.GAE/g DW. These variations in total phenol content could be due to the various factors. One such factor may be the genetic potential of individual species for polyphenol biosynthesis [TOUNSI, M.S., 2009]. Apart from the genetic (varietal) background, maturation stage may also be critical in this respect [DE FREITAS, V.A.P., 1999]. It was observed that amounts of all seed polyphenols in grape decline considerably, and this fact was attributed to an initiation of oxidative phenomena, which appeared to follow second order kinetics [KENNEDY, J.A., 2000].

Winery waste was ground before extraction in order to reduce particle size and increase the yield of extracted phenols and their antioxidant activities. Ethanol gave extracts with lower phenol content than those of methanol. Extraction with aqueous acetone gave a yield of 43.5%, compared to extraction with methanol.

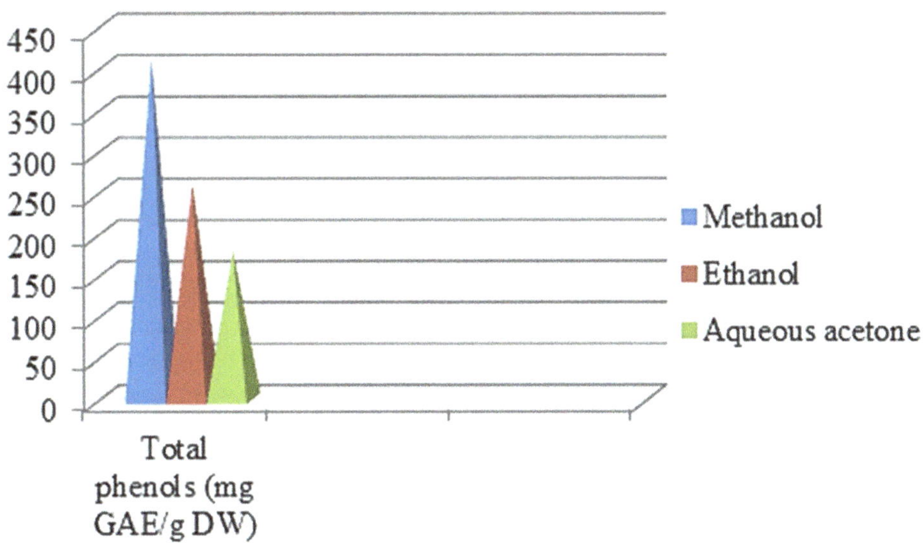

Figure 3. *The effect of solvent on the quantity of extracted phenols from Babeasca winery waste.*

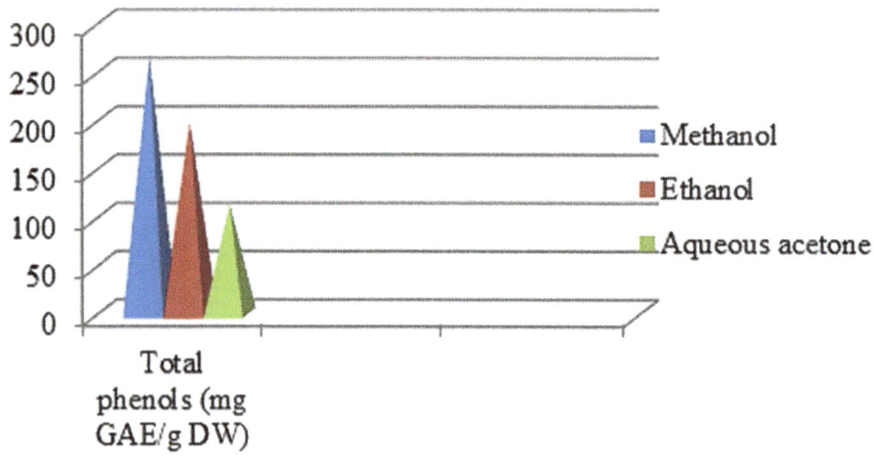

Figure 4. *The effect of solvent on the quantity of extracted phenols from Fraguta winery waste.*

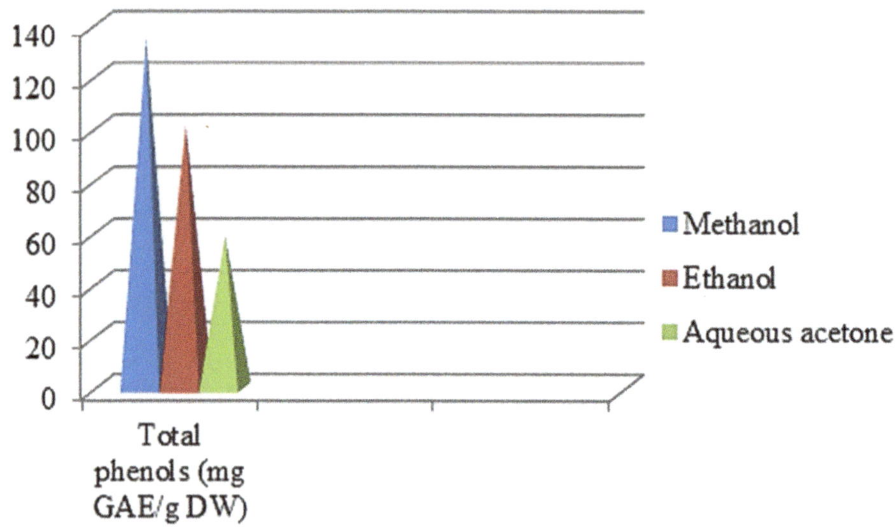

Figure 5. *The effect of solvent on the quantity of extracted phenols from Tamaioasa winery waste.*

3.2. Antioxidant Activity

The assessment of antioxidant activity showed that the examined seeds and peels were able to scavenge this radical (Table 1). Tamaioasa seeds displayed the highest activity compared to Fraguta seeds and Tamaioasa seeds (IC50 values were: 2.8, 7.9 and 27 µg/mL, respectively). The synergic effect of the antioxidants in the extracts should be considered. The values of IC50 for the grape peels show that are no significant differences between Babeasca, Fraguta and Tamaioasa.

Table 1. *Antiradical activity (DPPH) of Babeasca, Fraguta and Tamaioasa seeds and peels.*

Samples	IC50(µg/mL) DPPH
Babeasca seeds	2.8 ± 0.12
Fraguta seeds	7.9 ± 0.67
Tamaioasa seeds	27± 0.33
Babeasca peels	1.096 ± 1.11
Fraguta peels	1.107 ± 2.87
Tamaioasa peels	1.088 ± 1.65

Values are means ± standard deviation, *n= 3*

A good correlation was found for the total polyphenols and DPPH method ($r = 0.87$). Although researchers often reached different conclusions, the issue of relation of the antioxidant activity and the concentration of polyphenolic compounds in foods has been repeatedly undertaken in the literature. Some did not discover the correlation between polyphenol content and the antioxidant activity of plant extracts while others showed a strong relationship between them [MARKOWSKI, J., 1998, WILL, F., 2008].

4. Conclusion

The results of this work indicate the presence of compounds possessing high antioxidant activity in grape seeds and peels (Vitis vinifera L.). In addition, the obtained results showed large differences found among the varieties in relation to the polyphenol content.

The present study provides data for supporting the use of Babeasca, Fraguta and Tamaioasa seeds and peels as natural antioxidant agents, and confirms that these extracts represent a significant source of phenolic compounds.

References

[1] Ahmad, W., Khan, M.I., Waqar, M., Khan, M.A., Khan, A., Ramazan, R., Wali, S., Ahmad, F., Khan, N., Yousaf, S., Zeb, M., Khan, A.U., Rahman, M.U., Faisal, S., In vitro antibacterial activity of vitis vinifera leaf extracts against some pathogenic bacterial strains, Advances in biological research, 8(2), 2014, 62-67.

[2] Cce, comissão das comunidadeseuropeias. Para um sector vitivinícolasustentável. Comunicação da comissãoaoconselho e aoparlamentoeuropeu (pt), 319, Bruxelles.

[3] De freitas, V.A.P., Glories, Y., Concentration and compositional changes of procyanidins in grape seeds and skin of vitisvinifera varieties, J sci food agric, 79, 1999, 1601-1606.

[4] Deliormanorhan, D., Orhan, N., Özçelik, B., Ergun, F., Biological activities of vitis vinifera l. Leaves, Turk j biol, 33, 2009, 311-348.

[5] Dewanto, V., Wu, X., Adom, K., Liu, R.H., Thermal processing enhancing the nutritional value of tomatoes by increasing total antioxidant activity, J agric food chem, 50, 2002, 3010-3014.

[6] Duh, P.D., Antioxidant activity of burdock (arctiumlappalinne): its scavenging effect on free radical and active oxygen, Jaocs, 75, 1998, 455-465.

[7] Fernández, C.M., Ramos, M.J., Pérez, A., Rodríguez, J.F.,Production of biodiesel from winery waste: extraction, refining and transesterification of grape seed oil, Bioresourtechnol, 101, 2010, 7030-7035.

[8] Fiori, L., Grape seed oil supercritical extraction kinetic and solubility data: critical approach and modeling, J supercrit fluids, 43, 2007, 43 54.

[9] Habeebullah, S.F.K., Nielsen, N.S., Jacobsen, C., Antioxidant activity of potato peel extracts in a fish-rapeseed oil mixture and in oil-in-water emulsions, J am oil chemsoc, 87, 2010, 1319-1332.

[10] Hanato, T., Kagawa, H., Yasuhara, T., Okuda, T., Two new flavonoids and other constituents in licorice root: their relative astringency and radical scavenging effect, Chem pharm bull, 36, 1988, 1090-1097.

[11] Igartuburu, J.M., Pando, E., Rodríguez-Luis, F., Gil-Serrano, A., Structure of a hemicellulose b fraction in dietary fiber from the seed of grape variety palomino (vitisvinifera cv. Palomino), J nat prod, 61, 1998, 881-886.

[12] Kennedy, J.A., Troup, G., Pilbrow, J.R., Hutton, D.R., Hewitt, D., Hunter, C.R., Ristic, R., Iland, P.G., Jones, G.P., Development of seed polyphenols in berries from vitis vinifera l. Cv. Shiraz, Ajgwr, 6, 2000, 244-254.

[13] Lafka, T.I., Sinanoglou, V., Lazos, E.S., On the extraction and antioxidant activity of phenolic compounds from winery wastes, Food chemistry, 104, 2007, 1206-1214.

[14] Markowski, Jaroslaw, and Witold Plocharski. Determination of phenolic compounds in apples and processed apple products, Journal of fruit and ornamental plant research 14, 2006, 133-142.

[15] Mau, J.L., Chao, J.R., Wu, G.R., Wu, K.T., Antioxidant properties of methanolic extracts from several ear mushrooms, J agric food chem, 49, 2001, 5461-5467.

[16] Mazza, G., Miniati, E., 1993. Grapes. In anthocyanins in fruits, vegetables and grains, Bocaraton, Ann Harbor, London, Tokyo: CRC press, pp. 149-199.

[17] Mendes, J.A.S., Prozil, S.O., Evtuguin, D.V., Cruz Lopes, L.P., Towards comprehensive utilization of winemaking residues: characterization of grape skins from red grape pomaces of variety touriganacional, Industrial crops and products, 43, 2013, 25-32.

[18] Mohamed, H.M.A., Awatif, I.I., The use of sesame oil unsaponifiable matter as a natural antioxidant, Food chemistry, 1998, 62.3: 269-276.

[19] Muñoz-González, C., Rodríguez-Gencomo, J.J., Martín-Álvarez, P.J., Moreno-Arribas, M.V., Pozo-Bayón, M.Á. Recovery of aromatic aglycones from grape pomace winemaking by-products by using liquid-liquid and pressurized-liquid extraction, Food analytical methods 7.1, 2014, 47-57.

[20] Pinelo, M., Rubilar, M., Jerez, M., Sineiro, J., Núñez, M.J., Effect of solvent, temperature, and solvent-to-solid ratio on the total phenolic content and antiradical activity of extracts from different components of grape pomace, J agric food chem, 53, 2005, 2111-2117.

[21] Rzepecka-Stojko, A., Maciejewska-Paszek, I., Stec, M., Kurzeja, E., Kaska, A., Pawlowska-Góral, K., The influence of extraction method on obtaining polyphenolic compounds from bee pollen, Farmaceutycznyprzegladnaukowy, 7, issue 1, 2010, 38-41.

[22] Sánchez-Moreno, C., Review: methods used to evaluate the free radical scavenging activity in foods and biological systems, Food science and technology international, 8(3), 2002, 121-137.

[23] Tounsi, M.S., Ouerghemmi, I., Wannes, W.A., Ksouri, R., Zemni, H., Marzouk, B., Kchouk, M.E., Valorization of three varieties of grape, Industrial crops and products, 30, 2009, 292-296.

[24] Zheng, W., Wang, S.Y., Antioxidant activity and phenolic compounds in selected herbs, J agric food chem, 49, 2001, 5165-5170.

[25] Will, F., Roth, M., Olk, M., Ludwig, M., Dietrich, H., Processing and analytical characterisation of pulp-enriched cloudy apple juices. LWT-food science and technology 41.10, 2008, 2057-2063.

20

Physico-Chemical and Bacteriological Characterization of Spring and Well Water in Bamenda III (NW Region, Cameroon)

Alice Magha[1, 2, *], Margaret Tita Awah[1], Gus Djibril Kouankap Nono[1], Pierre Wotchoko[1], Mispa Ayuk Tabot[1], Veronique Kamgang Kabeyene[2]

[1]Department of Geology, HTTC, The University of Bamenda, Bambili; Bamenda, Cameroon
[2]Department of Earth Sciences, Faculty of Sciences, University of Yaoundé 1, Yaoundé, Cameroon

Email address:
amufur@yahoo.com (A. Magha), mtitawa2000@yahoo.com (M. T. Awah), kouankap@yahoo.fr (G. D. K. Nono), pierrewotchoko@yahoo.fr (P. Wotchoko), ayuktabot24@yahoo.com (M. A. Tabot), gemskruy@yahoo.fr (V. K. Kabeyene)

Abstract: Bamenda is a major town in the NW Region of Cameroon and constitutes part of the Cameroon Volcanic line. In the frame of this work, four springs and three wells in the Bamenda III council area were investigated to evaluate the physico-chemical and bacteriological characteristics of their waters. All the water points yielded pH values below 6.5 and were classified as acidic water. Electrical conductivity ranged between 0.04-2.38μm/cm. The relative abundance of major ions (meq/l) was $Ca^{2+} > Mg^{2+} > K^{+} > Na^{+}$ for cations and $HCO_3^{-} > SO_4^{2-} > NO_3^{-} > PO_4^{3-}$ for anions. Major ion concentrations were within the WHO guidelines for drinking water, but Ca^{2+} (110-1770mg/l) and Mg^{2+} (97.05-236.85mg/l) concentrations were much higher and above WHO limits. Main water types in the dry season were Ca-Mg-HCO3 while in rainy season the main water types were Ca-Mg-HCO3 and a mixed Ca-Mg-SO4. Bacterial analysis revealed that the water samples were highly polluted and classified B, C and D according to Cheesbrough classification. The spring and well waters analysed are not suitable for drinking and domestic purposes therefore the population of Bamenda III which depends on the mentioned water for drinking and domestic purposes may face serious health problems if appropriate actions are not taken to prevent and mitigate the problems.

Keywords: Bamenda III, Faecal Bacterial Analysis, Physico-Chemical Analysis, Water Quality

1. Introduction

Cameroon is endowed with numerous water sources [1,2]. It is a country with the second highest volume of available water in Africa [3]. Groundwater constitutes 21.5% (57 billion m3) of this resource and plays a very important role in the socio-economic life of the country.

The setting and evolution of most forms and pattern of settlements is linked to the availability of reliable and substantial water sources, and the fertility of volcanic soils have brought through ancient times the intense settlement and agricultural exploitation of the lower flanks of volcanoes [4]. This is the case of Bamenda town, a rapidly growing town which belongs to the Cameroon Volcanic Line (CVL) the principal watershed of the country [5]. Many thermal and mineral springs typified by the Fe - Mg - Ca $-HCO_3^-$ or Na $-$ HCO_3^- compositions are found along the CVL [6]. The chemical composition of surface and groundwater is controlled by many factors that include composition of precipitation, mineralogy of watershed and aquifers, climate and topography as well as anthropogenic influence such as urban, industrial and agricultural activites. These factors combine to create diverse water types that change spacially and temporally [3]. When water perculates through formations, it comes in contact with geologic formations and ions are exchanged especially during the process of weathering and alteration. These rocks release minerals whose ions enter aquifers and alter the chemical composition.

Geochemical studies from groundwater and surface water can provide a better understanding of potential water quality variations due to geology and land use practices [7-9].

Knowledge of the levels of microorganisms in water is also important for indexing the health harzard associated with its use, adopting meaningful interventions and improving bacterial water quality [10]. The Bamenda III Council Area is made up of many springs and wells on which about 80% of the population depends for both drinking and domestic purposes. Characterizing the mentioned water is of great importance to determine if the population is at risk. In this paper we present the evaluation of the physico-chemical and bacteriological qualities of some springs and wells in Bamenda III.

2. Natural Setting of the Study Area

Bamenda III is found in Mezam Division in the North West Region of Cameroon between latitudes 06°00'N and 06°89'N and longitudes 10°13'E and 10°28'E (Figure 1). This town is located at the central part of the Cameroon Volcanic Line, and is implanted below the high lava plateau, at an altitude of about 1500m. Bamenda III Subdivision is made up of two villages: Nkwen and Ndzah with about 250,000 inhabitants and a surface area of 74.61km2.

The climate is mainly of the equatorial type with two main seasons, the rainy season (April to October) and the dry season (November to March). Absolute annual average precipitation ranges from 1700mm to 2824mm and exhibits fluctuations [11].

The study area forms part of the Bamenda Mountains which lie along the Cameroon Volcanic Line. It is made up of a basement rock which is mainly composed of leucogranites of Precambrian age [12]. The basement rock is overlain by volcanic materials composed of mafic and felsic lavas. The mafic lavas include basalts which outcrop as lava flows and pyroclastic deposits respectively, while the felsic lavas include trachytes and rhyolites [13].

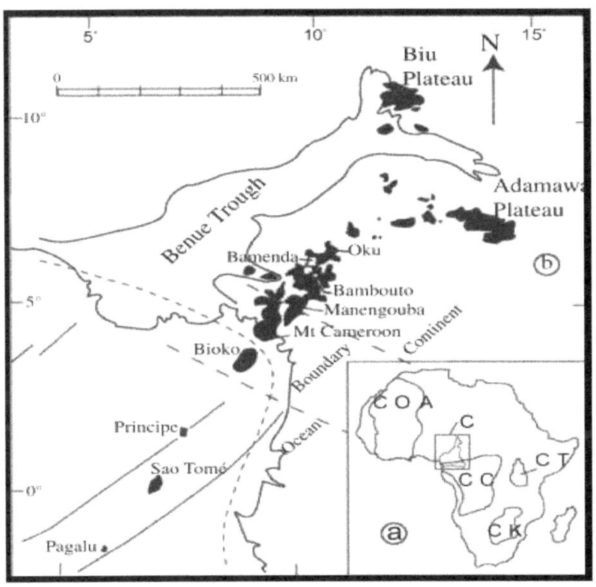

Figure 1. Location of Bamenda along the Cameroon Volcanic Line. From [14].

3. Methods

3.1. Field Work

Field studies were carried out in two phases and at different seasons of the year; the first phase took place in the rainy season (September) and the second phase in the dry season (February). Water samples were collected from 4 springs and 3 wells and at each sampling point, two sets of samples were collected and in one of the sets of samples 5ml of HCl were added to stabilise the heavy metal concentration. Sampling points were chosen in relation to their uses by the population.

3.2. Laboratory Analysis

Physico-chemical and bacteriological analyses were performed in the laboratory of the Faculty of Agronomy and Agricultural Sciences (FASA) in the University of Dschang, Cameroon.

3.3. Physical Analysis

For physical analysis, Total Dissolved Solids was determined. An NTU Turbidimeter (DRT 100B HF model type) was used to measure turbidity. Electrical Conductivity was determined using a Conductivity meter (WTW LF 521 model type).

3.4. Major Cations

Sodium (Na^+) and potassium (K^+) were determined by flame photometry. The analysis of Calcium (Ca^{2+}) was done by titration with 0.02M solution of Ca- EDTA together with 1ml of TEA and 1ml of 5% KCN. The determination of magnesium (Mg^{2+}) was done in a similar manner; ammonium (NH_4^+) was analysed using the colorimetric method.

3.5. Major Anions

Chloride (Cl^-) was analysed by titration. Nitrates (NO_3^-) and phosphate (PO_4^{3-}) were determined using colorimetry. Sulphates (SO_4^{2-}) were determined by turbidimetry and Bicarbonates (HCO_3^-) concentrations by titration.

3.6. Bacteriological Analysis

This was done on three sets of samples (two springs and one well) to determine the presence of faecal streptococcus, Escherichia coli, faecal and total coliform using the membrane filter procedure. 1ml of water sample was added to 9ml of distilled water. Each sample was diluted three times. A membrane was placed on a sterilized Wheaton Filtration funnel used to filter 20ml of undiluted sample. The funnel was sterilized after each filtration to avoid interferences. Several diluted samples were then processed so as to get filter plates with appropriate range of colonies. These filter plates were placed in an incubator at different temperature conditions for different bacteria. These were: 44oC for E-coli and feacal coliform, 35oC for streptococcus and total coliform.

4. Results

The physico-chemical characteristics of the studied water were similar, whether it was from well or from spring but significant differences were noted with the different seasons.

4.1. Physical Characteristics

Water temperatures were higher during the dry season and lower in the rainy season for all sampling points with the average of 22.6°C in the rainy season and of 26.31°C in the dry season (Figure 2). The average pH value was higher in the rainy season (5.71) than in the dry season (5.47) as shown in Figure 2, but all water points yielded acidic to barely acidic water (Table 1).

The values for EC were very low regardless of the sampling point and season but in the dry season a highest value (2.38µS/cm) was recorded in AMW3.

Turbidity was higher in the rainy season than in the dry season with the respective average value recorded of 8.14 and 2.40 (Table 1).

Table 1. *Variation of Temperature, PH, Electrical Conductivity, and Turbidity in the rainy season and dry season.*

	Temperature (T°)		pH		Electrical conductivity (EC)		Turbidity	
Sampling points	Rainy Season	Dry Season	Rainy Season	Dry Season	Rainy Season	Dry Season	Rainy Season	Dry Season
AMS_1	24.7	27.7	5.4	6.5	0.03	0.05	7.3	2.8
AMS_2	21.6	26.4	5.8	6	0.04	0.04	5.6	2.2
AMS_3	21.2	24	6.0	5.4	0.03	0.05	9.6	2
AMS_4	22.4	26.2	5.2	6.3	0.20	0.03	7.6	4.4
AMW_1	24	27.9	6.7	6.2	0.29	0.11	13.5	1.9
AMW_2	22.3	27	5.3	5.4	0.16	0.13	6.9	1.8
AMW_3	22	25	5.6	2.5	0.05	4.7	6.5	1.7
Average	22,60	26,31	5.71	5.47	0.11	0.73	8.14	2.40

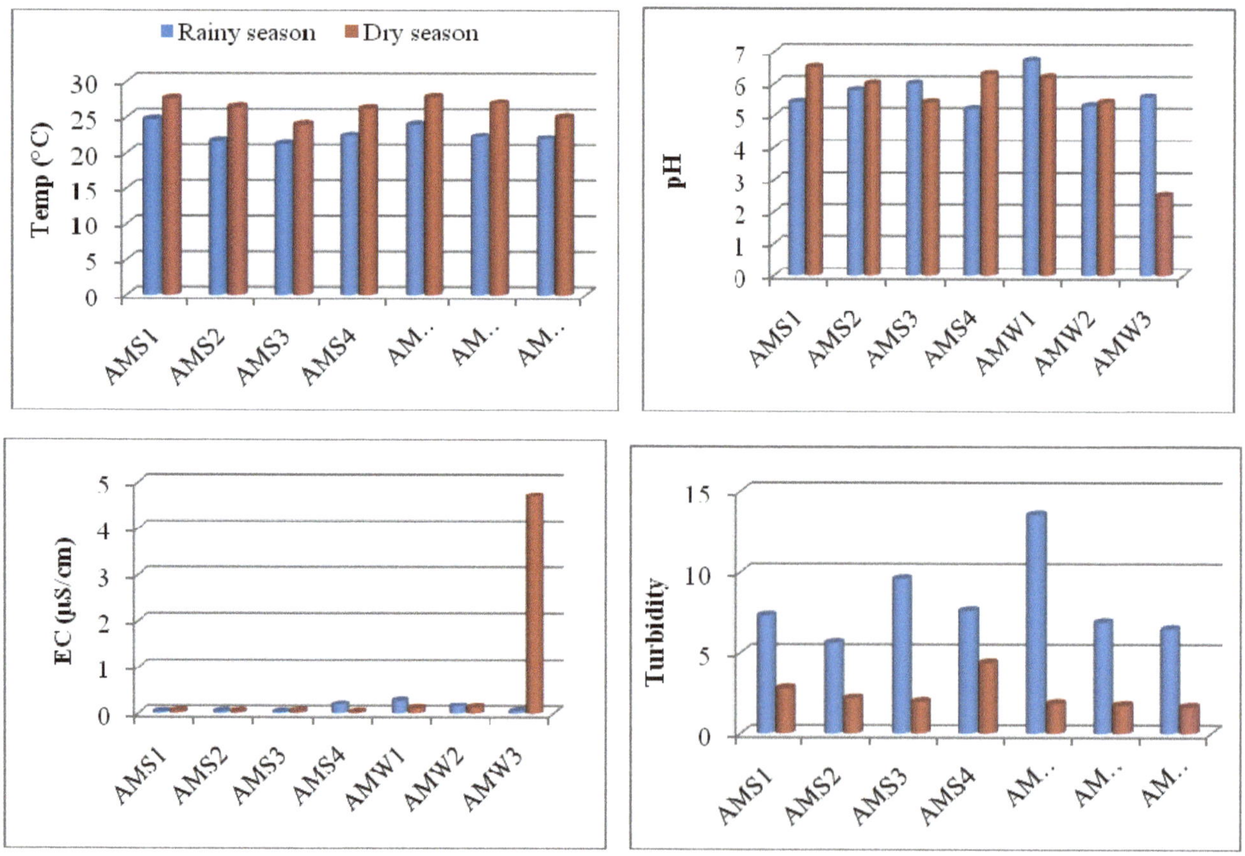

Figure 2. *Seasonal variation of physical parameters (Temperature, pH, Electrical conductivity, Turbidity).*

4.2. Chemical Characteristics of Cations

K^+ and Na^+ concentrations were high in the dry season and just barely detectable in the rainy season. There were no differences for the different sampling points, their respective average values for the rainy season were 0.39mg/l and

0.30mg/l, these values for dry season were 24.72mg/l and 11.37mg/l.

Regarding the average concentration in the dry season, Ca^{2+} was the most abundant cation in all the water samples with an average value above 500 mg/l (Table 2). Ca^{2+} was also higher in the dry season than in the rainy season for all samples.

Mg^{2+} was the second abundant cation present in the analysed water with an average of 232.3 mg/l in the rainy season and 126.57 mg/l in the dry season.

The NH_4^+ values were higher in the rainy season and barely detectable in the dry season (Figure 3) with average values ranging from 12.06 mg/l in the rainy season to 0.0073mg/l in the dry season.

Fe^{2+} was the least abundant cation in the water samples with the highest concentration of 1.14 mg/l in AMW_2 during the dry season. In the rainy season Fe^{2+} was not detected in AMS_3, AMS_4, AMW_1, AMW_2 and AMW_3 while in the dry season it was not detected in AMS_1 and AMS_2 (Table 2).

Table 2. *Variation of Cations in the rainy season and dry season.*

	K+		Na+		Ca2+		Mg2+		NH4+		Fe2+	
Sampling points	**Rainy Season**	**Dry Season**	**Rainy Season**	**Dry Season**	**Rainy Season**	**Dry Season**	**Rainy Season**	**Dry Season**	**Rainy Season**	**Dry Season**	**Rainy Season**	**Dry Season**
AMS_1	0.39	23.64	0.11	7.94	240	320	194.4	194	8.40	0.0062	0.89	-
AMS_2	0.40	25.39	0.37	8.01	280	400	109.4	219	12.88	0.0048	0.83	-
AMS_3	0.39	25.55	0.37	12.38	60	160	145.8	255	14	0.0070	-	0.39
AMS_4	0.40	24.68	0.37	8.38	260	500	206.6	170	16.80	0.0083	-	0.05
AMW_1	0.39	24.89	0.37	5.54	256	980	328.1	12	10.36	0.0084	-	0.05
AMW_2	0.39	24.72	0.37	26.67	340	800	461.7	12	12.77	0.0076	-	1.14
AMW_3	0.39	24.18	0.11	10.64	180	660	180	24	9.24	0.0087	-	0.18
Average	0.39	24.72	0.30	11.37	230.9	545.7	232.3	126.57	12.06	0.0073	0.25	0.26

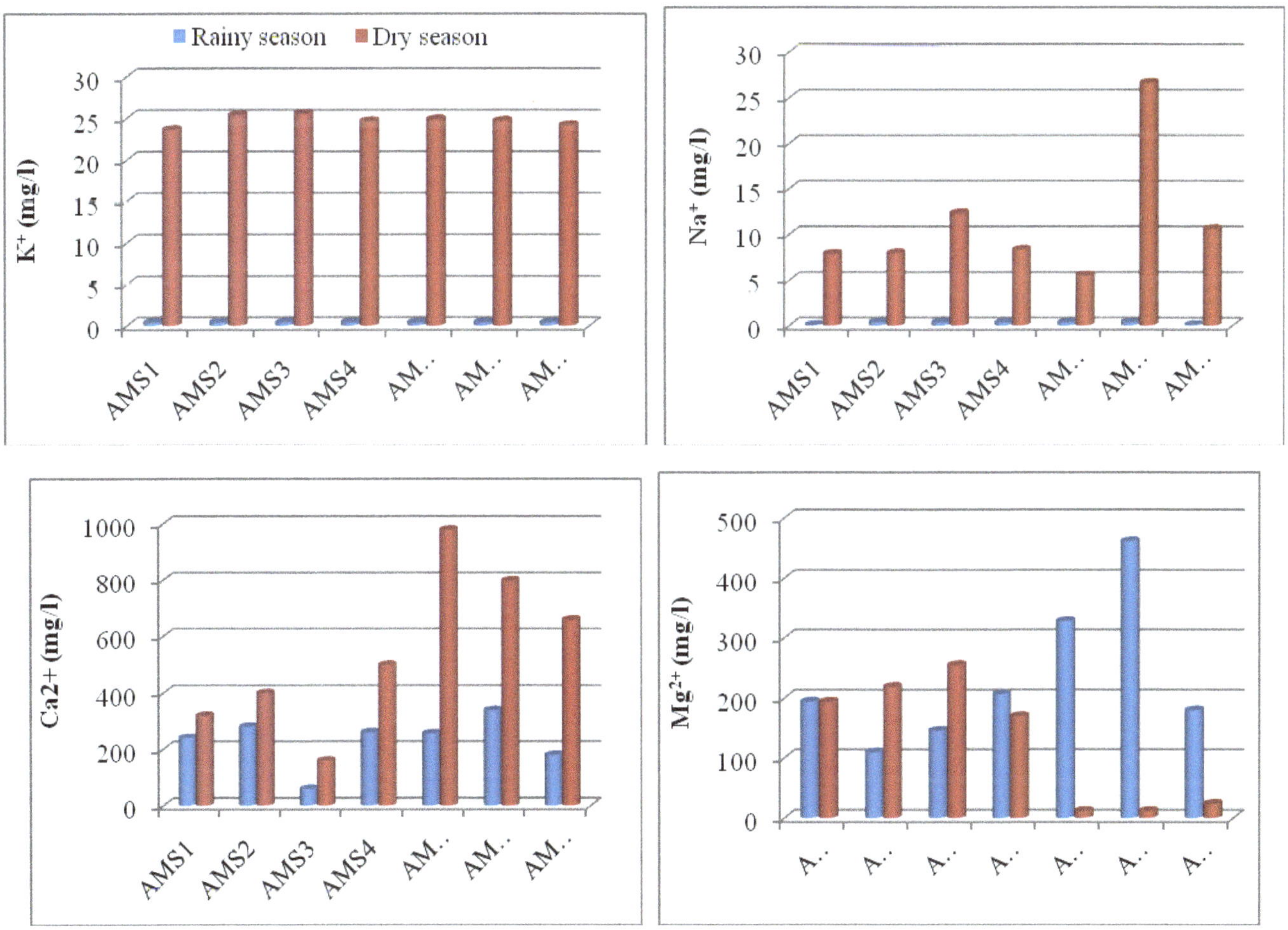

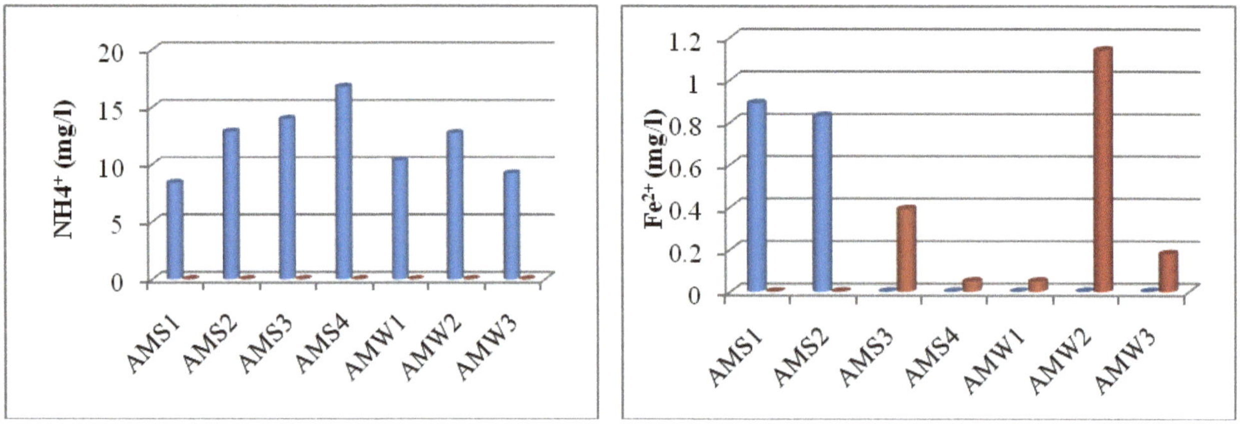

Figure 3. *Seasonal variation of cations (K^+, Na^+, Ca^{2+}, Mg^{2+}, NH_4^+, Fe^{2+}).*

4.2.1. Anions

Cl^- was completely absent in all the samples in the dry season. In the rainy season, high Cl^- values were recorded only in AMS_4 (158mg/l) and AMW_2 (270mg/l) as shown in Table 3.

Average values for HCO_3^- were higher in the rainy season (65.36mg/l) than in the dry season (50.13mg/l).

The values of PO_4^{3-} were higher in the dry season than in the rainy season but all the values were less than 2mg/l with an average of 0.16mg/l in the dry season and of 1.11mg/l in the rainy season (Table 3). NO_3^- and SO_4^{2-}were barely detectable in the dry season with the respective average values of 0.0033mg/l and 0.05mg/l. In the rainy season for each cation there were just minor differences in the sampling points, NO_3^- concentration yielded an average of 7.68mg/l and SO_4^{2-} an average of 17.61mg/l in the rainy season as presented in table 3 and Figure 4.

Table 3. *Variation of Anions in the rainy season and dry season.*

	Cl^-		HCO_3^-		PO_4^{3-}		NO_3^-		SO_4^{2-}	
Sampling points	**Rainy Season**	**Dry Season**	**Rainy Season**	**Dry Season**	**Rainy Season**	**Dry Season**	**Rainy Season**	**Dry Season**	**Rainy Season**	**Dry Season**
AMS_1	-	-	57.95	75.6	0.25	1.67	11.20	0.0028	15	0.05
AMS_2	-	-	30.50	40.32	0.19	1.23	8.40	0.0006	12	0.05
AMS_3	-	-	39.65	42.84	0.19	1.23	5.60	0.0022	18	0.06
AMS_4	158	-	33.55	37.8	0.13	0.80	7.28	0.0034	12	0.05
AMW_1	-	-	231.80	116.55	0.13	1.23	5.88	0.0048	26	0.06
AMW_2	270	-	30.50	37.8	0.13	0.80	7.00	0.0045	23	0.05
AMW_3	-	-	33.55	-	0.13	0.80	8.40	0.0048	17.3	0.05
Average	*61.14*	*0.00*	*65.36*	*50.13*	*0.16*	*1.11*	*7.68*	*0.0033*	*17.61*	*0.05*

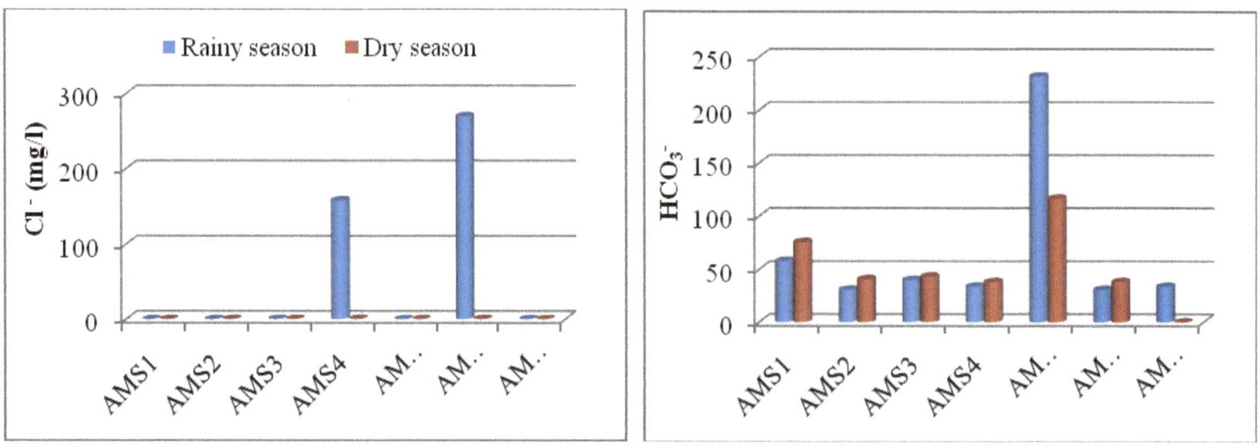

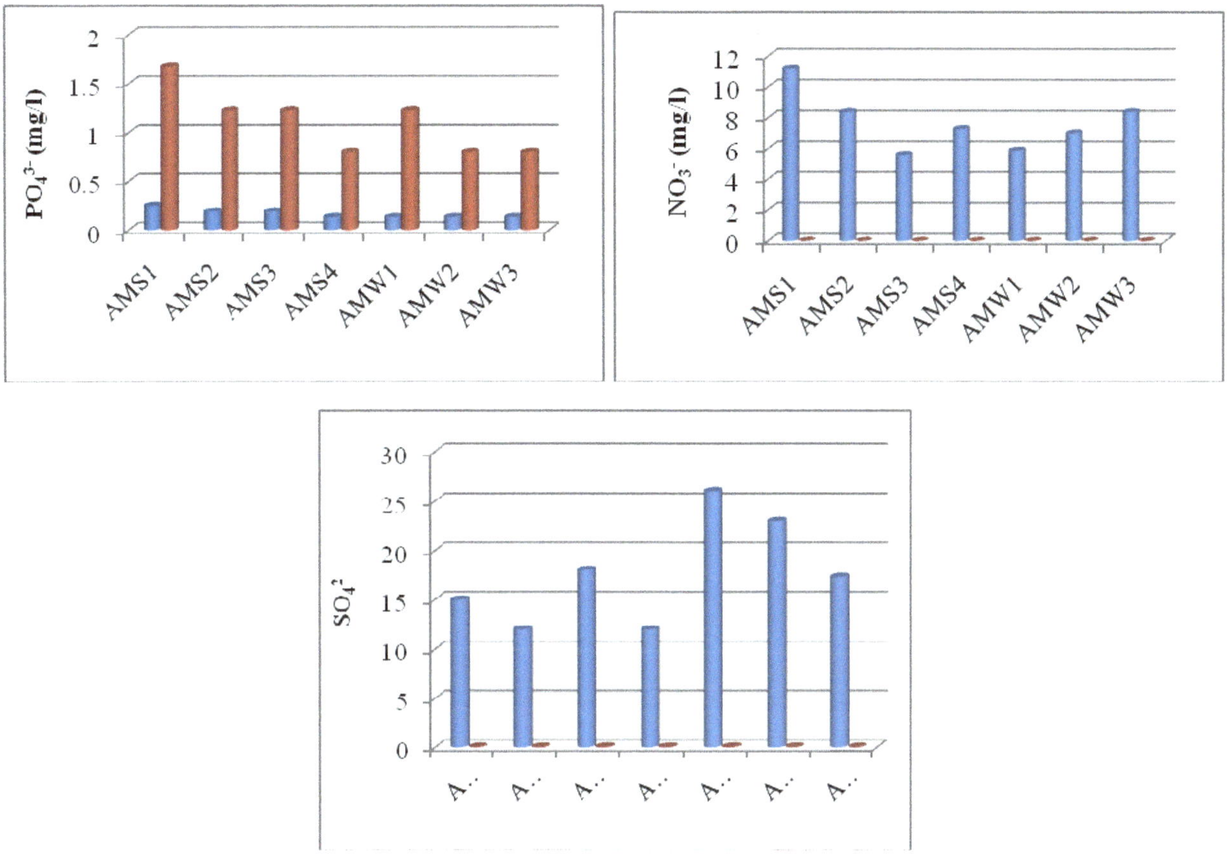

Figure 4. *Seasonal variation of anions (Cl^-, HCO_3^-, PO_4^{3-}, NO_3^-, SO_4^{2-}).*

4.2.2. Hydrochemical Facies

The piper diagram provides information on the different chemical facies of the analysed water.

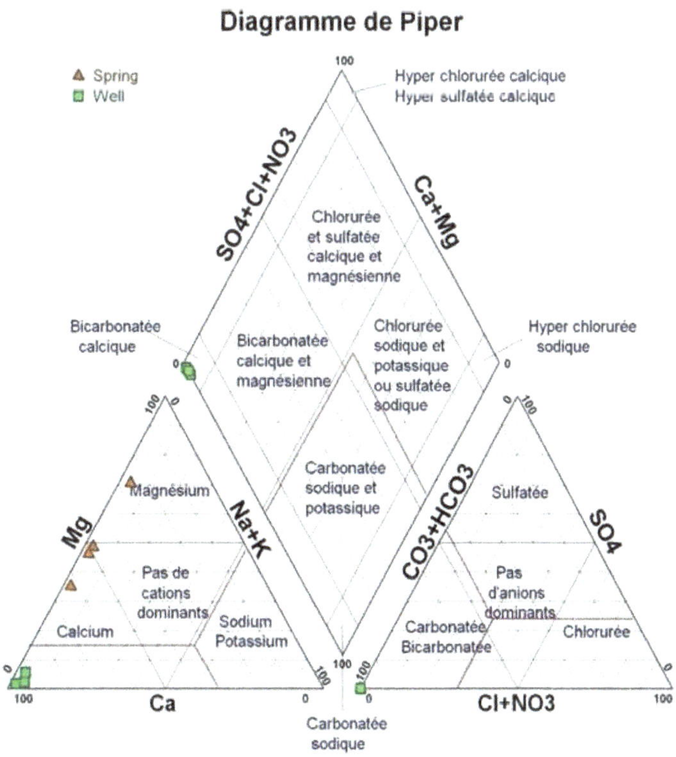

a

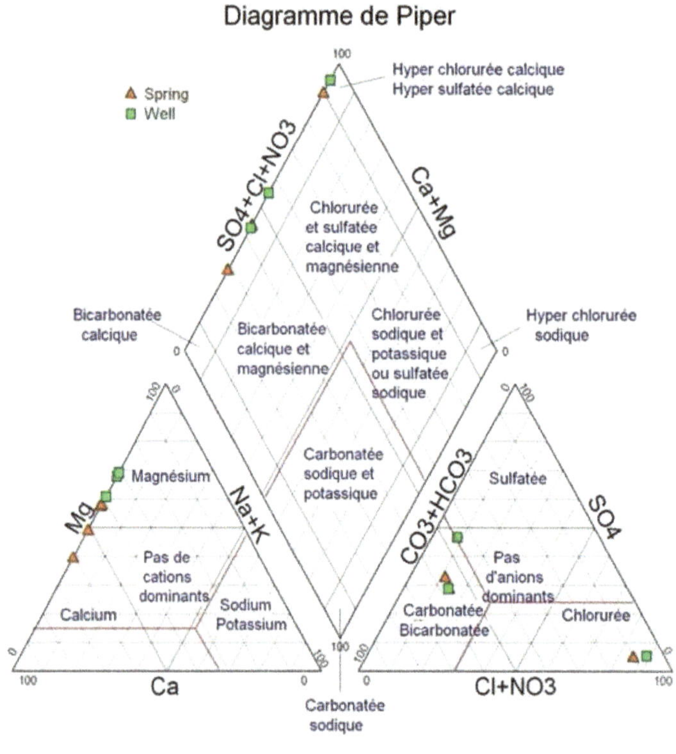

Figure 5. Piper's diagram [15], showing the water types in Bamenda III, a) dry season b) for rainy season.

In the dry season, the major cations and anions presented one facies of the Ca- Mg-HCO3 type (Figure 5a) whereas in the rainy season the major cations and anions presented two facies of Ca-Mg-HCO3 type and Ca-Mg-SO4 type (Figure 5b). This significant difference may be due to the hydrochemical processes taking place within the aquifer systems in different seasons. The mineralization process is more homogeneous in the dry season while in the rainy season mineralization is heterogeneous and influenced by the weathering of rocks.

4.3. Bacteriological Characteristics

Three (03) water points (AMS1, AMS2 and AMW2) were analysed for bacteria indicators of pollution as well as for specific microbes (Table 4).

Table 4. Bacterial counts of specific microbes isolated in springs and well in the rainy and dry seasons.

	Enterobacteria (cfu/100ml)		E. coli (cfu/100ml)		Streptococcus (cfu/100ml)		Salmonella (cfu/100ml)		Proteus (cfu/100ml)	
	Rainy Season	Dry Season	Rainy Season	Dry Season	Rainy Season	Dry Season	Rainy Season	Dry Season	Rainy Season	Dry Season
AMS_1	50	100	20	30	200	50	00	00	00	10
AMS_2	20	200	10	20	100	70	00	10	00	50
AMW_2	200	300	100	200	300	200	60	15	10	00
Average	90	200	43,33	83,33	200	106,67	20	8,33	3,33	20

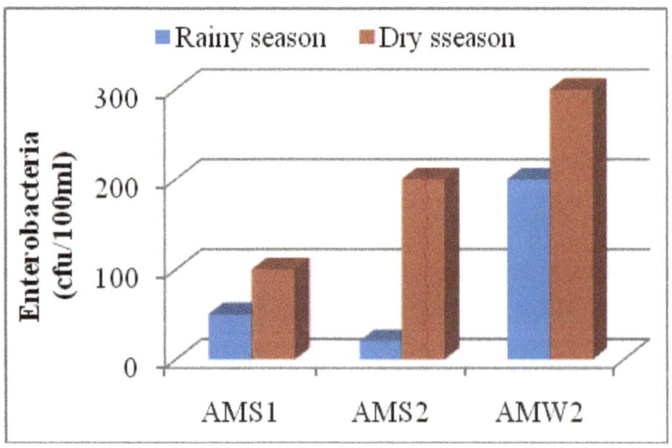

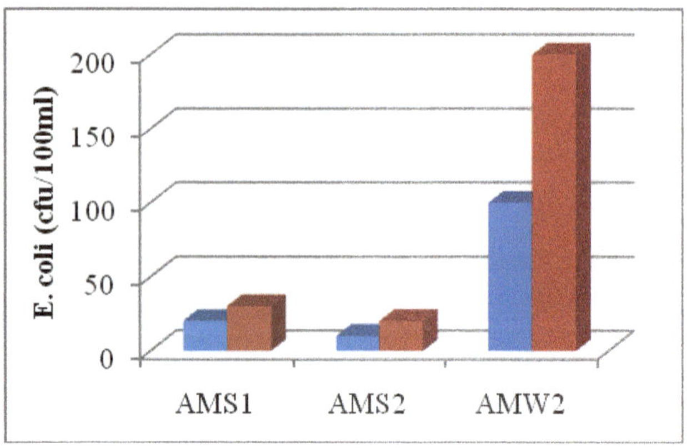

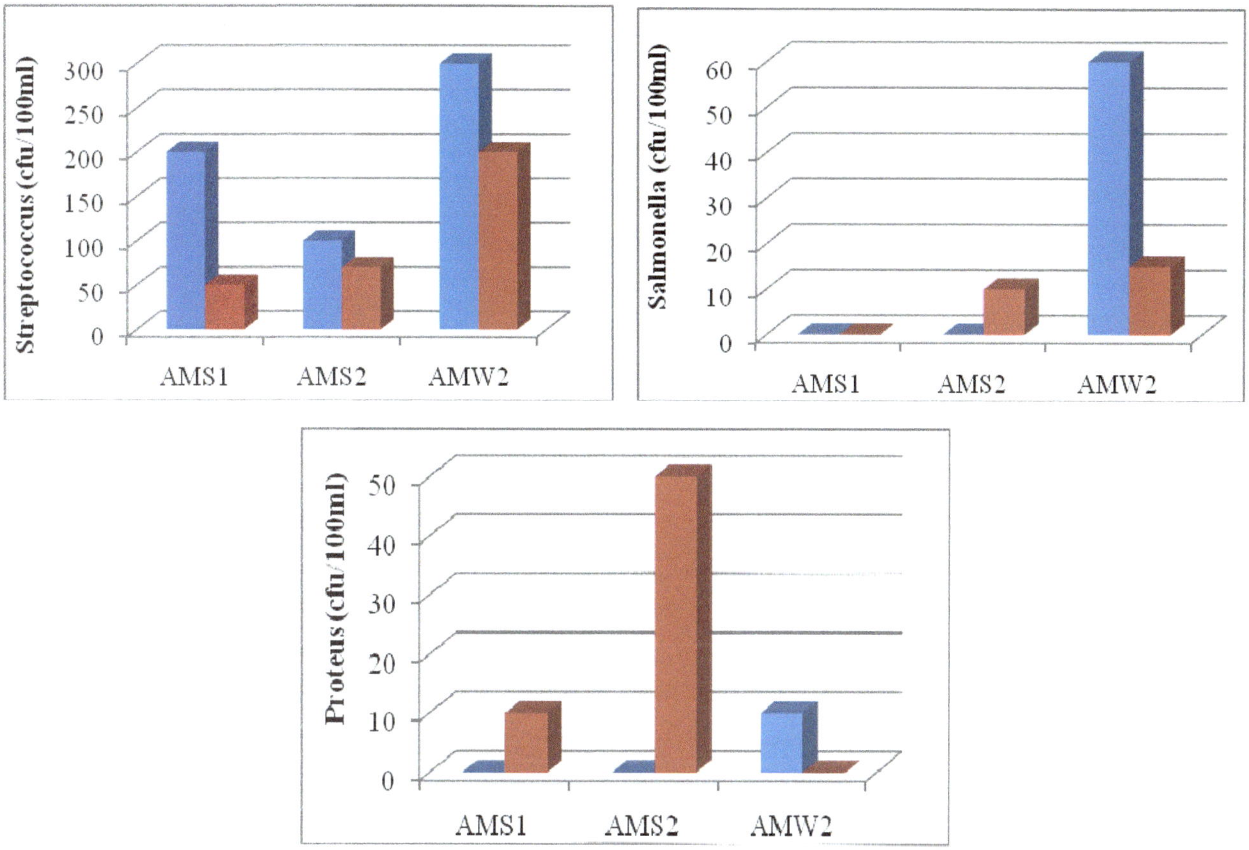

Figure 6. *Seasonal variation of bacteria indicators of pollution and specific microbes.*

Enterobacteria, Escherichia coli, Streptococcus, Salmonella and Proteus were detected in the water samples and the species richness varied from one water point to another: Enterobacteria, Escherichia coli and Streptococcus were present in all the sampling points whereas Salmonella and Proteus were present only in some sampling points. Higher counts were recorded in the dry season than in the rainy season with Streptococcus presenting an exception with higher counts in the rainy season (100 to 300 cfu/100ml) than in the dry season (50 to 200 cfu/100ml). Enterobacteria levels were particularly high during the rainy season (20 to 200 cfu/100ml) but increased greatly during the dry season (100 to 300cfu/100ml). On the other hand Salmonella and Proteus were absent in springs (AMS1 and AMS2) in the rainy season but registered an increase in the dry season.

5. Discussion

The major ions chemistry of groundwater and compositional relations among ionic species can reveal the origin of solutes and processes that generate water composition [16-18].

The acidic to barely acidic water may be an indication that the dissolve carbonates were predorminantly in the form of HCO_3; In Ghana [19] and in Nigeria [20] low pH has also been reported from granitic basement complex aquifers and were related to the same processes.

5.1. Cations

The relative abundance of the cations was as follows $Ca^{2+} > Mg^{2+} > K^+ > Na^+$, this shows that

Ca^{2+} is the most abundant cation followed by Mg^{2+} in the springs and wells of the study area. Their main source may be from the weathering of calco-sodico feldspaths and hornblende in the basic rock. [21] explains that these minerals are found in basic rocks.

2 Hornblende $+ 30H^+ + 9H_2O \rightarrow$ Kaolinite $+ 12H_4SiO_4 + 4Ca^{2+} + 10Mg^{2+} + 2Na^+$.

The most common source of K^+ is the silicate minerals: orthoclase, nepheline, leucite and biotite [22]. The basement rock of the study area is composed of granite and the K^+ is probably from the dissolution of orthoclase (KAlSi3O8)) and Fe Mg minerals like biotite ($K(Mg\ Fe)_3(AlSi_3)O_{10}(OH, Fe)_2$ present in the granites.

2 Orthoclase $+2H^+ + 9H_2O \rightarrow$ Kaolinite $+ 4H_4SiO_4 + 2K^+$

2 Biotite $+ 13H^+ + CO_2 + 2H_2O \rightarrow$ Kaolinite $+ 4H_4SiO_4 + 6Fe^{2+} + 6Fe^{3+} + HCO_3^- + 2K^+$

Albite in the granites is the probable source of the Na^+

2 Albite $+2H + 9H_2O \rightarrow$ kaolinite $+ 4H_4SiO_4 + 2Na^+$

Meanwhile the main source of NH_4^+ is probably from fertilizers used in farms which is being leached into the water while natural sources such as organic (metabolic processes) and inorganic (rock weathering and hydrothermal activity) could also contribute to the presence of NH_4^+.

The source of Fe^{2+} is from the hydrolysis of Fe Mg rich minerals such as biotite in the granite and diorite of the study area. The low concentration of Fe^{2+} is due to its being retained in the soil for the formation of clays, oxides and hydroxides during weathering [23,24].

5.2. Anions

The relative abundances of anions (mg/l) in different samples were as follows: $HCO_3^- > SO_4^{2-} > NO_3^- > PO_4^{3-} > Cl^-$.

HCO_3^- concentrations were high with average values ranging from 16.78 – 174.18mg/l. All the samples had values below the WHO value (200mg/l). The dorminance of HCO_3^- is consistent with most natural waters along the CVL [25]. The primary source is the dissolved CO_2 in rain water and the decay of organic matter at the surface which may release CO_2 for dissolution [22]. Weathering, precipitation and organic decay probably account for the dominance of this ion, $CO_2 + H_2O \rightarrow H^+ + HCO_3^-$ or $CO_2 + H_2O + CO_2 \rightarrow 2HCO_3^-$.

SO_4^{2-}could derive from natural source such as sulphate minerals common in igneous rocks [22]. It can also be from the oxidation of pyrite; $Fe\ S_2 + 2O_2 + H_2O \rightarrow 2\ SO_4^{2-} + H_2$. The low concentrations may be a consequence of gradual dissolution [26].

The presence of PO_4^{3-} probably indicates leaching of phosphate fertilizers from nearby farmlands.

Low concentrations of NO_3^- in the dry season suggest their adsorption in organic colloids [27]. The potential source may also be shallow pit toilets and oxidation of organic matters in biochemical processes as well as the oxidation of ammonia by Nitrosomonas and Nitrobacter bacteria.

Cl^- was absent in the rainy season and according to [28], at high altitude the decrease in Cl^- concentration is due to smaller rainfall inputs caused by rapid decrease of the absolute humidity and temperature with elevation resulting in a smaller condensation. An anormalous Cl^- concentration of 158mg/l and 270mg/l were obtained in the rainy season which might have been due to disinfection through chlorination.

6. Groundwater Suitability for Drinking and Domestic use in Bamenda III

The analytical results were evaluated to ascertain the suitability of groundwater in the study area for drinking and domestic purposes based on the WHO, 2004 standard (Table 5) and the [28, 29] classification (Table 6).

Table 5. *Groundwater quality in Bamenda III in compliance with WHO (2004) drinking water standard.*

Parameter	Range in study area	WHO limit (2004)
pH	4.1 – 6.5*	6.5 – 8.5
T (C°)	20.7 – 25.9*	15
EC (ųs/cm)	0.04 – 2.38	750
Ca2+ (mg/l)	110 – 1770*	75
Mg2+ (mg/l)	97.05 – 236.85*	30
Na+ (mg/l)	4.02 – 13.52	200
K+ (mg/l)	12.02 – 12.97	100
Cl- (mg/l)	0 – 270*	250
NO3⁻ (mg/l)	2.801 – 5.601	10
SO42⁻ (mg/l)	6.025 – 11.525	250
HCO3- (mg/l)	16.78 – 174.18	200

* values above WHO limit

EC, Na^+, K^+, NO_3^-, SO_4^{2-}, and HCO_3^- concentrations of the analysed water were under the limits required by the WHO (2004) drinking water norms whereas pH and T (C°) values, Ca^+, Mg^+, and Cl^- concentration values were above the WHO limits.

pH values for the study area ranged from 4.1 to 6.5 and were below the WHO guideline limit (Table 6). Even though pH has no effect on human health, it is closely related to other chemical constituents of water [3].

The temperature values were high which favour bacterial growth and may increase taste, odour, colour and corrosion problems.

Ca^+ and Mg^+ values were much higher than the WHO norms. The presence of these alkaline earth metals determines water hardness. [30] expresses Hardness of water as Total Hardness (TH) and is given by $TH = 2.5Ca + 4.1Mg$. Though hardness has no adverse effect on human health, there is an esthetic concern because of the unpleasant taste that it impacts. It can prevent formation of lather with soap, increase the boiling point of water and may cause encrustation in water supply distribution system. According to [31], there is some suggestive evidence that long term comsumption of extremely hard water might lead to increase urolithiasis, ancephaly, parental mortality and some kind of cancer and cardiovascular disorders. On the other hand water softness (low in Ca^{2+} and Mg^{2+}) could be a health problem since soft water has been linked to cardiovascular ailments [32].

WHO limit for Cl^- is 250mg/l and sample AMW_2 exceed this limit in the rainy season. Higher concentration may cause gastrointestinal irritation when associated with Na^+ and Mg^{2+} [33].

According to the [29] classification of drinking water, two of the sampling points (AMS_1 and AMW_2) could be placed in category C and one (AMS_2) under category B in the rainy season (Table 5). In the dry season, AMS_1 and AMS_2 fell under category C and AMW_2 under category D (Table 5).

Table 6. *Category of water samples according to Cheesbrough classification of drinking water.*

Sampling points	Seasons	Mean count	Category
AMS_1	Rainy	30	C
	Dry	35	C
AMS_2	Rainy	10	B
	Dry	30	C
AMW_2	Rainy	40	C
	Dry	90	D

These bacterial loads indicate an excessive level of pollution of the water samples [01]. Feacal coliforms such as E. coli are prevalent in the digestive tracts of warm – blooded animals and their presence in water is definite evidence of feacal contamination.

Contaminations of water points are much higher in the dry season than in the rainy season. This may be feacal bacteria accumulate and settle as a result of greater contact between water and sediments which enables significant sediment–water exchange [34].

7. Conclusion

Bamenda III, which is petrographicaly made up of two units, plutonic (granite) and volcanic (Rhyolite, basalt and trachyte), was subjected to physico-chemical and bacteriological analyses of its 7 water points: 03 wells and 04 springs.

All water samples had pH range between 4.1 and 6.5 which is below the WHO standard for drinking water. The low EC and TDS classified the water as low-mineralised and freshwater. The relative abundance of major cations and anions in the water (mg//l) was $Ca^{2+} > Mg^{2+} > K^{+} > Na^{+}$ and $HCO_3^{-} > SO_4^{2-} > NO_3^{-} > PO_4^{3-}$ respectively. Major ion concentrations were low, and below maximum values of the WHO standards for drinking water, except for Ca^{2+} and Mg^{2+} which were higher classifying the water as hardwater.

The variety and level of faecal bacteria recorded indicate that this area is under anthropogenic influence and receives feacal contamination. None of the water points are suitable for drinking. For this reason, the population of Bamenda III, 80% of which depends on the well and spring water for drinking and domestic purposes, may face serious health problems if appropriate actions are not taken to prevent and mitigate the problems.

References

[1] Katte V.Y., Fonteh M.F., Guemuh G.N. (2003): Domestic water quality in urban centres in Cameroon: A case study of Dschang in the West Province. Afr. Water. J. Pilot ed.: 43-54.

[2] Molua E.L., Lambi C.M. (2006) : Climate hydrology and water resources in Cameroon. CEEPA.

[3] Ako A.A. (2011): Hydrogeological study on groundwater in the Banana Plain and Mount Cameroon area – Cameroon Volcanic Line. Kumamoto University.

[4] Auippa A., Bellomoa S., Bruscab L., Alessandrolo W., Federico C. (2003): Natural and Anthropogenic factors affecting ground water quality of an Active volcano (Mt. Etna, Italy): Applied Geochemistry 18: 863-882.

[5] Mafany G.T., Fantong W.Y., Nkeng G.E. (2006): groundwater quality in Cameroon and its vulnerability to pollution. In Groundwater pollution in Africa. Edited by Yongxin Xu and Brent Usher. Taylor and Francis London

[6] Ako A.A., Shimada J., Hosono T., Kagabu M., Akoachere R.A., Nkeng G.E., Eneke G.T., Fouepe Appelo C., Postma D. (2005): Geochemistry, groundwater, and pollution (2nd ed.). Balkema, Amsterdam p. 635.

[7] Edmunds W.M., Smedley P.L. (1996): Groundwater geochemistry and health: An overview. In: Appleton JD, Fuge R, McCall GJH (eds.) Environmental geochemistry and health with special reference to developing countries. Geological Society Special Publication No.113, pp. 91-105.

[8] Deutsch W.J. (1997): Groundwater geochemistry: Fundamentals and applications to contamination. CRC Press, Florida p. 221

[9] Ramesh K., Elango L. (2012) : Groundwater quality and its suitability for domestic and agricultural use in Tondiar river basin, Tamil Nadu, India. Environ. Monit. Assess. 184:3887-3899.

[10] Entry J.A., Farmer N. (2001): Movement of coliform bacteria and nutrients in groundwater flowing through basalt and sand aquifers. J. Environ. Q. 30: 1533 – 1539.

[11] Acho C. (1998): Human interface and environmental instability: Addressing the environmental problems of rapid urban growth. In: Environment and urbanisation. Vol. 10. No. 2. London. P.415.

[12] Nzenti J.P., Abaga B., Suh E.C., Nzolang C., (2010): Petrogenesis of peraluminous magmas from the Akum-Bamenda massifs, Pan-African Fold Belt.International Geology Review. 1-29p.

[13] Njilah I.K., (1991): Geochemistry and petrogenesis of the Tertiary-Quaternary volcanic rocks from Oku Ndu area, North West Cameroon. PhD Thesis, Leeds University, 350pp+1map.

[14] Kamgang P., Njonfang E., Nono A., Gountie D.M., Tchoua F. (2010):Petrogenesis of a silicic magma system: geochemical evidence from Bamenda Mountains, NW Cameroon, Cameroon Volcanic Line; Journal of African Earth Sciences 58: 285–304

[15] Piper A.M. (1944): A graphic procedure in the geochemical interpretation of water analyses. Transactions American Geophysicists Un. Papers, Hydrology, 914-929.

[16] Hounslow A.W. (1995): Water quality data: Analysis and interpretation. Lewis Publishers, Boca Raton.

[17] Demlie M., Wohnlich S., Wisotzky F., Gizaw B. (2007) : Groundwater recharge, flow and hydrogeochemical evolution in a complex volcanic aquifer system, central Ethiopia. Hydrogeol. J. 15: 1169-1181.

[18] Tay C.K. (2012) : Hydrochemistry of groundwater in the Savelugu-Nanton District, Northern Ghana. Environ. Earth Sci. 67: 2077-2087.

[19] Adomako D., Osae S., Akiti T.T., Faye S., Maloszewski P. (2011): Geochemical and isotopic studies of groundwater conditions in the Densu River Basin of Ghana. Environ. Earth Sci. 62:1071–1084.

[20] Edet A., Nganje T.N., Ukpong A.L., Ekwere A.S. (2011) : Groundwater chemistry and quality of Nigeria: A status review. Afr. J. Environ. Sci Technol. 5 (13): 1152-1169.

[21] Marzoli A., Piccirillo E.M., Renne P.R., Bellien G., Facumin G.M., Nyobe J., Tongwa A.T. (2000): The CVL Revisited: petrogenesis of continental Basaltic mangmas form lithospheric and Asthenospheric Mantle sources; Journal of petrology; 41(1): 87-109.

[22] Karanth K.R. (1987): Groundwater Assessment: Development and management; Tatamc Grawl Hill publishing company limited; 7 west patel Nagar, New Delhi; 488pp.

[23] Goldschmidt (1954): Geochemistry. A Muir (éd) clarendo press, Oxford 730p

[24] Millot G. (1964): Geologie des argiles Masson et Cie (ed) Paris , 499 p

[25] Tanyileke G. Z., Kusakabe M., Evans W. C., (1996): Chemical and isotopic characteristics of fluids along the CVL, Cameroon – Journal of African Earth Sciences, 22. (4): 433-441.

[26] Ngwa M. (2013): Hydrochemical variation of groundwater in Bafut Subdivision (North West Region, Cameroon).

[27] Fonge B.A., Egbe E.A., Fongod A.N., Tening A.S., Achu R.M., Yinda G.S., Mvondo Z. (2012): Effects of land use on macrophyte communities and water quality in the Ndop wetland plain, Cameroon. J. Agric. Soc. Sci. 12: 41-49.

[28] Benedetti M.F., Dia A., Riotte J., Chabaux F., Gerald M. Boulegue J., Fritz B., Chauvel C., Bulourde M., Deruelle B., Ildefonse P. (2003): Chemical weathering of basaltic lava flows undergoing extreme climatic conditions: the water geochemistry record. Chemical geology 2011- 17.

[29] Cheesbrough M. (1991). Medical Laboratory manual for tropical countries. Low priced edition, Daddington, Cambridgshire.

[30] Todd D. k. (1980): Goundwater Hydrogeology. New York : Wiley.

[31] Durvey V.S., Sharma L.L., Saini V.P., Shama B.K. (1991): Handbook on the methodology of water quality assessment in India: Rajasthan Agriculture University.

[32] Kozisek F. (2005): Health risks from drinking demineralised water. In: Nutrients in drinking water. World Health Organization, Geneva. pp. 148-163.

[33] Garg V.K., Suthar S., Singh S., Sheoran A., Garima M., Jain S. (2009): Drinking water quality of Southwestern Hanyana India : Assessing human health risks associated with hydrochemistry. Environmental Geology. Dio: 10. 1007/s00254 – 008 1636 – y.

[34] Mitsch W.J., Gosselink J.G. (2000): Wetlands. 920 p. wiley, New York, USA.

Levels of Selected Essential and Nonessential Metals in Roasted Coffee Beans of Yirgacheffe and Sidama, Ethiopia

A. Tesfay Gebretsadik[1], Tarekegn Berhanu[2], Belete Kefarge[3]

[1]Department of Industrial Chemistry, Addis Ababa Science and Technology University, Addis Ababa, Ethiopia
[2]Strengthening of Agricultural Pesticide Residue Analysis System Project, JICA, Addis Ababa, Ethiopia
[3]Department of Chemistry, Jigjiga University, College of Natural and Computational Science, Jigjiga, Ethiopia

Email address:
tgtesfageb@gmail.com (A. T. Gebretsadik)

Abstract: The study was conducted to assess the contents of essential and non-essential metals in coffee beans. For this matter, seven essential metals such as K, Mg, Ca, Na, Mn, Cu and Zn and two nonessential metals (Cd and Pb) in four roasted coffee samples (washed Yirgacheffe, unwashed Yirgacheffe, washed Sidama and unwashed Sidama) were determined by FAAS. Closed microwave assisted wet digestion method with addition of concentrated (69-70%) HNO_3 and 30% H_2O_2 were selected for decomposition of ground roasted coffee samples. Generally, the levels of metals in all roasted coffee samples were found: K > Mg > Ca > Na >Mn> Zn > Cu, but the non-essential metals Pb and Cd were found to be below method detection limit. The digestion method was evaluated by spiking roasted coffee samples and their percentage recoveries were in the range of 95 −104 %. It is suggested that the consumption of roasted coffee beans could be a source of dietary essential metals and a possible entrance path way for trace metals to the food chain.

Keywords: Essential and Non-essential Metals, FAAS, Micro-wave Digestion, Roasted Coffee Beans

1. Introduction

Currently, one of the greatest challenges is the production of food, not only in sufficient amounts, but also with adequate nutritional quality, assuring nutritional balanced and a contaminant free food supply. Moreover, the increasing awareness of the important role of trace elements in biological systems has stimulated the extension and refinement of studies in this field which have provided a great deal of information. Accordingly, the content of mineral micronutrients and some other trace elements in coffee beans is of great general and special interest [1].

Over two-thirds of all the research literature on geographic origin commodities involves the analysis of vitamins or other organic molecules (amino acids, triglycerides, volatile aromatic compounds, etc.) [2]. However, a shortcoming of using vitamins (or other organic compounds) is their susceptibility to degradation those organic chemicals including enzymatic changes from the time of harvest through storage to the time of analysis [3]. This is also important because coffee beans are processed and organic chemical profiles are likely to be especially susceptible. The use of minerals and trace elements is therefore powerful because trace elements are significantly more stable in the commodity versus vitamins or some other types of organic compounds [4].

For determination of metals in raw or roasted coffees have used different analytical techniques like flame atomic absorption spectrometry (FAAS) [5], inductively coupled plasma optical emission spectrometry (ICP OES) [6] and neutron activation analysis (NAA) [7], results of these studies have revealed that the levels of metals differs among coffee species and varieties growing in the different parts of the world [8, 9]. This study aimed to determine the levels of essential and non-essential metals in washed and unwashed Yirgacheffe and Sidama roasted coffee beans which grow in Ethiopia.

2. Experimental

2.1. Instruments and Apparatus

Coffee Roasting Machine (PROBAT-WERKE, BRZ-2 type

Emmerich am Rhein, 2006, Germany), Microwave oven (Buck Scientific, inc. Type MWS-1, East Norwalk, USA) equipped with digestion vessels made of Teflon Fluor Modified (TFM), Flame atomic absorption spectrophotometer (FAAS) (Buck Scientific Model 210VGP AAS, East Norwalk, USA) fitted with deuterium background corrector and air acetylene flame atomizer.

2.2. Chemicals and Reagents

All the chemicals used were of analytical grade: 69−70% HNO_3 (Uni-Chem^R, Chemical reagents.), 30% H_2O_2 (Uni-Chem^R, Chemical reagents.), Lanthanum nitrate hydrate, 99.9% (Aldrich, USA), Stock standard solutions of concentration 1000 mg/L in 2% HNO_3 of the metals K, Mg, Ca, Na, Mn, Zn, Cu, Cd and Pb (Inorganic^{TM} Ventures.) and Deionized water chemically pure with <1.5 µs/cm.

2.3. Sample Preparation

Both washed and unwashed coffee samples were collected using plastic bags from Ethiopia Commodity Exchange (ECX, Hawassa, Ethiopia) and Yirgacheffe Woreda marketing and cooperative bureau. The pre-treated (cleaned) green or raw coffee bean samples were roasted in 150 g batches by using coffee roasting machine (PROBAT-WERKE, BRZ-2 type Emmerich am Rhein, 2006, Germany) at the Ethiopian Commercial Exchange (ECX, Hawassa, Ethiopia) Coffee Quality and Flavor Inspection laboratory Center. All the roasting process was carried out by an expert on coffee roasting and the temperature and time needed for roasting of all coffee samples were 180 °C and 8-12 min respectively, medium roasting was selected among the degree of roasting coffee beans (light, medium and dark). Finally the roasted coffee samples were ground in to fine powder using mortar and pestle (to protect metal contamination).

2.4. Digestion of Ground Roasted Coffee Samples

In this study a method developed by Oleszczuk*et al.* [14] with a mass of 0.3 g powder of each roasted coffee sample was transferred to TFM digestion vessels followed by addition of 5 mL of concentrated (69-70%) HNO_3 and 3 mL of 30% H_2O_2 was selected for decomposition of ground roasted coffee samples. In the first step the temperature was linearly increased to 145 °C in 5 min with power of the rotating 80 W. In the second step the temperature was kept 200 °C for 10 min with power of the rotating 90 W and in the third step the temperature was linearly decreased to 50 °C in 10 min with rotating power of 0 W. For digestion of blank samples, eight reagent blanks were digested following the same procedure used for the ground roasted coffee samples. Finally, the concentrations of metals (K, Na, Ca, Mg, Mn, Cu, Zn, Cd and Pd) were determined directly in the extract solution by flame atomic absorption spectrometry (FAAS).

2.5. Digestion the Spiked Roasted Coffee Samples

A spiking experiment was done to evaluate the efficiency and accuracy of the the method used. Known amounts of standard metal solutions were added to the roasted coffee samples taking care of the dilution of the final solution. Aliquots of 0.75 mL of 1000 mg/L K, 0.75 mL of 100 mg/L Mg, Ca and Na and 0.3mL of 10 mg/L Mn, Zn, Cu, Pb and Cd were spiked at once in to a 0.3 g of ground roasted coffee samples in the TFM digestion vessel and digested following the same optimized procedure as the unspiked samples.

2.6. Statistical Analysis

The significance of variation between samples was analyzed by student's t-test and one-way ANOVA using Microsoft Excel (Microsoft Excel 13.0 Work Sheet, The Apache Software Foundation, 2007) was used.

3. Results and Discussions

3.1. Method Validation

The method detection limit for the metals in roasted coffee samples (Table 1) were calculated as as given below in equation 1:

$$MDL = t(n-1, 1-\mu = 0.99)(S) \quad (1)$$

Where: MDL = method detection limit t (n-1, 1-µ= 0.99) = students' t value appropriate for a 99% confidence level with (t_7= 2.99, for n= 8) S = standard deviation estimate with n-1 degrees of freedom (for n= 8).

Table 1. *Method detection limit (MDL) for roasted coffee bean samples.*

Elements	K	Mg	Ca	Na	Mn	Zn	Cu	Cd	Pd
MDL(µgg^{-1})	0.07	0.05	0.04	0.06	0.02	0.03	0.02	0.01	0.04

The percentage recoveries were calculated using equation 2 given below.

$$R\,(\%) = \frac{\text{metalconc.inspicked sample} - \text{metal conc.in unspicked sample}}{\text{Conc.of metal spiked}}\, x\, 100 \quad (2)$$

The percentage recoveries of metals in the spiked roasted coffee samples were found in the range of 95 %−104 % with RSD 3−6% indicated in Table 2.

Table 2. *Recovery test results for the analysis of metals in roasted coffee samples.*

Element	Concentration in the sample (μgg^{-1})^a	Amount added (μgg^{-1})	Concentration in the spiked sample (μgg^{-1})^b	Recovery (%)^c
K	18563	2500	21013±21	98±6
Mg	1992	250	2234±19	97±4
Ca	965	250	1213±12	95±3
Na	446	250	706±20	104±6
Mn	19	10	29±0.8	96±5
Zn	16	10	26±0.6	103±4
Cu	10	10	21±0.5	101±3
Cd	ND	10	10.3±0.3	103±4
Pb	ND	10	10.2±0.2	102±5

^{a}Mean concentration of samples, ^{b}Mean concentration ± SD of spiked samples, ^{c}Mean recovery ± SD of percentage recoveries, ND= Not detected

3.2. Distribution of Metals in Roasted Coffee Samples

The concentrations of macro-elements (K, Mg, Ca and Na), micro-elements (Mn, Cu and Zn) and non-essential elements (Cd and Pb) determined in four types of roasted coffee samples summarized in Table 3.

Generally, the mean concentration of metals in all roasted coffee samples follows the order: K > Mg > Ca > Na >Mn> Zn > Cu, but the non-essential metals Pb and Cd were found to be below detection limit.

Table 3. *Concentration (mean ± SD, n = 3, μgg^{-1}) of elements in the roasted coffee samples.*

Element	Concentration of metals (μgg-1)a			
	Washed Yirgacheffe roasted coffee	Unwashed Yirgacheffe roasted coffee	Washed Sidama roasted coffee	Unwashed Sidama roasted coffee
K	19610±343	19471±392	18754±462	18563±477
Mg	1970±41	1943±45	2030±98	1992±93
Ca	943±29	931±17	1009±18	976±24
Na	484±85	463±69	475±65	446±73
Mn	23±0.5	21±0.3	20±0.1	19±0.2
Zn	14±0.3	15±0.2	17±0.2	18±0.3
Cu	12±0.3	13±0.2	9±0.3	10±0.1
Cd	ND	ND	ND	ND
Pb	ND	ND	ND	ND

^{a}Mean concentration ± standard deviation of each sample analysed.

Washed Yirgacheffe roasted coffee: contains K in highest amount of the macro-elements with concentration 19610±343 μgg^{-1}followed by Mg (1970±41 μgg^{-1}) and Ca (943±29 μgg^{-1}). Na (484±85 μgg^{-1}) was found to be present at the lowest concentration of the macro-elements analyzed (Table 3). Among the concentration of trace micro-elements, Mn (23±0.5 μgg^{-1}) was found in higher amount than others followed by Zn (14±0.3 μgg^{-1}) and Cu (12±0.3 μgg^{-1}).

Unwashed Yirgacheffe roasted coffee: of the macro-elements K with concentration 19471±392 μgg^{-1} was found to be present in the highest level followed by Mg (1943±45 μgg^{-1}), Ca (931±17 μgg^{-1}) and Na (463±69 μgg^{-1}) with the lowest concentration of the macro-elements analyzed (Table 3). Unwashed Yirgacheffe roasted coffee type also contains, Mn with concentration 21±0.3 μgg^{-1}followed by Zn (15±0.2 μgg^{-1}) and Cu (13±0.2 μgg^{-1}).

Washed Sidama roasted coffee: among the macro--elements, K was found in highest amount with concentration 18754±462 μgg^{-1} than Mg (2030±98 μgg^{-1}) and Ca (1009±18 μgg^{-1}). Na with concentration 475±65 μgg^{-1}was found to be present at the lowest concentration of the macro-elements analyzed (Table 3). Washed Sidama roasted coffee also contains Mn with concentration 20±0.1 μgg^{-1}, Zn (17±0.2 μgg^{-1}) and Cu (9±0.3 μgg^{-1}) slightly lower than the micro-elements.

Unwashed Sidama roasted coffee. the fourth which is washed Sidama roasted coffee type also contains K with concentration 18563±477 μgg^{-1} in the highest level followed by Mg (1992±93 μgg^{-1}), Ca (976±24 μgg^{-1}) and Na (446±73 μgg^{-1}) with the lowest concentration of the macro-elements analyzed (Table 3). In this roasted coffee sample Mn with concentration 19±0.2 μgg^{-1}, Zn (18±0.3 μgg^{-1}) and Cu (10±0.1 μgg^{-1}) slightly lower than the microelements.

3.3. Statistical Analysis

The significance of variation in metals concentration between samples was analysed by students t-test and one-way ANOVA. Accordingly, students' t-test shows, for concentration of; K, Mg, Ca and Na no significant difference at $p = 0.05$ and for concentrations of; Mn, Zn and Cu a significant difference at $p = 0.05$ was observed between washed and unwashed Yirgacheffe as well as between washed and unwashed Sidama roasted coffee samples respectively.

From one way ANOVA test for significant difference; for concentrations of Mg and Na no significant difference at $p = 0.05$ and for Mn, Zn and Cu, a significant difference at $p = 0.05$ was observed between all four roasted coffee samples.

However, for concentrations of K and Ca; a significant difference at p = 0.05 was observed only for both washed Yirgacheffe and Sidama as well as for both unwashed Yirgacheffe and Sidama roasted coffee samples respectively.

3.4. Comparison of the Metal Content of Roasted Coffee Samples with Other Reported Values

In different studies the metal content of roasted coffee samples have been analyzed for both Arabica and Robusta coffee types [9-11]. The composition of Ba, Ca, Cu, Fe, K, Mg, Mn, Na, P, Sr and Zn have been determined and used as chemical descriptions to differentiate between roasted coffee samples from Arabica and Robusta coffee varieties. According to the report on roasted coffee varieties present in southeast Brazilian market, there is a difference in metal content among the samples of roasted coffee varieties. These variations in metal compositions among samples roasted coffee can indicate the differences in the factors that influence the cultivation of the coffee plant; such as the type of soil, the use of fertilizers with different chemical compositions and the ambient conditions [9].

Although various chemical analyses target to a similar objective, there may also be a difference in sampling, sample preparation and analytical techniques. Considering all these, the result of the present study can be compared to the findings of other authors. Martin *et al.*[11] have determined the metal content of roasted coffee samples of various origin and the concentrations have been used as chemical descriptors to differentiate between roasted coffee samples from the Arabica and Robusta varieties. The Arabica coffee variety from Nicaragua contained (μgg^{-1}): Ca (970), Cu (13.4), Fe (46.2), K (14930), Mg (1740), Mn (14.1), Na (33.4) and Zn (19.3) while the Robusta variety from Ivory Cost contained Ca (940), Cu (16.1), Fe (56.1), K (14480), Mg (1610), Mn (12.2), Na (21.4) and Zn (13.9) comparable to the values reported in this paper. Furthermore, they used the P, Mn and Cu content to discriminate between the Arabica and Robusta roasted coffee varieties. Ashu and Chandravanshi [13] also determine the concentrations of metals (K, Mg, Ca, Na, Mn, Fe, Cu, Zn, Co, Pb, Cd) in three brands of commercially available roasted Ethiopian coffee powders (Abyssinia, Alem and Pride). The mean concentration of each metal in the three brands of coffee powder samples was (μgg^{-1}): K (14488±467), Mg (1964±78), Ca (945±65), Na (484±12), Fe (52.0±4.0), Mn (23.0±0.9), Cu (14.0±0.6), Zn (15.0±0.8) and Co (1.60±0.05) in which both those results are comparable with the present study.

The results of the present study are in good agreement with the most of reported values. Moreover, the general trend of the metal concentration for macro-elements K > Mg > Ca > Na is in good agreements. Regarding the trace microelements, Mn> Zn > Cu > Co in roasted coffee powder is followed for most reports. Similarly, reports by Ashu and Chandravanshi [13] and Grembecka *et al.* [16-17] shows that toxic elements Pb and Cd were not detectable under their analysis conditions using the FAAS which is in a good agreement with the present study. Generally, the concentration of the macro and micro-elements in both washed and unwashed Yirgacheffe and Sidama roasted coffee samples (present study) were compared to other reported values in Table 4 and 5.

Table 4. *Comparison of the concentrations of macro-elements in roasted coffee beans samples with other reported values.*

Origin	Concentration (μgg^{-1}) of metal in roasted coffee beans				Reference
	K	Mg	Ca	Na	
Ethiopia	18563–19610	1943–2030	931–1009	446–484	Present study
Various origins	11750–15850	800–2840	513–1620	382–459	[16]
Brazil	32500–39800	2120–4150	1110–1890	274–665	[10]
Ivory Coast	14010–14480	1610–1820	940–1220	15.9–21.4	[11]
Ethiopia	14310–19400	5470–6080	500–990	–	[12]
Uganda	14080–14990	1670–1780	1170–1230	11–14	[11]
Ethiopia	14361–14583	1959–1968	843–1045	468–501	[13]

Table 5. *Comparison of the concentrations micro-elements in the roasted coffee bean samples with literature values.*

Origin	Concentration (μgg^{-1})^b of metal in roasted coffee beans					Reference
	Mn	Zn	Cu	Cd	Pb	
Ethiopia	19–23	14–18	9–13	ND	ND	Present study
Brazil	29–40	3.9–9.8	49–88	0.7–0.9	0.015–0.058	[10]
Ethiopia	22–24	12–19	11–17	ND	ND	[13]
Various origins	16.5–40.6	3.2–16.2	12.1–20.1	ND	ND	[16]
Ethiopia	15–20	6–30	13–28	ND	ND	[12]

^{b}Their concentrations were below detection limit, ND = Not detected

4. Conclusions

The closed microwave assisted wet digestion method and determination of selected metals in roasted coffee samples by flame atomic absorption spectroscopic method was found to be efficient. K has high concentration among the macro-elements (Mg, Na and Ca). Moreover, Mn found high level comparing with other micro-elements (Mn, Zn and Cu). In all roasted coffee bean samples the level of toxic elements (Cd and Pb) were avail below the method detection limits. The levels of both essential and non-essential metals investigated in roasted coffee beans were comparable with that of the rest of the world. It could be suggested that the roasted coffee beans under investigation could be source of dietary minerals and trace metals and could be valuable in complementing available food composition data and estimating dietary intakes of essential and nonessential metals in Ethiopia through coffee consumption.

Acknowledgments

Ministry of Education (MOE) for financial support and Department of Chemistry, Hawassa University (Ethiopia) for providing the laboratory facilities are gratefully acknowledged.

References

[1] V. Krivan, P. Barth, A. Feria and M. Mikrochim (1993) Multielement Analysis of Green Coffee and Its Possible Use for the Determination of Origin. *MikrochimicaActa* 110: 217-236

[2] J. A. DeSousa (1996) Verifying wine origin: a neural network approach. *American Journal of Enology and Viticulture* 47: 410-414

[3] F. Lavedrine, A. Ravel, A. Poupard, and J. Alary (1997) Effect of geographic origin, variety and storage on tocopherol concentrations in walnuts by HPLC. *Food Chemistry* 58:135-140

[4] A. Kim, Anderson, and W. B. Smith (2002) Chemical Profiling To Differentiate Geographic Growing Origins of Coffee. *Journal of Agriculture and Food Chemistry* 50:2068–2075

[5] M. Grembecka, E. Malinowska and P. Szefer (2007) Differentiation of market coffee and its infusions in view of their mineral composition. *Science of Total Environment* 383:59-69

[6] R. J. Clarker and L. J. Walker (1974) Potassium and Other Mineral Contents of Green, Roasted and Instant Coffee. *Journal of the Science of Food and Agriculture* 25:1389 –1404

[7] J. H. Zaidi, I. Fatima, M. Arif and I. H. Qureshi (2006) Determination of trace elements in coffee beans andinstant coffee of various origins by INAA.*Journal of Radioanalytical and Nuclear Chemistry* 267:109–112

[8] E. E. Santos, D. C. Lauria and C. L. Porto da Silveira (2004) Assessment of daily intake of trace elements due to consumption of foodstuffs by adult inhabitants of Rio de Janeiro city. *Science of the* Total Environment 327:69-79

[9] M. J. Martin, F. Pablos and A. G. Gonzalez (1998) Characterization of Green Coffee Varieties According to Their Metal Content. *AnalyticaChimicaActa* 358:177-183

[10] E. J. dos Santos and E. de Oliveira (2001) Determination of Mineral Nutrients and Toxic Elements in Brazilian Soluble Coffee by ICP –AES. *Journal of food Composition and Analysis* 14:523-531

[11] M. J. Martin, F. Pablos and A. G. Gonzalez (1999) Characterization of Arabica and Robusta Roasted Coffee Varieties and Mixture Resolution According to Their Metal Content. *Food Chemistry* 66:365-370

[12] A. Gure (2006) Investigation of Metals in Raw and Roasted Indigenous Coffee Varieties in Ethiopia. M.Sc. Project, Addis Ababa University, Addis Ababa, Ethiopia 1-53

[13] R. Ashu and B. S. Chandravanshi (2011) Concentration levels of metals in commercially available Ethiopian roasted coffee powders and their infusions. *Bulletin of Chemical Society of Ethiopia* 25:11-24

[14] N. Oleszczuk, J. T. Castro, M. M. da Silva, M. G. A. Korn, B. Welz, and M. G. R., Vale (2007) Method development for the determination of manganese, cobalt and copper in green coffee comparing direct solid sampling electrothermal atomic absorption spectrometry and inductively coupled plasma optical emission spectrometry. *Talanta* 73:862–869

[15] D. Perrone, C. M. Donangelo and A. Farah (2008) Fast simultaneous analysis of caffeine, trigonelline, nicotinic acid and sucrose in coffee by liquid chromatography-mass spectrometry. *Food Chemistry* 110:1030-1035

[16] M. Grembecka, E. Malinowska and P. Szefer (2007) Differentiation of market coffee and its infusions in view of their mineral composition. *Science of Total Environment* 383:59-69.

[17] J. B. Edward, E. O. Idowu, J. A. Oso, O. R. Ibidapo (2013) Determination of Heavy Metal Concentration in Fish Samples, Sediment and Water from Odo-Ayo River in Ado-Ekiti, Ekiti-State, Nigeria, International Journal of Environmental Monitoring and Analysis. 1:27-33

Involvement in Tourism Activities and Perceived Benefits in Communities around Udzungwa Mountain National Park in Tanzania

John Safari*, Vicent Gowele, James Lwelamira

Institute of Rural Development Planning (IRDP), Dodoma, Tanzania

Email address:
jsafari@irdp.ac.tz (J. Safari)

Abstract: Tourism constitutes one of the fastest growing sectors in Tanzania. The industry is placed as one of the major means through which development of local communities can be achieved. There is, however, paucity of information on how local communities living close to the national parks are involved in tourism. A cross-sectional study was carried out in February 2013 in six villages adjacent to Udzungwa Mountain National Park (UMNP). The objectives of the study were (i) to examine the extent to which people living close to UMNP participate in tourism, (ii) to identify the determinants of individual involvement in tourism activities, and (iii) to assess the benefits accrued from tourism by local communities. Data were collected from 150 households using a semi-structured questionnaire. A 3-point Likert scale was used to obtain information on residents' involvement and perceived benefits in tourism activities. Data were then analyzed using Statistical Package for Social Sciences (SPSS) program version 20. Results show that almost all respondents (97%) were involved in at least one activity related to tourism. The activities include hotel and local restaurant services, agro-tourism, environmental protection, transport services, adventure tourism, local handcrafting, revenue sharing and walking safari. In most of these activities, residents' involvement was rather moderate. Overall, findings show that tourism provides array of benefits that contribute to the wellbeing of people. However, lack of transparency on revenues from UMNP was identified as ingrained challenge that affected cooperation and trust between residents and leaders. These results underline the need for the Local Government Authority to contextualize governance challenges taking into account the social-political environment. These results underline the need for changes in social-political environment with emphasis on increased accountability and transparency. Addressing these challenges will build greater trust between community members, village leaders and UMNP authority in pursuit of economic development, conservation requirements and sustainability in tourism industry.

Keywords: Benefits, Community Development, Conservation, Tourism

1. Introduction

Many developing countries now regard tourism as an important and integral part of their economic development strategies. For sub-Saharan African countries, the industry presents an opportunity to diversify export earnings and contribute to economic growth from sources other than primary commodities [24]. The industry is placed as one of the major means through which development of local communities can be achieved [3, 4, 25, 30, 32, 35]. Tanzania is endowed with various natural resources that form a main stay of tourist attractions. During the past decade, tourism increased its contribution to GDP, from US $615 million in 2005 to US$1.75 billion in 2010, making it the largest source of foreign exchange and constituting about 8% of the Tanzanian GDP [31].

Udzungwa Mountains in Tanzania's southern highlands are internationally recognised as an area of rich biodiversity and a hotspot for a range of unique endemic species. Tourism activities in these areas are mainly nature-based and recreation which are categorized as cultural ecosystem services [12, 17]. These present high potential for conservation of protected areas if local communities gain a sense of ownership of the natural resources. Effectiveness of nature-based tourism in providing incentives for sustainable ecosystem management often depends *inter alia* on sufficient returns to neighbouring communities through elaborate profit-sharing mechanisms. Community involvement in the tourism development process can be viewed from at least two perspectives: in the decision-making process and in the benefits of tourism development [39, 40]. It is widely acknowledged that participation of local communities in enterprises linked to natural resources determines the extent to which such strategies achieve conservation and economic development goals [5, 8, 13, 23, 26, 33]. Thus, community involvement is indispensable if tourism is to contribute to economic development and attain sustainable levels of growth [7, 16, 19, 21, 25, 27, 36]. Despite the growing significance of tourism in the country, there is limited information on the level at which local communities are integrated in the industry. The objectives of this study were (i) to examine the extent to which communities living close to Udzungwa Mountain National Park participate in tourism, (ii) to identify the determinants of individual involvement in tourism activities, and (iii) to assess the benefits accrued from tourism by local communities.

2. Materials and Methods

2.1. Study Area

This study was conducted in villages surrounding Udzungwa Mountain National Park (UMNP). This park is located at 7°46'S and 36°43'E with a size of 1990 km^2 in Kilombero District, Morogoro Region. The park is famous for high species diversity and endemism. Tourism in UMNP revolves around hiking and trekking. Kilombero district is situated in a vast floodplain, between the Kilombero River in the south-east and the Udzungwa-Mountains in the north-west. The area is predominantly rural with the semi-urban district headquarters in Ifakara. The majority of the villagers are subsistence farmers of maize and rice. The district has an estimated 407,880 inhabitants of whom, 205, 091(50.3%) are females and 202,789 (49.7%) are males [41].

2.2. Data Collection

The field surveys were conducted in February, 2013 in six villages namely MsolwaIteja, SanjeBarabarani, SanjeShuleni, Mang'ula A, Mang'ula B and Mwaya (Fig 1). The villages were selected on the basis of proximity to the UMNP. A cross-sectional design was applied to enroll 150 randomly selected respondents. Approximately equal sample size (average 25) was drawn from each village. Data were collected using a semi-structured questionnaire. The questionnaire covered the following aspects: social-demographic characteristics, personal involvement in tourism activities and benefits accrued from tourism. A 3-point Likert scale (i.e. 1 = "not at all", 2 = "somewhat", and 3 = "a lot") was used to rate the level of involvement and benefits in tourism activities. Mean scores were generated and categorized into three levels (low, ≤1.5; moderate, 1.6-2.5 and high, 2.6-3.0). In addition, qualitative data were obtained from four focus group discussions (FGDs) – each consisting of 8 to 12 members. Each group meeting involved, among other issues, discussion on stakeholders' involvement in tourism activities and benefit sharing in the community.

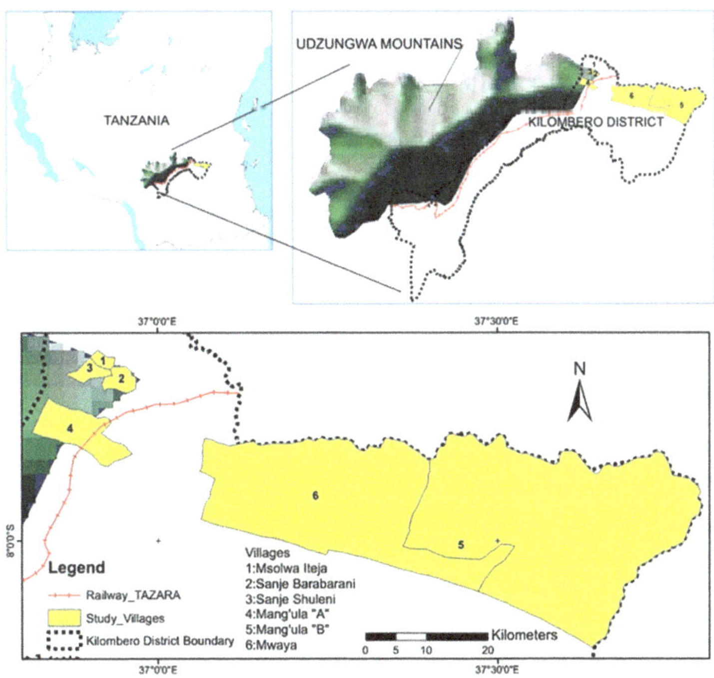

Fig. 1. *Map showing surveyed villages around the Udzungwa Mountains, Tanzania.*

2.3. Statistical Analysis

Data were entered, edited and analyzed in Statistical Package for Social Science (SPSS) version 20. The analysis included descriptive statistics to determine distributions of means, standard deviations, frequencies and percentages. To identify the factors that may influence involvement of individuals in tourism activities, linear regression analysis was performed. Individual scores for each activity were summed to form a composite index for the overall involvement in tourism activities. During regression analysis, the index was used as a dependent variable against age, sex, education level, marital status, distance to the park, length of residence, and employment status of the respondent as independent variables.

3. Results and Discussion

3.1. Characteristics of Respondents

Socio-demographic characteristics of the respondents are presented in Table 1. Age of the respondents ranged from 18 to 50 years. Mean age was 30.5 ± 6.9 years (29.1±7.1, females and 31.4±6.7, males). Majority of the respondents (55.4%) were in the age of 25 to 45 years. More males (62.7%) than females (37.3%) were involved in tourism activities. The average household size was 4.5 members, which is slightly higher than the district average of 4.3 but lower than the national average of 4.8 [41]. With regard to education level, 38% of the respondents had completed primary education while 50.7% and 17% had completed secondary and tertiary studies, respectively. Most families depend on agriculture (50.7%) and small business (36.7%) for their livelihood.

3.2. Involvement of Local Communities in Tourism related Activities

Tourism in UMNP involves various attractions including nature, views, climate, recreation and culture. Ninety-seven percent of the respondents were involved in at least one activity related to tourism. The activities include hotel and local restaurant services, agro-tourism, environmental protection, transport services and adventure tourism (means score 1.5-2.0 or moderate involvement). Others were local handcrafting, revenue sharing and walking safari (means score <1.5 or low involvement). Since agriculture is the mainstay in this community, supply of agricultural produce to hotel and restaurants is of high significance. In Udzungwa Mountains, adventure tourism is a central activity. It entails viewing waterfalls and primate species such as Crested Mangabey (*Cercocebusgaleritussanjei*) and the Udzungwa red colobus (*Procolobusbadiusgordonorum)* which are only found in UMNP. Through adventure tourism local residents (usually young men) were hired as touring guides.

Table 1. Characteristics of respondents (N=150).

Variables	n	%
Age (Years)		
18-24	37	24.7
25-35	64	42.7
36-45	19	12.6
46+	30	20.0
Sex		
Female	56	37.3
Male	94	62.7
Education level		
Primary	57	38.0
Secondary	76	50.7
Tertiary	17	11.3
Marital status		
Single	55	36.7
Married	64	42.7
Divorced/ Widow/widower	31	20.6
Main source of income		
Agriculture	76	50.7
Small business	55	36.7
Employment (civil service)	19	12.6

Results from a 3-point Likert scale show that the overall mean score for involvement in tourism related activities was 1.6 (Table 2). This means that respondents scored 'somewhat' for most activities. Further, results show that none of the activity scored 'high' in involvement. In general, tourism in this area is still limited in scale partly because the industry itself is not well developed. Deficiency in infrastructure, travel facilities and service industries present a major challenge for the delivery of world-class service quality. This explains why the clientele of the UMNP tourism are mainly budget and backpacker tourists [34]. Thus, a considerable amount of investment will be required for the improvement of transportation infrastructures and networks necessary for tourism to occupy a market niche. During focus group discussions, lack of transparency on revenues gained from UMNP was reported as ingrained challenge that needed attention. There were suspicions on how well funds from UMNP were used by the village leaders. Village members did not have influence or control over village resources.

Table 2. Community involvement in tourism related activities (N=150).

Variable	Not at all n(%)	Somewhat n (%)	A lot n (%)	Mean	S.D
Hotel & local restaurant services	5(12.5)	12(30.0)	23(57.5)	2.0	0.8
Agro-tourism	7(13.7)	30(58.8)	14(27.5)	1.8	0.8
Environmental protection	8(19.5)	20(48.8)	13(31.7)	1.7	1.0
Transport services	13(28.9)	18(40.0)	14(31.1)	1.6	1.0
Adventure tourism	14(35.9)	18(46.2)	7(17.9)	1.5	0.7
Local handcrafting	10(41.7)	12(50.0)	2(8.3)	1.4	0.7
Revenue sharing	9(45.0)	8(40.0)	3(15.0)	1.4	0.7
Walking safari	13(46.4)	14(50.0)	1(3.6)	1.3	0.6
Overall mean				1.6	1.1

Participants of the focus groups claimed that planning and decisions regarding resources and community benefits associated with the park remained entirely in the hands of leaders. Cooperation and trust between citizens and leaders were limited although these two are necessary to enhance participation in collective activities [29]. This scenario presents a clear example of marginalized voices of community members. Indeed, it shows failure of the ostensible aim of participatory approaches to development that advocates for increased people's influence or control over development initiatives [10,18]. On the other hand, experience shows that protected areas can deliver multiple benefits that contribute to wellbeing but require a strong social and political framework in which to spread benefits equitably [11]. In this regard, success cannot be achieved in conditions of weak governance, unrealistic expectation and inadequate engagement of stakeholders and mechanisms to transfer benefits [20]. Poor governance has negative consequences both on community development and conservation of resources. There is compelling evidence that communities that lack sense of ownership or of benefit-sharing are less inclined in looking after the natural resources [1, 4]. Experience from Uganda, for example, shows four key components of successful revenue-sharing programmes are needed: long-term institutional support, appropriate identification of the target community and project type, transparency and accountability, and adequate funding [4]. Thus, these aspects need to be taken into account for efficient utilization of resources and local economic development.

3.3. Determinants of Individual Involvement in Tourism Activities

Table 3 shows that age, sex and education level of an individual were determinants of involvement in tourism related activities ($p < 0.05$). More men than women were involved in tourism related activities ($p = 0.01$). It has been documented that women are often less engaged in business because of barriers associated with socialization practices, stereotypes, educational experiences, household chores and lack of networks of business contacts [9, 15, 22, 28]. It is possible that some of these factors were responsible for the observed pattern in our study. Findings also show that respondents in their middle age (25-35 years) were more likely than those in younger or older groups to engage in tourism activities ($p = 0.024$). On the other hand, having tertiary education was associated with high level of involvement in tourism activities ($p = 0.02$). Respondents with lower level of education, especially women, often remarked that they did not have skills and capacities to engage in tourism activities. This challenge is not exceptional to this community. Elsewhere, lack of education and the wide range of business and marketing skills necessary to manage day-to-day operations have been identified as significant barriers to engagement in tourism [37].

Table 3. *Linear regression analysis for determinants of involvement in tourism related activities (N=150).*

Potential determinant	β	SE	T	p-value
Constant	4.14	0.87	4.76	0.00
Respondent is in middle age (25-45)	1.14	0.62	2.27	0.02
Sex of respondent is male	0.96	0.37	2.59	0.01
Respondent is married	0.22	0.14	0.76	0.45
Distance to national park is <15 km	0.01	0.01	1.53	0.13
Length of residence is >10 yrs	0.00	0.04	0.06	0.96
Respondent has tertiary level of education	1.26	0.53	2.36	0.02
Respondent is not formally employed	-0.41	0.48	-0.86	0.39

$R^2 = 0.42$, Significant at $p< 0.05$

3.4. Benefits of Tourism

Respondents were inquired whether tourism had any benefit at the family level. The question was not simply whether they enjoyed personal benefits from tourism but the degree to which each of the potential benefit contributed to household living condition. Table 4 shows a wide range of acquired benefits. These include acquisition of both productive assets (e.g. increased value of land) and non-productive assets (e.g. television) and access to basic needs (e.g. food supply) and other services (e.g. electricity). Overall, data show that tourism has important role on improving standard of living of community members (mean score 1.9). Increase in food supply, paying school fees and enhancing small business were among the most frequently mentioned benefits. At community level, the benefits included supporting schools, health facilities and supply of potable water. Community members also received environmental education which was designed as a strategy for conserving biodiversity. Through this program, UMNP authority raises environmental awareness of community members. The program works with community groups, such as youth and women, and provides support in planting trees for supply of fruits or delimitating the farms.

Table 4. *Benefits of tourism at household level.(N=150).*

Attribute	Not at all n(%)	Somewhat n (%)	A lot n (%)	Mean	S.D
Improves standard of living	6(7.5)	43(53.8)	31(38.8)	1.9	0.3
I own a bicycle because of tourism	7(21.2)		18(54.5)	1.9	0.7
Increases food supply	2(5.7)	29(82.9)	4(11.4)	1.8	0.2
Helps me to pay school fees	4(10.3)		7(17.9)	1.7	0.2
Enhances small business	13(24.5)	30(56.6)	10(18.9)	1.6	0.5
Increases value of house and land	26(33.8)	38(49.4)	13(16.9)	1.5	0.6
Led to installation of electricity	23(38.3)	20(33.3)	17(28.3)	1.5	0.8
I own a car because of tourism	3(42.9)		2(28.6)	1.4	0.9
I own a generator because of tourism	3(33.3)		1(11.1)	1.4	0.6
I own a motorbike because of tourism	4(40.0)		2(20.0)	1.4	0.8
Helps savings income	19(38.8)	26(53.1)	4(8.2)	1.4	0.6
Helps to improve farming skills	15(46.9)	13(40.6)	4(12.5)	1.3	0.7
I own a Television because of tourism	5(50.0)		5(50.0)	1.2	0.6

However, environmental education was not perceived as real benefit by some members of the village as they associated it with punitive and restrictive policies pursued by the park. Others had the view that tourism provided minimal economic benefits to the local community and so they demanded more support than they had received. The above argument suggests that creating a platform that is more inclusive in planning and decision making processes is likely to result in reduced tensions among different actors in the community. Through such a platform, stakeholders will have a chance to negotiate their demands and seek to agree upon. Limited participation of local communities in decision making process in the sampled villages is a reflection of what is referred in [40]:

"...participation in tourism development process in many developing countries has been recognized as helping people get more economic benefits...rather than creating opportunities for local people to have a say in decision-making process of the tourism development"

In this regard, it is important to note that while the role of tourism has been critical for achieving sufficient support and legitimacy among local people in many places [14], it is widely posited that, increased involvement of local community increases trust, positive future biodiversity outcomes and can promote conservation of protected areas [2, 38, 42]. Indeed, considering the fact that at least 96 vertebrate species are endemic in these mountain ranges [6], increasing stakeholder involvement is critical for sustainable management of resources and the overall tourism development.

4. Conclusion and Recommendations

This study shows moderate involvement in tourism activities among people living close to UMNP. Through this involvement, the industry provides array of benefits that contribute to the wellbeing of people. However, residents' involvement and benefits in tourism activities is limited by underdeveloped infrastructures necessary to attract international tourism and increase service industries. Inadequate representation of community members in planning and decision making process poses a challenge to growth in tourism. Further, women and individuals with low level of formal education seemed to be disfranchised groups in tourism activities. The described limits to community participation may not be only specific to tourism development but rather a common problem in day-to-day community activities – a situation that could undermine collective actions for economic and social development in the area. These results underline the need for changes in social-political environment with emphasis on increased accountability and transparency. Addressing these challenges will build greater trust between community members, village leaders and UMNP authority in pursuit of economic development, conservation requirements and sustainability in tourism industry.

References

[1] Agrawal, A and Gibson, C. C. (Eds.). 2001. Communities and the environment: ethnicity, gender, and the state in community-based conservation. Rutgers University Press.

[2] Agrawal, A and Gibson, C. C. 1999. Enchantment and disenchantment: the role of community in natural resource conservation. *World development*, 27(4), 629-649.

[3] Akama, J. S and Kieti, D. 2007. Tourism and socio-economic development in developing countries: A case study of Mombasa Resort in Kenya. *Sustainable tourism*, 15(6), 735-748.

[4] Archabald, K and Naughton-Treves, L. 2001. Tourism revenue-sharing around national parks in Western Uganda: early efforts to identify and reward local communities. *Environmental conservation*, 28(02), 135-149.

[5] Blom, B., Sunderland, T and Murdiyarso, D. 2010. Getting REDD to work locally: lessons learned from integrated conservation and development projects. *Environmental Science and Policy*, 13(2), 164-172.

[6] Burgess, N.D., Butynski, T.M, Cordeiro, N.J, Doggart, N.H., Fjeldså, J. Howell, K.M, Kilahama, F.B. Loader, S.P, Loader, S.P, Lovett, J.C, Mbilinyi, B, Menegon, M., Moyer, D.C, Nashanda, E., Perkin, A., Rovero, F., Stanley, W.T and Stuar, S.N. 2007. The biological importance of the Eastern Arc Mountains of Tanzania and Kenya, *Biological Conservation*, 134(2), 209-231

[7] Camargo, C., Maldonado, J. H., Alvarado, E., Moreno-Sánchez, R., Mendoza, S., Manrique, N and Sánchez, J.A. 2009. Community involvement in management for maintaining coral reef resilience and biodiversity in southern Caribbean marine protected areas. *Biodiversity and Conservation*, 18(4), 935-956.

[8] Campbell, L. M and Vainio-Mattila, A. 2003. Participatory development and community-based conservation: opportunities missed for lessons learned?. *Human Ecology*, 31(3), 417-437.

[9] Chitsike, C. 2000. Culture as a barrier to rural women's entrepreneurship: Experience from Zimbabwe. *Gender and Development*, 8(1), 71-77.

[10] Cooke, B and Kothari, U. (Eds.). (2001). *Participation: The new tyranny?*. Zed Books.

[11] Dudley, N., Mansourian, S., Stolton, S and Suksuwan, S. 2010. Do protected areas contribute to poverty reduction? *Biodiversity*, 11(3-4), 5-7.

[12] Egoh, B. N., O'Farrell, P. J., Charef, A., Gurney, L. J., Koellner, T., Abi, H. N and Willemen, L. 2012. An African account of ecosystem service provision: use, threats and policy options for sustainable livelihoods. *Ecosystem services*, 2, 71-81.

[13] Fabricius, C., Koch, E., Turner, S and Magome, H. (Eds.). 2013. Rights resources and rural development: Community-based natural resource management in Southern Africa. Routledge University Press.

[14] Ghimire, K. and Pimbert, P. 2000. Conservation in the Context of Rural Development. In: Ghimire, K. and Pimbert, P. (eds.) *Social Change and Conservation*. London: Earthscan, 1-45.

[15] Gupta, V.K., Turban, D.B, Wasti, S.A. and Sikdar, A. 2009.The role of gender stereotypes in perceptions of entrepreneurs and intentions to become an entrepreneur. *Entrepreneurship theory and practice*, 33(2), 397-417.

[16] Harris, R., Griffin, T. and Williams, P. (Eds.). 2002. *Sustainable tourism: a global perspective*. Routledge University Press.

[17] Hernández-Morcillo, M., Plieninger, T and Bieling, C. 2013. An empirical review of cultural ecosystem service indicators. *Ecological Indicators*, 29, 434-444.

[18] Inskeep, E. (1991). Tourism planning: an integrated and sustainable development approach. Van Nostrand Reinhold.

[19] Isaac, R. 2014. Responsible Tourism and Development in the Context of Peace-building and Peace. In International Handbook on Tourism and Peace, pp 387.

[20] Janishevski, L., Noonan-Mooney, K., Gidda, S.B. and Mulongoy, K.J. (2008). Protected areas in today's world: their values and benefits for the welfare of the planet. *CBD Technical Series*, (36).

[21] Kalemo, Z.R. (2011). Understanding the role of tourism in poverty reduction, the case of communities adjacent to Mikumi National Park in Tanzania, MSc Thesis, University of Bedfordshire, pp 72.

[22] Kalleberg, A.L. and Leicht, K.T. 1991. Gender and organizational performance: Determinants of small business survival and success. *Academy of Management*, 34(1), 136-161.

[23] Kiss, A. 2004. Is community-based ecotourism a good use of biodiversity conservation funds?. *Trends in Ecology and Evolution*, 19(5), 232-237.

[24] Kweka, J., Morrissey, O and Blake, A. 2003. The economic potential of tourism in Tanzania. *International Development*, 15(3), 335-351.

[25] Lee, C.C. and Chang, C.P. 2008. Tourism development and economic growth: a closer look at panels. *Tourism management*, 29(1), 180-192.

[26] Lele, S., Wilshusen, P., Brockington, D., Seidler, R. and Bawa, K. 2010. Beyond exclusion: alternative approaches to biodiversity conservation in the developing tropics. *Current Opinion in Environmental Sustainability*, 2(1), 94-100.

[27] Manyara, G and Jones, E. 2007. Community-based tourism enterprises development in Kenya: An exploration of their potential as avenues of poverty reduction. *Sustainable tourism*, 15(6), 628-644.

[28] Mariki, S.B., Hassan, S. N., Maganga, S.L.S., Modest, R.B. and Salehe, F.S. 2011. Wildlife-based domestic tourism in Tanzania: experiences from northern tourist circuit. *Ethiopian Environmental Studies and Management*, 4(4).

[29] Mendoza-Botelho, M. 2013. Social capital and institutional trust: Evidence from Bolivia's popular participation decentralisation reforms. *Development Studies*, 49(9), 1219-1237.

[30] Milne, S and Ateljevic, I. 2001. Tourism, economic development and the global-local nexus: theory embracing complexity. *Tourism geographies*, 3(4), 369-393.

[31] Mitchell, J., Keane, J., Laidlaw, J., 2009. Making success work for the poor: package tourism in Northern Tanzania. Overseas Development Institute. In '*Tracing the Tourism Dollar' Stakeholder Workshop* Held at KiboPalace Hotel in Arusha, 29 October 2008

[32] Murphy, P. E. (2013). Tourism: A community approach. Routledge University Press.

[33] Naughton-Treves, L., Holland, M. B and Brandon, K. 2005. The role of protected areas in conserving biodiversity and sustaining local livelihoods. *Annual Review of Environment and Resources*, 30, 219-252.

[34] Okayasu, S. 2008. The status of Ecotourism in the Eastern Arc Mountains of Tanzania, Msc Thesis, Imperial College London, p. 79

[35] Salazar, N. B. 2012. Community-based cultural tourism: Issues, threats and opportunities. *Sustainable Tourism*, 20(1), 9-22.

[36] Sarrasin, B and Tardif, J. 2012. Ecotourism and natural resources in Dominica: co-management as an innovative practice. *Téoros*, Special Issue, 85-90

[37] Scheyvens, R. 2002. Tourism for development: empowering communities: Pearson Education Limited.

[38] Stem, C.J., Lassoie, J.P., Lee, D.R., Deshler, D.D. and Schelhas, J.W. 2003. Community participation in ecotourism benefits: The link to conservation practices and perspectives. *Society and Natural Resources*, 16(5), 387-413.

[39] Timothy, D. 1999. Participatory planning: A view of tourism in Indonesia. *Annals of Tourism Research*, 26(2), 371-391.

[40] Tosun, C. 2000. Limits to community participation in the tourism development process in developing countries. *Tourism management*, 21(6), 613-633.

[41] United Republic of Tanzania. 2013. Population and Housing Census. National Bureau of Statistics (Tanzania).

[42] Young, J.C., Jordan, A., Searle, K.R., Butler, A., Chapman, D.S., Simmons, P. and Watt, A.D. 2013. Does stakeholder involvement really benefit biodiversity conservation?. *Biological Conservation*, 158, 359-370.

Low level fluoride content in groundwater of the Younger Granite aquifers in parts of the Jos Plateau, Nigeria

Dibal Hyeladi[1], Shem Auta Dahilo[2], Daspan Raymond[1], Lar Uriah[1]

[1]Department of Geology University of Jos, Jos, Nigeria
[2]Federal College of Education, Akwanga, Nasarawa State, Nigeria

Email address:
hassandibal@yahoo.com (Dibal H. U.), sdahilo@yahoo.com (Shem A.), daspanr@yahoo.com (Daspan R. I.), ualexanderlar@yahoo.co.uk (Lar U. A.)

Abstract: Fluoride content in water has received a world-wide attention due to its importance to health. This study attempts to trace the factors responsible for low fluoride levels in the aquifers of the Jos Younger Granites despite the high fluorine content in the rocks. A total of 41groundwater samples (1 mining pond, 2 hand pumps and 38 hand dug wells), 7 rock samples from the various lithological units, and 13 soil sections from two exposed locations were collected and analyzed for their fluorine content. Analysis of major cations was carried out using ICP-OES; the anions were analyzed using the UV multi-ion parameter and bicarbonates by titration method. Fluoride in underground water was determined by multi-ion parameter. Fluorine in rocks and soil were analyzed by the fusion method. The rock samples show variations in their fluorine content (Jos – Bukuru Biotite Granite 6,231, aplo-pegmatic granite-gneiss (basement rock) 4,864, Quartz-pyroxenes-fayalite porphyry 1,280, Dilimi-Biotite Granite 258 and Ngeil Biotite Granite-162 ppm). The soil sections from different locations also show variations of fluorine with depth of sampling. Cumulatively, the fluorine content in the sections and fluoride content in water do not correlate with fluorine in rock in the rock units. The low content in the two media, indicate that: 1. bulk fluorine have not been released from minerals in the host rocks and those retained in soils have not been mobilized to the groundwater. 2. Fluorine have formed complexes with other ions and occurring in compound form rather than ionic form. Apart from low fluoride in about 70% of the water samples all other parameters are within the WHO recommended limit for house hold uses. Although there are no records on the effect of low consumption of fluoride in water, inferences from the data show that most areas with low fluoride level should have dental caries.

Keywords: Fluoride, Groundwater, Younger Granites, Aquifers, Jos Plateau

1. Introduction

Fluoride content in groundwater is of concern because of two reasons – 1. Low content of fluoride (< 0.5 mg/l) may results in the development of dental caries. 2. Concentration levels (> 1.5 mg/l) may result in development of dental fluorosis, skeletal fluorosis, neurological effect, reproductive effect, genetic effect, low intelligent quotient (IQ) (WHO 2004; Hussain et al, 2004). Studies on fluoride in groundwater and effects on health of humans and animal have been reported in different parts of the world both in epidimeological study and the earth sciences for well over a hundred years now. However, in Nigeria, geo-medical studies on fluoride in groundwater and health received attention in the last fifteen years in the earth sciences (Lar and Dibal, 2013). Earlier studies in the area of epidemiology were carried out by Alakija, (1983), Bano et al, (1987) and Wongdem et al, (2002). Before now, data exist as part of water quality parameter, but none has linked the fluoride to source and or health effect. Thompson, 1957 reported fluoride level in waters of the Gombe area to range from 0.0 – 2.1 mg/l. In the Pleistocene sediments of the Chad Formation, Barber, (1965) found fluoride level of 2.0 mg/l. Schoeneich and Mbonu, (1990) reported very low fluoride in groundwater in parts of Jos and Bukuru area. Oteze and Ayegbusi, (2004) recorded value of 5.00 mg/l in the Chad Formation at Maiduguri. The link between fluoride occurrence in groundwater and health was first reported by

Dibal and Lar in 2005. They reported the existence of high fluoride in groundwater in the basement aquifer of (Kaltungo area) as the causal factor for the high incidence of mottled teeth in the area and environs. Lar et al, (2007) linked the source of the high fluoride in groundwater in Dorong area and the Furzi spring on the Jos Plateau, central Nigeria to fluorite minerals in the host granite rocks where it has caused serious dental fluorosis in children. Lar and Tejan, (2008) highlighted the human health effect related to the consumption of fluoride in groundwater particularly in Nigeria. Dibal, et al, (2008) reported the high incidence of dental fluorosis to consumption of high fluoride in groundwater of Langtang area. Dibal and Schoeneich, (2009) reported the occurrence of high fluoride in groundwater in several rural communities in some parts of northern Nigeria. Other studies on fluoride in groundwater are those of Bale and Goni, 2012. Although the Younger Granite lithologies which constitute the major aquifer units in the Jos area have been reported to have high fluorine content (ribeckite and biotite granite) (Bowden, 1965), fluoride in groundwater in these aquifers have been reported to be generally low (< 0.5 mg/l) in many areas (Schoeneich and Mbonu, 1990; Chilota, 2010, Solomon and Piwuna, 2011 and Dibal et al, 2012). Causes of high and low fluoride in Indian groundwater have been reported by Handa in 1975. He found out that the chemical composition of ground water from confined aquifers shows fluoride content below 1.0 mg/l and generally below 0.5 mg/l. He also reported positive correlation between fluoride and silica as well as between fluoride and sodium in groundwater, which indicate a silica-mineral source of fluoride. Hitchon, (1995) found out that fluoride is adsorbed on clay minerals, where ionic exchange takes place, and hence fluoride ions partly replaced which may lead to low fluoride in groundwater. According to (Abu Rukah and Alsokhny, 2004; Jacks et al, 2004), the concentration of fluoride in groundwater depends on the geological, chemical and physical characteristics of aquifers (e.g, porosity and acidity of soils, rocks, temperature and depth).

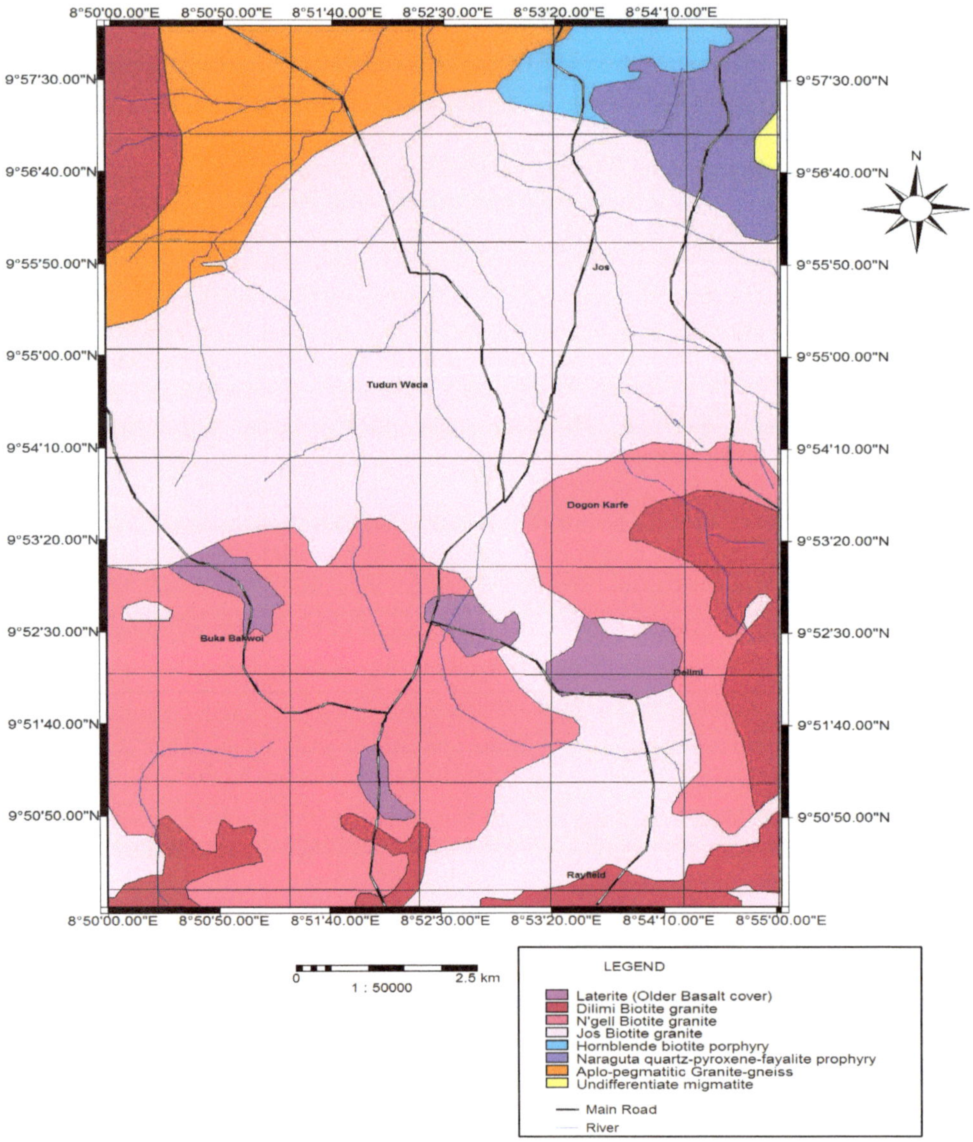

Fig 1. Geologic Map of the Study Area

Rock chemistry, groundwater age, well depth, hydrologic condition, residence time and geologic structure are important factors of fluoride rich groundwater (Edmunds, 1994; Kim and Jeong, 2005). The aim of this study is an attempt to trace the factors responsible for low level fluoride content in the Younger Granite Aquifer.

1.1. Description of the Study Area

Geographically the area lies between Latitudes $9^0$48'9 and $9^0$58'308" and Longitudes $8^0$50'.00" and $8^0$55'00.00", with aerial extent of about 106 km^2 within the Younger Granite Province of Nigeria. The Younger Granites are characterized by circular intrusions and represent one of the classical areas of occurrences of ring-complexes in the world. They are non - orogenic rocks, which have intruded the Late Precambrian to Lower Paleozioc Basement Complex of Northern Nigeria in a N - S trend (Turner, 1983). Ring faulting and cauldron subsidence are the major tectonic controls governing the emplacement of the Younger Granite (Macleod et al, 1971). These controls have operated during the volcanic and the stages of the emplacement cycles. The pattern of initial volcanism has been determined largely by ring fracturing which have extended to the surface. These fractures controlled the alignment and distribution of the vents and the major cauldron of the same mechanism at greater depth, beneath the lava accumulation, which localized the peripheral ring dykes and the granitic plutons. The Younger Granites exhibit a variety of rock types, which present similar characteristics throughout the whole province. These rock types include; biotite granite, hornblende biotite granites, rhyolite, riebeckite - biotite granite, syenite, gabbro, pyroxene -fayalite granite (Macleod et al. 1971). However, the following rock types characterized the study area; Jos Bukuru Biotite Granite, N'geil Biotite Granite, hornblende biotite granite, Dilimi Biotite Granite, aplo-pegmatitic granite gneiss (Basement rock), Naraguta Quartz Pyroxene Fayalite Porphyry and the weathered Older Basalt (Laterite) (Fig 1).

2. Methodology

Water samples were collected from different sources (hand dug wells, hand pumps and mining ponds) at intervals of 200 meters depending on availability of sources. 250 ml plastic bottles were used to collect water samples. Two samples each were collected at every sampling point. One sample from each sample point was then acidified with two drops of concentrated pure nitric acid to a pH of <2 for the samples that will be used for the detection of cation. Physical parameters such (temperature, Total Dissolved Solids (TDS), Electrical Conductivity (EC), and pH) for all the samples were determined in the field with the Oakton 5/6 pH/temperature meter and TDS/EC meter. The Inductive Coupled Plasma- Optical Emission Spectrophotometry (ICP-OES) was used to analyse the major cations and trace elements at the ACME Laboratory in Canada. The Multi-Ion Parameter Bench Photometer (HI 83200) was used to determined fluoride at the University of Jos, Faculty of Pharnaceutical Sciences. While the UV Multi-Ion Parameter Model 721 was used to determine SO_4 and bicarbonate and chloride were by titration method at the Univesity of Jos, Department of Geology. Soil samples were collected at intervals of 1 meter from weathered section of the Jos - Bukuru biotite granite the largest rock type which constitutes an aquifer (Plates). Fresh rock samples were also collected from the different rock types. Soil samples were dried at room temperature and both (rock and soil) samples were pulverized with FRITSCH PLANETARY MICRO MILL (Pulverisette 7). 1.5 grammes of each were weighed using AINSWORTH electric balance for analysis. Fluorine in rocks and soils were analysed by the Fusion method at the ACME Laboratory.

3. Results and Discussion

Table1 shows the concentration of fluorine in the rocks sampled at 7 locations on different rock types within the study area (Dillimi-Biotite –Granite, aplo- pegmatitic granite –gneiss, Jos Biotite Granite and Quartz-pyroxene-fayalite porphyry). From the Table 1, The Jos - Bukuru Biotite Granite shows the highest fluorine concentration of 6,231ppm at location 1 and 914 ppm at location 6, this is followed by the aplo-pegmatitic granite –gneiss (basement rock) with 4,864, Dillimi Biotite –Granie, 2,587, Quartz-Pyroxenes-Fayalite Porphyry 1,280. The Ngeil Biotite Granite has the lowest concentration of fluorine of 47ppm.

Table 1. *Concentration of fluorine in rocks at various locations in the study area*

SID	Longitude	Latitude	Date	Rock type	F(ppm)
LR1	8 54 38	9 48.9	11/4/2012	Dillimi-Biotite -Granite	2,587
LR2	8 50 39.3	9 56 49.6	12/4/2012	Aplo-pegmatitic granite-gneiss	4,864
LR3	8 51 53.2	9 55 49.7	12/4/2012	Jos - Bukuru Biotite Granite	6,231
LR4	8 50 06.7	9 51 53.1	12/4/2012	Ngell Biotite Granite	162
LR5	8 54 30.8	9 56 27.6	13/4/2012	Quartz -pyroxene-fayalite pophyry	1,280
LR6	8 53 28.6	9 54 15.6	13/4/2012	Jos - Bukuru Biotite Granite	914
LR7	8 50 25.3	9 51 53.2	13/4/2012	Ngell Biotite Granite	47

The concentration of fluorine in the soil samples taken from weathered sections of the Jos-Bukuru Biotite Granite at two locations (only rock type with exposed weathered section) is shown in Table 2. From (Table 2), it shows that the soils generally have moderate concentration of fluorine however; there are variations of fluorine content in the different sections with the bottom layers from the two locations having the highest concentrations and the top have the list concentration. The sample LS1-4 a sample collected at the bottom of the soil section at location 1 has fluorine concentration of 598 ppm. The sample LS2-5 a sample taken at the bottom of the soil section at location 2 has fluorine concentration of 375 ppm. The top soils represented by samples LST-2 and LST-1 from locations 2 and 1 respectively have fluorine concentrations of 155 ppm and 188 ppm probably indicating leaching of fluorine from the top to bottom. The samples LLSM-1 and LSV1 are soils from weathered pegmatite vein with fluorine concentration of 246 and 263 ppm respectively. Generally it can be observed that all the soil sections have lower fluorine content compared to the parent rock. The top soils appear to have lower fluorine concentration than those at the bottom which could probably explains the high concentration at the bottom. Probably this is because they are friable and have acted as a sink for the fluorine to be leached down to the bottom.

Table 3, shows the concentrations of major cations, anions and fluoride in water samples of the study area and their physical parameters and Table 4 shows the summary of the data. Temperature ranges from 23 to 25.50 °C with a mean of 24.5°C. pH, ranges from 7 to 11.1 with an average pH of 8.2 indicating neutral to alkaline water in the area .Total Dissolved Solids (TDS) ranges from 3 to 598 mg/l with few wells having TDS > 500 mg/l indicating such hand dug wells may probably have been affected by residential pollution. Calcium level is relatively high except at locations 1 an 2 which have 0.57 and 0.81 ppm respectively. Magnesium ranges from <0.05 to 14.18 ppm on average 2.15 ppm. Sodium concentration is relatively high except at locations 1 and 50 which recorded 0.63 and 1.09 ppm respectively. Potassium is generally low except at location 30 where the concentration is 44.9 ppm. Sulphate concentration is generally low ranging from 6.5 to 42.5 on average 12.68, chloride ranges from 0.00 to 206 on average 30.95. The absence of chloride in some locations could only be explained on the basis of analytical error, since it conservatively behaves in the geochemical environment. Bicarbonates concentration ranges from 8.8 to 328 ppm on average 68.27. Fluoride concentration level is very low; it ranges from 0.01 to 1.56. 97% of the locations have fluoride level of less than 1.5. These areas include Rayfield, Zaramaganda, Baza Bakwoi and Tudun Wada

Table 2. *Concentration of fluorine in the soil samples at two locations in the study area: Location: Ray Field Weathered Sections*

SID	Date	Depth (m)	Texture	colour	Soil type	F ppm
LST2	11/4/2012	surface	coarse-grained	Reddish-Brown	lateritic	155
LS37	11/4/2012	0 - 1	fine- grained	Brownish	clay	288
LS1-3	11/4/2012	1 - 2	Medium grained	Reddish-Brown	lateritic	170
LS1-4	11/4/2012	2 – 3	fine- grained	Reddish-Brown	clay	598
LSM-1	11/4/2012	3 - 4	Medium grained	Brownish	clay	246
LS2-3	11/4/2012	4 - 5	fine- grained	Pinkish white	clay	202
LS1-2	11/4/2012	5 - 6	fine- grained	Brownish	clay	302
Location: Air Force Base Weathered Sections						
LS2-4	11/4/2012	surface	fine- grained	Pinkish white	clay	212
LSV1	11/4/2012	0 – 1	Medium grained	Pinkish white	clay	263
LST1	11/4/2012	1 - 2	fine- grained	Reddish-Brown	lateritic	188
LS15	11/4/2012	2 - 3	fine- grained	Reddish-Brown	clay	231
LS2-2	11/4/2012	3 - 4	coarse-grained	Brownish	clay	200
LS2-5	11/4/2012	4 - 5	fine-grained	Brownish	clay	375

Table 3. *Compositions of groundwater in the study area*

SID	Longititude	Latitudes	Date	Source	Depth	Elev(m)	EC	T (°c)	pH
SD1	08°54'41.7"	9°49'6"	11/4/2012	MP	1.5m	4241	6	22.5	11.1
SD2	08°54'38"	09°48'9"	11/4/2012	well	3.5m	4218	17	23.2	9.7
SD3	08°53'17.6"	09°52'37.1"	11/4/2012	well	3.5m	4241	203	23	9.1
SD4	08°53'17.3"	09°52'42.0"	11/4/2012	well	5m	4152	11	26	9.1
SD5	08°52'51.4"	09°52'57"	11/4/2012	HP	None	4204	40	27	8
SD6	08°52'49.5"	09°53'04.9"	11/4/2012	well	7m	4172	149	23	8.4
SD7	08°53'06.4"	09°53'03.2"	11/4/2012	well	6m	4208	437	25	7.9
SD9	08°54'11.9"	09°52'33.6"	11/4/2012	well	10m	4218	136	24	8.2
SD10	08°53'58.1"	09°52'38.1"	11/4/2012	well	1.3m	4214	40	24.5	9.1
SD11	08°53'36.4"	09°53'53.5"	11/4/2012	well	0.5m	3985	289	24	8.1
SD12	08°51'25.9"	09°57'22.6"	12/4/2012	well	7m	4142	351	24.2	8.3
SD13	08°50'27"	09°58'17.8"	12/4/2012	HP	10.6m	3850	67	25	9.1
SD14	08°50'42.2"	09°56'50.6"	12/4/2012	well	10m	3808	86	24	9.1
SD15	08°50'52.3"	09°56'54"	12/4/2012	well	5.5m	3781.8	146	23.2	8.8
SD16	08°51'15.2"	09°56'53.4"	12/4/2012	well	8m	3768.7	102	24	8.8
SD18	08°50'47.5"	09°57'57.8"	12/4/2012	well	7.8m	3693	68	23.5	8.8
SD19	08°51'32.9"	09°56'28.6"	12/4/2012	well	8.8m	3775	72	24	8.8
SD20	08°51'53.2"	09°55'49.7"	12/4/2012	well	3.8m	3857	284	25	8.1
SD21	08°50'51.5"	09°55'38.4"	12/4/2012	well	3.7m	3873	387	25	8.2
SD22	08°53'21.9"	09°55'04.2	12/4/2012	well	3m	3988	415	24	8.5
SD23	08°50'06.7"	09°51'53.1"	12/4/2012	well	10m	4283	7	25	9.3
SD24	08°52'14.3"	09°54'21.5"	13/4/2012	well	5m	4063	81	24	9.7
SD27	080 5140.6	090 54 05.3	13/04/2012	well	5m	4080	1138	24	7.5
SD29	08°51'48.1"	09°54'23.1"	13/4/2012	well	4.2m	4001	408	24.5	8
SD30	08°59'30.4"	09°54'32.1"	13/4/2012	well	4.2m	4044	1178	24	7.3
SD31	08°51'57.9"	09°53'53.7"	13/4/2012	well	1m	4077	126	25.5	8.2
SD32	08°51'10.6"	09°53'10.6"	13/4/2012	well	6.4m	4100	79	23.5	8.6
SD33	08°50'57.8"	09°52'02.6"	13/4/2012	well	3m	4296.8	76	23.5	8.5
SD36	08°50'09.3"	09°50'29.5"	13/4/2012	well	5.8m	4319.7	9	24	9
SD37	08°50'25.2"	09°50'53.7"	13/4/2012	well	8.9m	4303	12	24.5	9
SD38	08°50'32.6"	09°50'50.0"	13/4/2012	well	3.2m	4286	8	25	9
SD41	08°50'48.0"	09°52'57.3"	13/4/2012	well	5.5m	4346	29	23	8.4
SD43	08°52'11.3"	09°52'54.0"	13/4/2012	well	6.8m	4195	81	25	8.7
SD45	08°54'42.5"	09°54'15.2"	14/4/2012	well	4.4m	4031	94	23	8.5
SD46	08°50'51.9"	09°55'00.6"	14/4/2012	well	5.7m	4011	339	24	8
SD47	08°54'53.6"	09°56'00.2"	14/4/2012	well	5.2m	3919.6	235	25	7.8
SD48	08°54'30.8"	09°56'27.6"	14/4/2012	well	5.9m	3880	166	25	7.3
SD49	08°54'16.1"	09°57'14.2"	14/4/2012	well	6.1m	3922.8	30	25	7.9
SD50	08°53'28.6"	09°54'15.6"	14/4/2012	well	5m	4096.7	117	24.5	8.7
SD29	08°51'48.1"	09°54'23.1"	13/4/2012	well	4.2m	4001	408	24.5	8
SD30	08°59'30.4"	09°54'32.1"	13/4/2012	well	4.2m	4044	1178	24	7.3
SD31	08°51'57.9"	09°53'53.7"	13/4/2012	well	1m	4077	126	25.5	8.2
SD32	08°51'10.6"	09°53'10.6"	13/4/2012	well	6.4m	4100	79	23.5	8.6
SD33	08°50'57.8"	09°52'02.6"	13/4/2012	well	3m	4296.8	76	23.5	8.5
SD36	08°50'09.3"	09°50'29.5"	13/4/2012	well	5.8m	4319.7	9	24	9
SD37	08°50'25.2"	09°50'53.7"	13/4/2012	well	8.9m	4303	12	24.5	9
SD38	08°50'32.6"	09°50'50.0"	13/4/2012	well	3.2m	4286	8	25	9
SD41	08°50'48.0"	09°52'57.3"	13/4/2012	well	5.5m	4346	29	23	8.4
SD43	08°52'11.3"	09°52'54.0"	13/4/2012	well	6.8m	4195	81	25	8.7
SD45	08°54'42.5"	09°54'15.2"	14/4/2012	well	4.4m	4031	94	23	8.5
SD46	08°50'51.9"	09°55'00.6"	14/4/2012	well	5.7m	4011	339	24	8
SD47	08°54'53.6"	09°56'00.2"	14/4/2012	well	5.2m	3919.6	235	25	7.8
SD48	08°54'30.8"	09°56'27.6"	14/4/2012	well	5.9m	3880	166	25	7.3
SD49	08°54'16.1"	09°57'14.2"	14/4/2012	well	6.1m	3922.8	30	25	7.9
SD50	08°53'28.6"	09°54'15.6"	14/4/2012	well	5m	4096.7	117	24.5	8.7

***Table 3.** Continued*

SID	TDS	Ca	Mg	Na	K	SO4	Cl	HCO3	F
SD1	3	0.57	0.07	1.09	0.44	6.5	ND	8.8	0.12
SD2	8	0.81	0.09	2.66	1.72	7.9	0	10.1	0.02
SD3	104	11.09	1.38	26.51	7.25	10.7	21.27	64	0.21
SD4	6	1.06	0.08	2.34	0.61	7.5	7.09	120	0.07
SD5	20	4.36	1.24	2.18	1.22	8	21.27	18.18	0.02
SD6	75	10.87	1.87	13.58	6.93	12.2	14.18	40.4	0.01
SD7	221	31.50	5.01	39.26	12.96	11.8	63.81	30.3	1.27
SD9	68	25.96	0.21	2.72	2.91	19.6	92.17	44.44	0.12
SD10	20	2.48	0.47	4.40	2.82	12.1	0	62.62	1.18
SD11	149	34.02	14.18	7.60	1.87	14	14.18	144	1.22
SD12	177	33.03	5.11	28.29	6.16	16.3	14.18	141.4	0.16
SD13	33	3.14	1.54	10.62	2.07	14.8	28.36	76.76	0.12
SD14	43	9.53	0.33	7.90	3.20	8.45	2.5	42.42	0.18
SD15	72	16.40	0.55	5.93	11.36	9	4.2	38.38	1.19
SD16	52	15.93	0.48	3.50	3.21	13.5	0	58.58	0.16
SD18	34	4.13	1.30	8.72	2.83	9.2	49.63	50.5	0.51
SD19	36	4.98	0.83	6.84	5.00	13	14.18	54.54	0.2
SD20	145	22.52	3.68	25.11	7.05	18.6	28.36	84	0.09
SD21	191	25.63	2.55	39.95	10.42	9.75	7.09	200	0.06
SD22	208	14.83	1.61	51.50	10.03	19.4	35.45	52.52	1.31
SD23	3	0.27	<0.05	1.68	0.74	37.5	42.54	60.6	0.38
SD24	41	6.67	0.68	6.49	4.56	8.9	42.54	12.12	0.49
SD27	181	73.95	12.92	127.73	27.34	8.0	42.54	480	0.26
SD29	206	27.04	3.17	29.81	25.47	11.5	42.54	158	0.41
SD30	598	71.51	10.14	113.05	44.97	12.6	77.99	328	0.11
SD31	61	8.02	1.44	11.56	6.68	42.5	205.6	129.28	0.34
SD32	38	5.56	0.50	6.48	6.30	8.8	42.54	10.1	0.19
SD33	38	4.18	0.81	8.70	2.79	10.5	42.54	14.14	1.14
SD36	4	0.55	<0.05	2.11	0.24	20.5	184.3	20.2	0.27
SD37	7	L.N.R.	L.N.R.	L.N.R.	L.N.R.	15.3	21.27	20.2	1.36
SD38	4	0.37	0.05	1.28	0.56	14.4	7.09	20.2	1.01
SD41	14	2.92	0.27	2.58	1.48	12.8	21.27	20.2	1.04
SD43	41	2.67	0.48	10.57	7.45	8.1	21.27	20.2	0.13
SD45	47	7.50	0.63	9.18	4.38	9.7	77.99	10.1	1.11
SD46	173	16.28	3.11	34.42	9.78	8.8	56.72	14.14	0.1
SD47	118	10.75	1.18	30.73	8.22	8.9	56.72	16.16	1.15
SD48	84	6.70	1.15	21.50	8.90	8.49	63.81	10.1	1.14
SD49	15	1.91	0.31	2.81	3.42	9.75	28.36	18.18	1.56
SD50	48	1.38	0.12	0.63	0.24	7.98	63.81	14.14	1.42
SD29	206	27.04	3.17	29.81	25.47	11.5	42.54	158	0.41
SD30	598	71.51	10.14	113.05	44.97	12.6	77.99	328	0.11
SD31	61	8.02	1.44	11.56	6.68	42.5	205.6	129.28	0.34
SD32	38	5.56	0.50	6.48	6.30	8.8	42.54	10.1	0.19
SD33	38	4.18	0.81	8.70	2.79	10.5	42.54	14.14	1.14
SD36	4	0.55	<0.05	2.11	0.24	20.5	184.3	20.2	0.27
SD37	7	L.N.R.	L.N.R.	L.N.R.	L.N.R.	15.3	21.27	20.2	1.36
SD38	4	0.37	0.05	1.28	0.56	14.4	7.09	20.2	1.01
SD41	14	2.92	0.27	2.58	1.48	12.8	21.27	20.2	1.04
SD43	41	2.67	0.48	10.57	7.45	8.1	21.27	20.2	0.13
SD45	47	7.50	0.63	9.18	4.38	9.7	77.99	10.1	1.11
SD46	173	16.28	3.11	34.42	9.78	8.8	56.72	14.14	0.1
SD47	118	10.75	1.18	30.73	8.22	8.9	56.72	16.16	1.15
SD48	84	6.70	1.15	21.50	8.90	8.49	63.81	10.1	1.14
SD49	15	1.91	0.31	2.81	3.42	9.75	28.36	18.18	1.56
SD50	48	1.38	0.12	0.63	0.24	7.98	63.81	14.14	1.42

Table 4. *Minima, maxima, mean and standard deviations of analyzed water samples in the study area.*

	N	Minimum	Maximum	Mean	Std. Deviation
EC	41	6	94	35.68	27.595
Temp	41	23	27	24.25	.923
pH	41	7	11	8.20	.782
TDS	41	3	598	85.98	105.487
Ca	40	0	74	13.36	16.899
Mg	38	0	14	2.15	3.318
Na	40	1	127	18.35	26.880
K	40	0	44	6.48	8.476
SO4	40	7	42	12.69	7.138
Cl	39	0	92	30.95	23.943
HCO3	40	9	480	68.27	92.913
F	40	0	1	.35	.483
Valid N (listwise)	36				

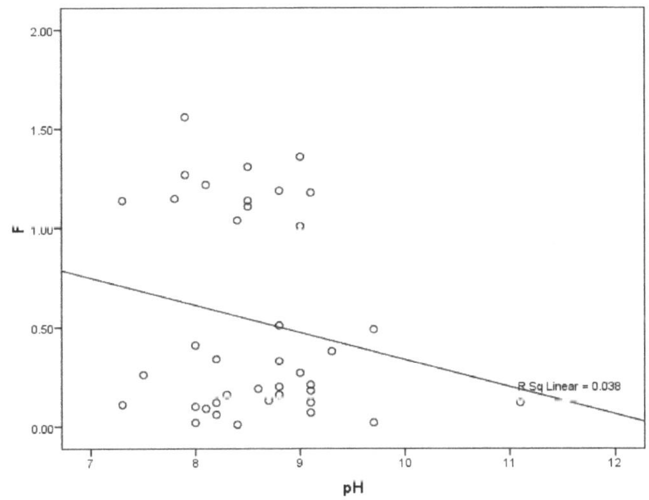

2a

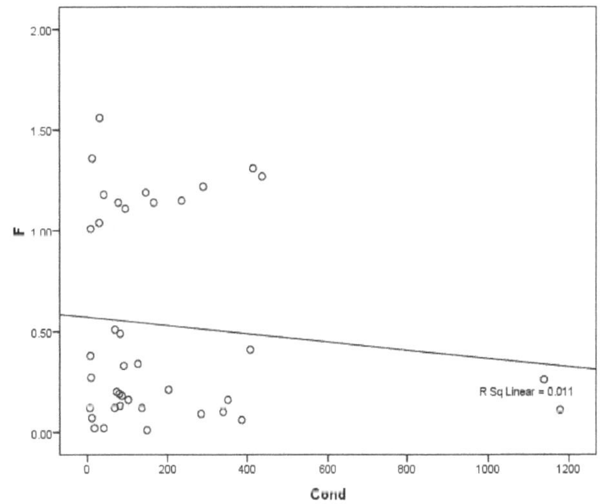

b

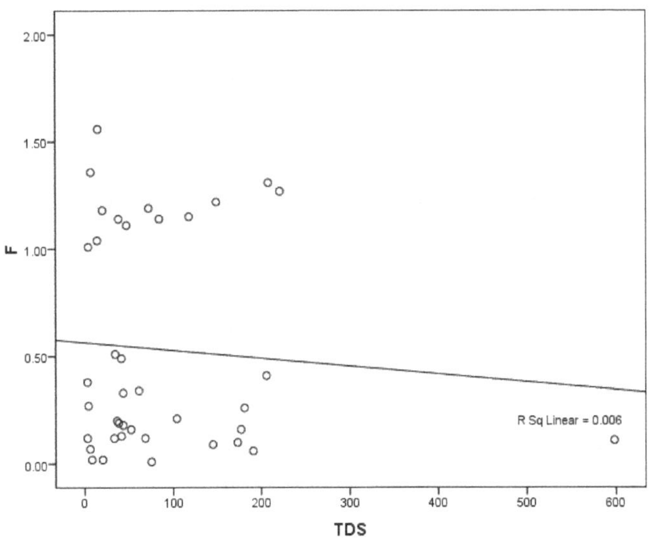

c

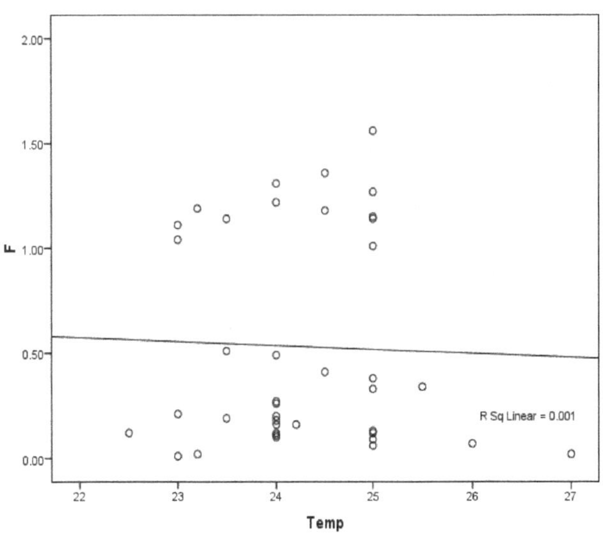

d

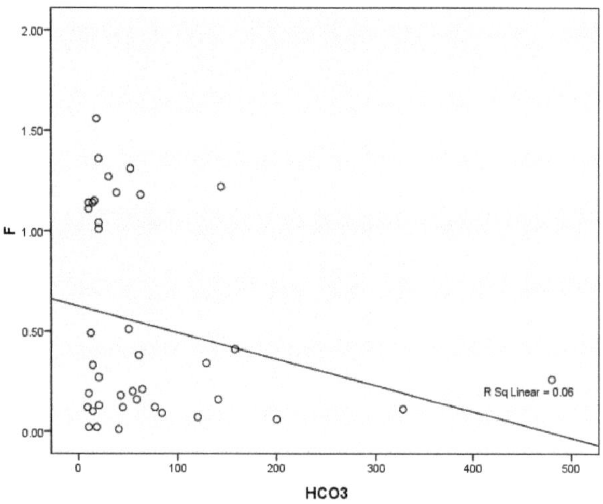
F
HCO3
R Sq Linear = 0.06

e

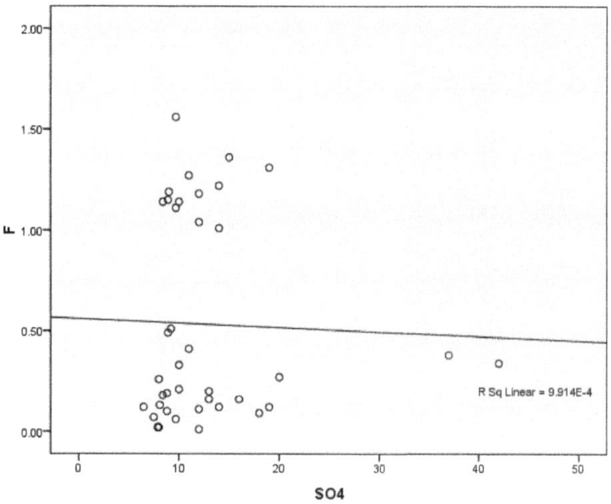
F
SO4
R Sq Linear = 9.914E-4

f

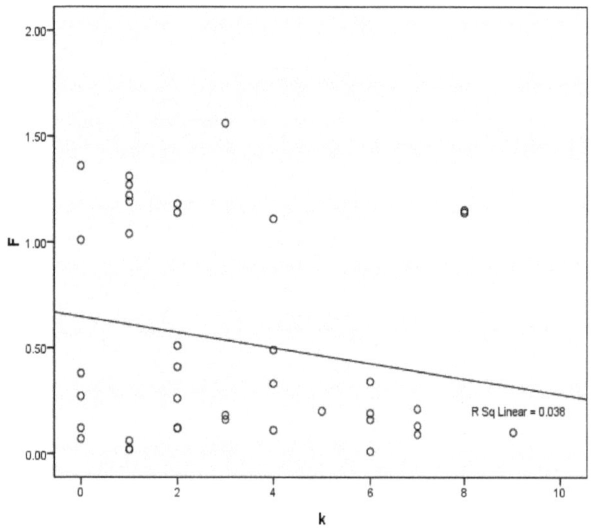
F
k
R Sq Linear = 0.038

g

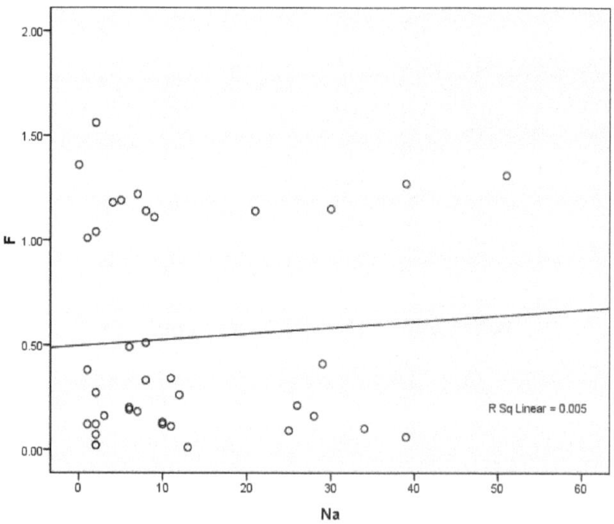
F
Na
R Sq Linear = 0.005

h

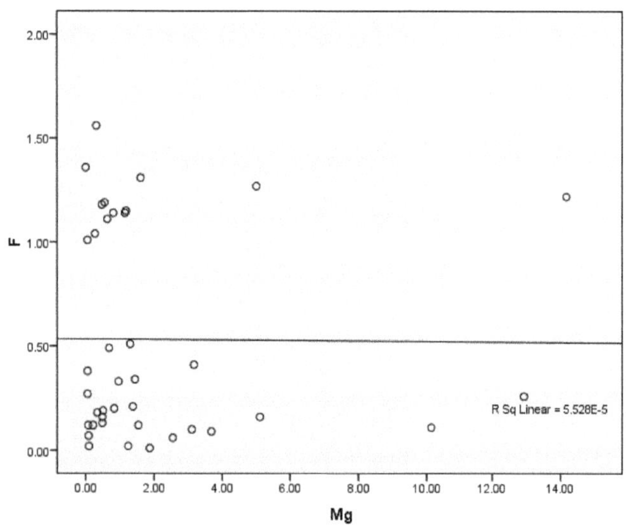
F
Mg
R Sq Linear = 5.528E-5

i

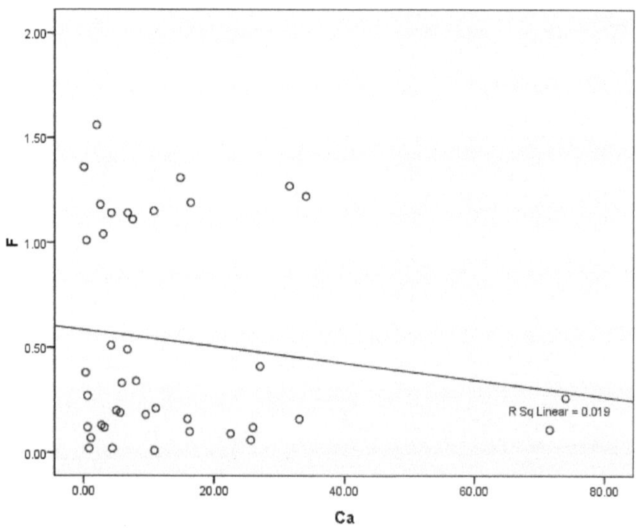
F
Ca
R Sq Linear = 0.019

j

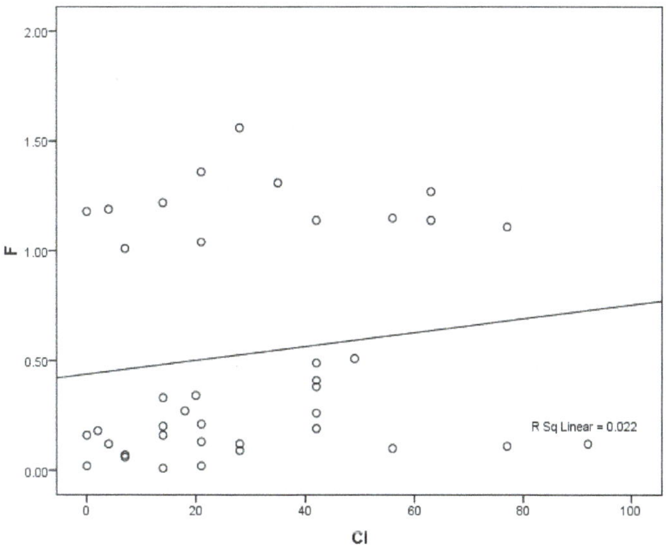

k

Fig 2 (a – k). *Relationship between Fluoride and some Parameters in the Study Area*

The Figures 2 (a -k) show scatter plots between F^-, and the physical parameters and other ionic constituents in the study area. There seem not to be a clear relationship between F and Temperature, TDS, Conductivity, Mg and SO_4 in the water. However, F exhibit negative relationship with pH, Ca, K and HCO_3 and a poor but positive relationship with Na and Cl.

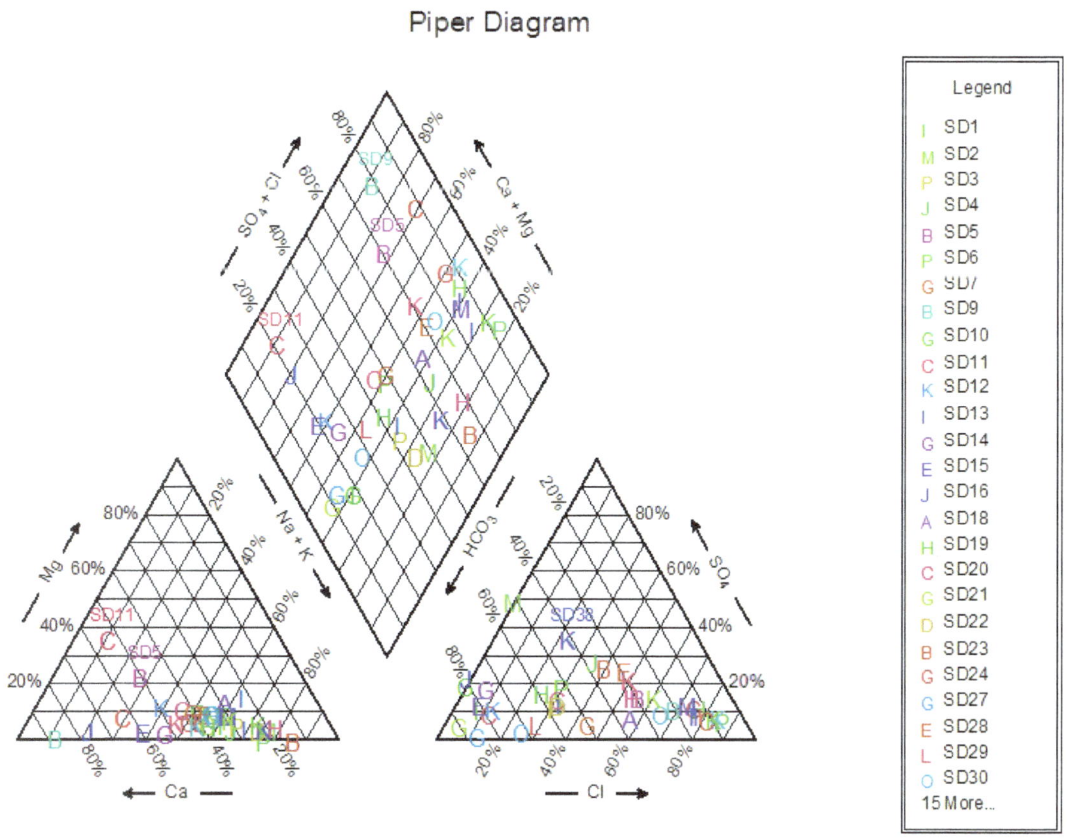

90%

Fig 3. *Piper Diagram showing Major Ion Chemistry of Groundwater in the Study Area.*

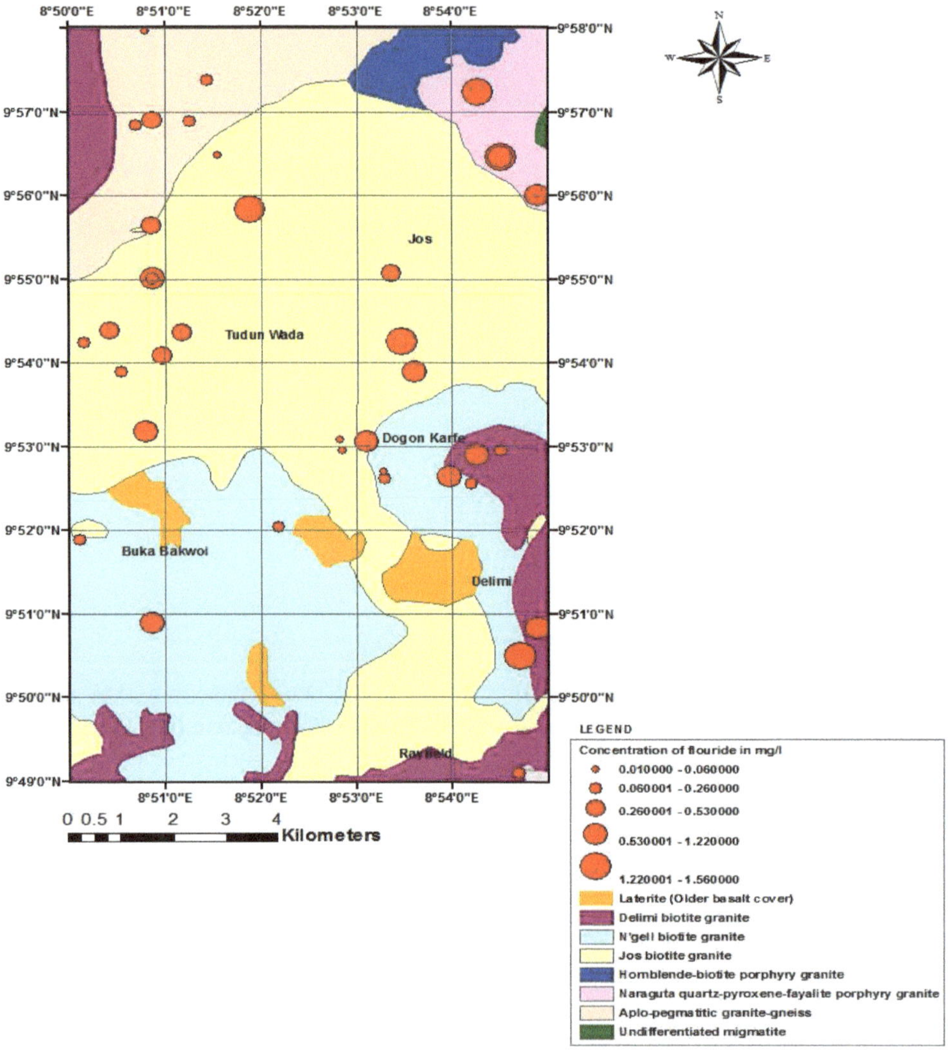

Fig 4. *Fluoride symbol map of the study area*

3.1. Groundwater Types

Groundwater type is useful in determining groundwater flow history, groundwater contact time and residency. High fluoride groundwaters have been associated to the Na-HCO_3 water type (Handa, 1975, Gaciri and Davies, 1993; Edmunds, 1994 Ampambire et al, 1997; Fordyce, 2007). Low level fluoride groundwaters have been associated to Ca-HCO_3 water type (Handa, 1975, Gaciri and Davies, 1993; Edmunds, 1995). The ionic concentration of major cations and anions in groundwater of the study area are plotted in Piper's Trillinear Diagram (Fig. 3). The analysis of Piper diagram suggests the majority of groundwater samples belong to the Na-Ca-HCO_3-Cl (mixed) water type with very few which is of the SO_4-Cl water type.

In order to examine the relationship between F^- levels in groundwater and geology, the fluoride values were plotted on geological map of the study area in the form of symbol map (Figure 4). It is observed that, of the 40 wells located on the different lithological units which make up the aquifer in the area, 13 (32%) have F^- concentration between 0.53 and 1.56 mg/l and 5 (38%) out of these 13, are located on the Jos-Bukuru Biotite Granite, 3 (23%) are located on the applo – pegmatitic granite terrain, 2 (15%) in the Dilimi Biotite-Granite area and 3 (23%) are located on the Naraguta quartz-pyroxene-fayalite-porphry granite terrain. The wells with lower F^- content occur mostly in the Ngeil-Biotite Granite. High concentration of F^- in groundwater occurs in the Jos-Bukuru -Biotite granite, Naraguta quartz –pyroxene-fayalite-porphry granite and applo-pegmatitic granite as earlier mentioned. The highest concentration of F- in groundwater occurs in the Jos-Bukuru -Biotite granite. It appears that the concentration of F^- in groundwater is proportional to the concentration of fluorine in the rocks (Fig 2).

4. Discussion of Results

The fluorine content in rocks of the study area is high except that of the Ngeil Biotite Granite area. The soils from the different sections of the weathered Jos - Bukuru Biotite Granite show variations in the content of fluorine, but appear to have higher fluorine content at the sections at the bottom. Cumulative (from top to bottom) value of fluorine collected from a section in the Ray Field area show slightly lower than half the value obtained in the rock mass and around the Air Force Base it has almost the same value in the rock mass. There is a clear indication that fluorine has been leached, evidenced from the low values obtained from the top soil. However, the values obtained from the different sections and the cumulative value especially that of the Ray Field areas do not correlate with the values obtained in the rock units. The high pH, low TDS, High bicarbonate and Na, low Ca content of the water are factors which favour the accumulation of fluoride in groundwater (Handa, 1975; Edmunds, 1994, Gaciri and Davies, 1993). However, the existence of these factors does not justify the content of fluoride in the groundwater of the study area. We therefore believe that apart from fluorine that have been adsorbed to the bodies of clay, bulk of the fluorine have not been leached from the biotite and other variety of minerals in the soils (Pickering, 1995) or probably fluorine may have formed complexes with other ions in soils or in the waters, rather than occurring in ionic form.

4.1. Groundwater Quality

The groundwater quality of the study area is evaluated by comparing the range of values of different geochemical parameters with the standard of the World Health Organization (WHO 1984). Physico-chemical properties of analysed water samples show considerable variation in the water quality with respect to their chemical composition. The waters in the study area are neutral to alkaline in nature. Nine (22%) out of 40 water samples from the study area have pH values above the WHO standard of 6.5 to 8.5 for human consumption. The electrical conductivity (EC) values are found to be within the range of 6 to 94 with mean value of 35.68. 91% of the samples were found to have EC values less than the maximum of 1,400 ppm at 25^0C specified by WHO (2004). With the concentrations of Total Dissolved Solids (TDS) ranging from 3 mg/l to 598 mg/l most of the samples are classified as soft as mean values for TDS is 85.98 less than the prescribed limit of 500 mg/l WHO, (2004) for drinking water. All the other ions found in the groundwater of the study area are within the WHO (2004) standard for house hold uses. With only 38% of water in the study area having fluoride level between 0.5 – 1.5 mg/l, most areas are imbibing water with fluoride below the recommended level of 0.5 mg/l (WHO 2004). Although there is no record to show the manifestation of dental caries in the study area, inferences can be drawn for the occurrence of dental caries owing to the low level of fluoride in the waters.

5. Summary and Conclusion

The Younger Granites rocks of the Jos, particularly the biotite granites as revealed in the study area, have considerable amounts of fluorine, but generally, fluoride levels in the groundwater is low. There is clear indication that fluorine has been leached as evidenced from the low content in the top soils. Also, factors which favour the accumulation of fluoride in groundwater in the study area have been satisfied, but it seems they have not been mobilized to the water. This low level fluoride in the waters may probably be due non release of fluorine in the biotite minerals and other minerals which fluorine may be residing in, apart from those that have been adsorbed to the bodies of clays. Probably also, fluorine may have formed complexes with other ions in the water and the soils and may be occurring in compound from rather than in ionic form.

About 90% of the concentration of cations, anions are within the permissible standard limit prescribed by WHO (2004) for drinking water..

38% of the F^- content in the water samples are within the permissible standard limit prescribed by WHO (2004) for drinking water. Some locations, particularly the Rayfield area, have F^- levels below the prescribed permissible standard limit by WHO 2004 for drinking water. People living in these areas may likely suffer from dental carries.

References

[1] Alakija, W (1983): Dental caries in Primary School Children in Nigeria. In: Lar and Tejan (Ed) Highlights of some environmental problems of geomedical significance in Nigeria. Environmental Geochemisty and Health, Springer, pp 383 – 389.

[2] Apambire, W.B.,Boyle, D.R. and Micheal, F.A., (1997): Geochemistry, genesis and ealth implication of fluoriferous groundwater in the upper regions, Ghana. Environ Geology. 33:13-24.

[3] Abu Rukah, Y. and Alsonkhny, K., (2004): Geochemical assessment of groundwater contamination with special emphasis on fluoride concentration, North Jordan. Chemie der Erde- Geochemistry. 64(2): 174-181.

[4] Bano, A.I., Isichei, W.P., Das, S.U., Morimoto, L and Nagataki, S (1987): Common trace elements in potable water in Plateau State of Nigeria and their impact on goiter prevalence in the state. In: Lar, and Tejan (Ed) Highlights of some environmental problems of geomedical significance in Nigeria. Environmental Geochemistry and Health, Springer, pp 383 – 389.

[5] Barber, W. (1958): Pressure water in the Chad Formation of Bornu and Dikwa Emirate, Northeast Nigeria, Geological Survey of Nigeria Bulletin 35, pp 62- 68.

[6] Bowden, P. (1966): Lithium in the Younger Granites of Northern Nigeria. In: Kogbe, C. A (Ed) Geology of Nigeria. pp, 207- 225 Rock View, Nigeria Limited, Jos, Nigeria

[7] Chilota, E.R. (2011): Geology and trace elements content in abandoned mined ponds waters of Bargada Area, Plateau State Unpublished B.Sc Project, Department of Geology and Mining, University of Jos

[8] Dibal, H.U. and Lar, U.A. (2005): Preliminary survey of fluoride concentrations in the groundwaters of Kaltungo Town and Environs. Gombe State. Northeastren Nigeria: Water quality and health implications. Journal of Environmental Sciences, Vol 9, no 2.

[9] Dibal, H.U., Lekmang, I.C and Lar, U.A (2008) Dental fluorosis from Drinking water consumption in Langtang Town, Plateau State. Continental Journal of Earth scinces, Vol 3, 77-82

[10] Dibal, H.U and Schoeneich, K (2009): Fluoride concentration in the groundwaters of Northern Nigeria. Nigerian Association of Hydrogeologist. Conference paper page 22

[11] Dibal H, U, Schoeneich, K, Garba I, Lar U.A. and Bala E.A (2012): Occurrence of fluoride in the drinking waters of Langtang Area, North Central Nigeria. Health, vol.4, No11, 1116

[12] Dibal, H. U., Schoeneich, K Garba, I., Lar U.A. and Bala E.A. (2012): Overview of Fluoride distribution in major aquifer Units of Northern Nigeria. Health vol 4, No 12, 1987-1294

[13] Edmunds, W.M. (1994): Characterization of Groundwaters in Semi - Arid and Arid Zones Using MinorElements, In: Nash, I.I and McCall, G.J.H (Ed), Groundwater Quality, pp 19 – 30. Chapman and Hall, London.

[14] Fordyce, F. M., Vrana, K., Zhovinsky, E., Povoroznuk, V., Toth, G., Hope, B. C., Iljinsky, U and Baker, J. (2007): A Health Risk Assessment for Fluoride in Central Europe. Journal of Environmental Geochemistry and Health, vol 29, no 2. pp 83 - 102.

[15] Gaciri, S. J. and Davies, T.C. (1993): The occurrence and geochemistry of fluoride in some natural waters of Kenya. Journal of Hydrology 143, pp 395-412.

[16] Hitchon, B., (1995): Fluorine in formation water, Alberta basin, Canada. Applied Geochemistry. 10:357-367.

[17] Hussain, J., Sharma, K.C and Hussain, I (2004): Fluoride in the groundwaters of Rajastan. Indian Journal Environmental Health. Pp. 84 - 87.

[18] Handa, B.K., (1975). Geochemistry and genesis of fluoride containing groundwater in India. Soil research laboratory, 1147, sector 21-B, Chandigarh, India vol.13, No.3.

[19] Jack, G., Bhattacharya, p., Chaudhary, V. and Singh, K.P., (2004): Controls on the genesis of some high-fluoride groundwater in India. Applied Geochemistry, 20(2): 221-228.

[20] Kim, K. and Jeong,G.Y., (2005): Factors influencing natural occurrence of fluoride-rich groundwater: a case study in the south eastern part of the Korean Peninsula. Chemosphere. 58(10): 1399-1408.

[21] Lar, U.A and Tejan, A.B (2008): Highlights of some environmental problems of geomedical significance in Nigeria. Environmental Geochemisty and Health, Springer, pp 383 – 389.

[22] Lar, U.A. and Dibal, H. U. (2014): The status of medical geology in Nigeria: challenges, prospects and benefits. 1st International Symposium on Medical Geology in Africa (ISMGAf)University of Johannesburg, Republic of South Africa "Health and Earth – Building a Safer Environment" Program and Abstract Volume

[23] Lar,U.A., Dibal,H. U., Daspan, R. and Jaryum, S.W. (2007): Fluoride occurrence in the surface and groundwater of Fobur area of Jos East LGA of Plateau States. Journal of Environmental Sciences II, (2) pp 99-105

[24] Macleod, W.N., Turner, D.C. and Wright, E.P. (1971): The Geology of Jos Plateau. Geological Survey of Nigeria, Bulletin No 32.

[25] Pickering, W.F. (1983): The Mobility of Soluble Fluoride in Soils. Environmental Pollution (Series B) 9, pp281 -308.

[26] Schoeneich, K. and Mbonu, W.C. (1991): Excursion Guide Book: On the Geology of the Jos Plateau and the Yankari Game Reserve. The Nigerian Mining and Geosciences SocietyAnnual Conference and 30th Anniversary Celebration held at Jos. Bros - way Press 79/1 Sarki Street, Jos.

[27] Thompson, J.H. (1965): The Geology and Hydrogeology of Gombe, Bauchi Province. Records of the Geology Survey, Pp 46 -56.

[28] WHO (2004) Guidelines for drinking water quality, Vol.2 health criteria and other supporting information. World Health Organization, Geneva.

[29] Wongdem, J.G., Aderinokun, G.A., Sridhar, M.R. and Selkur. S. (2002): Prevalence and distribution pattern of enamel fluorosis in Langtang Town. African Journal of Medical Science. Fluoride 35. pp. 120-135.

Debates on Land Titling

Ayele Behaylu

Department of Geography and Environmental Studies, Collage of Social Science and Humanities, Adigrat University, Adigrat, Ethiopia

Email address:
ayeleby@gmail.com

Abstract: Countries like Ethiopia emphasis that certification process guarantees land use and transfer rights ensures tenure security, and is a vital incentive for farmers to invest in their land and improve production. The program is assumed to addresses the issue of tenure insecurity which is a chronic problem in most developing countries. Moreover, land titling has a good effect in reducing of both border and inheritance disputes. However, there are debating issues on the effect of land titling in tenure security, land management and land disputes. The aim of this study is to investigate whether land titling has resulted in insuring tenure security, assuring land management, and reducing land related disputes or not. Though land titling is not a sufficient factor for sustainable land management, it is a necessary condition so that it enables farmers to have confidence on their tenures which is a preliminary condition to start land management activities. In addition to this, land titling is vital for minimizing of land disputes.

Keywords: Land Disputes, Land Management, Land Policy, Land Titling, Tenure Security

1. Introduction

Land is one of the most important natural resource in which we can get different uses from it like forest products, production of crops, construction materials, and what not (Olsson and Lundmark, 2014). Though there are various factors like climate change for the deterioration of environment in general and land in particular (Towfiqul Islam, *et al*, 2014), tenure insecurity has a lion share for land degradation.. Tenure insecurity creates doubt on farmers t their tenure so that they will not manage properly. For the proper utilization of this resource without degradation, land titling is vital. Land titling is a process of recording rights on land which provides safe and certain foundation of acquisition and disposal of rights in land, where disposition includes transfer, leasing or mortgage of the holding rights (Pankhurst,1966). For African countries registration of land rights is a continuing issue regarding debate in land policies. Many West African countries have instituted the process of land registration through their land legislation. Land registration and bestowing of land certificates is assumed to have many effects on different land related issues. Some of these are tenure security, sustainable land management practices, and reducing of land disputes. These issues are intended to be discussed in the following sections.

2. Land Titling and Tenure Security

Land titling is assumed to have a good effect in resource management and planning (Djiré, 2007). However, there are conflicting ideas about the issue of land registration and its impact in ensuring land tenure security and assuring sustainable land management. There are many studies supporting the neoclassical hypothesis that tenure insecurity is correlated negatively with the quality of resource management. In some countries certain results do not show a positive link between tenure security and land related investments.

For most African countries land tenure, land registration and title certification do not necessarily ensure tenure security. For example, comparative studies conducted by Migot–Adhola *et al* (1991) in African countries such as Ghana, Rwanda and Kenya showed that land registration and title certification do not have a clear impact on land improvements. Lack of formal land titles may not be the cause of tenure insecurity. Rather physical infrastructures, effective credit systems and marketing institutions or access to credits had a greater impact on land improvements (Migot-Adhola *et al,* 1991). In similar way Feder and Nishio (1998) revealed that land registration is unlikely to be economically worthwhile for much of sub Saharan African countries. Place and Migot-Adholla (1994) with their studies in four Kenyan sites showed that land registration and certification showed a very weak impact on perceived tenure rights of farmers, land improvements crop yields, or concentration of landholdings.

Broegaard (2002) with her study in Nicaragua stated that in spite of the fact that most farmers in Nicaragua posses a legal title to their land, they still feel tenure insecurity and there is no improvement in land management as a result of land titling. According to Broegaard (2002) more attention should be given to aspects other than land certification like inequality of wealth and power, lack of enforcement, lack of impartiality on the part of formal institutions when addressing an institutional unstable setting. In her conclusion Broegaard (2002) explained that even though titles are much desired to secure property rights, they did not necessarily create tenure security. Similarly, Dickerman (1989) stated that registration and title certification is no guarantees of success in bringing tenure security and land improvement but merely the necessary foundation (Dickerman, 1989 cited in AAE, 2006). Lemel (1988) also stated that lack of individual titles does not necessarily mean that there is tenure insecurity, and individual titles do not significantly ensure security of tenure.

A study conducted by Zikhalui (2008) in the central province of Zimbabwe called Mashona-land stated that, in order to investigate the impact of Zimbabwe's "fast track land reform program" on the perception of tenure security and conservation of natural resources including investments on land management showed that the program did not ensure tenure security among the farmers and had no positive impacts on soil conservation practices. It is assumed that this program might have failed to provide tenure security to farmers who had got land under reform program (Zikhalui, 2008). Adams (2001) argues that unless the land administration is accompanied by other development activities, the land reform in general may unlikely to make much difference to the rural poor. Unless cadastral surveys and modern system of records are keeping essential, land registration and title certification could not to be effective to bring tenure security, and even it may lead to conflict and uncertainty (Dessalegn, 1994).

There are also other researchers who argued that tenure insecurity does not prohibit land investment activities. A study by Neef *et al* (2001) in Niger, Benin, Thailand and Vietnam show that tenure insecurity does not always lead to decreasing investments on land. They explain that sustaining environmental resources does not primarily depend on whether farmers have titled land or not. Tenure security did not necessarily bring as a result of certification (Neef *et al*, 2001). The same case also found in Burkina Faso where land related investments are made to increase tenure security rather than as a consequence of more secure rights. They stated that land registration and title certification enable security of tenure by stimulating long-term investments to improve soil fertility and land management and to increase productivity (ibid).

On the other hand, Cotula *et al* (2004) stated that land registration and title certification increases tenure security, and hence stimulates land investment in addition to improving farmers' access to credit in which it serve as a collateral to financial institutions. In most African countries land management activities and investments made to bring tenure security rather than as a consequence of more secure rights .Whereas, in case of Ethiopia the reverse is true, i.e. land tenure security enables the land management (Brasselle *et al.*, 2002). Land registration and certification stimulates a more efficient use of land, enables creation of land market, and it provides farmers with a title that can be offered as collateral for financial institutions. Tenure insecurity which in turn caused by absence of land registration is the main reason why developing countries did not developed as western countries. There are other paramount of research results which show the positive relationship between land registration and certification and tenure security as well as land improvement and productivity (ibid). There are quite a number of case studies conducted in different countries that show the role of land registration and title certification in ensuring tenure security. For example, Sikor (2005) in his study in the northwest Vietnam assured that land registration and title certification is counterproductive to rural peoples of the country by avoiding the economic and political uncertainties and providing tenure security to the region. A study by Kabubo-Mariana and Linderhof (2009) in two provinces of Kenya (Naro and Kajado) show that land management practices strengthened as a result of land tenure security which was ensured due to land registration and title certification in the region as the cost of other incentives. Deininger *et al* (2003b) in their study at Nicaragua also founded that land titling and administration project has great contribution to reduce the pre-existing uncertainties on land tenure; and enable poor people to have a guarantee on their holdings. Their study and finding was supported by another study conducted by Binswanger *et al* (1995) in that landholder can reasonably expect that if his/her property rights are well enforced, a farmer will have strong incentives to make long-term land investments, when these long-term investments combined with better formal credit access it will led to higher land price and better functioning of land markets (Binswanger *et al,* 1995 cited in Broegaard, 2002).

According to Holden *et al* (2007), important policy concerns are whether the land reform in form of registration and certification has contributed to increased tenure security, especially for the poor, including women that women think differently about their land certificates than men do as their tenure rights have been less secure than that of men, and the certificates may therefore have a higher value to them than they have to men"

3. Land Titling and Land Resource Management

Land tenure security cannot be directly measured, and to a larger extent it is what peoples perceive it to be, and may change from one context to another context (FAO, 2005).

Therefore, it is vital to use other cross-checking mechanisms. One of the mechanisms used to measure whether there is tenure security among farmers or not is land management. Since it increases security of tenure rural land titling enable landholders to have incentives to investment in land and long-term protection of the environment (Solomon, 2006). Peoples having certificates are more interested in making long-term land related investments (Berhanu *et al*, 2003; Holden *et al*, 2008b). Insecure of tenure on land and other natural resources is the main reason for the natural resources end-users to be unwilling to invest in long-term sustainable land management (GEF, 2005). Proper land management and investment on land productivity are expected outcomes of land registration and certification of farmlands (Charles, 2004). Feder and Nishio (1998) from their findings in their empirical evidence in Thailand concluded that in addition to increased access to formal credit opportunities, formal land registration has contributed to greater land investment which in turn has resulted in increment in production. According to the report of Orgut (2010) in Amhara state of Ethiopia, land certification increased land investment, investment in farming equipments, and improvement in agriculture. Similarly, Berhanu and Fayera (2005) in their studies in four districts of Amhara state founded that more than 59% of farmers perceived that land registration and certification had assured improvement in sustainable land management. Another study by Sabita (2010) in Ethiopian rift valley region showed that more than 70% of the respondents (farmers) completely agree that certification had increased investment in sustainable land management. The results from the study conducted by Holden *et al* (2008a) in Tigray State also showed that land certification has contributed to better improvement in soil conservation structures, increased investments in trees, and enhancement of land productivity. According to the findings of these researchers, people having certificates are more interested in making long term land-related investments and also the high use of chemical as well as organic fertilizers. On the other hand, Berhanu *et al* (2003) from their study in Tigray State of Ethiopia founded that land certification has brought short-term land investments. Another study conducted by Kabubo- Mariana and Linderhof (2009) in Kenya long-term investments in land management technologies were brought as a result of land registration and certification. In their findings they had explained that the three most frequently observed land investments in their survey were tree planting (28%), terracing with grass strips (26%) and grass strips (23%). Increasing of land tenure security and associated property rights as a result of land certification in china has brought land investments, and hence farmers applied somewhat more labor (18%), nitrogen fertilizer (15%), animal traction (14%), organic manure (32%), and phosphates (20%) (Rozelle *et al*, 1998).

According to Laiglesia (2004) in Nicaragua possession of legal property documents/ registries in enhancing agricultural investment incentives increased the probability of carrying out land-attached investments by 35%. Bruce and Migot-Adholla (1994) also investigated that land titles and clarity of land rights play an important role for providing incentives for investment in land use. Besides to this, an influential study was made by Feder *et al* (1988) in Thailand show that certification of land titles to farmers provide not only tenure security but also higher level of land investment and higher land price. Titled land had higher capital stocks of 56-250% and use of labor (increased by 8-15%), draft power (increased by 25-39 %), and fertilizers and pesticides (increased by 23-34%) and compared to untitled land, resulting in higher output and productivity.

In Costa Rica ownership of tenure security as a result of protected land right had brought farm investment per unit of land. Increased investment and higher output and income were also realized on titled land in Costa Rica, Brazil, Ecuador and Paraguay (Salas *et al* 1970; Feder and Nishio 1998). Analogously, the study by Vilamizar (1984) at three Brazilian states shows that investment per hectare is substantially greater on titled land than untitled land. Similarly, a survey study made by IDB (1986) in Jamaica revealed that there was a greater incidence of permanent and semi permanent crops among farmers with titled lands than untitled lands. According to the IDB's report, following the bestowing of certificates farmers planted more permanent and semi permanent crops. Moor (1996) in his study in Manincaland province of Zimbabwe argued that there is a positive impact of land rights status on land investment activities and productivity (Moor, 1996). The study demonstrated that tenure security has significant positive effect on farmers' long-term land investments and increment in yield. Alston *et al* (1996) in their studies at Brazil show that although there is difference with location from market center, the titled land is more valuable than untitled land. Without taking in to consideration of market distance, titled land of a given agricultural quality has a 189% higher value than untitled land at the same location. Berhanu and Swinton (2003) also founded that land titling in Tigray region has brought durable long-term land investments such as stone terraces and tree planting than short- term low cost investments like soil bunds.

4. Land Titling and Land Disputes

Dispute over land has been the root cause for civil wars because land had been the primary means of generating livelihood for majority of rural population of many countries (Holden and Yohannes, 2001). In most countries of the world disputes over land rights and property boundaries make up the largest percentage of court cases. Land disputes resulted in social unrest, expensive litigation and breakdown in law and order, and hence courts spend much of their time in resolving these matters (Marquardt, 2006). With reference to smallholder farmers, the commonest land-related dispute is related to inheritance and boundary (Gizachew, 2006). In addition to land redistribution and expropriation of land by government agencies with or without adequate compensation, disputes with neighbors over boundaries and disputes with relatives over inheritance in Ethiopia under successive governments are major sources of insecurity (Bruce and Migot-Adholla, 1994).

Dispute over land can be manifested in many basic ways. They may be occurred when there is lack of adequate demarcation, registration and record keeping which led to overlapping land claims stemming from inheritance that is beginning to result in conflict, when the law is weak in protecting the rights of individual landholders, in which violence and intimidation are used against women who attempt to use the law to establish and defend their right to landholdings, and the feeling of tenure insecurity (Brasselle *et al*, 2002; Holden *et al*, 2008b). However, many of the land disputes could be resolved through access to accurate land ownership information of land holders which may be useful for the design of environmentally sound development programs for the provision of incentives to land holders who practice sound land management, or the imposition of sanctions on those perpetuating environmentally destructive land use activities. (Marquardt, 2006).

Holden *et al* (2007b) in their study in Tigray region stated that in addition to improving tenure security and land-related investments land certification contributed in reducing land disputes/boarder conflicts among the households. Land registration and certification in the region revealed that the number of border conflicts during and after the reform (registration and certification) were more likely to have decreased than increased. They stated that clarification of boundaries and field-based adjudication that were undertaken in the context of the certification program had contributed towards a marked reduction in land related conflicts because formalization of land rights and better demarcation of land borders enable farmers to have security of tenures. Land registration and certification was more successful in solving land disputes that arose during the last land redistributions in Tigray state (ibid). In addition to being a proof of ownership, land registration documents/certificates are compulsory by provide documentary evidence that is necessary for resolving property disputes as well as information for a wide variety of public information such as land valuation (Dale and Mclaughlin, 1999).

Sikor (2005) in his study in the northwest Vietnam stated that land registration and certification reduced border disputes in the region. In addition to this, Deininger *et al* (2007) in their study in Tigray regional state founded that 66% of land disputes decreased after certification. In Amhara region, in addition to increased the perception of tenure security among farmers which improved tree plantation and other SLM activities, land titling and bestowing of certificates resulted in the decline of incidence of land disputes. This is because land registration and title certification enables the proper demarcation of land boundaries; hence, resulted in lesser land disputes (Palm, 2010).

5. Conclusion

From this review paper one can understand that there are divergent views about the fruitfulness of land titling regarding tenure security, sustainable use of our limited natural resources, and reducing of land disputes. Even though there are other factors that have a negative impact on tenure security, land titling is a means by which farmers' confidence on their tenures can be increased so that they can practice whatever land management activities on the land. Moreover, land titling enables the reduction of land related conflicts among neighbors and relatives.

References

[1] AAE (Action Aid Ethiopia) .2006. Case Study on Policies and Practices for Securing and Improving Access to and Control over Land in Ethiopia. Prepared for International Conference on Agrarian Reform and Rural Development (ICARRD) January 2006. Addis Ababa, Ethiopia.

[2] Abu Reza Md. Towfiqul Islam, Anjum Tasnuva, Soniya Sultana , Shifat Rumana .2014. Climate change impact: Food production and local Perception. Science Publishing Group. American Journal of Environmental Protection 2014; 3(2): 45-50.

[3] Adams, M. 2001. Tenure security livelihoods and sustainable land use in southern Africa. Paper presented at SARNP. Conference on land reform and poverty alleviation in southern Africa June 4th -5th.

[4] Alston, L., Libeecap, G.D. and Schneider, R. 1996. The determinants and impact of property rights: land titles on the Brazilian frontier. The Journal of law, economics and organization, 12(1).

[5] Anna Olsson, Robert Lundmark .2014. Modelling the competition for forest resources: The case of Sweden. Science Publishing Group. Journal of Energy and Natural Resources 2014; 3(2): 11-19.

[6] Berhanu Adenew and Fayera Abdi. 2005. Land Registration in Amhara Region, Ethiopia. Securing Land Rights in Africa. Research Report 3. IIED: London.

[7] Berhanu Gebremedhin and Swinton, S.M .2003. "Investment in soil conservation in northern Ethiopia: the role of tenure security and public programs." Agricultural Economics, 29, 69–84.

[8] Berhanu Gebremedhin, Pender, J. and Ehui, S. 2003. Land Tenure and Land Management in the Highlands of Northern Ethiopia. International livestock Research Institute (ILRI) http://homepages.wmich.edu/sefa/Conference%20.

[9] Brasselle, A.S., Gaspart, F. and Platteau J.P., 2002. Land Tenure Security and Investment Incentives: Puzzling Evidence from Burkina Faso. Journal of Development Economics, 67, pp. 373-418. University of Namur, Rempart, de la Vierge.

[10] Broegaard, R. J .2002. Land Tenure Insecurity and Inequality in Nicaragua. Institute of Geography Copenhagen: University of Copenhagen.

[11] Bruce, J. W. and Migot-Adholla, S. E .1994. Searching for Land Tenure Security in Africa: Review, Kendall-Hunt Publishing Company, Dubuque, Iowa.

[12] Charles, C.G. 2004. The Impact of Land Titling on Land Transaction Activity and Registration System Sustainability: The Case of St. Lucia (PHD Dissertation) University of Florida.

[13] Cotula L., Toulmin C., and Hesse, C.2004. Land Tenure and Administration in Africa: Lessons of Experience and Emerging Issues. IIED.

[14] Dale, P.F. and Mclaughlin, J.D. 1999. Land administration systems. USA, New York UK.

[15] Deininger, K., Daniel Ayalew Ali, Holden, S. and Zevenbergen, J. (2007) 'Rural land certification in Ethiopia: Process, initial impact, and implications for other African countries', in World Bank Policy Research Working Paper 4218: The World Bank.

[16] Deininger, K., E. Zegarra and Lavadenz, I. 2003b. 'Determinants and Impacts of Rural Land Market Activity: Evidence from Nicaragua', World Development 31(8): 1385–1404.

[17] Dessalegn Rahmato .1994. Land policy in Ethiopia at Cross-roads. Dessalegn, R. (ed.).Tenure and Land Policy in Ethiopia after Derg: proceedings of the second workshop of land tenure project. University of Trondheim: Norway. Center for Environment and Development. Norway: Reprocentralen ANH. Pp 1-34.

[18] Djiré, M. 2007. Land registration in Mali– No land ownership for farmers? Observations from peri-urban Bamako. http://books.google.com.ct /books? Id =r s3BWtm.

[19] FAO .2005. "Access to rural land and land administration after violent conflicts. Food and Agricultural Organization of United Nations, Rome. ISBN 9-25-105343-x".

[20] Feder G. and A. Nishio .1998. The Benefits of Land Registration and Titling: Economic and Social Perspectives, Land Use Policy, Vol. 15:1. Pp25-43. PII: S0264-8377/97.

[21] Feder, G., T. Onchan, Chalamwong, Y. and Hongladarom , C. 1988. Land Policies and Farm Productivity in Thailand. Baltimore, MD: The Johns Hopkins University Press in association with the World Bank.

[22] GEF council .2005. Scientific and Technical Advisory Panel to the Global Environment Facility: Land management and its benefits – the challenge, and the rationale for Sustainable management of dry lands. (Global Environment Facility).

[23] Gizachew Abegaz .2006. Basic Definitions of Land Registration Terms. A Review. http://eltap net/ download/NationalCoferenceOnStandardization.

[24] Holden, S.T. and Yohannes Hailu. (2001). "Land Distribution, Tenure Insecurity, and Intensity of Production: A Study of Farm Households in Southern Ethiopia" CAPRi Working Paper No. 21. International Food Policy Institute Washington DC USA: CGIAR.

[25] Holden S.T., Deininger, K. and Hosaena Ghebru .2008a. Land Certification, Land -Related Investment and Land Productivity Impacts. Research Report1..http://www sta tk ar t .no / Engelsk and Land Registry.

[26] Holden, S.T., Deininger, K. and Hosa'ena Ghebru .2007. Impact of Land Certification on Land Rental Market Participation in Tigray Region, Northern Ethiopia. Paper submitted for the Nordic Development Economics Conference in Copenhagen June 18-19, 2007. Department of Economics and Resource Management Norwegian University of Life Sciences, Norway. (2007:13).

[27] Holden, S.T., Deininger, K. and Hosaena Ghebru .2008b. Impact of Land Certification on and related Conflicts in Ethiopia. Working Paper, Norwegian University of Life Sciences, Norway.

[28] Holden, S.T., Deininger, K. and Hosaena, Ghebru. 2007b. Impact of land registration and certification on land border conflicts in Ethiopia. http://siteresources.World bank.org/E/ Resources.

[29] Inter-American Development Bank (IDB).1986. Jamaica land titling project: feasibility report. Washington, DC.

[30] Kabubo-Mariana and Linderhof.2009. Land Tenure, Sustainable Land Management and poverty: Case Studies from Kenya. School of Economics, University of Nirobi, Kenya.

[31] Laiglesia J. R. 2004. Investment and credit effects of land titling and registration: evidence from Nicaragua. London School of Economics.

[32] Lemel, H. 1988. 'Land Titling: Conceptual, Empirical and Policy Issues', Land Use Policy 5(3): 273–90.

[33] Marquardt, M. A. 2006. Global Experiences in Land Registration and Titling Global Experiences in Land Registration and Titling. http://eltap.net/dhttp://eltap.net/download /proceeding/ T1P 1 % 20 Marquardt .

[34] Migot-Adhola, S., Hazell, P., Blarel, B. and Place, F., 1991. Indigenous Land Rights Systems in Sub-Saharan Africa: A Constraint on Productivity? The World Bank Economic Review, Rev., 5: 155-175, January 1991.

[35] Moor, G.M. 1996. Tenure Security and Productivity in the Zimbabwean Small Farm Sector: Implications for South Africa. Department of Agricultural Economics, University of Natal, Pietermaritzburg.

[36] Neef, A., Sangkapitux C. and Kirchmann K .2001. Does Land Tenure Security Enhance SLM? Evidence from Mountainous Regions of Thailand and Vietnam.

[37] Orgut .2010. Land Registration and Certification: Experiences from the Amhara National Regional State in Ethiopia. Report by the Bureau of Environment Protection, Land Administration and Use and Orgut Consulting. http://www.orgut.se/files/L and 20 %20Registration.

[38] Palm, L. 2010. Quick and Cheap Mass Land Registration and computerization in Ethiopia. Stockholm, Sweden.

[39] Pankhurst, R. 1966. State and land in Ethiopian history. Addis Ababa: Institute of Ethiopian Studies, in association with Oxford University Press. http: //www.africanidea.org Richard_Pankhurst.pdf.

[40] Place, F. and Migot-Adholla, S.1994.The Economic Effects of Land Registration on Smallholder Farms In Kenya: Evidence from Nyeri and Kaakmega Districts. Land economics. Vol. 74. No. 3. University of Wisconsin press.

[41] Rozelle, S., Li, G. and Brand, L.1998. Tenure, land rights, and farmer investment incentives in China. Agricultural Economics 19 (1998) 63±71. Elsevier Science B.V.

[42] Sabita Giri. 2010. The Effect of Rural Land Registration and Certification Program on Farmers' Investments in Soil Conservation and Land Management in the Central Rift Valley of Ethiopia. (MSc Thesis). Wageningen University, The Netherland.

[43] Salas, O., Knight, F. And Saenz, C. 1970. Land Titling in Costa Rica: a Legal and Economic Policy, USAID Contract 515-178—T.USAID And University of Costa Rica Law School, San Jose.

[44] Sikor, T.2005. Politics of rural land registration in post-socialist societies: Contested titling in villages of Northwest Vietnam. Junior Research Group on Post socialist Land Relations, Luisenstr. Humboldt University, Berlin, Germany.

[45] Solomon Abebe .2006. Land Registration System in Ethiopia. Comparative Analysis of Amhara, Oromia, SNPP and Tigray Regional States. http://www.Goog le.c/ search = Solomon % 2 C +A.+%282006%.

[46] Vilamizar, M.-F. (1984) IBD financing in land administration programs. Paper presented at the International Workshop on Land Tenure Administration, August 20-24, Salvador, Brazil.

[47] Zikhalui, P. 2008. Fast Track Land Reform, Tenure Security and Investments in Zimbabwe. Environmental Economics Unit. University of Gothenburg, Sweden.

25

Diversity of Vascular and Insects Canopy Epiphytes on Palm Oil in North Sumatra, Indonesia

Fitra Suzanti[1], Retna Astuti Kuswardani[2], Suci Rahayu[3], Agus Susanto[4]

[1]Biology Education Study Program, University of Riau, Riau, Indonesia
[2]Faculty of Agriculture, University of Medan Area, Medan, Indonesia
[3]Faculty of Mathematics and Natural Sciences, University of North Sumatra, Medan, Indonesia
[4]Palm Research Center of Marihat, North Sumatra, Indonesia

Email address:
fitrausu@gmail.com (F. Susanti)

Abstract: This paper describes the role of the insect, which is essential for human survival, but still people are not familiar with widely even some people consider insects no meaning or viewed from the downside. Diversity of epiphytes and insects in the canopy of palm oil plantations in North Sumatra assessed by field observations at five locations, namely, Bukit Sentang, Padang Mandarsyah, Ajamu, Sijambu-jambu, and Aek Pancur. Data were analyzed using qualitative and quantitative. Based on the results obtained by analysis of the biological diversity of insects that affect the quality and quantity of farm products. In natural ecosystems, generally there has been a stable population of pests and their natural enemies so that the presence of insect pests in cropping is no longer harmful. Thereby, the conditions need to be developed so as to reduce the use of pesticides to control pests in field crops, especially in export-oriented and has high economic value.

Keywords: Diversity, Vascular Epiphytes, Insects Canopy, Oil Palm Plantations

1. Introduction

Instability of the agro-ecosystem and increasing of pest infestation is caused by changing in natural ecosystems, making the main crop plantation ecosystem type (monoculture), so the diversity of micro-habitats to be reduced. Agriculture sustainable can be realized by reducing the action of pests through the use of natural enemies of insects and increasing the diversity of plants. This needs to be done to improve the stability of the ecosystem and reduce the risk of pests [1]. One of the diversity of plants in the ecosystem of oil palm plantations is a vascular epiphytes. These weeds are grouped in the plantation, because their numbers are small (1.5%) compared with other weeds so often overlooked existence [2]. The type is ferns, ficus and other wood. It provides an important resource and habitat for many animals and microorganisms including eating insects, and vertebrates nectarivorous fruit-eating bats and birds in particular. [3] found the presence of epiphytes in a coffee plantation turned out to support more insects compared with no epiphytes. The presence of epiphytic provide a valuable resource for arthropods.

As long as this is done in oil palm plantations is controlling epiphyte epiphytic ferns woody and let live, as long as their presence does not interfere with harvesting and plant maintenance. Information about epiphytes in oil palm plantations, both in terms of ecology, insects contribution to the canopy and also the microclimate created is still small. It is necessary for better understanding of the ecology on human modified ecosystems to support the conservation of the landscape level [4]. Natural ecosystem with a simpler artificial ecosystem is more easily shaken by pests caused fewer species [5, 6]. Biodiversity in monoculture plantations are low, may result in lower complexity compared to other natural habitats. It seemed from the tree whose age is uniform, lower canopy, and almost no bush. This is a result of human disorders that are high in oil palm plantations [7, 8]. It is part of the ecosystem, and has a function in the cycle of nutrients and water, which changes the composition of epiphytes can have negative effects, affecting the ecosystem processes in general [9, 10].

The survival and distribution of epiphytes is dependent on the density of porofit, microclimate, the distance from the seed source, the type and size of the tree, the type and history of the disorder, epiphytic population dynamics of trees and epiphytes physiology [11]. Bryophyta and epiphytes richness of coverage decreased from the tree to the top of the stem, while increasing the wealth of moss and lack of coverage of the base to the top of the rod [12]. There are 136 species of weeds, which consists of 33 families dikotil, 3 families of monocots and 8 families Pterydophyta of weeds [13].

Weed of the family *Poaceae* and *Asteraceae* is very invasive. *Chromolaena odorata, Aspillia africana* and *Melanthera* scandens of *Asteraceae*, *maximum Panicum* and *Imperata cylindrica of Poaceae* and *Euphorbiaceae Mallotus oppositifolius* of widespread and problematic. [14] stating there is a correlation of epiphytic vegetation on the palm *Socratea Exorrhiza* with increased diameter, the age of the tree and *Bryophyta* cover. The most common epiphytes are three species of fern, which *Ananthacorus angustifolius, Elaphoglossum sporadolepis* and *Dicranoglossum panamense.* [1] found that weeds affect diversity and the existence of herbivorous insects and their natural enemies in the farm system. Flowers certain weeds (mostly *Umbelliferae, Leguminosae* and *Compositae*) plays an important role as a source of adult parasitoids feed which can suppress the insect pest population. Natural mechanisms such as predatisme, parasitism, pathogenicity, and interspesies intraspesies competition, succession, productivity, stability and biological diversity can be utilized to achieve sustainable agriculture.

The expansion of monoculture crops at the expense of natural vegetation is the result in reduced diversity of local habitats, eventually leading agro-ecosystem instability and increased pest attacks. Commodity crops modified to meet human needs will be damaged due to high pest infestation. Generally, the more intensive the plant is modified it will be more intensive pest attack him anyway. Characteristic properties of its own settings commodity will naturally disappear when humans modify these commodities by breaking the interaction of plant life and eventually becomes brittle. This solution can be improved with the recovery of commodity components through the addition or enhancement of biodiversity. [15] stated polyculture in annual plants is more supportive of herbivores to a level lower than monocultures, in terms of the dynamics of insects. This trend is caused by the stability of the natural enemy population with the availability of food sources and habitat micro continuously. Another possibility is certain herbivores prefer similar habitats on the plant, which provides centralized all its needs, and the physical conditions are always the same. Manipulation using cover crops (cover crops) also affect insect pests and their natural enemies. Orchard with wild plants underneath a lower damage by insect attacks than cultivated orchards free of other plants (clean cultivated), because of the abundant amount and efficiency of predators and parasitoids.

Plants are the main source of feed for herbivorous insects, because it provides a compound of the main nutrients, namely carbohydrates, proteins, amino acids, vitamins, minerals, and water. Nitrogen (N) in the form of protein and amino acids in the plant tissue is one of the reasons herbivorous insects choosing one type of plant as their food source [16]. The main role of nitrogen in the metabolic processes that improve the health, growth, proliferation, and resistance organism to make this element as a factor limiting in nature. Insects are generally grouped into insect pests, useful insects and insect neutral from the standpoint of farming. As useful organisms, insects on the role of natural enemies either as parasitoids and predators, insect pollinators and decomposers. While neuters often fall prey to predators, so a very large role in maintaining the balance of the ecosystem. Insect biodiversity affects the quality and quantity of product produced. In natural ecosystems, generally there has been a stable population of pests and their natural enemies so that the presence of insect pests in cropping is no longer harmful. These conditions need to be developed so as to reduce the use of pesticides to control pests in field crops, especially in export-oriented and has high economic value. Insects have a role that is essential for human survival, but the role of these insects is still less known by the public and even some people consider insects no meaning or viewed from the downside. This became the basis of a multifaceted assessment canopy epiphytes and insects in oil palm plantations in North Sumatra.

2. Method

The study was conducted at five locations in North Sumatra oil palm plantations, namely on the Bukit Sentang, Padang Mandarsah, Ajamu, Sijambu-jambu and Aek Pancur. Identification of epiphytes and insects do in Herbarium Medanense, Biology laboratory of North Sumatra University, PPKS Marihat, Laboratory and LIPI Cibinong.

2.1. Equipment and Materials

The tools used in the study are shears, knife, hatchet, yarn, pencils, books field (collector book), etiquette hanging, binoculars, plastic bags, stairs, Egrek, mistblowwer Maruyama MD 1800X, Knapsact spayer Solo 425, bottle collection, magnifying glass, tool documentation, digital termohygrometer TFA, bouffant/ herbarium presses, newsprint, fabric insect container, cardboard insulation, rubber/ rope/ wire bond, a razor and a location map. The materials used are 70% alcohol, the herbicide paraquat dichloride and 276 g/l, insecticide Deltamethrin 25 g/l.

2.2. Research Procedure

The location is determined through field observation, and secondary data collected to get a diversity of epiphytes and insects. Identification carried out on the park boundaries, limits each location, topography of the region, the oil age, location maps, soil conditions, climate and average annual secondary data. Palm oil grouped into three age groups, namely the age of 1-5 years, 6-15 years and over 15 years at each study site and calculated diversity epifitnya. The observation is made with a size of 20m x 20m, with each age group as many as five plots. The laying of the plot is done by purposive sampling, with Objective for the entire community of the various existing topography can be represented. The number of trees was observed in one observation plot is 3 trees. Observations epiphytes that grow on the trunk and canopy carried by climbing poles or using outreach and assistance binoculars.

The type and amount of epiphytes attached to the palm tree stands were observed as well as sorting plant species of vascular epiphytes. For the kind already known are immediately noticed, and that is not known to be made of specimens sampled herbariumnya. Herbarium made wet and dry. Herbarium specimens are used to wet watery, mushy and hard dried, such as fruits and orchids pseudobulb. Dried herbarium specimens are used to form sheets or rods can be dried.

Identification of epiphytic plants is done by reference to the Taxonomy of Vascular Plants [17], Type of Paku Indonesia [18], Fern of Malaysia In Colour [19], Comparative Morphology of Plants Vasculat [20]. The data were analyzed qualitatively and quantitatively. Quantitative analysis was carried out on the kinds of epiphytic plants to determine the density and frequency of species at each location. Epiphyte communities steadiness level at each location is calculated using diversity index and Evenness Index Shannon-Wiener type [21] and a kind of similarity between locations tested with Sorensen similarity index [22].

Observations and insect diversity calculation is done in the age group of trees. Spraying insect was conducted in the morning around 09.00 pm and evenings 7pm. Spraying insecticide active ingredient Deltamethrin (synthetic pyrethroid compounds) with the recommended dose is was done on each of 10 trees per age group. Each of these insect tree container fitted cloth measuring 1x1 m by the two individual trees. After 30 minutes fabric pitcher taken, and insects that falls is collected and put into a bottle containing 70% alcohol to be identified.

Differences in the structure of insects among the research sites is done by calculating the diversity index of each location and compare them. Diversity index calculated by the formula Shannon-Wiener. Then calculated the index evenness (E), the similarity of Sorensen index (ISS). Further views the relationship between the diversity of epiphytes and diversity of insects with a simple correlation, and the relationship between the diversity of epiphytes and diversity of insects with climate [23].

3. Results and Discussion

At five locations are obtained 87 species of vascular epiphytes with the overall number of 65 751 people, included into 37 tribes. Epiphytes are grouped into three classes, namely class Filicopsida with 11 tribes, Liliopsida class with five tribes, and Class Magnoliopsida with 20 tribes as in Table 1. Based on the nature of his group can not be considered a true epiphytes, but rather tends included hemiepifit.

Table 1. *Composition of vascular epiphytes in North Sumatra oil palm plantations.*

The composition of epiphytes	Location					Amount
	I	II	III	IV	V	
Number of Individuals	13.380	11.013	7.989	11.625	21.744	65.751
Number of Species	21	37	23	62	34	177
Number of Genus	18	31	21	47	29	146
Total of Family	16	19	16	30	23	104

Description:
I = Bukit Sentang Gardens; II = Padang Mandarsyah Gardens; III = Ajamu Gardens; IV = Sijambu-jambu Gardens; V = Aek Pancur Gardens

Overall the number of species of epiphytes in the fifth location is more than the epiphytes in natural forests. This is in line with the findings [24], ie there are 47 kinds of macro epiphytes, and 26 kinds of macro epiphytes in Forest Tourism Tangkahan TNGL Langkat. However epiphytes are found largely not true epiphytes (holoepiphytes) as found in natural forests. In general, a shortage of plantation species are protected, but is dominated by species that are not protected [25, 26].

The number of epiphytic vascular found is relatively large, due to the presence of plant species that are not included epiphytic true but the chance to grow on palm trunks like *Asystasia gangetica, Elaeis guineensis, Clidemia hirta* and other types of *Liliopsida* nations and *magnoliopsida* also called *epiphytic accidental*, there is also hemiepifit and facultative epiphytes. It was found that hemiepifit more resistant than holoepifit to habitat change because they are rooted in the ground (most of his life) in line with the findings [27, 28]. It's hard to find true epiphytes in oil palm plantations, almost all epiphytes are found in palm trees are also found living in the soil. Group of ferns *(Pteridophyta)* is most prevalent, namely *family of davalliaceae, Nephrolepidaceae, Polypodiaceae, Vittariaceae, Adiantaceae, Pteridaceae, Dryopteridaceae, Atthyriaceae, Aspleniaceae, Thelypteridaceae, Hymenophyllaceae* and *Ophioglosaceae.*

The longer a plantation allowed, then a chance for epiphytes to grow and grow bigger. This is a reflection of the nature of the ecosystem that always moves towards equilibrium (homeostasis). Vascular types found are shown in Table 2.

Table 2. Vascular epiphytes in North Sumatra oil palm plantations.

No	Spesies	Location					Number
		I	II	III	IV	V	
1	*Nephrolepis biserrata*	9486	2658	3567	3795	7176	26682
2	*G. verrucosum*	1782	1389	1539	1731	4518	10959
3	*Vittaria elongata*	324	537	1041	1143	6678	9723
4	*Asystasia gangetica*	126	1668	84	819	633	3330
5	*Davallia divaricata*	66	1689	330	117	621	2823
6	*Elaeis guineensis*	552	591	297	783	531	2754
7	*Peperomia pellucida*	357	642	36	681	318	2034
8	*Vittaria ensiformis*	48	306	24	723	210	1311
9	*G. integrifolium*	9	6	6	3	21	45

Type *Nephrolepis biserrata, Goniophlebium verrucosum*, and *Davallia divaricata* similar to those found by the [29] in the oil palm plantations of Malaysia and Singapore. *Vittaria elongata* and *Vittaria ensiformis* is the type often found in old oil palm plantations. *Asystasia gangetica* is a terrestrial shrubs and plants, including invasive especially in disturbed habitats. *Gymnopetalum integrifolium* and *Peperomia pellucida* including hemiepifit, while *Elaeis guineensis* is a plantation crop. Vascular epiphytes are most commonly found is *Nephrolepis biserrata* and *Goniophlebium verrucosum* as shown in Figure 1a and 1b. If at first glance these two types seems to be almost the same as growing mixes in the palm tree. To distinguish to note the shape and position of the spores. Both types of epiphytes is found almost exclusively in tree eraser even sometimes cover the surface of the rod.

Figure 1a. *Nephrolepis biserrata.*

Figure 1b. *Goniophlebium verrucosum.*

Type Davallia divaricata (Figure 2a) are found in trees exposed to sunlight. On the inside of the garden that is less sunlight of this type are rare, if there is only a little. Elaeis guineensis (Figure 2b) excluding epiphytes because this plant is used to grow here because of their palm kernel spilled at harvest time and germinate in the palm trunks. At first, its presence does not interfere as well as other epiphytes, but if left will be parasite. The roots of this plant will try to look for land to meet the needs of the water so it would be a competitor for its host tree.

Figure 2a. *Davalia divaricate.*

Figure 2b. *Elaeis guineensis.*

Epiphytes usually take advantage of the remaining pieces of leaf midrib as a place to grow, because this is where a lot of accumulated dust, litter and other nutrients needed by epiphytes to grow and develop. Thus the age difference palm will automatically affect the existence of epiphytes. The age difference will affect the stems and fronds as a place to grow epiphytes. Palm oil was aged 1-5 years have a short tree and its trunk is intact or not cut pelepahnya. Even if there is cut pelepahnya only a few are usually not more than 10 midrib. Oil palm frond aged 6-15 years who cut many with stem height of about 1-10 m while the palm over the age of 15 years is usually pelepahnya started off so the stems partly bald look. Number of species of epiphytes at five locations for the three age groups are the most common type found in the age group of 6-15 years unless Padang Mandarsyah Gardens location where the highest number of species that are in the age group 1-5 years and in the Bukit Sentang garden, the number of species in the group age 1-5 years is equal to 6-15 years.

Many types of vascular epiphytes in the age group 6-15 years due to oil palm fronds so epiphytes have been a lot of media to grow a lot. The remaining pieces of this frond will hold a lot of dust, debris and other nutrients and water so as to create a good habitat for the growth of several types of vascular epiphytes. This is in line with the findings [30], that the vertical stratification is one of the factors that affect the distribution of life niches (niche) on epiphytic plants. This vertical stratification generated by differences in height different oil palm planting year. Differences in plant height is related to the possibility of catching the seeds of epiphytes that can be spread by wind or insects further that not only affect plant height, age also has an effect in terms of providing the right layer in helping the growth of epiphytes. This situation is illustrated shown in Figure 3.

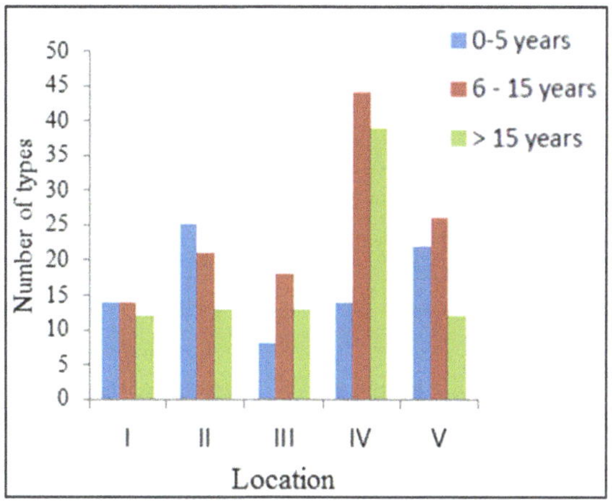

Figure 3. *The number of vascular epiphytes in three age groups in five locations of oil palm plantations of North Sumatra.*

Description: I = Bukit Sentang Gardens
II = Padang Mandarsyah Gardens
III = Ajamu Gardens
IV = Sijambu-jambu Gardens
V = Aek Pancur Gardens

Relations epiphytes number of individuals in each age group indicate a trend, namely the older the age of the tree vascular epiphytes growing number of individuals who inhabit it. Except in Ajamu Gardens seen that in the age group above 15 years epifitnya less than the age group 6-15 years. Means the type of epiphytic win the competition with other species growing breed and number of individuals more so that this type become the dominant species. This is in line with the findings of [31], that the bigger or getting older a tree then the carrying capacity for the growth of epiphytes are also getting bigger and so are the findings [32] and [33] which explains that the larger the tree will be more extensive surfaces inhabited by epiphytes. Number of individuals of epiphytes at five locations shown in Figure 4.

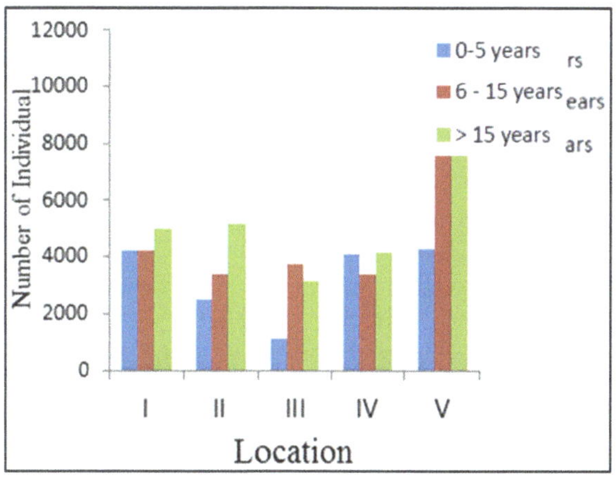

Figure 4. *Number of Individuals of vascular epiphytes in three age groups in five locations.*

Description: I = Bukit Sentang Gardens
II = Padang Mandarsyah Gardens
III = Ajamu Gardens
IV = Sijambu-jambu Gardens
V = Aek Pancur Gardens

The number of individuals of the epiphytic contained in oil palm over the age of 15 years Ajamu Gardens is lower than the age of 6-15 years, because the sample for the age group above 15 years of residual pelepahnya many are separated from the tree, so that the media place to grow epiphytes also the less. Oil palm is old, it's midrib start off one by one. Midrib detachment process was followed by the fall of epiphytic riding her life. This resulted in a reduced number of epiphytes contained in palm tree. Epiphytes overgrown palm only on the top of the stem, while the stem lower part relatively clean of epiphytes. This is in line with the findings [34] that the reduction porofit parts will result in decreased diversity of epiphytes.

The highest relative density which is kind of Nephrolepis biserrata at five locations. This shows that this type of match live on oil palm plantations, and are not influenced by geographical location, temperature, humidity and other environmental factors. Possible nutrients and a place to grow palm oil supplied to meet the needs of this plant. This type of relative control of most palm trunks and got rid of the other species that are less resistant. This is in line with the findings [35] that the spikes of the family Polypodiaceae including vascular epiphytes like habitat disturbed, so he can take advantage of anthropogenic disturbance, and it might be better if he lives in a managed environment of the human being in the natural environment. Diversity index illustrates the stability of vegetation as outlined in Table 3.

Table 3. *Diversity Index (H') of vascular epiphytes in five locations for the three age groups of palm oil in North Sumatra.*

Location	Age		
	1-5 Years	6-15 Years	> 15 Years
Bukit Sentang	1.06	1.16	0.97
Padang Mandarsyah	2.11	2.00	1.92
Ajamu	0.72	1.78	1.7
Sijambu-jambu	1.55	2.19	2.45
Aek Pancur	1.5	1.46	0.95

Diversity of vascular epiphytes were highest in the age group above 15 years old in the Sijambu-jambu garden is by 2,45 index. Because of at this location, the community are quite old and not too many human disorders. Based on the information and conditions in the field this location it's is epifitnya never discarded unless the condition is very disturbing harvest. While in other locations epifitnya more often disposed hygiene purposes and policies maisng garden of the gardens. Based on this, the more often a disturbance, the lower the diversity epifitnya. This is in line with the opinion of the [28] which states that if there is a reduction of midrib means a place to grow epiphytes is also reduced so that the diversity epifitnya was reduced. Action to reduce the parts porofit created the conditions for dry and sunny can result in decreased diversity of epiphytes. Distribution of individuals of each species of epiphytic described by evenness index (E) as shown in Table 4.

Table 4. *Evenness Index (E) of the vascular epiphytes in three age groups of palm oil in North Sumatra.*

	Years	6-15 Years	>15 years
Bukit Sentang	0.4	0.42	0.39
Padang Mandarsyah	0.66	0.66	0.75
Ajamu	0.35	0.62	0.66
Sijambu-jambu	0.59	0.58	0.67
Aek Pancur	0.50	0.44	0.38

The above data shows that the highest evenness when all of the same species have the same number of individuals or equally abundant. The similarity of organisms that can live in two different places is showing its spread, and can also be used to determine its spread. And the similarities between the existing types of epiphytes are shown in Table 5.

Table 5. *Similarity Index (ISS) of vascular epiphytes on five oil palm plantations in North Sumatra (%).*

Location	I	II	III	IV	V
I	100	41	50	42	51
II	41	100	52	46	42
III	50	52	100	38	49
IV	42	46	38	100	40
V	51	42	49	40	100

Description:
I = Bukit Sentang; II = Padang Mandarsyah; III = Ajamu; IV = Sijambu-jambu; V = Aek Pancur

Index of similarity of the five research sites ranging from 38% to 52%, where the location of which has an index of similarity is the smallest comparison site III (Ajamu) conveniently IV (Sijambu-guava) and the highest is a comparison site II (Padang Mandarsyah) with location III (Ajamu). This is in line with the findings of [36] that the greater the similarity index then it means the same type at different locations each. The composition of the canopy insects include only the insect class by the number of 10 264 individuals were included in the Order 17, 147 families as shown in Table 6.

Table 6. *The composition of canopy insects in North Sumatra oil palm plantations.*

The composition of Insects	Location				
	I	II	III	IV	V
Number of Individuals	1056	1332	1539	2222	4115
Number of Family	74	67	75	102	78
Number of Order	10	10	9	11	11

Description:
I= Bukit Sentang garden; II = Padang Mandarsyah garden; III = Ajamu garden; IV = Sijambu-jambu garden; V = Aek Pancur garden

Based on data from the composition and structure of vascular epiphytes, it appears that Sijambu-jambu gardens has a number of species, family and genus of epiphytic most, demikianjuga by the number of families of insects most. This shows that the presence of epiphytic provide homes and food for many insects canopy. In line with the findings of [37], the more diverse epiphytes that grow on the tree canopy also increasingly diverse micro-habitats which can be occupied by insects so the variety of all insects that live in the tree canopy. Epiphytes also increases diversity by creating an environment arthropod soil and litter as habitat and foliage consumed by herbivorous insects. Based on the data of the composition of canopy insects in three age groups and five locations, obtained diversity index, which is one indicator of the stability of an ecosystem. The higher the diversity index showed more complicated or complex the food web, which means also more stable or better an environment. Index canopy insect diversity in North Sumatra oil palm plantations are shown in Table 7.

Table 7. *Diversity Index (H ') insect canopy in three age groups of palm oil in North Sumatra.*

Location	Age		
	1-5 Years	Years	>15 Years
Bukit Sentang	1,81	2,39	2,29
Padang Mandarsah	2,61	2,05	2,20
Ajamu	1,68	2,43	2,78
Sijambu-jambu	2,33	1,64	1,97
Aek Pancur	1,39	0,42	1,92

Insect diversity of the five locations that are classified as being in all age groups, except in the garden Aek Pancur 6-15 years age group is low. This shows that the ecosystem is in a state of almost balanced. While in Aek Pancur age group 6-15 yrs unbalanced state of the ecosystem, because at this location has often done pest control with pesticides. When H '<1, it means low diversity, where the presence of insect pests and natural enemies are not balanced to make the damage to the plant, when H' 1-3 means diversity medium, which leads to the well where the presence of pests and natural enemies in the field almost balanced, and if H '> 3 means higher insect diversity, where the state of the ecosystem in the field is balanced, ie between pests and their natural enemies in balance so it does not need to be treated to kill insect pests [38]. There is no one location that high insect diversity index. Whilst in the natural forest insect diversity index is usually high [39]. It is associated with a lower diversity of vascular epiphytes in oil palm plantations. Overall it appears that the vegetation structure and floristic diversity complex will support a high diversity of insects. This is confirmed by [40] and [41], that the high degree of variation or diversity of plants affects the diversity of insects and insect populations.

[42] also stated that the plant diversity 75% less in oil palm plantations than natural forests. This resulted in a wealth of species in primary forest sites are on average five times higher than the plantation site [43]. In natural ecosystems all living things are in balance and control each other so there is no threat. In natural ecosystems, species diversity is very high, which means there is a space in any unity of diverse flora and fauna. The level of crop diversity affect the incidence of the pest problem. Diverse cropping systems affect the populations of pest species [44].

Distribution of canopy insects can be explained by the evenness index (E), which ranges from 0:12 to 0.67. The lowest scores were found in the age group of 6-15 years in the Aek Pancur garden and the highest in the age group> 15 years in the garden Ajamu. In the Bukit Sentang garden and Ajamu garden seen that the mature age of the plantation the higher the index Evenness. But in the garden Padang Mandarsah and Sijambu-guava, Evenness index is highest in the age group 1-5 yrs and 6-15 yrs lowest age group, as shown in Table 8.

Table 8. *Evenness index (E) of insects canopy in three age groups of palm oil in North Sumatra.*

	1-5 Years	6-15 years	> 15 Years
Bukit Sentang	0.51	0.61	0.63
Padang Mandarsyah	0.64	0.40	0.56
Ajamu	0.45	0.64	0.67
Sijambu-jambu	0.63	0.48	0.58
Aek Pancur	0.36	0.12	0.47

Similarities canopy insects is determined by the similarity index, which is the study ranged from 56% - 94% or including similar criteria [45]. The sites have the same major crops and species composition of epiphytic dominant also relatively similar, so micro-habitat conditions provided also almost the same as shown in Table 9. It is a major cause of canopy insects found also almost the same (similar).

Table 9. *Similarity Index (IS) insect canopy on five oil palm plantations in North Sumatra (%).*

Location	I	II	III	IV	V
I	100	70	67	56	64
II	70	100	76	94	68
III	67	76	100	64	69
IV	56	94	64	100	61
V	64	68	69	61	100

Description:
I = Bukit Sentang; II = Padang Mandarsyah;III = Ajamu; IV = Sijambu-jambu; V = Aek Pancur

The relationship between the diversity of epiphytes and canopy insect diversity in North Sumatra oil palm plantations is shown in the model equation, $Y = 0.176x + 1717$, with a value of $R2 = 0.024$. This shows that the diversity of insects can only be explained as much as 2% and as much as 98% more there are other factors that are not revealed. In Figure 5 also shows that the diversity of epiphytes and insect diversity are positively correlated, meaning the higher the diversity of its epiphytic insect diversity also increased.

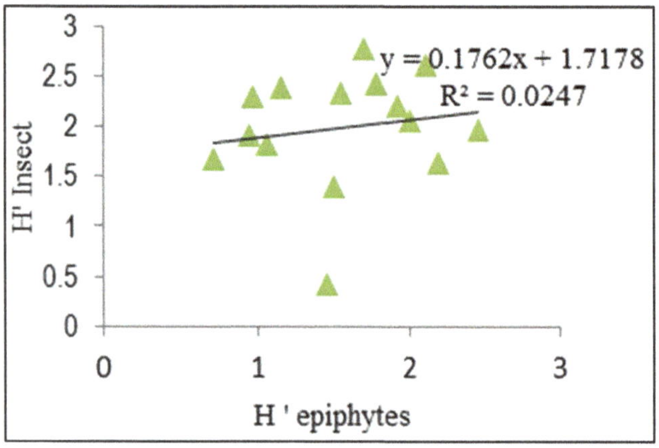

Figure 5. The relationship between the diversity of epiphytes with insect diversity in North Sumatra Oil Palm Plantation.

This relationship indicates that epiphytic vegetation provide important resources and habitat for many animals and microorganisms. Epiphytes that grow on palm trees will be a source of food, shelter and shelter canopy insects and other small animals. The more epiphytes that grow on the palm tree will be more and more varied food and a place that can be occupied by insects. The high diversity of epiphytes will increase the number of niche and micro habitat for insects. This resulted in a positive correlation between epiphytes and insects. However, many other factors that come into play has not been revealed in this study. In an ecosystem is always a mutual relationship between the constituent components of these ecosystems. Likewise in oil palm plantations, where epiphytes and insects will affect the ecosystems they occupy, reverse vascular epiphytes was also influenced by the physical state of the host environment. Epiphytic plants are very sensitive to changes in the composition of plants and micro-climate in an ecosystem so that the vascular epiphytes are often used as an indicator of disruption and change in an ecosystem [46] and [47].

The abundance of vascular epiphytes contribution Against Insects Canopies In Oil Palm in North Sumatra obtained by the ANOVA results shown in Table 10.

Table 10. *The effect of age differences epiphytes and oil palm on the number of individual insects canopy (tail) in North Sumatra.*

Age Main plot	Subplot of	Number of insects (tail)					Total	Average
(A)	Epiphytes (B)	I	II	III	IV	V		
A1	B1(100%)	30	84	15	161	74	364	72,8
(1-5 Years)	B2(50%)	11	53	10	106	21	201	40,2
	B3(0%)	21	47	8	42	29	147	29,4
Total of A1		62	184	33	309	124	712	
A2	B1(100%)	44	75	121	163	49	452	90,4
(6-15 Years)	B2(50%)	23	60	37	23	21	164	32,8
	B3(0%)	18	71	11	30	25	155	31
Total of A2		85	206	169	216	95	771	
A3	B1(100%)	88	111	36	94	71	400	80
(> 15 Yrars)	B2(50%)	80	49	34	55	17	235	47
	B3(0%)	79	35	24	60	37	235	47
Total of A3		247	195	94	209	125	870	
Total		394	585	296	734	220	2229	

Descriptions:
I = Bukit Sentang; A = Group of Age
II= Padang Mandarsyah; B = Epiphyte
III= Ajamu
IV = Sijambu-jambu
V = Aek Pancur

The analysis showed that the age difference does not affect the oil palm canopy number of individual insects. Epiphytes is very real existence affects the number of individual insects canopy, differences in study site is also real influence, but influence the interaction of different age groups and the presence of epiphytic no effect. DNMRT further test with confidence interval of 0.5% showed that the presence of epiphytic real 100% higher than the number of insects epiphytic 50% and 0% with an average increase of 81.0667. While the location of Padang Mandarsah and Sijambu-jambu has a number of real insects more to the value of the average increment of 65 and 81.556 comparison site Ajamu Gardens, Aek Pancur Gardens and the Bukit Sentang Gardens.

Table 11. Effect of epiphytes and the age difference of the amount of oil palm canopy insect family in North Sumatra.

Age main plot	Subplot of	Family number of insect					Total	Average
Umur (A)	Epiphytes (B)	I	II	III	IV	V		
A1	B1	11	22	4	26	21	84	16,8
	B2	6	20	7	10	12	55	11
	B3	5	24	7	16	12	64	12,8
Total of A1		22	66	18	52	45	203	
A2	B1	15	8	10	18	11	62	12,4
	B2	6	16	8	8	7	45	9
	B3	11	12	7	11	9	50	10
Total of A2		32	36	25	37	27	157	
A3	B1	16	16	11	18	7	68	13,6
	B2	19	17	11	12	6	65	13
	B3	13	16	15	22	5	71	14,2
Total of A3		48	49	37	52	18	204	
Total		102	151	80	141	90	564	

Description:
I = Bukit Sentang; A = Group of Age Kelompok Umur
II = Padang Mandarsyah; B = Epiphytes
III = Ajamu
IV = Sijambu-jambu
V = Aek Pancur

The results of these analyzes showed that trees with epiphytes have a number of species and number of individuals were significantly higher than the tree without epiphytes. The insects can be grouped according to their role in the ecosystem, which is useful insects, insect pests, insects and insect neutral unknown role. Detailed data is the role of insects in oil palm plantations are shown in Table 12.

Table 12. Number of canopy insects caught in oil palm plantations in North Sumatra in accordance with its role in the ecosystem.

Location	Insect role			
	Useful	Pest	Netral	Unknown
Bukit Sentang	180	23	180	18
Padang Mandarsyah	306	53	366	31
Ajamu	155	47	189	33
Sijambu-jambu	227	145	112	23
Aek Pancur	453	150	343	56

Overall insects fact that this estimate is the most useful insects and neuters. When viewed by age group and presence epifitnya palm turned out beneficial insects and insect remains neutral for more, both in epiphytes condition of 100%, 50% or 0%. Nevertheless, there have also been found insect pests more than useful insects, which in Aek Pancur Gardens, Gardens Sijambu-pink at the age> 15 years old and 50% epiphytes.

As useful organisms, insects on the role of natural enemies either as parasitoids and predators, insect pollinators and decomposers. While neuters often fall prey to predators, so a very large role in maintaining the balance of the ecosystem.

One insect predators that are found are spiders (Order Araneae). Insects found useful is that acts as a predator, for example of the order Araneae (spiders) and Dermaptera. Family Formicidae also many who acted as a predator for example Oecophyla smaragdina, Camponotus sp., Hypoponera sp., Pheidole sp., Tetramorium sp., Strumigenys sp., Technomyrmex sp., Odontomachus sp., And others. Other relatives who act as predators among others Reduviidae, Coccinelidae, Cantharidae and Anthicidae. There is also a role as a biocontrol like Aphytis sp., Nipponochalcidia sp., Parasitoid as Apantheles sp., Gavrana sp., Goryphus sp., Spathius sp., Orgilus sp., Phanerotoma sp.

Species Dermaptera, some Coleoptera, especially from the family Coccinellidae and some ant species of the genera Pachicondyla, Odontomachus (Ponerinae) and Crematogaster (Myrmicinae) are predators and can help to protect epiphytic herbivores. Spiders (Order Araneae) is a member of the arthropod phylum which has high adaptability to different types of environments. Spiders are predators for insects in the vicinity. Thus, the Spiders have an important role in the food chain. Spiders also have a role in agriculture, plantations, and housing is to protect from insects destroyer. Spiders are cosmopolitan animals that can live anywhere. Spiders can be found in terrestrial habitats, arboreal, and some aquatic. One habitat is oil palm plantations.

4. Conclusion

The results of this study suggest that it is very important to maintain the diversity of vascular epiphytes in an oil palm plantation in an effort that can be done to increase the diversity of insects canopy. Preservation of biodiversity in and around the oil palm plantations can help to ensure that the checks and balances that normally regulate invasive species and pest outbreaks in their natural habitat be kept functioning.

Acknowledgement

The authors are grateful to Hamonangan Tambunan, for his help with the authors.

References

[1] Altieri, M. A. and C. I. Nicholls. (1999). Biodiversity, Ecosystem Function, and Insect Pest Management in Agricultural System, in *Biodiversity in Agroecosystems,* Eds. W. W. Collins & C. O. Qualset. Lwis Publ. New York. pp. 69-84.

[2] Ginting, K., E. S. Sutarta dan R. Y. Purba, 2004. Pengendalian Gulma Epifit Pada Kelapa Sawit. *Warta PPKS,* 12 (2-3): 23-27.

[3] Cruz-Ang et al. (2009). The contribution of epiphytes to the abundance and species richness of canopy insects in a Mexican coffee plantation. *Journal of Tropical Ecology,* 25, 453–463. doi: 10.1017/S0266467409990125.

[4] Brown, S., & Lugo, A. E. (1990). Tropical secondary forests. *Journal of Tropical Ecology,* 6(01), 1. doi: 10.1017/s0266467400003989.

[5] Fitzherbert, E. B., M. J. Struebig, A. Morel, F. Danielsen, C. A. Bruhl, P. F. Donald dan B. Phalan. (2008). "How will oil palm expansion affect biodiversity?" *Trends in Ecology and Evolution,* 23, 538-45.

[6] Mackinnon, P. J., & Alderton, M. R. (2000). An investigation of the degradation of the plant toxin, ricin, by sodium hypochlorite. *Toxicon,* 38(2), 287–291. doi: 10.1016/s0041-0101(99)00142-7.

[7] Corley, R. H. V., & Tinker, P. B. (Eds.). (2003). *The Oil Palm.* doi: 10.1002/978047075097.

[8] Danielsen, F., Beukema, H., Burgess, N., Parish, F., Brühl, C., Donald, P., Fitzherbert, E. (2009). Biofuel plantations on forested lands: Double jeopardy for biodiversity and climate. IOP Conference Series: *Earth and Environmental Science,* 6(24). doi: 10.1088/1755-1307/6/4/242014.

[9] Hanski, I., & Gilpin, M. E. (1997). *Preface. Metapopulation Biology, xv–xvi.* doi: 10.1016/b978-012323445-2/50001-8.

[10] Li, Q., Moorhead, D. L., DeForest, J. L., Henderson, R., Chen, J., & Jensen, R. (2009). Mixed litter decomposition in a managed Missouri Ozark forest ecosystem. *Forest Ecology and Management,* 257(2), 688–694. doi: 10.1016/j.foreco.2008.09.043.

[11] Hietz, P. 1999. Diversity and conservation of epiphytes in a changing environment. Pure Appl. Chem.70:2114. http://uipac.Org/symposia/proceedings/phuket97/hietz. (diakses: 26 Oktober 2013).

[12] Green, N. D. C., Brown, M. D., & Coote, J. H. (2007). Failure of vascular autoregulation in the upper limb with increased +Gz acceleration. *Eur J Appl Physiol,* 100(6), 621–626. doi: 10.1007/s00421-007-0465-0.

[13] Essandoh, P. K., Armah, F. A., Odoi, J. O., Yawson, D. O. and Afrifa, E. K. A. 2011. Floristic Composition and Abundance of Weeds in an Oil Palm Plantation in Ghana.ARPN. *Journal of Agriculture and Biological Science,* 6(1).

[14] Zotz, G., & Vollrath, B. (2003). The epiphyte vegetation of the palm Socratea exorrhiza - correlations with tree size, tree age and bryophyte cover. *Journal of Tropical Ecology,* 19(01). doi: 10.1017/s0266467403003092.

[15] Southwood, T. R. E. and Way. M. J. (19700. Ecological background to pest management. *in Concepts of Pest Management,* pp.7-13. R. L. Rabb & F. E. Guthrie, eds. North Carolina State University, Raleigh. Van Driesche, R. G. & T. S. Bellows Jr. 1996.*Biological Control.*Chapman and Hall. New York.

[16] Mattson, W. J. J. (1980). Herbivory in relation to plant nitrogen content. *Annual Reviewof Ecology & Systematics,* 11, 119-161.

[17] Lawrence, G. H. M. 1951. *Taxonomy of Vascular Plants.* The Macmillan Company, New York.

[18] Sastrapradja, S., J. J. Afriastini, D. Darmedi dan Elizabeth. (1980). *Jenis Paku Indonesia.* Lembaga Biologi Nasional, Bogor.

[19] Piggott, A. G. (1984). *Fern of Malaysia in colour.* Tropical Press, ISBN 9677300296. Kualalumpur. Malaysia.

[20] Foster, A. S. and E. M. Gifford, Jr. (1967). *Comparative Morphology of Vascular Plants.* Second Indian Edition.Vaklis, Feffer and Simons Private Ltd. Bombay.

[21] Magurran, A. E. (1988). *Ecological Diversity and Its Measurement.* New Jersey: Princeton University Press.

[22] Mueller-Dombois, D. and Ellenberg, H. (1974) Aims and methods ofvegetation ecology.John Wiley & Sons, New York, USA.Nation, J. L., 2002. Insect physiology and Biochemistry. CRC Press, Boca Raton.

[23] Michael, P. (1995). *Metode ekologi untuk penyelidikan lading dan laboratorium.* Jakarta: UI Press.

[24] Aththorick, T. A., E. S. Siregar dan S. Hartati. 2007. Kekayaan Jenis Makroepifit Di Hutan Telaga Taman Nasional Gunung Leuser (Tngl) KabupatenLangkat. *Jurnal Biologi Sumatera,* Vol. 2(1): 12–16.

[25] Danielsen, F. and M. Heegaard, 1995. *Impact of logging and plantation development onspecies diversity: a case study from Sumatra. Management of tropical forests: towards an integrated perspective,* Centre for Development and the Environment, Universityof Oslo.Norway.

[26] Fitzherbert, E. B., M. J. Struebig, A. Morel, F. Danielsen, C. A. Bruhl, P. F. Donald dan B. Phalan. (2008). "How will oil palm expansion affect biodiversity?" *Trends in Ecology and Evolution,* 23: 538-45.

[27] Benzing, D. H. (2004). Vascular Epiphytes. *Forest Canopies,* 175–211. doi:10.1016/b978-012457553-0/50014-9.

[28] Nieder, J., dan W. Barthlott. (2001). Epiphytes and their role in the tropical forest canopy. In J. Nieder and W. Barthlott (Eds.).*Epiphytes and canopy fauna of the Otonga Rain Forest (Ecuador),* vol. 2, pp. 23–86. Results of the Bonn-Quito epiphyte project funded by the Volkswagen Foundation, Bonn, Germany.

[29] Yusuf, F. B., Tan, B. C, and Turner, I. M. (2003). Species Richness of Pteridophytes in Natural Versus Man-Made Lowland Forests in Malaysia and Singapore. *Pteridology in The New Millenium.* 283-298. Kluwer Academic Publisher.

[30] Ruiz-Cordova, J. P., Toledo-Hernández, V. H., & Flores-Palacios, A. (2014). The effect of substrate abundance in the vertical stratification of bromeliad epiphytes in a tropical dry forest (Mexico). *Flora - Morphology, Distribution, Functional Ecology of Plants*, 209(8), 375–384. doi: 10.1016/j.flora.2014.06.003.

[31] Johansson, D. (1974). Ecology of vascular epiphytes in West Africanrain forest. *Acta Phytogeographica Suecica*, 59: 1–136.

[32] Hietz, P. and U. Hietz-Seifert. (1995). Intra- and interspecific relations within an epiphyte community in a Mexican humid montane forest. *Selbyana*, 16, 135–140.

[33] Haro-Carrión, X., Lozada, T., Navarrete, H., de Koning, G. H. J. (2009). Conservation of the vascular epiphyte diversity in shade cacao plantations in the Chocó region in northwestern Ecuador. *Biotropica*, 41: 520–529.

[34] Nadkarni, N. M. (2002). Potential effect of climate change on canopy communities in a tropical cloud forest: An experimental approach. *Oecologia*, 131: 580-586.

[35] Barthlott, W., V. Schmit-Neuerburg, J. Nieder, dan S. Engwald. (2001). Diversity and abundance of vascular epiphytes; A comparison of secondary vegetation and primary montane rain forest in the Venezuelan Andes. *Plant Ecology*. 152: 145-156.

[36] Krebs, C. J. (1985). Ecology. *The Experimental Analisys of Distribution and Abundance*. Third Edition.Harper & Row Publisher. New York.

[37] Stuntz, S., C. Ziegler, U. Simon, and G. Zotz. (2002). Diversity and structure of the arthropod fauna within three canopy epiphyte species in central Panama. *Journal of Tropical Ecology*, 18, 161-176.

[38] Michael, P. 1995. *Metode ekologi untuk penyelidikan lading dan laboratorium*. Jakarta: UI Press.

[39] Pelawi, A. P. (2009). *Indeks Keanekaragaman Jenis Serangga Pada Beberapa Ekosistem di Areal Perkebunan PT. Umbul Mas Wisesa Kabupaten Labuhanbatu*. Skripsi. Universitas Sumatera Utara, Medan Indonesia.

[40] Haddad, N. M., M. C. Gregory, G. Kevin, H. John, T. David. (2011). Plant Diversity and The Stability of Foodwebs. *Ecol. Lett*, 14, 42-46.

[41] Mulder, C. P. H., J. Koricheva, K. Huss-Danell, P. Hoegberg, J.Joshi. (1999). *Ecol. Lett.* 2, 237-246.

[42] Gillison A and Liswanti N. (1999). *Impact of oil palm plantations on biodiversity in Jambi, Central Sumatra, Indonesia*. Bogor, Indonesia: Center for International Forestry Research.

[43] Brühl, C. A. and T. Eltz, (2009). *Fuelling the crisis:* Species loss of ground-dwelling forest ants in oil palm plantations in Sabah, Malaysia (Borneo)."Biodiversity & Conservation.

[44] Oka, I. N., (1995). *Pengelolaan Hama Terpadu dan Implementasinya di Indonesia*. UGM-Press, Yogyakarta.

[45] Suin, N. M. (1997). *Ekologi Hewan Tanah*. Bumi Aksara. Bandung.

[46] Padmawathe, R., Qureshi, Q., & Rawat, G. S. (2004). Effects of selective logging on vascular epiphyte diversity in a moist lowland forest of Eastern Himalaya, India. *Biological Conservation*, 119(1), 81–92. doi: 10.1016/j.biocon.2003.10.024.

[47] Larrea, M. L., and Werner, F. A. 2010. Response of vascular epiphyte diversity to different land-use intensities in a neotropicalmontane wet forest. *Forest Ecology and Management*, *260*(11): 1950–1955. doi: 10.1016/j.foreco.2010.08.029.

Radon, Thoron and Progeny Measured in Urban Health Centres and the Resulting Radiation Doses to Doctors, Nurses and Patients from the Inhalation of Air

M. A. Misdaq[1, *], A. Matrane[1, 2], J. Ouguidi[1]

[1]Nuclear Physics and Techniques Laboratory, Faculty of Sciences Semlalia, University of Cadi Ayyad, Marrakech, Morocco (URAC-15 Research Unit Associated to the CNRST, Rabat, Morocco)

[2]Nuclear Medicine Service, Mohamed VI University Hospital Centre, Faculty of Medicine and Pharmacy, University of Cadi Ayyad, Marrakech, Morocco

Email address:

misdaq@uca.ma (M. A. Misdaq)

Abstract: Doctors and nurses spent about 8 hours a day inside urban health centres examining a large number of patients. To assess radiation dose due to the attached and unattached fractions of the short-lived alpha-emitting radon decay products from the inhalation of air by working personnel and patients, concentrations of these radionuclides as well as those of radon and thoron gases were measured in indoor air of different health centres in the city of Marrakech (Morocco) by means of CR-39 and LR-115 type II solid state nuclear track detectors (SSNTDs). Committed equivalent doses per hour of exposure due to the attached and unattached fractions of ^{218}Po and ^{214}Po radon short-lived progeny were evaluated in different tissues of the respiratory tract of individuals from the inhalation of air inside the studied health centres. The influence of the activity of the attached and unattached fractions of ^{218}Po and ^{214}Po and mass of the tissue on the committed equivalent doses per hour of exposure was investigated. Annual committed effective doses due to the attached and unattached fractions of ^{218}Po and ^{214}Po radon short-lived progeny from the inhalation of air by doctors, nurses and patients inside the studied hospitals were determined. A maximum value of 7.1 mSv y^{-1} was found for doctors working 40 hours per week.

Keywords: Radon, Thoron and Their Progenies, Attached and Unattached Fractions, Health Centres, Nuclear Track Detectors, Human Respiratory Tract, Radiation Dose Assessment

1. Introduction

Humans are continuously exposed in homes and workplaces to some ionising radiation in form of alpha- and beta- and gamma-photons emitted by radon, thoron and their corresponding decay products coming from building materials, thermal water and air pollution (cigarette smoke, fly ashes and dusts) [1-5]. Radon and its short-lived progeny in dwellings and workplaces represent the main source of public exposure from natural radiation. They contribute to nearly 50% of the global effective dose to population [6]. Lung cancer hazard from inhalation exposure to radon is due to alpha-dose deposited by short-lived radon progeny. Being inert gases, the main characteristic of radon (^{222}Rn) and thoron (^{220}Rn) among other natural radioelements is the fact that their behaviour is not affected by chemical processes. In addition, their concentration levels depend strongly on geological and geophysical conditions, as well as on atmospheric influences such as barometric pressure and rainfall. Formed as a result of ^{238}U and ^{232}Th natural radioactive series in the earth's crust they are free to move through soil pores and rock fractures; then to escape into the atmosphere. Once in the air atmosphere, the ^{222}Rn and ^{220}Rn atoms decay producing isotopes of polonium, lead, and bismuth- also thallium for ^{220}Rn. These elements are heavy metals chemically very active, which may exist briefly as ions/ or free atoms before forming molecules in condensed phase or attached to airborne dust particles, typically to those with a sub-micron range sizes, forming radioactive aerosols. A variable proportion of airborne ^{222}Rn and ^{220}Rn decay products remains unattached and is referred as the airborne-unattached fraction. This fraction may be inhaled and deposited in the respiratory tract, in which they release all their alpha-emissions. The particle distribution of the aerosol-attached fraction in the inhaled air also influences the dose to the airways, because particles of different sizes deposit preferentially in different areas of the respiratory tract.In controlling radiation exposure for the working personnel and patients inside hospitals, it is necessary to measure the concentrations of radon, thoron and their progenies in the air of these locations.^{222}Rn concentration was measured in dwellings and workplaces in different countries by using LR-115 type films[7], Pylon AB-5 scintillometer with a continuous passive radon detector [8], AlphaGUARDPQ 2000PRO [8]and Radim 2P and Radim 3P with a semiconductor detector [8,9].

In the work described here, a technique based on using CR-39 and LR-115 type II SSNTDs was used for measuring the concentration of radon, thoron and their progenies inside different urban health centres. Annual committed effective doses due to the inhalation of radon short- lived decay products by doctors, nurses and patientsinside the studied health centres were also evaluated.

2. Materials and Methods

2.1. Description of the Studied Health Centres

The studied health centres are situated in different districts of the city of Marrakech (Morocco) in which thousands of patients received health cares all over the year. They are generally built with concrete, cement, bricks and granite rocks as shown in Table 1. The surface (S) to volume (V) ratio is equal to 1.5 m^{-1}for doctors' and nurses' offices and 2.0 m^{-1} for patients' rooms for the studied urban health centres.

Table 1. Description of the studied health centres.

Health centre	Building materials		
	Walls	Floor	Ceiling
HC1	Concrete, cement	Concrete, cement, granite	Concrete, cement
HC2	Concrete, cement	Concrete, cement, granite	Concrete, cement
HC3	Concrete, cement	Concrete, cement, granite	Concrete, cement
HC4	Bricks, cement	Concrete, cement, granite	Concrete, cement
HC5	Cement, granite	Concrete, cement, granite	Zinc plaques, cement
HC6	Bricks, cement	Concrete, cement, granite	Concrete, cement
HC7	Bricks, cement	Concrete, cement, granite	Concrete, cement
HC8	Bricks, cement	Concrete, cement, granite	Concrete, cement
HC9	Bricks, cement	Concrete, cement, granite	Concrete, cement
HC10	Bricks, cement	Concrete, cement, granite	Concrete, cement

2.2. Determination of Alpha- and Beta- Activities Per Unit Volume Due to Radon, Thoron and Their Decay Products in the Air of Urban Health Centres

The alpha-activities of ^{222}Rn, ^{220}Rn and their decay products were measured using the following types of solid state nuclear track detectors (SSNTDs):

(i) CR-39 discs, 2 cm in radius and 500 μm thick, manufactured by Pershore Mouldings Ltd, United Kingdom;

(ii) LR-115 type II discs, 2 cm in radius, comprising 12 μm of cellulose nitrate on a 100 μm thick polyester base, manufactured by Kodak Pathé, France, and marketed by Dosirad, France.

The detectors were hung in different doctors' and nurses' offices and patients' rooms inside health centres in the city of Marrakech (Morocco) for 24 hours in May 2014(Figure 1). During the exposure time, α-particles emitted by radon, thoron and their progenies bombarded the SSNTD films. After the irradiation, the exposed SSNTDs were etched in two NaOH solutions: one was of 2.5 normality at 60°C during 2 hours for the LR-115 II films, andthe other of 6.25 normality at 70°C for 7 hours for the CR-39 detectors [10]. After chemical treatment, the track densities registered on the CR-39 and LR-115 II SSNTDs were determined by means of an ordinary microscope with magnification 40x. Backgrounds of the CR-39 and LR-115II SSNTDs have been evaluated by placing five unexposed films inside small well closed plastic pockets during 24 hours in each location of the studied hospitals and counting the resulting track density rates. The average track density rates were found to be equal to ρ_G^{CR} = (1.58 ±0.01) 10^{-6} tracks cm^{-2} s^{-1} and ρ_G^{LR} = (0.535 ± 0.005) 10^{-6} tracks cm^{-2} s^{-1}, respectively. The reproducibility of the method was checked by placing four couples of CR-39 and LR-115 type II SSNTDs at the same room in the studied health centres for 24 hours. The distance between two successive couples of detectors is of 30 cm. The average track density rates registered on the CR-39 and LR-115 type II SSNTDs were evaluated. The relative uncertainty of the average track density rate determination is smaller than 1%.For the experimental etching conditions, the residual thickness of the LR-115 type II SSNTD is 5 μm which corresponds to the lower (E_{min}=1.6 MeV) and upper (E_{max}=4.7 MeV) energy limits for the registration of tracks of alpha-particles in the LR-115 II films [11]. All alpha-particles emitted by the radon and thoron series that reach the LR-115 II detector surface under an angle lower than its critical angle of etching, θ_c' with a residual energy between 1.6 MeV and 4.7 MeV are registered as bright track-holes. The CR-39 detector is sensitive to all alpha-particles reaching its surface under an angle smaller than its critical angle of etching θ_c. The critical angles of etching θ_c' and θ_c were calculated using a method described in detail by Misdaq et al. [12].

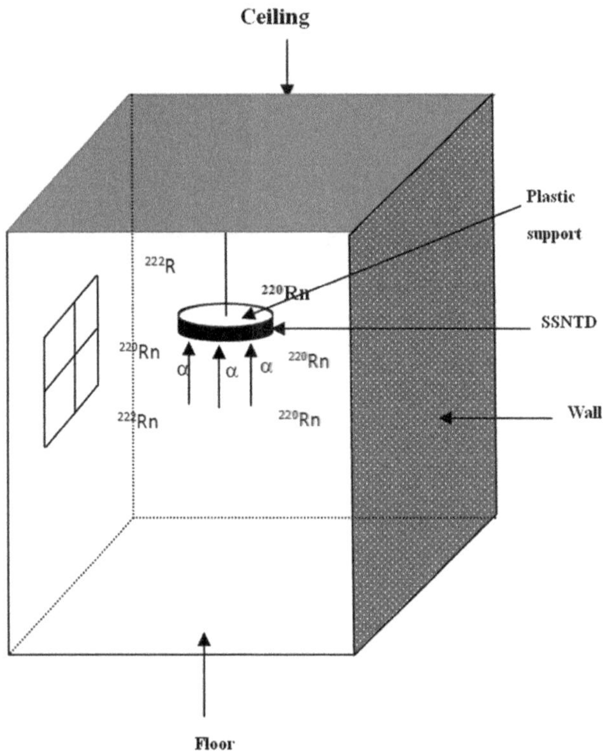

Fig. 1. Sketch of the arrangement of the CR-39 and LR-115 type II solid state nuclear track detectors inside a room in a health centre. The detectors are placed at a distance of 80 cm from the ceiling and 220 cm from the floor.

The global track density rates (tracks cm^{-2} s^{-1}), due to α-particles emitted by the ^{222}Rn (three α-emitters) and ^{220}Rn (four α-emitters) series inside a location, registered on the CR-39 (ρ_G^{CR}) and LR-115 II (ρ_G^{LR}) detectors, after subtracting the corresponding backgrounds, were obtained as described by [13]:

$$\rho_G^{CR} = \frac{\pi q^2}{2S_d'}\left\{\sum_1^3 A_c(j) K_j \varepsilon_j^{CR} R_j + \sum_1^4 A_c(j) K_j' \varepsilon_j'^{CR} R_j'\right\} \quad (1)$$

and

$$\rho_G^{LR} = \frac{\pi q^2}{2S_d'}\left\{\sum_1^3 A_c(j) K_j \varepsilon_j^{LR} R_j + \sum_1^4 A_c(j) K_j' \varepsilon_j'^{LR} R_j'\right\} \quad (2)$$

where q is the radius of the detector films (2 cm), S_d and S_d' are respectively the surface areas of the CR-39 and LR-115 II films, $A_c(j)$ (Bq cm^{-3}) is the alpha-activity of the jth alpha-emitter, R_j and R_j' are the ranges in air of an alpha-particle of index j and initial energy E_j emitted by the nuclei of the ^{222}Rn and ^{220}Rn series inside the air of the studied cave, respectively, K_j and K_j' are, respectively, the branching ratios corresponding to the disintegration of the nuclei of the ^{222}Rn and ^{220}Rn series and ε_j^{CR}, $\varepsilon_j'^{CR}$, ε_j^{LR} and $\varepsilon_j'^{LR}$ are respectively the detection efficiencies of the CR-39 and LR-115 II detectors for the emitted α-particles [13]. The first terms (right of Equations (1) and (2)) correspond to the number of alpha-particles emitted by the radon series, when the second terms correspond to the number of alpha-particles emitted by the thoron group.

The relationship between the activity concentrations of the unattached fraction $A_{c,u}(j-1)$ and attached fraction $A_{c,a}(j-1)$ of a (j-1)th radionuclide and those of its jth unattached progeny $A_{c,u}(j)$ and attached progeny $A_{c,a}(j)$ is given by the following [14]:

$$A_c(j) = A_{c,a}(j) + A_{c,u}(j) \quad (3)$$

$$A_{c,u}(j) = \frac{\lambda_j A_{c,u}(j-1) + r_{j-1}\lambda_j A_{c,a}(j-1)}{V+\lambda_j+q^u+X} \quad (4)$$

and

$$A_{c,a}(j) = \frac{(1-r_{j-1})\lambda_j A_{c,a}(j-1) + XA_{c,u}(j)}{V+\lambda_j+q^a} \quad (5)$$

where λ_j (s^{-1}) is the decay constant of the jth radon or thoron progeny, r_j is the recoil factor of the aerosol-attached radon or thoron progeny j, V is the ventilation rate inside a location which is measured by using a CO_2 tracer method and was found to be in the range of 0.26-0.57 h^{-1}) (Table 2), X is the mean attachment rate [14], q^u is the plate-out rate of the unattached radon or thoron daughters [14] and q^a is the plate-out rate of the aerosol-attached radon or thoron progenies [14].

The values of j for radon and its progeny are:

- ^{222}Rn: j = 0;
- ^{218}Po: j = 1;
- ^{214}Pb: j=2;
- ^{214}Bi: j=3;
- ^{214}Po: j = 4.

The corresponding values for ^{220}Rn and its progeny are:

- ^{220}Rn: j = 0;
- ^{216}Po/ j = 1;
- ^{212}Pb: j = 2;
- ^{212}Bi: j = 3;
- ^{212}Po: j = 4.

Since ^{222}Rn and ^{220}Rn are gaseous, all of the activity is unattached, that is:

- $A_{c,u}(0) = A_c(0)$;
- $A_{c,a}(0) = 0$.

The recoil factors for the attached ^{222}Rn and ^{220}Rn progeny are as follows [15]:

- $r_1=r_4=0.8$;
- $r_2=r_3=0$.

The plate-out rate of the unattached radon or thoron daughters' q^u is equal to 22.5 h^{-1} for doctors' and nurses' offices and 30.0 h^{-1} for patients' rooms in the studied health centres. The plate-out rate of the aerosol-attached radon or thoron progenies q^a is equal to 0.15 h^{-1} for doctors' and nurses' offices and 0.2 h^{-1} for patients' rooms in the studied health centres. The mean attachment rate X was adjusted to get values of the radon and thoron concentrations comparable to those obtained by using a method previously developed by Misdaq et al. [13], within the uncertainty interval, for doctors' and nurses offices and patients' rooms inside the studied health centres (Table 3).

2.3. Evaluation of Annual Committed Effective Doses Due to the Inhalation of Radon Short-Lived Decay Products in the Respiratory Tract of Working Personnel and Patients Inside a Health Centre

In terms of the International Commission on Radiological Protection [16,17], there are ten compartments in the thoracic region of the human respiratory tract numbered from 1 to 10and, respectively, named AI_1, AI_2, AI_3, bb_1, bb_2, bb_{seq}, BB_1, BB_2, BB_{seq} and LN_{TH}. The extrathoracic region contains four compartments numbered from 11 to 14 and, respectively named ET_2, ET_{seq}, LN_{ET} and ET_1.As radon is inert, nearly all the gas inhaled is subsequently exhaled. However, when inhaled, the radon short-lived progeny are assumed to be attached to particles of an activity median aerodynamic diameter (AMAD) of 200nm with a geometric standard deviation σ_g=2.35 [16] in indoor air.The rate of change of the alpha-activity of the attached fraction of a j'th radon decay product in a compartment i of the respiratory tract [16] at any time is given by:

$$\frac{dA^i_{c,a}(j')}{dt} = F_d(i) I_0(j') + \sum_n \lambda_{n,i} A^n_{c,a}(j') - \left(\sum_n \lambda_{i,n} + \lambda_{j'}\right) A^i_{c,a}(j') \tag{6}$$

where F_d (i) is the fractional deposition in the compartment i of the respiratory tract of individuals[16], $I_0(j') = B A_{c,a}(j')$. B is the average breathing rate for workers and patients in a health center. $A_{c,a}$(j') ($Bq.m^{-3}$) is the alpha-activity of the attached fraction of a j'th radon decay product in a health centre atmosphere, $\lambda_{n,i} = m_{n,i} + S_s$ where $m_{n,i}$ is the clearance rate from region n to region i due to particle transport and S_s is the clearance rate due to particle absorption into blood [16]. The rate of absorption of a material into blood is the same in all regions of the respiratory tract, except in the anterior nasal passages (ET_1), where no absorption occurs [16]. $\lambda_{n,i}$ =0 for i=1,2,3,5,6,8,9,12 and 14 [16], $\lambda_{i,n} = m_{i,n} + S_s$ where $m_{i,n}$ is the clearance rate from region i to region n due to particle transport [16] and $\lambda_{j'}$ is the radioactive constant of the j'th radon decay product.

Alpha-activities corresponding to the attached fraction of the j'th radon decay product in each of the 1-14 compartments of the respiratory tract as functions of time are obtained by solving Equation [6]. Alpha-equivalent dose rate (Sv s^{-1}) in a tissue T of the respiratory tract of an individual due to the inhalation of the attached fraction of the j'th radon daughter is given by:

$$\dot{H}^a_T(j')(t) = A^T_{c,a}(j')(t)\, D^T_{SP}(j')\, W_R \tag{7}$$

where $A^T_{c,a}(j')(t)$ (Bq) is the alpha-activity of the attached fraction of the j'th radon decay product in the tissue T of the respiratory tract, $D^T_{SP}(j')$ is the specific alpha-dose (Gy) deposited by alpha-particles emitted by 1 Bq of the j'th radon daughter inside the tissue T and W_R is the radiation weighting factor which is equal to 20 for alpha-particles [18].

The $D^T_{SP}(j')$ specific alpha-dose is given by:

$$D^T_{SP}(j') = k\frac{K_{j'} R_{j'} S_{j'}}{m_T} \tag{8}$$

where m_T is the mass of the target tissue T (18), $K_{j'}$ is the branching ratio, $R_{j'}$ is the range of the alpha-particle emitted by the j'th radon decay product, $S_{j'}$ is the stopping power of the tissue T for the emitted alpha-particle and k=1.6 10^{-10} is a conversion factor. R_j and S_j were calculated by using the TRIM program [19] (using the elemental chemical composition of tissues given in the ICRP publication 89 [20]).

The equivalent dose in the tissue T of the respiratory tract for the attached fraction of the j' radon decay product is given by:

$$H^a_T(j') = \int_0^{t'_e} \dot{H}^a_T(j')(t)\, dt \tag{9}$$

where t_e' is the exposure time of the tissue T.

The rate of change of the alpha-activity of the unattached fraction of the j'th radon decay product (AMAD=1nm) (14)in a compartment i of the respiratory tract [16] at any time is given by:

$$\frac{dA^i_{c,u}(j')}{dt} = F_d(i) I_0(j') + \sum \lambda_{n,i} A^n_{c,u}(j') - (\sum_n \lambda_{i,n} + \lambda_{j'}) A^i_{c,u}(j') \tag{10}$$

where F_d (i) is the fractional deposition in the compartment i of the respiratory tract of individuals(16), $I_0(j') = B A_{c,u}(j')$. B is the average breathing rate for working personnel and patients in a health centre, $A_{c,u}$(j')(Bq m^{-3}) is the alpha-activity of the unattached fraction of the j'th radon decay product in a health centre atmosphere.

Alpha-activities corresponding to the unattached fraction of the j'th radon decay product in each of the 1-14 compartments of the respiratory tract as functions of time are obtained by solving Equation [10].

Alpha-equivalent dose rate (Sv s^{-1}) in a tissue T of the respiratory tract of an individual due to the inhalation of the unattached fraction of the j'th radon daughter is given by:

$$\dot{H}_T^u(j')(t) = A_{c,u}^T(j')(t) D_{SP}^T(j') W_R \quad (11)$$

where $A_{c,u}^T(j')(t)$ (Bq) is the alpha-activity of the unattached fraction of a j'th radon decay product in the tissue T of the respiratory tract.

The equivalent dose in the tissue T of the respiratory tract for the unattached fraction of the j' radon decay product is given by:

$$H_T^u(j') = \int_0^{t'_e} \dot{H}_T^u(j')(t)dt \quad (12)$$

Regional doses, weighted with factors assigned for the partition of radiation detriment, are summed to give a value of committed equivalent dose for the thoracic $H_{TH}(j')$ and extrathoracic $H_{ET}(j')$ regions. Indeed, we have:

$$H_{TH}(j') = A_{BB}(H_{BB}^a(j') + H_{BB}^u(j')) + A_{bb}(H_{bb}^a(j') + H_{bb}^u(j')) + A_{AI}(H_{AI}^a(j') + H_{AI}^u(j')) + A_{LNTH}(H_{LNTH}^a(j') + H_{LNTH}^u(j')) \quad (13)$$

and

$$H_{ET}(j') = A_{ET1}(H_{ET1}^a(j') + H_{ET1}^u(j')) + A_{ET2}(H_{ET2}^a(j') + H_{ET2}^u(j')) + A_{LNET}(H_{LNET}^a(j') + H_{LNET}^u(j')) \quad (14)$$

Where

$H_{BB}^a(j'), H_{BB}^u(j'), H_{bb}^a(j'), H_{bb}^u(j'), H_{AI}^a(j'), H_{AI}^u(j'), H_{LNTH}^a(j')$ and $H_{LNTH}^u(j')$ are respectively the equivalent doses in the BB, bb, AI and LN_{TH} tissues of the thoracic region, $H_{ET1}^a(j'), H_{ET1}^u(j'), H_{ET2}^a(j'), H_{ET2}^u(j'), H_{LNET}^a(j')$ and $H_{LNET}^u(j')$) are respectively the equivalent doses in the ET_1, ET_2 and LN_{ET} tissues of the extrathoracic region, A_{BB}=0.333, A_{bb}=0.333, A_{AI}=0.333 and $A_{LN_{TH}}$=0.001 are respectively the weighting factors for the partition of radiation detriment for the BB, bb and AI tissues of the thoracic region and A_{ET_1}=0.001, A_{ET_2}=1 and $A_{LN_{ET}}$=0.001 are the weighting factors for the partition of radiation detriment for the ET_1, ET_2 and LN_{ET} extrathoracic region [16].

The global effective dose (mSv y^{-1} per hour of exposure) due to unattached and attached fractions of the ^{218}Po and ^{214}Po radon short-lived decay products from the inhalation of air by patients and working personnel inside a health center is given by:

$$E = E^u + E^a \quad (15)$$

where

$$E^u = 0.12\left[H_{TH}^u(^{218}Po) + H_{TH}^u(^{214}Po)\right] + 0.025\left[H_{ET}^u(^{218}Po) + H_{ET}^u(^{214}Po)\right] \quad (16)$$

is the committed effective dose due to the unattached fractions of the ^{218}Po and ^{214}Po radon progeny and

$$E^a = 0.12\left[H_{TH}^a(^{218}Po) + H_{TH}^a(^{214}Po)\right] + 0.025\left[H_{ET}^a(^{218}Po) + H_{ET}^a(^{214}Po)\right] \quad (17)$$

is the committed effective dose due to the attached fractions of the ^{218}Po and ^{214}Po radon progeny.

3. Results and Discussion

3.1. Alpha- and Beta-Activities per Unit Volume Due to ^{222}Rn, ^{220}Rn and Their Progenies in the Air of Different Urban Health Centres

Alpha-activities per unit volume due to radon (^{222}Rn) and thoron (^{220}Rn) have been measured in May 2014 in the air of doctors' office, nurses' office and patients' room of various health centres in the city of Marrakech (Morocco).

Table 2. *Data obtained for ^{222}Rn and ^{220}Rn concentrations in doctor's offices, nurses' offices and patients' rooms of different health centres. V is the ventilation rate. Given errors correspond to one-sigma uncertainties.*

Health centre	Doctor's office			Nurse's office			Patients' room		
	V (h^{-1})	$A_c(^{222}Rn)$ (Bq.m^{-3})	$A_c(^{220}Rn)$ (Bq.m^{-3})	V (h^{-1})	$A_c(^{222}Rn)$ (Bq.m^{-3})	$A_c(^{220}Rn)$ (Bq.m^{-3})	V (h^{-1})	$A_c(^{222}Rn)$ (Bq.m^{-3})	$A_c(^{220}Rn)$ (Bq.m^{-3})
HC1	0.50	74±5	6.8±0.5	0.45	78±5	7.9±0.5	0.40	88±6	8.7±0.6
HC2	0.50	74±5	6.8±0.5	0.46	77±5	7.8±0.5	0.46	78±5	7.6±0.5
HC3	0.29	131±10	13±1	0.28	138±10	13.0±0.8	0.30	127±9	12.3±0.8
HC4	0.26	165±11	14±1	0.40	88±6	9.1±0.6	0.48	76±5	6.8±0.5
HC5	0.30	125±9	12.0±0.8	0.39	90±6	9.5±0.6	0.48	76±5	6.8±0.5
HC6	0.31	119±9	12.0±0.8	0.34	105±8	11.1±0.8	0.37	95±6	9.9±0.6
HC7	0.47	77±6	7.4±0.5	0.44	80±5	8.0±0.5	0.57	71±5	6.2±0.4
HC8	0.42	84±6	8.4±0.6	0.46	78±5	7.6±0.5	0.53	73±5	6.4±0.4
HC9	0.27	153±10	13.3±0.8	0.28	140±9	13.1±0.8	0.30	128±9	12.6±0.8
HC10	0.36	99±8	10.3±0.8	0.40	88±6	9.5±0.6	0.43	81±5	8.2±0.6

Data obtained are shown in Table 2. From the statistical error on track counting one can determine the error on track density rate and then evaluate the relative uncertainty of the radon and thoron determination which is of 8%. It was noted that alpha-activities due to radon are higher than those due to thoron for the health centres studied. Indeed, due to its too short half-life (55s) thoron (^{220}Rn) has a diffusion length in air smaller than that of radon (^{222}Rn) which has longer half-life (3.82d). Note that since the studied health centres are practically built with the same materials (concrete, cement, granite and bricks), ^{222}Rn and ^{220}Rn concentrations increase when the ventilation rate inside the studied locations decreases. According to the ICRP publication 65 recommendations [21] the radon action level is situated between 500Bq m^{-3} and 1500 Bq m^{-3} for workplaces. The action level advised by the European Union [22], accepting the recommendations of the ICRP Publication 65 [21], is also the same: between 500Bq m^{-3} and 1500Bq m^{-3}. In the United Kingdom, the Health and Safety Executive (HSE) has adopted a radon action level of 400Bq m^{-3} for workplaces based on advice from the National Radiological Protection Board (NRPB) [23]. The USA has one reference level (150 Bq m^{-3}) which is applied for workplaces [24]. In Ireland, the advisory reference level is 200 Bq m^{-3} for workplaces [25]. In the European Union Directive 2013/59/EURATOM, the radon reference level is of 300 Bq m^{-3} in workplaces [26].In Morocco, the ICRP Publication 65 [21] recommendations are applied for workplaces.

Alpha-activities per unit volume air due to the attached and unattached fractions of the radon short-lived decay products were determined in the air of the studied health centres by using Equations (1)-(5). Data obtained are shown in Table 3. The relative uncertainty of the radon progenies determination is of 8%. It is to be noted that the concentrations of the attached fraction of the radon short-lived decay products are clearly higher than those of the unattached fraction in the air of the studied health centres.

Table 3. *Data obtained for the alpha-activities due to the unattached ($A_{c,u}$ (^{218}Po) and $A_{c,u}$ (^{214}Po)) and attached ($A_{c,a}$ (^{218}Po) and $A_{c,a}$ (^{214}Po)) fractions of the ^{218}Po and ^{214}Po radon short-lived progeny in nurses' office (a), doctors' office (b) and patients' room (c) of different health centres. Given errors correspond to one-sigma uncertainties.*

Table 3 (a)

Health centre	V (h^{-1})	$A_{c,u}$ (^{218}Po) (Bq.m^{-3})	$A_{c,a}$ (^{218}Po) (Bq.m^{-3})	$A_{c,u}$ (^{214}Po) (Bq.m^{-3})	$A_{c,a}$ (^{214}Po) (Bq.m^{-3})
HC1	0.45	8.0±0.6	56±4	0.0120±0.0007	30±2
HC2	0.46	7.9±0.5	55±4	0.0120±0.0007	29±2
HC3	0.28	14.1±0.8	99±8	0.021±0.001	61±4
HC4	0.40	8.9±0.6	63±5	0.0130±0.0007	35±2
HC5	0.39	9.2±0.6	65±5	0.0140±0.0007	36±2
HC6	0.34	10.7±0.6	7564	0.016±0.001	44±3
HC7	0.44	8.1±0.6	57±4	0.012±0.001	31±2
HC8	0.46	8.0±0.6	56±4	0.0120±0.00107	30±2
HC9	0.28	14.3±0.8	101±8	0.022±0.001	62±4
HC10	0.40	9.0±0.6	63±5	0.0130±0.0007	35±2

Table 3 (b)

Health centre	V(h^{-1})	$A_{c,u}$ (^{218}Po) (Bq.m^{-3})	$A_{c,a}$ (^{218}Po) (Bq.m^{-3})	$A_{c,u}$ (^{214}Po)(Bq.m^{-3})	$A_{c,a}$ (^{214}Po) (Bq.m^{-3})
HC1	0.50	7.5±0.5	53±4	0.0110±0.0007	27±2
HC2	0.50	7.5±0.5	53±4	0.0110±0.0007	27±2
HC3	0.29	13.4±0.7	95±7	0.020±0.001	57±4
HC4	0.26	17.0±0.8	119±9	0.025±0.001	74±5
HC5	0.30	13.0±0.7	90±7	0.019±0.001	54±4
HC6	0.31	12.1±0.7	86±6	0.018±0.001	51±4
HC7	0.47	7.8±0.5	54±4	0.0120±0.0007	29±2
HC8	0.42	8.5±0.5	60±4	0.0130±0.0007	33±2
HC9	0.27	15.6±0.8	110±8	0.024±0.001	68±5
HC10	0.36	10.0±0.6	71±5	0.0150±0.0007	41±3

Table 3 (c)

Health centre	V(h^{-1})	$A_{c,u}$ (^{218}Po) (Bq.m^{-3})	$A_{c,a}$ (^{218}Po) (Bq.m^{-3})	$A_{c,u}$ (^{214}Po)(Bq.m^{-3})	$A_{c,a}$ (^{214}Po) (Bq.m^{-3})
HC1	0.40	8.9±0.5	63±4	0.0130±0.0007	35±2
HC2	0.46	7.9±0.5	56±4	0.0120±0.0007	30±2
HC3	0.30	13.0±0.7	92±7	0.020±0.001	55±4
HC4	0.48	7.7±0.5	54±4	0.0120±0.0007	28±2
HC5	0.48	7.7±0.5	54±4	0.0120±0.0007	28±2
HC6	0.37	9.7±0.5	68±4	0.0150±0.0007	39±2
HC7	0.57	7.2±0.5	50±4	0.0110±0.0007	25±2
HC8	0.53	7.4±0.5	52±4	0.0110±0.0007	26±2
HC9	0.30	13.0±0.7	92±7	0.020±0.001	56±4
HC10	0.43	8.3±0.5	58±4	0.0120±0.0007	31±2

3.2. Committed Effective Doses Due to Short-Lived Alpha-Emitting ^{222}Rn Decay Products in the Respiratory Tract of Doctors, Nurses and Patients

Committed equivalent doses per hour of exposure due to the attached and unattached fractions of the short-lived alpha-emitting ^{222}Rn daughters ^{218}Po ($H_T^a(^{218}Po)$ and $H_T^u(^{218}Po)$) and ^{214}Po ($H_T^a(^{214}Po)$ and $H_T^u(^{214}Po)$) were evaluated in the respiratory tract of doctors, nurses and patients from the inhalation of air in the studied health centres (Tables 4-8). The statistical relative uncertainty of the committed equivalent dose determination is about 10%.

Table 4. *Data obtained for committed equivalent doses due to the unattached (a) and attached (b) fractions of the ^{218}Po and ^{214}Po radon progeny in the tissues of the respiratory tract from the inhalation of air by male doctors in different health centres. Given errors correspond to one-sigma uncertainties.*

Table 4 (a)

Health centre	^{218}Po					^{214}Po				
	Thoracic region			Extrathoracic region		Thoracic region			Extrathoracic region	
	H^u_{AI} (x10^{10}Sv.y^{-1})	H^u_{bb} (x10^7Sv.y^{-1})	H^u_{BB} (x10^7Sv.y^{-1})	H^u_{ET2} (x10^5Sv.y^{-1})	H^u_{ET1} (x10^7Sv.y^{-1})	H^u_{AI} (x10^{19}Sv.y^{-1})	H^u_{bb} (x10^{15}Sv.y^{-1})	H^u_{BB} (x10^{15}Sv.y^{-1})	H^u_{ET2} (x10^{13}Sv.y^{-1})	H^u_{ET1} (x10^{15}Sv.y^{-1})
HC1	0.32±0.03	4.8±0.4	5.8±0.5	2.0±0.2	5.7±0. 5	0.46±0.04	0.69±0.06	0.85±0.07	0.36±0.03	0.82±0.07
HC2	0.32±0.03	4.8±0.4	5.8±0.5	2.0±0.2	5.7±0.5	0.46±0.04	0.69±0.06	0.85±0.07	0.36±0.03	0.82±0.07
HC3	0.57±0.06	8.5±0.8	10±1	3.5±0.3	10±1	0.83±0.07	1.2±0.1	1.5±0.1	0.66±0.06	1.5±0.1
HC4	0.71±0.07	11±1	13±1	4.4±0.4	13±1	1.0±0.1	1.5±0.1	1.9±0.1	0.83±0.07	1.8±0.1
HC5	0.54±0.06	8.1±0.8	10±1	3.4±0.3	9.7±0.7	0.79±0.07	1.2±0.1	1.4±0.1	0.63±0.06	1.4±0.1
HC6	0.51±0.05	7.7±0.7	9.4±0.9	3.2±0.3	9.1±0.7	0.75±0.07	1.1±0.1	1.4±0.1	0.59±0.05	1.3±0.1
HC7	0.33±0.03	5.0±0.5	6.0±0.6	2.1±0.2	5.9±0.5	0.48±0.04	0.72±0.06	0.90±0.07	0.38±0.03	0.85±0.07
HC8	0.36±0.03	5.4±0.4	6.6±0.6	2.2±0.2	6.4±0.5	0.52±0.04	0.78±0.06	0.9±0.07	0.41±0.03	0.93±0.07
HC9	0.66±0.06	10±1	12±1	4.1±0.4	12±1	0.97±0.07	1.4±0.1	1.8±0.1	0.77±0.07	1.7±0.1
HC10	0.42±0.04	6.4±0.6	7.8±0.7	2.6±0.2	7.6±0.7	0.62±0.04	0.9±0.07	1.1±0.1	0.49±0.04	1.1±0.1

Table 4 (b)

Health centre	^{218}Po					^{214}Po				
	Thoracic region			Extrathoracic region		Thoracic region			Extrathoracic region	
	H^a_{AI} (x10^9Sv.y^{-1})	H^a_{bb} (x10^7Sv.y^{-1})	H^a_{BB} (x10^7Sv.y^{-1})	H^a_{ET2} (x10^5Sv.y^{-1})	H^a_{ET1} (x10^7Sv.y^{-1})	H^a_{AI} (x10^{15}Sv.y^{-1})	H^a_{bb} (x10^{13}Sv.y^{-1})	H^a_{BB} (x10^{13}Sv.y^{-1})	H^a_{ET2} (x10^{11}Sv.y^{-1})	H^a_{ET1} (x10^{13}Sv.y^{-1})
HC1	9.1±0.8	10±1	3.4±0.3	2.1±0.2	5.4±0.5	4.5±0.4	5.2±0.5	1.7±0.1	1.4±0.1	2.7±0.2
HC2	9.1±0.8	10±1	3.4±0.3	2.1±0.2	5.4±0.5	4.5±0.4	5.2±0.5	1.7±0.1	1.4±0.1	2.7±0.2
HC3	16±1	19±1	6.1±0.6	3.8±0.3	9.8±0.9	9.5±0.8	11±1	3.6±0.3	2.9±0.2	5.7±0.5
HC4	20±2	23±2	7.7±0.7	4.8±0.4	12±1	12±1	14±1	4.6±0.4	3.7±0.3	7.5±0.6
HC5	15±1	18±1	5.8±0.5	3.6±0.3	9.3±0.8	9.0±0.8	10±1	3.4±0.3	2.7±0.2	5.4±0.4
HC6	15±1	17±1	5.5±0.5	3.4±0.3	8.8±0.8	8.5±0.8	10±1	3.2±0.3	2.5±0.2	5.1±0.4
HC7	9.4±0.8	11±1	3.5±0.3	2.2±0.2	5.6±0.5	4.8±0.4	5.5±0.5	1.8±0.1	1.4±0.1	2.9±0.2
HC8	10±1	12±1	3.9±0.3	2.4±0.2	6.2±0.5	5.5±0.5	6.3±0.5	2.1±0.2	1.6±0.2	3.3±0.3
HC9	19±1	22±2	7.1±0.7	4.4±0.4	11±1	11±1	13±1	4.3±0.4	3.4±0.3	6.8±0.6
HC10	12±1	14±1	4.6±0.4	2.8±0.2	7.3±0.7	6.7±0.5	8.0±0.7	2.5±0.2	2.0±0.2	4.1±0.4

Table 5. *Data obtained for committed equivalent doses due to the unattached (a) and attached (b) fractions of the ^{218}Po and ^{214}Po radon progeny in the tissues of the respiratory tract from the inhalation of air by female nurses in different health centers. Given errors correspond to one-sigma uncertainties.*

Table 5(a)

Health centre	^{218}Po					^{214}Po				
	Thoracic region			Extrathoracic region		Thoracic region			Extrathoracic region	
	H^u_{AI} (x10^{10}Sv.y^{-1})	H^u_{bb} (x10^7Sv.y^{-1})	H^u_{BB} (x10^7Sv.y^{-1})	H^u_{ET2} (x10^5Sv.y^{-1})	H^u_{ET1} (x10^7Sv.y^{-1})	H^u_{AI} (x10^{19}Sv.y^{-1})	H^u_{bb} (x10^{15}Sv.y^{-1})	H^u_{BB} (x10^{15}Sv.y^{-1})	H^u_{ET2} (x10^{13}Sv.y^{-1})	H^u_{ET1} (x10^{15}Sv.y^{-1})
HC1	0.42±0.04	5.1±0.4	6.8±0.6	2.4±0.2	7.1±0.7	0.60±0.05	0.70±0.05	1.0±0.1	0.40±0.03	1.0±0.1
HC2	0.41±0.04	5.0±0.4	6.8±0.6	2.4±0.2	7.0±0.7	0.60±0.05	0.70±0.05	1.0±0.1	0.40±0.03	1.0±0.1
HC3	0.73±0.06	8.9±0.8	12±1	4.3±0.4	12±1	1.1±0.1	1.3±0.1	1.8±0.2	0.80±0.07	1.8±0.2
HC4	0.46±0.04	5.7±0.5	7.6±0.7	2.7±0.2	7.9±0.7	0.70±0.05	0.80±0.05	1.1±0.1	0.50±0.04	1.2±0.1
HC5	0.48±0.04	5.8±0.5	7.9±0.7	2.8±0.2	8.2±0.7	0.70±0.05	0.8±0.1	1.2±0.1	0.50±0.04	1.2±0.1
HC6	0.55±0.05	6.8±0.6	9.1±0.8	3.2±0.3	9.5±0.8	0.80±0.05	1.0±0.1	1.3±0.1	0.60±0.05	1.4±0.1
HC7	0.42±0.04	5.2±0.4	6.9±0.6	2.5±0.2	7.2±0.7	0.60±0.05	0.70±0.05	1.0±0.1	0.50±0.04	1.0±0.1
HC8	0.41±0.04	5.1±0.4	6.8±0.6	2.4±0.2	7.1±0.7	0.60±0.05	0.70±0.05	1.0±0.1	0.40±0.03	1.0±0.1
HC9	0.74±0.06	9.1±0.8	12±1	4.3±0.4	13±1	1.1±0.1	1.3±0.1	1.8±0.2	0.80±0.07	1.9±0.1
HC10	0.47±0.04	5.7±0.5	7.7±0.7	2.7±0.2	7.9±0.7	0.70±0.05	0.80±0.05	1.1±0.1	0.50±0.04	1.2±0.1

Table 5 (b)

Health centre	^{218}Po					^{214}Po				
	Thoracic region			Extrathoracic region		Thoracic region			Extrathoracic region	
	H^{a}_{AI} (x10^{9}Sv.y^{-1})	H^{a}_{bb} (x10^{7}Sv.y^{-1})	H^{a}_{BB} (x10^{7}Sv.y^{-1})	H^{a}_{ET2} (x10^{5}Sv.y^{-1})	H^{a}_{ET1} (x10^{7}Sv.y^{-1})	H^{a}_{AI} (x10^{15}Sv.y^{-1})	H^{a}_{bb} (x10^{13}Sv.y^{-1})	H^{a}_{BB} (x10^{13}Sv.y^{-1})	H^{a}_{ET2} (x10^{11}Sv.y^{-1})	H^{a}_{ET1} (x10^{13}Sv.y^{-1})
HC1	12±1	11±1	4.0±0.3	2.6±0.2	6.8±0.6	6.1±0.6	5.8±0.5	2.1±0.2	1.7±0.1	3.6±0.3
HC2	11±1	11±1	3.9±0.3	2.6±0.2	6.7±0.6	6.0±0.6	5.7±0.5	2.0±0.2	1.7±0.1	3.5±0.3
HC3	21±2	20±2	7.1±0.6	4.6±0.4	12±1	12±1	12±1	4.2±0.4	3.5±0.3	7.2±0.6
HC4	13±1	12±1	4.5±0.3	2.9±0.2	7.7±0.7	7.1±0.7	6.7±0.6	2.4±0.2	2.0±0.2	4.1±0.4
HC5	13±1	13±1	4.6±0.4	3.0±0.2	7.9±0.7	7.4±0.7	7.0±0.6	2.5±0.2	2.1±0.2	4.3±0.4
HC6	16±1	15±1	5.4±0.4	3.5±0.2	9.2±0.9	8.9±0.8	8.4±0.7	3.0±0.3	2.5±0.2	5.2±0.5
HC7	12±1	11±1	4.1±0.3	2.6±0.2	6.9±0.6	6.2±0.6	5.9±0.5	2.1±0.2	1.8±0.1	3.6±0.3
HC8	12±1	11±1	4.0±0.3	2.6±0.2	6.8±0.6	6.0±0.6	5.7±0.5	2.1±0.2	1.7±0.1	3.5±0.3
HC9	21±2	20±2	7.2±0.6	4.7±0.4	12±1	13±1	12±1	4.3±0.4	3.5±0.3	7.3±0.6
HC10	13±1	12±1	4.5±0.3	2.9±0.2	7.7±0.7	7.2±0.7	6.8±0.6	2.4±0.2	2.0±0.2	4.2±0.4

Table 6. *Data obtained for committed equivalent doses due to the unattached (a) and attached (b) fractions of the ^{218}Po and ^{214}Po radon progeny in the tissues of the respiratory tract from the inhalation of air by 15 y girls in different health centres. Given errors correspond to one-sigma uncertainties.*

Table 6 (a)

Health centre	^{218}Po					^{214}Po				
	Thoracic region			Extrathoracic region		Thoracic region			Extrathoracic region	
	H^{u}_{AI} (x10^{12}Sv.y^{-1})	H^{u}_{bb} (x10^{7}Sv.y^{-1})	H^{u}_{BB} (x10^{7}Sv.y^{-1})	H^{u}_{ET2} (x10^{5}Sv.y^{-1})	H^{u}_{ET1} (x10^{7}Sv.y^{-1})	H^{u}_{AI} (x10^{21}Sv.y^{-1})	H^{u}_{bb} (x10^{15}Sv.y^{-1})	H^{u}_{BB} (x10^{15}Sv.y^{-1})	H^{u}_{ET2} (x10^{13}Sv.y^{-1})	H^{u}_{ET1} (x10^{15}Sv.y^{-1})
HC1	0.39±0.03	0.80±0.07	3.0±0.3	0.90±0.07	2.8±0.2	0.56±0.04	0.12±0.01	0.45±0.03	0.18±0.01	0.40±0.04
HC2	0.35±0.03	0.70±0.06	2.7±0.2	0.80±0.07	2.5±0.2	0.50±0.04	0.11±0.01	0.40±0.03	0.16±0.01	0.36±0.03
HC3	0.56±0.05	1.2±0.1	4.4±0.3	1.4±0.1	4.0±0.3	0.82±0.07	0.18±0.01	0.65±0.05	0.26±0.02	0.59±0.05
HC4	0.34±0.03	0.70±0.06	2.6±0.2	0.80±0.07	2.4±0.2	0.48±0.04	0.10±0.01	0.38±0.03	0.15±0.01	0.35±0.03
HC5	0.34±0.03	0.70±0.06	2.6±0.2	0.80±0.07	2.4±0.2	0.48±0.04	0.10±0.01	0.38±0.03	0.15±0.01	0.35±0.03
HC6	0.42±0.04	0.90±0.07	3.3±0.3	1.0±0.1	3.0±0.3	0.61±0.05	0.13±0.01	0.48±0.04	0.19±0.01	0.44±0.03
HC7	0.31±0.03	0.70±0.06	2.5±0.2	0.80±0.07	2.2±0.2	0.45±0.04	0.10±0.01	0.36±0.03	0.14±0.01	0.32±0.02
HC8	0.32±0.03	0.70±0.06	2.5±0.2	0.80±0.07	2.3±0.2	0.46±0.04	0.10±0.01	0.37±0.03	0.14±0.01	0.33±0.02
HC9	0.57±0.05	1.2±0.1	4.4±0.3	1.4±0.1	4.1±0.3	0.82±0.07	0.18±0.02	0.66±0.05	0.26±0.02	0.59±0.05
HC10	0.36±0.03	0.80±0.07	2.8±0.2	0.90±0.07	2.6±0.2	0.52±0.04	0.11±0.01	0.41±0.03	0.16±0.01	0.37±0.03

Table 6 (b)

Health centre	^{218}Po					^{214}Po				
	Thoracic region			Extrathoracic region		Thoracic region			Extrathoracic region	
	H^{a}_{AI} (x10^{9}Sv.y^{-1})	H^{a}_{bb} (x10^{7}Sv.y^{-1})	H^{a}_{BB} (x10^{7}Sv.y^{-1})	H^{a}_{ET2} (x10^{5}Sv.y^{-1})	H^{a}_{ET1} (x10^{7}Sv.y^{-1})	H^{a}_{AI} (x10^{16}Sv.y^{-1})	H^{a}_{bb} (x10^{13}Sv.y^{-1})	H^{a}_{BB} (x10^{13}Sv.y^{-1})	H^{a}_{ET2} (x10^{11}Sv.y^{-1})	H^{a}_{ET1} (x10^{13}Sv.y^{-1})
HC1	3.8±0.3	6.1±0.6	1.9±0.1	0.50±0.04	1.5±0.1	20±2	3.3±0.3	5.2±0.5	0.40±0.04	0.80±0.07
HC2	3.4±0.3	5.4±0.5	1.7±0.1	0.50±0.04	1.3±0.1	17±1	2.8±0.2	4.4±0.4	0.30±0.03	0.70±0.06
HC3	5.6±0.5	8.9±0.8	2.8±0.2	0.80±0.06	2.2±0.2	32±3	5.2±0.4	8.3±0.8	0.60±0.05	1.3±0,.
HC4	3.3±0.3	5.3±0.5	1.7±0.1	0.50±0.04	1.3±0.1	17±1	2.7±0.2	4.2±0.4	0.30±0.03	0.70±006
HC5	3.3±0.3	5.3±0.5	1.7±0.1	0.50±0.04	1.3±0.1	17±1	2.7±0.2	4.2±0.4	0.30±0.03	0.70±0.06
HC6	4.1±0.4	6.6±0.5	2.1±0.2	0.60±0.05	1.6±0.1	23±2	3.6±0.3	5.8±0.5	0.40±0.04	0.90±0.08
HC7	3.1±0.3	4.9±0.4	1.6±0.1	0.40±0.03	1.2±0.1	14±1	2.3±0.2	3.7±0.3	0.30±0.03	0.60±0.05
HC8	3.1±0.3	5.0±0.4	1.6±0.1	0.40±0.03	1.2±0.1	15±1	2.5±0.2	3.9±0.3	0.30±0.03	0.60±0.05
HC9	5.6±0.5	9.0±0.8	2.9±0.2	0.80±0.06	2.2±0.2	32±3	5.2±0.4	8.3±0.8	0.60±0.05	1.3±0.1
HC10	3.5±0.3	5.6±0.5	1.8±0.1	0.50±0.04	1.4±0.1	18±1	3.0±0.3	4.7±0.4	0.30±0.03	0.70±0.06

Table 7. *Data obtained for committed equivalent doses due to the unattached (a) and attached (b) fractions of the ^{218}Po and ^{214}Po radon progeny in the tissues of the respiratory tract from the inhalation of air by 10 y child in different health centres. Given errors correspond to one-sigma uncertainties.*

Table 7 (a)

Health centre	^{218}Po					^{214}Po				
	Thoracic region			Extrathoracic region		Thoracic region			Extrathoracic region	
	H^{u}_{AI} (x10^{12}Sv.y^{-1})	H^{u}_{bb} (x10^{7}Sv.y^{-1})	H^{u}_{BB} (x10^{7}Sv.y^{-1})	H^{u}_{ET2} (x10^{5}Sv.y^{-1})	H^{u}_{ET1} (x10^{7}Sv.y^{-1})	H^{u}_{AI} (x10^{21}Sv.y^{-1})	H^{u}_{bb} (x10^{15}Sv.y^{-1})	H^{u}_{BB} (x10^{15}Sv.y^{-1})	H^{u}_{ET2} (x10^{13}Sv.y^{-1})	H^{u}_{ET1} (x10^{15}Sv.y^{-1})
HC1	0.90±0.08	1.2±0.1	3.7±0.3	1.3±0.1	3.4±0.3	1.3±0.1	0.20±0.01	0.50±0.05	0.20±0.02	0.50±0.05
HC2	0.80±0.07	1.1±0.1	3.3±0.3	1.1±0.1	3.1±0.3	1.2±0.1	0.16±0.01	0.48±0.04	0.21±0.02	0.44±0.04
HC3	1.3±0.1	1.8±0.1	5.4±0.5	1.8±0.1	5.0±0.5	2.0±0.2	0.27±0.02	0.80±0.08	0.34±0.03	0.74±0.07
HC4	0.80±0.07	1.1±0.1	3.2±0.3	1.1±0.1	3.0±0.3	1.2±0.1	0.16±0.01	0.47±0.04	0.20±0.02	0.43±0.04
HC5	0.80±0.07	1.1±0.1	3.2±0.3	1.1±0.1	3.0±0.3	1.2±0.1	0.16±0.01	0.47±0.04	0.20±0.02	0.43±0.04
HC6	1.0±0.1	1.4±0.1	4.0±0.4	1.4±0.1	3.7±0.3	1.5±0.1	0.20±0.02	0.59±0.05	0.25±0.02	0.54±0.05
HC7	0.70±0.06	1.0±0.1	3.0±0.3	1.0±0.1	2.8±0.2	1.1±0.1	0.15±0.01	0.44±0.04	0.19±0.01	0.40±0.04
HC8	0.70±0.06	1.0±0.1	3.1±0.3	1.0±0.1	2.9±0.2	1.1±0.1	0.15±0.01	0.45±0.04	0.19±0.01	0.41±0.04
HC9	1.3±0.1	1.8±0.1	5.4±0.5	1.8±0.1	5.0±0.5	2.0±0.2	0.27±0.02	0.80±0.08	0.34±0.03	0.74±0.07
HC10	0.80±0.07	1.2±0.1	3.4±0.3	1.2±0.1	3.2±0.3	1.2±0.1	0.17±0.01	0.50±0.05	0.21±0.02	0.46±0.04

Table 7 (b)

Health centre	^{218}Po					^{214}Po				
	Thoracic region			Extrathoracic region		Thoracic region			Extrathoracic region	
	H^{a}_{AI} (x10^{9}Sv.y^{-1})	H^{a}_{bb} (x10^{7}Sv.y^{-1})	H^{a}_{BB} (x10^{7}Sv.y^{-1})	H^{a}_{ET2} (x10^{5}Sv.y^{-1})	H^{a}_{ET1} (x10^{7}Sv.y^{-1})	H^{a}_{AI} (x10^{16}Sv.y^{-1})	H^{a}_{bb} (x10^{13}Sv.y^{-1})	H^{a}_{BB} (x10^{13}Sv.y^{-1})	H^{a}_{ET2} (x10^{11}Sv.y^{-1})	H^{a}_{ET1} (x10^{13}Sv.y^{-1})
HC1	7.0±0.7	8.4±0.8	2.4±0.2	0.90±0.08	2.3±0.2	38±3	4.5±0.4	1.3±0.1	0.60±0.05	1.2±0.1
HC2	6.2±0.6	7.5±0.7	2.1±0.2	0.80±0.07	2.1±0.2	32±3	3.9±0.3	1.1±0.1	0.50±0.04	1.1±0.1
HC3	10±1	12±1	3.5±0.3	1.3±0.1	3.4±0.3	60±5	7.2±0.7	2.0±0.2	1.0±0.1	2.0±0.2
HC4	6.0±0.6	7.2±0.7	2.0±0.2	0.80±0.07	2.0±0.2	30±3	3.7±0.3	1.0±0.1	0.50±0.04	1.0±0.1
HC5	6.0±0.6	7.2±0.7	2.0±0.2	0.80±0.07	2.0±0.2	30±3	3.7±0.3	1.0±0.1	0.50±0.04	1.0±0.1
HC6	7.6±0.7	9.1±0.8	2.6±0.2	1.0±0.1	2.5±0.2	42±4	5.0±0.5	1.4±0.1	0.70±0.06	1.4±0.1
HC7	5.6±0.5	6.8±0.6	1.9±0.1	0.70±0.06	1.9±0.1	27±2	3.2±0.3	0.9.±0.08	0.40±0.03	0.90±0.08
HC8	5.7±0.5	6.9±0.6	2.0±0.2	0.70±0.06	1.9±0.1	28±2	3.4±0.3	1.0±0.1	0.50±04	0.90±0.08
HC9	10±1	12±1	3.5±0.3	1.3±0.1	3.4±0.3	60±5	7.2±0.7	2.0±0.2	1.0±0.1	2.0±0.2
HC10	6.4±0.6	7.8±0.7	2.2±0.2	0.80±0.07	2.1±0.2	34±3	4.1±0.4	1.1±0.1	0.50±0.04	1.1±0.1

Table 8. *Data obtained for the committed equivalent doses due to the unattached (a) and attached (b) fractions of the ^{218}Po and ^{214}Po radon progeny in the tissues of the respiratory tract from the inhalation of air by 5 y child in different health centres. Given errors correspond to one-sigma uncertainties.*

Table 8 (a)

Health centre	^{218}Po					^{214}Po				
	Thoracic region			Extrathoracic region		Thoracic region			Extrathoracic region	
	H^{u}_{AI} (x10^{12}Sv.y^{-1})	H^{u}_{bb} (x10^{7}Sv.y^{-1})	H^{u}_{BB} (x10^{7}Sv.y^{-1})	H^{u}_{ET2} (x10^{5}Sv.y^{-1})	H^{u}_{ET1} (x10^{7}Sv.y^{-1})	H^{u}_{AI} (x10^{21}Sv.y^{-1})	H^{u}_{bb} (x10^{15}Sv.y^{-1})	H^{u}_{BB} (x10^{15}Sv.y^{-1})	H^{u}_{ET2} (x10^{13}Sv.y^{-1})	H^{u}_{ET1} (x10^{15}Sv.y^{-1})
HC1	1.4±0.1	1.4±0.1	4.4±0.4	1.6±0.1	4.5±0.4	2.2±0.2	0.20±0.02	0.60±0.05	0.29±0.02	0.70±0.06
HC2	1.3±0.1	1.2±0.1	3.9±0.3	1.4±0.1	4.0±0.4	1.9±0.1	0.18±0.01	0.60±0.05	0.26±0.02	0.60±0.05
HC3	2.1±0.2	2.0±0.2	6.4±0.6	2.3±0.2	6.6±0.6	3.2±0.3	0.29±0.02	0.90±0.08	0.42±0.04	1.0±0.1
HC4	1.2±0.1	1.2±0.1	3.8±0.3	1.4±0.1	3.9±0.3	1.9±0.1	0.17±0.01	0.60±0.05	0.25±0.02	0.60±0.05
HC5	1.2±0.1	1.2±0.1	3.8±0.3	1.4±0.1	3.9±0.3	1.9±0.1	0.17±0.01	0.60±0.05	0.25±0.02	0.60±0.05
HC6	1.5±0.1	1.5±0.1	4.8±0.4	1.7±0.1	4.9±0.4	2.4±0.2	0.22±0.02	0.70±0.06	0.31±0.03	0.70±0.06
HC7	1.2±0.1	1.1±0.1	3.6±0.3	1.3±0.1	3.7±0.3	1.7±0.1	0.16±0.01	0.50±0.04	0.23±0.02	0.50±0.04
HC8	1.2±0.1	1.1±0.1	3.7±0.3	1.3±0.1	3.8±0.3	1.8±0.1	0.17±0.01	0.50±0.04	0.24±0.02	0.50±0.04
HC9	2.1±0.2	2.0±0.2	6.4±0.6	2.3±0.2	6.6±0.5	3.2±0.3	0.30±0.03	1.0±0.1	0.42±0.04	1.0±0.1
HC10	1.3±0.1	1.3±0.1	4.1±0.4	1.4±0.1	4.2±0.4	2.0±0.2	0.19±0.01	0.60±0.05	0.27±002	0.60±0.05

Table 8 (b)

Health centre	^{218}Po					^{214}Po				
	Thoracic region		Extrathoracic region			Thoracic region		Extrathoracic region		
	H^a_{AI} (x10^{9}Sv.y^{-1})	H^a_{bb} (x10^{7}Sv.y^{-1})	H^a_{BB} (x10^{7}Sv.y^{-1})	H^a_{ET2} (x10^{5}Sv.y^{-1})	H^a_{ET1} (x10^{7}Sv.y^{-1})	H^a_{AI} (x10^{15}Sv.y^{-1})	H^a_{bb} (x10^{13}Sv.y^{-1})	H^a_{BB} (x10^{13}Sv.y^{-1})	H^a_{ET2} (x10^{11}Sv.y^{-1})	H^a_{ET1} (x10^{13}Sv.y^{-1})
CS1	11±1	9.3±0.8	2.8±0.2	1.5±0.1	3.9±0.3	6.0±0.6	5.0±0.5	1.5±0.1	1.0±0.1	2.1±0.2
CS2	10±1	8.3±0.8	2.5±0.2	1.3±0.1	3.5±0.3	5.1±0.5	4.3±0.4	1.3±0.1	0.90±0.08	1.8±0.1
CS3	16±1	14±1	4.1±0.4	2.2±0.2	5.7±0.5	9.6±0.8	7.9±0.7	2.4±0.2	1.6±0.1	3.3±0.3
CS4	9.7±0.8	8.0±0.8	2.4±0.2	1.3±0.1	3.3±0.3	4.9±0.4	4.0±0.4	1.2±0.1	0.80±0.07	1.7±0.1
CS5	9.7±0.8	8.0±0.8	2.4±0.2	1.3±0.1	3.3±0.3	4.9±0.4	4.0±0.4	1.2±0.1	0.80±0.07	1.7±0.1
CS6	12±1	10±1	3.0±0.3	1.6±0.1	4.2±0.4	6.7±0.6	5.5±0.5	1.7±0.1	1.1±0.1	2.3±0.2
CS7	9.0±0.8	7.5±0.7	2.2±0.2	1.2±0.1	3.1±0.3	4.3±0.4	3.5±0.3	1.1±0.1	0.70±0.06	1.5±0.1
CS8	9.3±0.8	7.7±0.7	2.3±0.2	1.2±0.1	3.2±0.3	4.5±0.4	3.7±0.3	1.1±0.1	0.80±0.07	1.6±0.1
CS9	16±1	14±1	4.1±0.4	2.2±0.2	5.7±0.5	9.6±0.8	8.0±0.8	2.4±0.2	1.6±0.1	3.3±0.3
CS10	10±1	8.6±0.7	2.6±0.2	1.4±0.1	3.6±0.3	5.5±0.5	4.5±0.4	1.3±0.1	0.90±0.08	1.9±0.1

After inhalation of air, ^{218}Po and ^{214}Po are deposited in the extrathoracic region ET and thoracic region TH of the lung. According to the ICRP compartmental model [16] a certain number of ^{218}Po atoms are transferred from the TH region to the ET region, since this radionuclide has a half-life of 3.05min. As a consequence, we found that annual committed equivalent doses due to ^{218}Po (attached and unattached fractions) are higher in the ET region than in the TH region for doctors, nurses and patients (Tables 4-8). It is noted that annual committed equivalent doses due to ^{214}Po (attached and unattached fractions) are smaller than those due to ^{218}Po in both ET and TH regions. This is due to fact that ^{214}Po has a very short half-life (1.64 10^{-4}s) compared to the exposure time of the tissues: this means that ^{214}Po comes essentially from the disintegration of ^{218}Po present in the ET and TH regions.

$H^a_T(^{218}Po)$, $H^u_T(^{218}Po)$, $H^a_T(^{214}Po)$ and $H^u_T(^{214}Po)$ are influenced by the integral of activity–time curves (the activity integrals)(Equations (9) and (12)) of ^{218}Po and ^{214}Po in a given tissue of the human body, the mass of the target tissue (m_T) and the weighting factor for the partition of radiation detriment (Equations(13) and (14)). Variation of the activities of ^{218}Po and ^{214}Po as functions of time in various tissues of the human body from the inhalation of air by different age groups of individuals are given in reference [27].

The results in Tables 4-8 show that:

- $H^a_T(^{218}Po)$, $H^u_T(^{218}Po)$, $H^a_T(^{214}Po)$ and $H^u_T(^{214}Po)$ are clearly higher in the bb and BB tissues than in the AI tissue of the thoracic region even though the latter tissue shows higher activity integral than the former ones[27]. This is due to the predominance of the mass tissue: the former tissues show lower masses than the latter.
- $H^a_T(^{218}Po)$, $H^u_T(^{218}Po)$, $H^a_T(^{214}Po)$ and $H^u_T(^{214}Po)$ are higher in the ET_2 tissue than in the ET_1 tissue of the extrathoracic region even though the latter tissue has smaller mass and higher activity integral than the former one. This is because the former tissue shows a higher weighting factor for the partition of radiation detriment than the latter.
- $H^a_T(^{218}Po)$, $H^u_T(^{218}Po)$, $H^a_T(^{214}Po)$ and $H^u_T(^{214}Po)$ in the bb tissue are higher for the adult male than for the adult female: it is due to the predominance of the activity integral [27].
- $H^a_T(^{218}Po)$, $H^u_T(^{218}Po)$, $H^a_T(^{214}Po)$ and $H^u_T(^{214}Po)$ are negligible in the LN_{TH} (thoracic region) and LN_{ET} (extrathoracic region) tissues. This is due to the fact that these tissues show lower activity integrals and weighting factors for the partition of radiation detriment.

Committed effective doses per hour of exposure due to the attached E^a and unattached E^u fractions of the ^{218}Po and ^{214}Po radon decay products and global committed effective dose E from the inhalation of air by doctors, nurses and patients in different health centres were evaluated by use of Equations 15-17. Data obtained are shown in Tables 9 and 10.The statistical relative uncertainty of the committed effective dose determination is about10%. It was noted that the global committed effective dose due to ^{218}Po and ^{214}Po radon progeny is higher for patients in the HC3 and HC9 health centres than in the others (Table 10). This is because activities due to these radionuclides are higher in these locations than in the others (Table 3(c)). It was also noted that global committed effective dose due to ^{218}Po and ^{214}Po is higher for 5y child than for the other age groups of patients. This is due to the fact that 5 y child have smaller lung mass than the others age groups. A maximum value of 7.1 mSv y^{-1} was found for female doctors spending 8 hours per day (2,080 hours per year) in the HC4 health centre which is within the (3-10 mSvy^{-1}) dose limit interval for workers [21].

Table 9. *Data obtained for the committed dose per hour of exposure due to the attached E^a and unattached E^u fractions of the ^{218}Po and ^{214}Po radon decay products and global committed effective dose E from the inhalation of air by doctors and nurses in different health centres.*

Health center	Committed effective dose (µSv y^{-1} h^{-1}exposure)											
	Doctors						Nurses					
	Male			Female			Male			Female		
	E^u	E^a	E	E^u	E^a	E	E^u	E^a	E	E^u	E^a	E
HC1	0.60±0.05	0.70±0.06	1.4±0.1	0.70±0.06	0.80±0.07	1.5±0.1	0.70±0.06	0.80±0.07	1.4±0.1	0.80±0.07	0.90±0.08	1.6±0.1
HC2	0.6±0.05	0.70±0.06	1.4±0.1	0.70±0.06	0.80±0.07	1.5±0.1	0.70±0.06	0.80±0.07	1.4±0.1	0.80±0.07	0.80±0.07	1.6±0.1
HC3	1.1±0.1	1.3±0.1	2.4±0.2	1.3±0.1	1.4±0.1	2.7±0.2	1.2±0.1	1.3±0.1	2.5±0.2	1.4±0.1	1.5±0.1	2.9±0.2
HC4	1.4±0.1	1.6±0.1	3.0±0.3	1.6±0.1	1.8±0.1	3.4±0.3	0.80±0.07	0.90±0.08	1.6±0.1	0.90±0.08	1.0±0.1	1.8±0.1
HC5	1.1±0.1	1.2±0.1	2.3±0.2	1.2±0.1	1.4±0.1	2.6±0.2	0.80±0.07	0.90±0.08	1.7±0.1	0.90±0.08	1.0±0.1	1.9±0.1
HC6	1.0±0.1	1.2±0.1	2.2±0.2	1.2±0.1	1.3±0.1	2.5±0.2	0.90±0.08	1.0±0.1	1.9±0.1	1.0±0.1	1.1±0.1	2.2±0.2
HC7	0.70±0.06	0.70±0.06	1.4±0.1	0.80±0.07	0.80±0.07	1.6±0.1	0.70±0.06	0.80±0.07	1.5±0.1	0.80±0.07	0.90±0.08	1.6±0.1
HC8	0.70±0.06	0.80±0.07	1.5±0.1	0.80±0.07	0.90±0.08	1.7±0.1	0.70±0.06	0.80±0.07	1.4±0.1	0.80±0.07	0.90±0.08	1.6±0.1
HC9	1.3±0.1	1.5±0.1	2.8±0.2	1.5±0.1	1.7±0.1	3.2±0.3	1.2±0.1	1.4±0.1	2.6±0.2	1.4±0.1	1.5±0.1	2.9±0.2
HC10	0.90±0.08	1.0±0.1	1.8±0.1	1.0±0.1	1.1±0.1	2.0±0.2	0.80±0.07	0.90±0.08	1.6±0.1	0.90±0.08	1.0±0.1	1.8±0.1

Table 10. *Data obtained for the committed effective doses per hour of exposure due to the attached E^a and unattached E^u fractions of the ^{218}Po and ^{214}Po radon decay products and global committed effective dose E from the inhalation of air by different age groups of patients in various health centres.*

Health centre	Committed effective dose (µSv y^{-1} h^{-1}exposure)								
	Adult male			Adult female			Male, 15 years		
	E^u	E^a	E	E^u	E^a	E	E^u	E^a	E
HC1	0.35±0.03	0.29±0.02	0.64±0.06	0.28±0.02	0.24±0.02	0.52±0.05	0.31±0.03	0.26±0.02	0.57±0.05
HC2	0.31±0.03	0.26±0.02	0.57±0.05	0.25±0.02	0.21±0.02	0.46±0.04	0.27±0.02	0.23±0.02	0.50±0.05
HC3	0.51±0.05	0.42±0.04	0.93±0.09	0.41±0.04	0.35±0.03	0.76±0.07	0.45±0.04	0.38±0.03	0.83±0.08
HC4	0.30±0.03	0.25±0.02	0.55±0.05	0.25±0.02	0.20±0.02	0.45±0.04	0.27±0.02	0.22±0.02	0.49±0.04
HC5	0.30±0.03	0.25±0.02	0.55±0.05	0.25±0.02	0.20±0.02	0.45±0.04	0.27±0.02	0.22±0.02	0.49±0.04
HC6	0.38±0.03	0.31±0.03	0.69±0.06	0.31±0.03	0.26±0.02	0.57±0.05	0.33±0.03	0.28±0.02	0.61±0.06
HC7	0.28±0.02	0.23±0.02	0.51±0.05	0.23±0.02	0.19±0.01	0.42±0.04	0.25±0.02	0.21±0.02	0.46±0.04
HC8	0.29±0.02	0.24±0.02	0.53±0.05	0.24±0.02	0.19±0.01	0.43±0.04	0.25±0.02	0.21±0.02	0.46±0.04
HC9	0.51±0.05	0.42±0.04	0.93±0.08	0.41±0.04	0.35±0.03	0.76±0.07	0.45±0.04	0.38±0.03	0.83±0.08
HC10	0.32±0.03	0.27±0.02	0.59±0.05	0.26±0.02	0.22±0.02	0.48±0.04	0.28±0.02	0.24±0.02	0.52±0.05

Table 10. *Continue.*

Health centre	Committed effective dose (µSv y^{-1} h^{-1}exposure)								
	Female, 15years			Chid, 10 years			Chid, 5 years		
	E^u	E^a	E	E^u	E^a	E	E^u	E^a	E
HC1	0.29±0.02	0.24±0.02	0.53±0.05	0.38±0.03	0.36±0.03	0.74±0.07	0.47±0.04	0.53±0.05	1.0.±0.1
HC2	0.26±0.02	0.21±0.02	0.47±0.04	0.34±0.03	0.32±0.03	0.66±0.06	0.42±0.04	0.47±0.04	0.90±0.09
HC3	0.42±0.04	0.35±0.03	0.77±0.07	0.56±0.05	0.53±0.05	1.1±0.1	0.69±0.06	0.78±0.07	1.5±0.1
HC4	0.25±0.02	0.20±0.02	0.45±0.04	0.33±0.03	0.31±0.03	0.64±0.06	0.41±0.04	0.46±0.04	0.87±0.08
HC5	0.25±0.02	0.20±0.02	0.45±0.04	0.33±0.03	0.31±0.03	0.64±0.06	0.41±0.04	0.46±0.04	0.90±0.09
HC6	0.32±0.03	0.26±0.03	0.58±0.05	0.41±0.04	0.39±0.03	0.80±0.08	0.51±0.05	0.57±0.05	1.1±0.1
HC7	0.24±0.02	0.19±0.01	0.43±0.04	0.31±0.03	0.29±0.02	0.60±0.06	0.38±0.03	0.42±0.04	0.80±0.08
HC8	0.24±0.02	0.19±0.01	0.43±0.04	0.32±0.03	0.29±0.02	0.61±0.06	0.39±0.03	0.44±0.04	0.83±0.08
HC9	0.43±0.04	0.35±0.03	0.78±0.07	0.56±0.05	0.53±0.05	1.1±0.1	0.69±0.06	0.78±0.07	1.5±0.1
HC10	0.27±0.02	0.22±0.02	0.49±0.04	0.35±0.03	0.33±0.03	0.68±0.06	0.44±0.04	0.49±0.04	0.93±0.09

Committed effective doses due to ^{222}Rn progeny from the inhalation of air by patients (at rest) in the studied health centres were evaluated by using this model and the ICRP [18] conversion dose coefficient for the members of the general public(10.5 ((nSvh^{-1})(Bqm^{-3}))). Data obtained for the average effective dose obtained by using this model for adult patients (at rest) were found in good agreement with those obtained by using the

ICRP model for the general public (Table 11).

Table 11. *Global committed effective dose (μSv y^{-1} h^{-1}of exposure)due to the attached and unattached fractions of the ^{218}Po and ^{214}Po radon short-lived progeny fromthe inhalation of air by adult patients inside different health centres by using the method described here and the ICRP dose conversion coefficient (18).*

Health centre	Committed effective dose (μSv y^{-1} h^{-1}of exposure)			
	This method			ICRP
	Adult male	Adultfemale	Average	
HC1	0.64±0.05	0.52±0.05	0.58±0.05	0.46
HC2	0.57±0.05	0.46±0.04	0.51±0.05	0.40
HC3	0.93±0.08	0.76±0.07	0.84±0.08	0.71
HC4	0.55±0.05	0.45±0.04	0.50±0.05	0.38
HC5	0.55±0.05	0.45±0.04	0.50±0.05	0.38
HC6	0.69±0.05	0.57±0.05	0.63±0.06	0.51
HC7	0.51±0.05	0.42±0.04	0.46±0.04	0.34
HC8	0.53±0.05	0.43±0.04	0.48±0.04	0.36
HC9	0.93±0.09	0.76±0.07	0.84±0.08	0.71
HC10	0.59±0.05	0.48±0.04	0.53±0.05	0.42

4. Conclusion

It has been shown by this study that by using CR-39 and LR-115 type II solid state nuclear track detectors (SSNTDs) one can evaluate alpha-activities per unit volume due to ^{222}Rn, ^{220}Rn and their decay products in the air of different health centres. Radiation doses to the respiratory tract of doctors, nurses and patients from the inhalation of air in the studied health centres were evaluated. It is concluded that global committed equivalent dose per hour of exposure due to the attached and unattached fractions of the ^{218}Po and ^{214}Po radon short-lived decay products increase when alpha-activities due to these radionuclides increase and the mass tissue decreases. Annual committed effective doses due^{218}Po and ^{214}Po radon short-lived progeny from the inhalation of air by working personnel and patients in the studied health centres were determined. It has been shown that the working personnel must spend less than 8 hours per day inside the studied health centres to avoid higher radiation dose exposure. Dose conversion coefficients obtained by using this model were found in good agreement with those of the ICRP for the inhalation of ^{222}Rn short-lived decay products. The SSNTD's method developed which has the advantage of being inexpensive, accurate, and sensitive and does not need the use of standard sources for its calibration is a good tool for measuring ^{222}Rn, ^{220}Rn and their progenies in the air of workplaces.

Funding

This work was performed under an URAC-15 research contract with the CNRST, Rabat, Morocco.

References

[1] M.A. Misdaq, H. Ezzahery, D. Elbboubi,"Determination of equivalent dose rates and committed effective doses in the respiratory system from the inhalation of radon decay products by using SSNTD and a dosimetric compartmental model", Radiat. Prot. Dosim., 2001, Vol. 93, No. 4, pp347-355.

[2] M.A. Misdaq, K. Flata,"The influence of the cigarette smoke pollution and ventilation rate on alpha-activities per unit volume due to radon and its progeny",J. Environ. Radioactivity, 2003, Vol. 67, pp207-218.

[3] M.A. Misdaq, M. Ghilane, J. Ouguidi, K. Outeqablit,"Radiation doses to individuals due to 238U, 232Th and 222Rn from the immersion in thermal waters and to radon progeny from the inhalation of air inside thermal stations", Radiat. Environ. Biophysics, 2012, Vol. 51, pp375-389.

[4] K.N. Yu, E.C.M. Young, M.J. Stokes, Z.J. Guan, K.W. Cho, "A survey of radon and thoron progeny for dwellings in Hong Kong", Health Phys.,1997,Vol. 73, No.2,pp373-377.

[5] C.M. Ha, S.Y. Chang, B. H. Lee,"Dose assessment to inhalation exposure of indoor ^{222}Rn daughters in Korea", Health Phys.,1999,Vol. 63, No. 4,pp453-456.

[6] United Nations Scientific Committee on the Effects of Atomic Radiation (UNSCEAR), Sources and effects of ionising radiation. Report to the General Assembly, New York, 1993.

[7] B. Lévesque, D. Gauvin, R.G. McGregor, R. Martel, S. Gingras, A. Dontigny A,"Radon in residences: Influence of geological and housing characteristics", Health Phys.,1997,Vol. 72, No. 6,pp907-914.

[8] N. Kávási, T. Kovács, C. Németh, T. Szabó, Z. Gorjánácz, A. Várhegyi A, "Difficulties in radon measurements at workplaces", Radiat. Meas., 2006,Vol. 41,pp229-234.

[9] T. Anastasiou, H. Tsertos, S. Christofides, G. Christodoulides, "Indoor radon (222Rn) concentration measurements in Cyprus using high-sensitivity portable detectors",J. Environ. Radioactivity, 2003, Vol. 68, pp159-169.

[10] M.A. Misdaq, H. Khajmi, F. Aitnouh, S. Berrazzouk, W. Bourzik, "A new method for evaluating uranium and thorium contents in different natural material samples by calculating the CR-39 and LR-115 type II SSNTD detection efficiencies for the emitted α-particles", Nucl. Instr. Meth. Phys. Res., 2000, Vol. B 171, pp350-359.

[11] A.F. Hafez, M.A. Naim,"Plastic nuclear track detection methods for estimation of thorium to uranium ratio in thick natural materials",Nucl. Instr. Meth. Phys. Res., 1992, Vol. 69B, pp373-381.

[12] M.A. Misdaq, A. Bakhchi, A. Ktata, A. Merzouki, N. Youbi, "Determination of uranium and thorium contents inside different materials using track detectors and mean critical angles", Appl. Radiat. Isot., 1999, Vol. 51,pp209-215.

[13] M.A. Misdaq, H. Ezzahery, J. Lamine,"Influence of the building material and ventilation rate on the concentration of radon, thoron and their progenies in dwelling rooms using SSNTD and Monte Carlo simulation" J. Radioanal. Nucl. Chem., 2002, Vol.252, pp67-74.

[14] J. Porstendörfer, "Properties and behaviour of radon and thoron and their decay products in the air" J. Aerosol Sci.,1994, Vol.25, No. 2,pp219-263.

[15] J. Porstendörfer, T. T. Mercer, "Influence of nuclei concentration and humidity upon the attachment rate of atoms in the atmosphere" Atmos. Environ., 1978, Vol. 12,pp2223-2238.

[16] International Commission on Radiological Protection (ICRP), Human respiratory tract model for radiological protection, ICRP Publication 66, Ann ICRP 24 (1-3), 1994.

[17] International Commission on Radiological Protection (ICRP), Human respiratory tract model for radiological protection, ICRP Publication 100, Ann ICRP 36 (1-2), 2006.

[18] International Commission on Radiological Protection (ICRP), Recommendations of the International Commission on Radiological Protection, ICRP Publication 103, Ann ICRP 37(2-4), 2007.

[19] J. P. Biersack, J.F. Ziegler, IBM Research, TRIM, Version 98, 1998.

[20] International Commission on Radiological Protection (ICRP), Basic anatomical and physiological data for use in radiological protection: reference values, ICRP Publication 89, Ann ICRP 32 (3.4), 2002.

[21] International Commission on Radiological Protection, Protection against radon-222 at home and at work,ICRP Publication 65, Ann ICRP 23 (2), 1993.

[22] European Commission, Office for Official Publications of the European Commission, Recommendations for the implementation of the Title VII of the European Basic Safety Standards concerning significant increase due to natural radiation sources, Radiation Protection Series, 1997.

[23] D.W. Dixon, T.D. Gooding, S. Mc Cready Shea, "Evaluation and significance of radon exposure in British workplace buildings", Environ. Internat., 1996, Vol. 22, S1079-S1082.

[24] US EPA Environments Division (6609J), A Citizen's Guide to Radon: the Guide to Protecting Yourself and Your Family From Radon, US EPA 402-K-006,Washington DC,2004.

[25] G. Åkerblom, "Radon Legislation and National Guidelines", Swedish Radiation Protection Institute, SSI Report 99-18, ISSN 0282-4434, 1999.

[26] EU, Directive 2013/59/EURATOM, Laying down basic safety standards for protection against the dangers arising from exposure to ionising radiation, 2013.

[27] M. A. Misdaq, J.Ouguidi, "Concentrations of radon, thoron and their decay products measured in natural caves and ancient mines by using solid state nuclear track detectors and resulting radiation dose to the members of the public", J. Radioanal. Nucl. Chem., 2011, Vol.287, 135-150.

Arc GIS Based Interpretation of Surface Sediment Heavy Metals Near Coastal Area of District Badin, Sindh, Pakistan

Mohammad Amin Qureshi[1, *], Ghulam Murtaza Mastoi[1], Mushtaque Ahmed Baloch[1], Muhammad Ali Bhatti[1], Sarfaraz Ali Mallah[2], Muhammad Waseem Junejo[1], Rashid Ali Malukhani[1], Imran Ali Abbasi[1]

[1]Centre for Environmental Sciences, University of Sindh, Jamshoro, Sindh, Pakistan
[2]M. A. Kazi Institute of Chemistry University of Sindh Jamshoro, Sindh, Pakistan

Email address:
m-ameen-q@hotmail.com (M. A. Qureshi)

Abstract: The present study investigates heavy metal contamination of sediments of the surface drains, lakes and lagoons of coastal district Badin. The accumulation of industrial, municipal waste water, pumped saline water and agricultural runoff through left bank outfall drain (LBOD) and other linking drains in the coastal lines of Badin, Sindh, Pakistan, have disturbed the environmental ecology of the area. The heavy metal characteristics of sediments were analyzed, i.e. ionic and molecular species of Hg,Ni, Fe, Cd, Pb, Cr, Co, Cu, Mn, and Zn. Arc GIS application was used for mapping of the studied area. The obtained results were represented by GIS based maps, with graduated colored circles. A total of 31 sediment samples were taken from drains, lakes and lagoons of coastal areas. The average results, analyzed for heavy metals, were found as Hg(mg/kg) 0.2032 ±0.173, Ni (mg/kg)1.455 ±1.449,Cd (Not Detected), Zn(mg/kg)2.789 ± 0.865, Cu(mg/kg)0.3050 ±0.3355, Fe(mg/kg)71.175±1.610, Mn (mg/kg)7.375±1.848, Co (mg/kg) 0.0548 ±0.0634,Pb(mg/kg)0.216 ± .114,Cr(mg/kg)0.7054 ±0.6949.The obtained results indicated that pollution of heavy metal concentration in surface sediments of lakes , lagoons, and drains near coastal lines of Badin Sindh, posed a serious question . The main contaminating sources in the area are LBOD and other link drains, carrier of industrial effluents of different districts. Poor infrastructure of LBOD facilitates backward flow of contaminated sea tidal water towards coastal saline lakes, also adds in sediment pollution.

Keywords: Heavy Metals, Sediments, Lakes, Lagoons, Badin Coastal Area, LBOD (Left Bank Outfall Drain), Kadhan Pateji Outfall Drain (KPOD), Dhoro Puran Outfall Drain (DPOD), GIS(Geographic Information System)

1. Introduction

Sediments consist of the loose sand, clay, silt, soil particles that settle at the bottom of a body of water. Soil particles transported by water are often deposited in rivers, streams, lakes, and wetlands. Water and wind brings these particles to rivers, lakes and other water bodies. They are repositories for a wide range of chemical and biological agents[1]. The heavy metal pollution of aquatic ecosystems has been a potential global problem. Heavy metals may enter into aquatic ecosystems from anthropogenic sources, such as industrial and sewage wastewater, fossil fuel combustion and atmospheric deposition [2].It is known that trace amounts of heavy metals are always present in fresh waters from weathering of rocks resulting into geo-chemical recycling of heavy metal elements in these ecosystems. These heavy metals in sediments may be found immobilized, and at the same time can be involved in absorption, co-precipitation, and complex formations[2].Heavy metals are among the most common environmental pollutants, and their occurrence in waters confirms the presence of natural or anthropogenic sources. The existence of heavy metals in sediments of water bodies can affect on plant and animal life [3, 4].

District Badin is situated between 24°-5` to 25°-25` north latitude and 68 21' to 69 20' east longitude [5]. This district is bordered with Arabian Sea in South, the Rann of Kutch, and desert in South-Eastern side. Coastal belt of this district is very complex phenomenon and appears very vulnerable due to multi-directional anthropogenic and climatic conditions. This coastal belt gets contaminated due to mega drainage projects and sea tidal water.

Prior to the construction of Ghulam Mohammad Barragein 1955, coastal area of Badin was one of fertile parts of the Indus Delta. Annual floods had not only maintained ecology but also kept sea water away. But after the construction of barrage (flood protected embankment), the area was completely cut off, and coastal ecosystem started degrading due to sea tidal water [5,6,7,8].New saline lakes appeared, on older deltic depressions' lands. Ecology of Badin further disturbed due to imbalance of polluted water supply, due to the introduction of two drainage systems in 1960s and 1980s respectively. The first one, Kotri barrage drain system comprises of Karo Gungro and Fuleli Guni Outfall drains, both drains, polluting coastal wetland like Sanhro, Mehro and Cholri shallow lakes. The second one is Left Bank Outfall Drain (LBOD) which ends into Arabian Sea (Fig1).The project LBOD was aimed to mitigate water logging and salinity of an area of 1.27 million acres of three Districts of Sindh province in 1986. This LBOD was initially aimed to drain out pumped saline water and agricultural runoff, but soon it carried municipal and industrial wastes along flood water towards the coastal areas of Badin. Badin coastal district became the final pathway of this contaminated water to Arabian Sea. LBOD has total discharge capacity of 4400 m3/s. It keeps flowing throughout the year. LBOD has other major components as Kadhan Pateji outfall drain (KPOD), Dhoro Puran Outfall Drain (DPOD), Tidal Link and Cholri weir (Fig1)[6,7]. The infrastructure of Tidal Link and Cholri weir found damaged completely after cyclone of 1999.This condition facilitated backflow of contaminated sea tidal water towards lakes and lagoons of coast (Fig1).The present study was focused to understand contamination of heavy metals of sediments in water bodies of coastal areas. The results were interpreted by using arc GIS application.

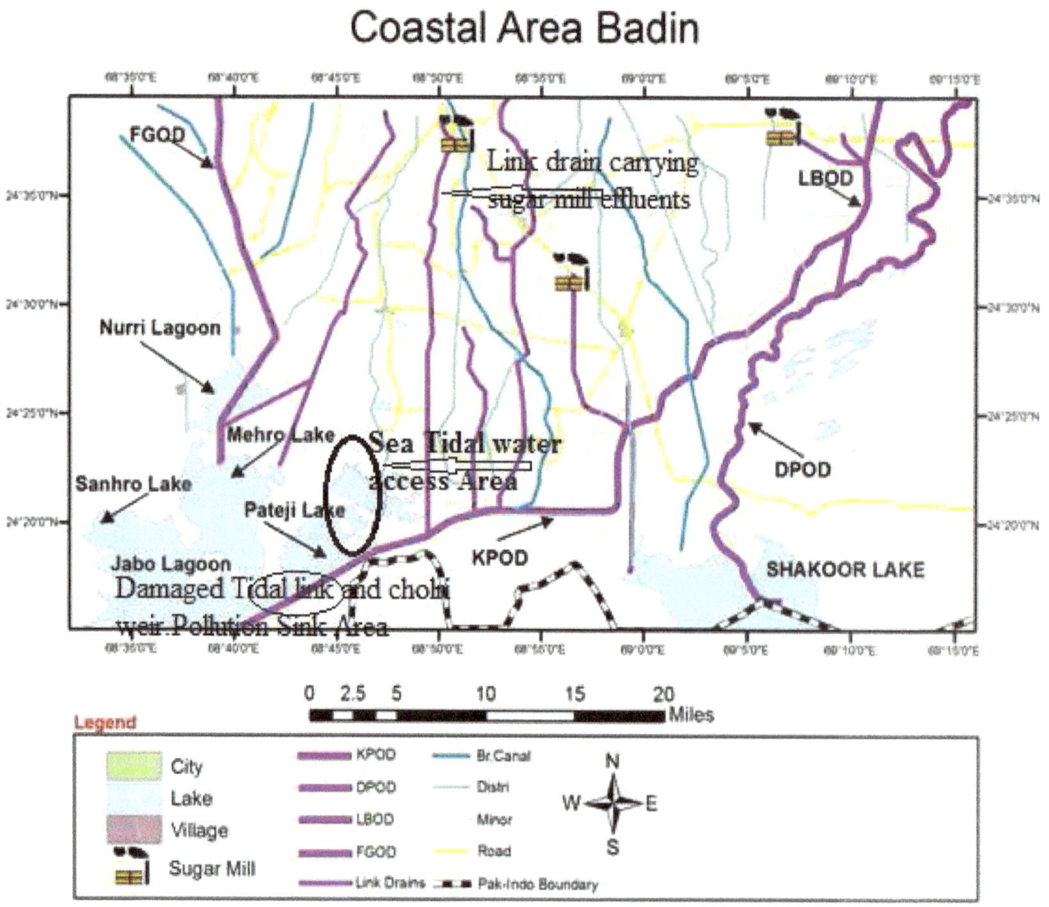

Fig.1. Base map ofstudied area.

2. Experimental Methodology

Sediment samples were collected from drains and shallow lakes and lagoons along with the coastal areas of Badin district (Fig 1).Samples were collected from 31 locations from 1st – 6th march, and 1st to 6th of September of 2013.The sediment samples were collected from the selected stations using a grab sampler at depths of 0–20 cm. The samples were kept tightly closed in polyethylene bags. All the sediment samples were stored in an insulated box containing ice and delivered to the laboratory. The samples were kept at 4°C until processing for analysis. Sediment samples were dried in an oven at 105oC overnight, sieved mechanically using a 0.5 mm sieve, homogenized and ground to 0.063 mm fine powder because metals are known to adhere to fine particles.

Sediment pollutants covering the total sorbed metals (cadmium, chromium, cobalt, copper, iron, lead, manganese, nickel, zinc) were determined by using strong acid extractants such as concentrated nitric acid (HNO_3) and aqua Regia usingUS-EPA method #3050 [9].10 ml of 1:1 HNO_3 was added to 2 g of air dried sediments in a 150ml beaker.The samples were placed on hotplate, and covered with watch glass and heat (reflux) at 950C for15 minutes.The digest was cooled, and 5 ml of concentrated HNO_3 added, reflux for an additional 30 minutes at 950C. The last step was repeated and solution was reduced near to 5 ml, without boiling, beaker was partially covered. The sample was cooled again, and 2 ml of deionized water and 3 ml of 30 % $H_2O_2$was added. With the beaker covered sample was heated gently to start the peroxide reaction. In case that effervescence became excessively vigorous the sample must be removed from the hot plate. 30 %$H_2O_2$was added, followed by gentle heating until the effervescence subsidized. Concentrated HCl and 10 ml deionized water was added, and samples were refluxed for an additional 15 min without boiling. The samples were cooled, and filtered through Whatman No 42. The samples were diluted to 50 ml with deinonized water for metal determination.The mercury was determined with coldvapour atomic absorption spectrometric method [9].

GIS was applied for collecting, storing, organizing, retrieving, transforming, and displaying data [10]. Arc GIS 9.2 was used for spatial analyses, and thematic mapsto map out the vector data[11]. GPS coordinates were usedto locate the sampling location.KMZ files were developed in Google earth and exported to Arc GIS for maps.The obtained results were used to make attribute table in arc GIS. By selecting symbology from properties, graduatedcircles and colors were selected to interpret the sediment results.The obtained results of parameters were divided into four classes, from lowest to highest ones.

3. Result and Discussions

Mercury has low solubility in aqueous solution; it is easily adsorbed on water-borne suspended particles. The water-borne Hg finally accumulates in the sediment, and the quantity of Hg contained in the sediment reflects the degree of pollution for the water body [12].The result of mercury concentration, as shown in Fig 2, is represented by graduated red circles, ranging in0.0428-0.8333(mg/kg). It was observed that mercury was in high concentration in the sediments of LBOD, DPOD, and KPOD, as compared to lakes and lagoons. Mercury was found reduced near damaged Cholri Weir areas of coastal waters.

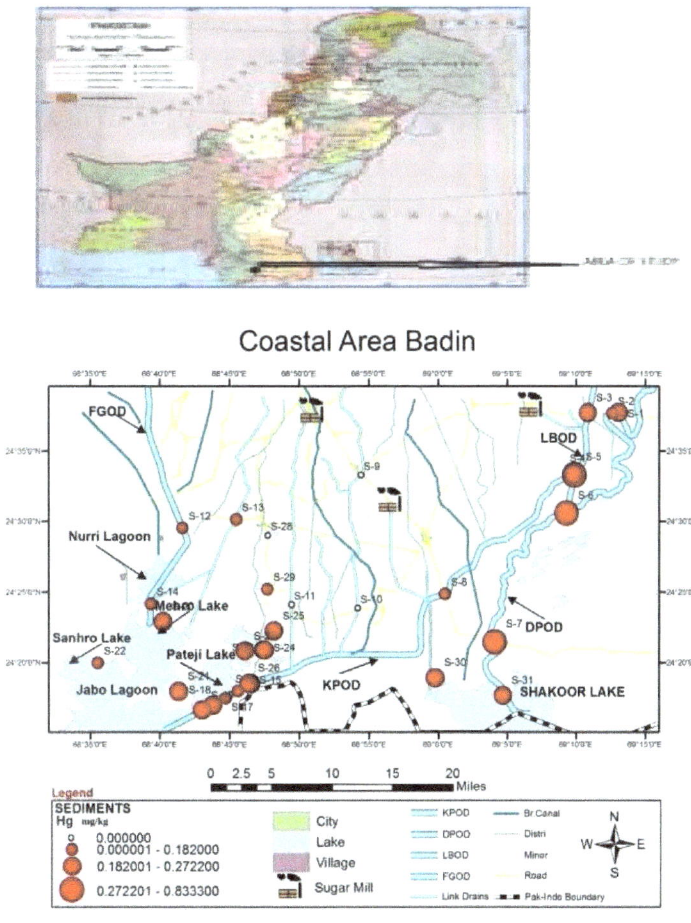

Fig. 2.Map of studied area showing mercury detection in sediment.

The results of nickel as shown in Fig 3 were ranged in 0.0256-4.528 mg/kg. It was observed that Ni was in high concentration in S-2, S-4, S-23, S-24, and S-25.Ni was found in reduced concentration near damaged Cholri Weir areas. The results were within guide line values (Nickel 20 mg/kg) of EPA for heavy metal in sediments [13].

The results of Zn concentrations, as shown in Fig 4,ranged in 1.2 -4.46mg/kg. The highest zinc range was found in S-23(table 1). The Zn was observed reduced as moving towards tidal water. Conducting the study of sediment heavy metals in three lakes of China, by Lu and Cheng (2011), it was reported the Zn range of58.67- 93.22 mg/kg [14].The results were within guide line values (Zinc 90 mg/kg) ofEPA for heavy metal in sediments [13].

The results of Cu concentrations as shown in Fig 5, with graduated red circles, ranged from 0.013-1.86mg/kg. The high concentration of Cu was observed at S-31 location. Chen et al. (2012), in a study conducted on distribution of copper in the sediments of Salt River Mouth, Taiwan were reported Cu range of 286–895 mg/kg [15].It was also reported, by Lu and Cheng (2011) while conducting study of sediments of heavy metals of three lakes of China, the Cu range 21.63- 35.64 mg/kg in lakes of the area [14]. The results were within guide line values (Cu 25 mg/kg) of EPA for heavy metal in sediments [13].

The result of iron content, as shown in Fig 6, was ranged 64.62-73.4mg/kg. It was observed that the iron concentration near coastal tidal water and damaged cholri weir was found reduced. The maximum range of Fe was observed in S-29.

The results of Mn concentration, as shown in Fig 7, ranged 3.28-9.134mg/kg. The maximum Mn was found in S-1, S-2, and S-4 locations (table 1). The results were within guide line values (Manganese300mg/kg) of EPA for heavy metal in sediments [13].

The results of cobalt concentration, as shown in Fig 8, were found in the range of 0.062-0.144mg/kg. The cobalt was not found detected in 17 sampling location s.

The results of Pb concentration, as shown in fig 9, were within range of 0.024-0.435mg/kg (table 2). The Pb was found maximum in the sample location S-1, S-2. The results were within guide line values (Lead 40mg/kg) of EPA for heavy metal in sediments [13].

The results of Cr concentration as shown in Fig 10, was ranging from0.562 to 1.983 mg/kg. The Cr was found high in S-23, S-24, S-25, S-26, and S-27 locations. The results were within guide line values (chromium25 mg/kg) of EPA for heavy metal in sediments. The high concentration of chromium may be due to industrial effluents brought by main LBOD network from different districts. [13].

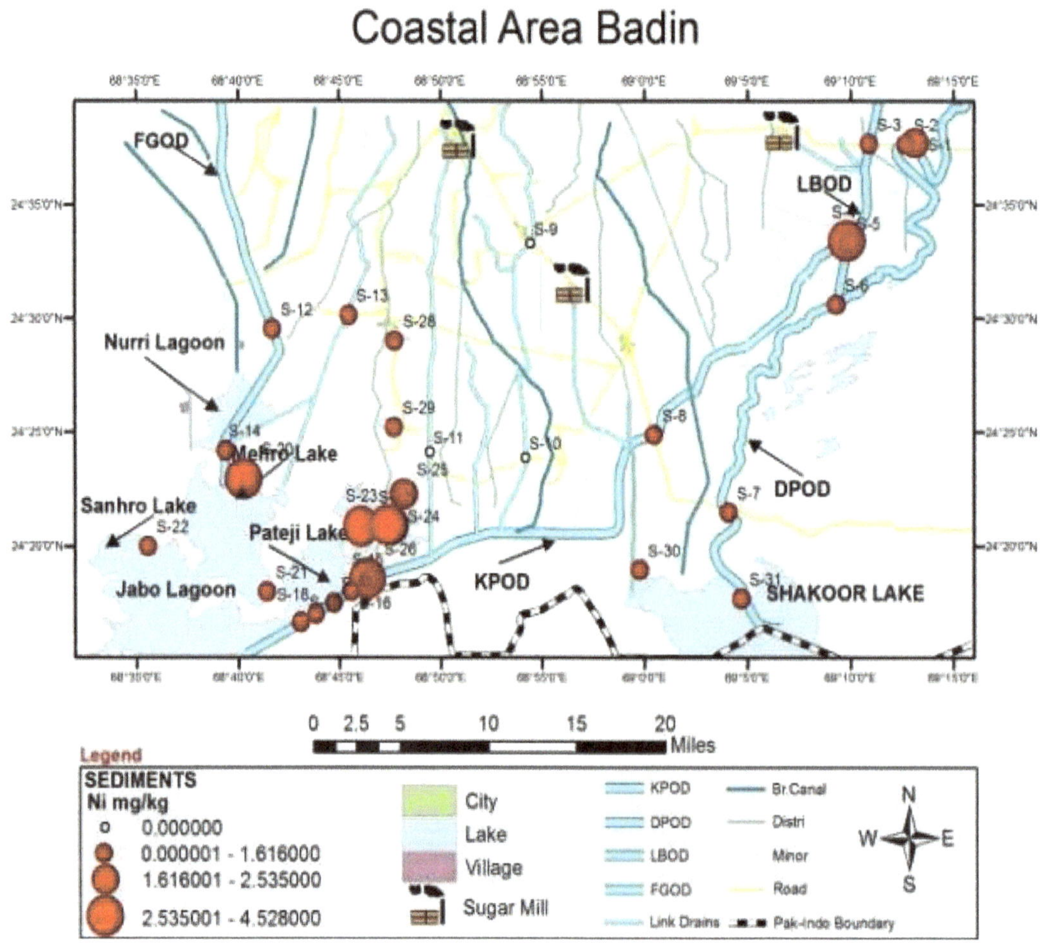

Fig. 3. Map of studied area showing detection of Ni in sediment.

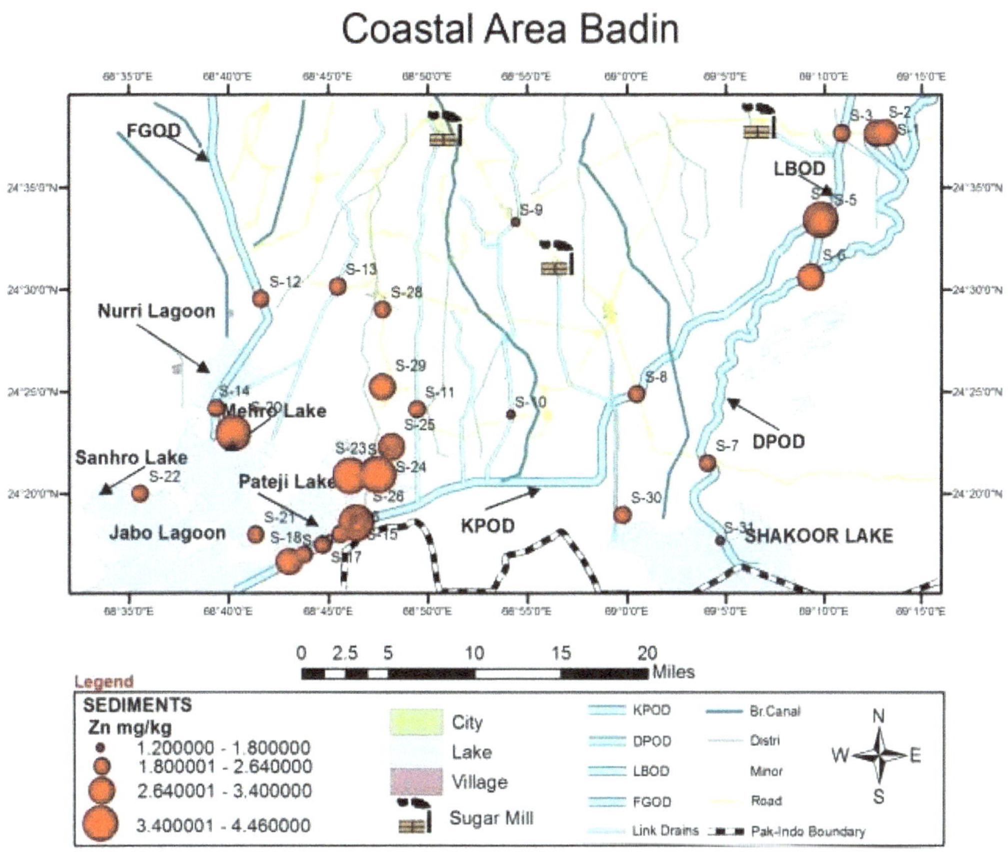

Fig. 4. Map of studied area showing Zinc detection in sediment.

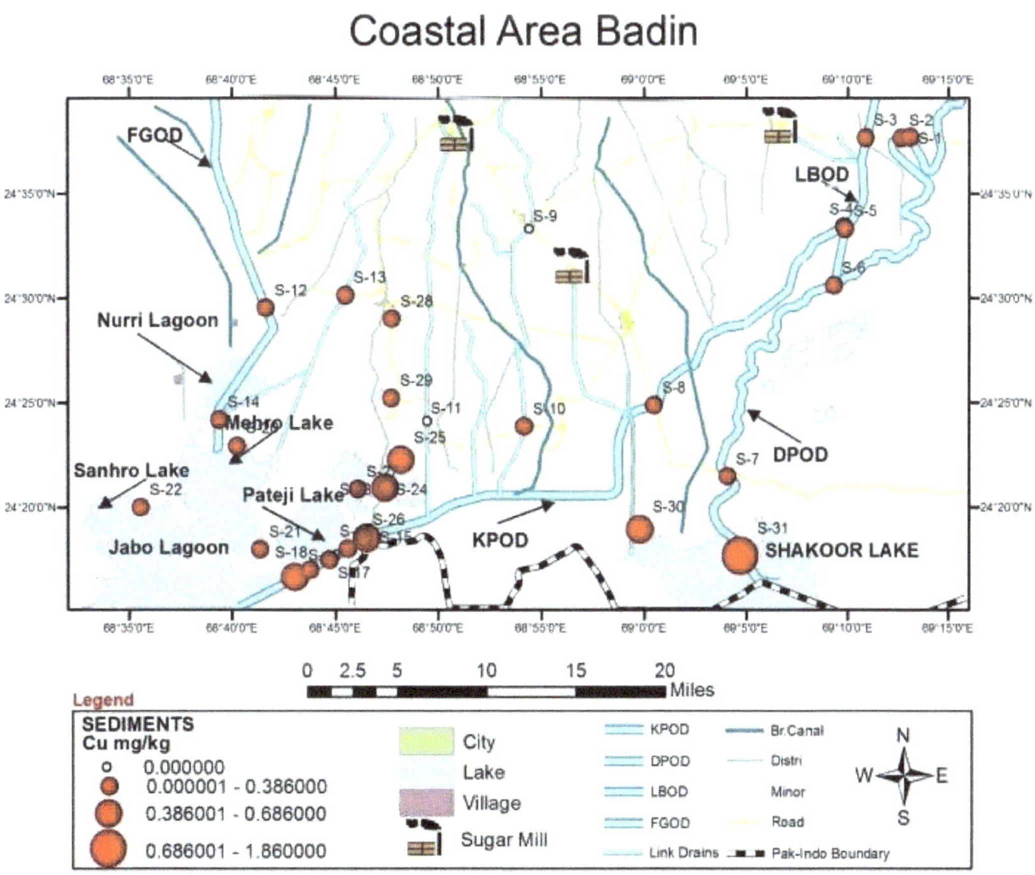

Fig. 5. Map of studied area showing Copper detection in sediment.

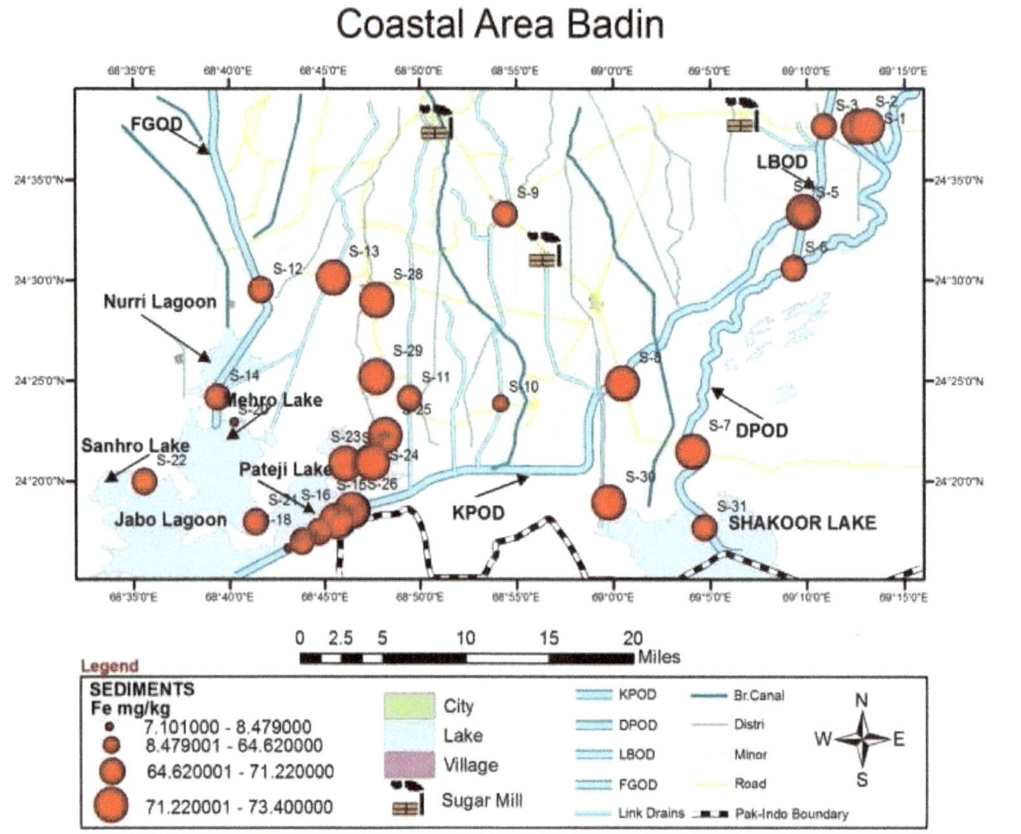

Fig. 6. Map s of studied area showing Fe detection in sediment.

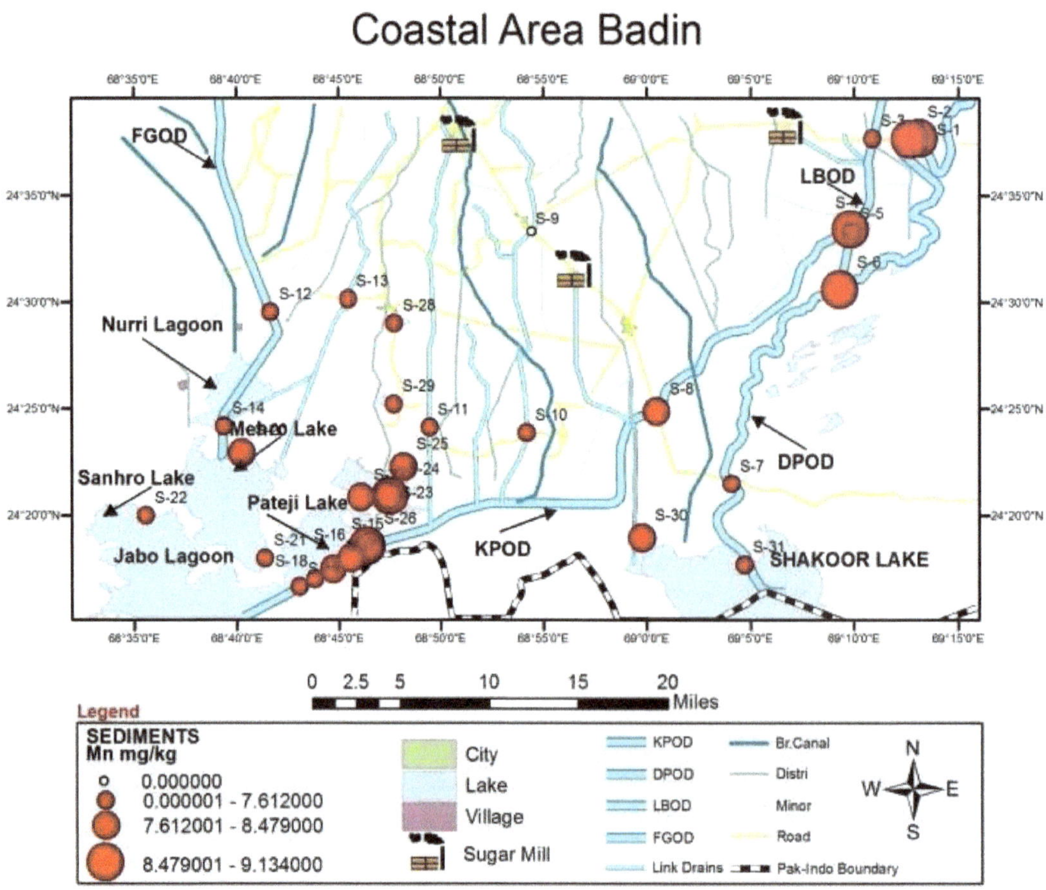

Fig. 7. Map of studied area showing Mn detection in sediment.

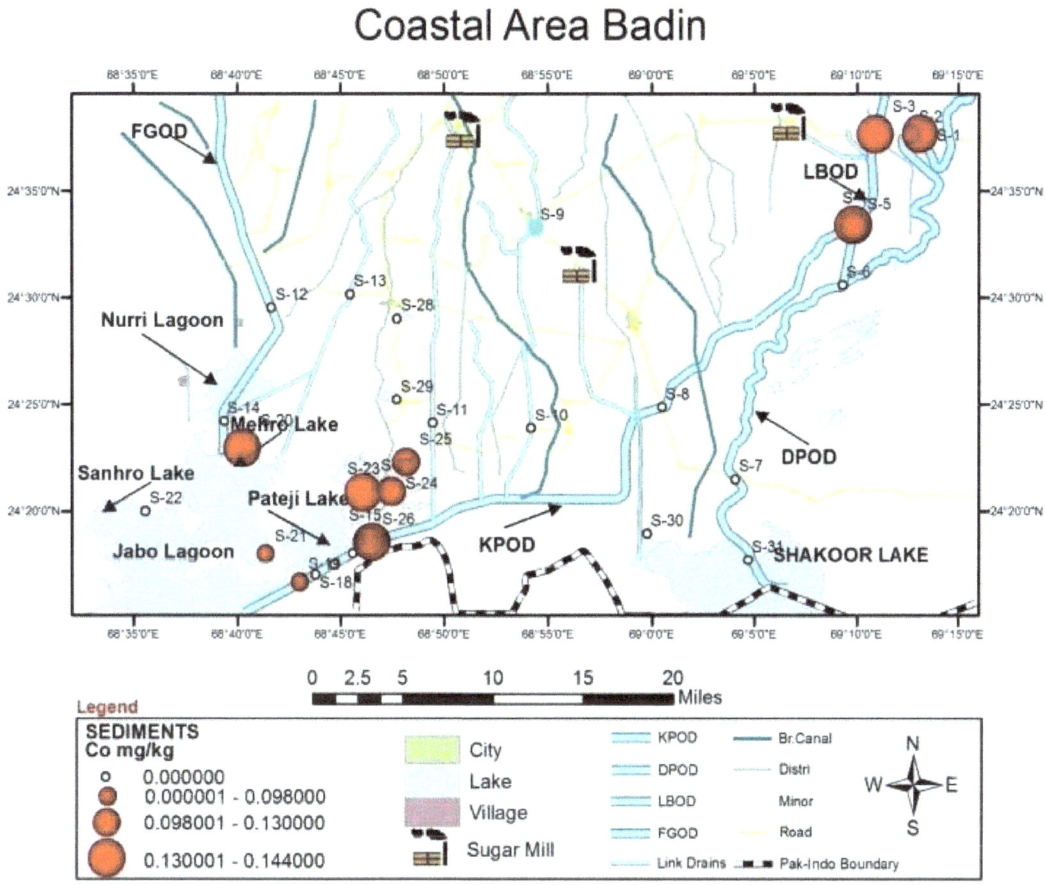

Fig. 8. Map of studied area showing Co detection in sediment.

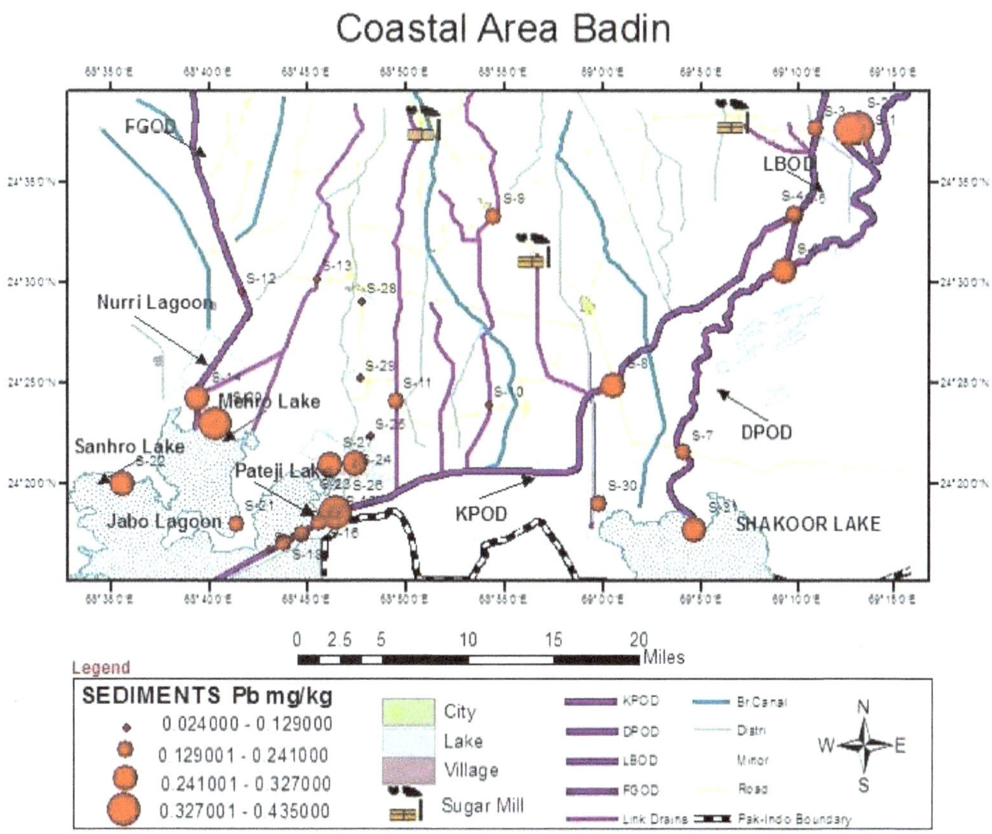

Fig. 9. Map of studied area showing Pb detection in sediment.

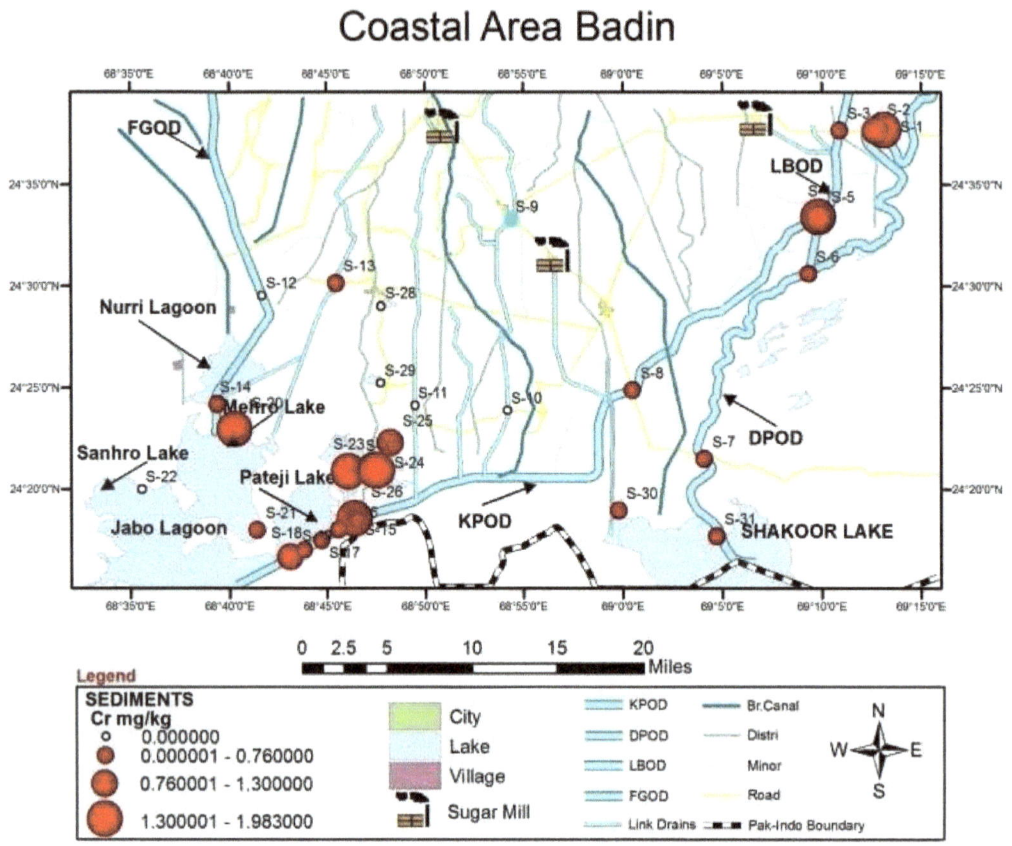

Fig. 10. *Map of studied area showing Cr detection in sediment.*

4. Conclusion

The coastal area of Badin is very vulnerable. The district is used as a pathway to dispose of saline pumped water, agriculture runoff along with the industrial effluents of three remote districts of province of Sindh. The waste water of main LBOD and damaged infrastructure of KPOD, tidal link and cholri weir has polluted the whole coastal area.

It was concluded that sediments of drains, lakes and lagoons of coastal areas have been more contaminated by anthropogenic activities. LBOD and other link drains carrying industrial and municipal wastes are major source of contamination of the areas.The mercury was in high concentration in the sediments of LBOD, DPOD, and KPOD, surface drains as compared to the areas of lakes and lagoons. This also confirms that Hg contamination source is main LBOD. High concentration of zinc was due to agriculture run off contributed by link drains near lakes. It was also observed that near Sea tidal water and damaged Cholri Weir area, heavy metal contents in sediments were in lowest concentration.

Table 1. *heavy metal concentrations of sediments of coastal area Badin.*

No	SamplingStations										
1	Sample codes	S-1	S-2	S-3	S-4	S-5	S-6	S-7	S-8	S-9	S-10
2	Location Names	Dhoro Puran(01)	Dhoro Puran (02)	LBOD at Shadi large	KPOD at bifurcation point	DPODat bifurcation point	DPOD at KathriJo Band	DPOD at Ali bander	KPODat kadhan	Luari Sharif Drain	Luari Drain near KPOD
3	Latitude	N 24° 37.665	N 24°37.660'	N 24° 37.698	N 24 33.406	N 24 33.325	N 24° 30.612	N 24° 21.490	N 24°24.888	N 24° 33.313	N 24° 23.871
4	Longitude	E069° 13.115	E069° 12.656	E069° 10.876	E069 09.822	E069 09.873	E069 °09.309	E069 04.090	E069° 00.473	E068°54.429	E068° 54.191
	Concentrations of heavy metals in sediments										
5	Hg(mg/kg)	0.1988	0.1540	0.2144	0.1900	0.8333	0.6412	0.4684	0.1720	N.D	N.D
6	Ni (mg/kg)	2.535	1.242	1.246	3.643	1.616	1.246	0.9642	0.8560	N.D	N.D
7	Cd (mg/kg)	N.D	N.D	N.D	N.D	N.D	N.D	N.D	N.D	N.D	N.D
8	Zn (mg/kg)	3.26	3.36	2.62	4.07	2.91	2.80	2.46	2.14	1.20	1.80

No	SamplingStations										
9	Cu (mg/L)	0.118	0.180	0.315	0.375	0.013	0.046	0.026	0.248	N.D	0.032
10	Fe (mg/kg)	72.56	73.0	70.96	71.79	70.97	71.0	72.0	71.50	68.20	64.62
11	Mn (mg/kg)	9.063	9.0	7.438	9.099	7.420	8.60	7.36	8.24	N.D	3.28
12	Co (mg/kg)	0.144	0.098	0.136	0.134	0.130	0	0	0	0	0
13	Pb (mg/kg)	0.435	0.389	0.204	0.210	0.223	0.327	0.201	0.321	0.189	0.102
14	Cr (mg/kg)	1.630	1.30	0.697	1.588	0.923	0.056	0.082	0.056	N.D	N.D

Table 01[1]*. heavy metal concentrations of sediments of coastal area Badin.*

No	Sampling stations										
1	Sample codes	S-11	S-12	S-13	S-14	S-15	S-16	S-17	S-18	S-19	S-20
2	Location Names	Serani Drain	Ameer Shah drain	Hamid drain	Nareri Zero Drain	KPOD AtZero point	KPOD Breach 1	KPOD Breach 2	KPOD Breach 3	KPOD Breach 4	Mehro lake
3	Latitude	N 24° 24.122	N 24°29.519	N 24°30.143	N 24 °24.204	N 24° 18.528	N 24°17.990	N 24°17.490	N 24°17.000	N 24°16.652	N 24° 22.953
4	Longitude	E068° 49.473	E068° 41.643	E068°45.4 47	E068° 39.332	E068° 46.514	E068° 45.600	E068°44.68 5	E068°43.79 1	E068°43.04 1	E068 °40.2 58
Concentrations of heavy metals in sediments											
5	Hg(mg/kg)	N.D	0.0506	0.0428	0.1404	0.2100	0.1806	0.1250	0.2340	0.2177	0.1988
6	Ni (mg/kg)	N.D	0.0684	0.0256	0.982	1.467	0.845	0.450	0.254	1.351	3.814
7	Cd (mg/kg)	N.D	N.D	N.D	N.D	N.D	N.D	N.D	N.D	N.D	N.D
8	Zn (mg/kg)	2.20	1.96	2.10	2.64	2.41	2.25	2.56	2.10	2.99	3.86
9	Cu (mg/L)	N.D	0.252	0.243	0.240	0.498	0.344	0.386	0.298	0.434	0.212
10	Fe (mg/kg)	69.3	71.2	72.6	70.5	72.19	71.40	71.22	70.50	70.45	71.73
11	Mn (mg/kg)	4.24	6.40	7.56	7.22	8.365	7.80	7.86	7.25	7.101	8.479
12	Co (mg/kg)	0	0	0	0	0.136	0	0	0	0.084	0.138
13	Pb (mg/kg)	0.156	0.129	0.105	0.280	0.170	0.190	0.182	0.156	0.112	0.382
14	Cr (mg/kg)	N.D	N.D	0.0842	0.562	0.684	0.720	0.426	0.760	0.920	1.543

Table 01[2]*. heavy metal concentrationsof sedments of coastal Badin.*

S.No	SamplingStations											
1	Sample codes	S-21	S-22	S-23	S-24	S-25	S-26	S-27	S-28	S-29	S-30	S-31
2	Location Names	Jubho lagoon	Sanhro lake	Sandho Dhandh	Waryaro Dhandh	Chorhadi Dhandh	Pateji lake.	Bakradi lake	Khana lake	Bajorerro lake	Kanderri lake	Shahkoor lake
3	Latitude	N 24° 17.992	N24° 19.989	N 24° 20.926	N24° 20.904	N 24° 22.281	N24°18 .543	N24°20.8 54	N24°29. 025	N24°25.21 6	N24°18.944	N24°17.7 71
4	Longitude	E068 °413 82	E068° 35.582	E068 °47.4 13	E068 °4 7.510	E068° 48.176	E068°4 6536	E068°46. 071	E068°47 .711	E068°47.7 25	E068°59.77 1	E069°.04. 756
Concentrations of heavy metals in sediments												
5	Hg(mg/kg)	0.1944	0.1820	0.2333	0.2133	0.2722	0.1922	0.2077	N.D	0.1426	0.2046	0.1860
6	Ni (mg/kg)	0.216	0.1528	3.619	4.453	2.177	4.291	4.528	0.5684	0.2460	1.026	1.246
7	Cd (mg/kg)	N.D	N.D	N.D	N.D	N.D	N.D	N.D	N.D	N.D	N.D	N.D
8	Zn (mg/kg)	2.33	2.60	4.46	4.24	3.06	4.22	4.31	2.33	3.40	2.56	1.26
9	Cu (mg/L)	0.365	0.150	0.593	0.328	0.415	0.199	0.298	0.180	0.124	0.686	1.860
10	Fe (mg/kg)	70.95	70.2	71.88	71.85	72.25	71.88	71.95	72.3	73.4	71.6	70.5
11	Mn (mg/kg)	7.193	7.440	8.243	8.755	8.018	9.134	8.115	7.612	7.324	7.820	7.210
12	Co (mg/kg)	0.062	0	0.129	0.121	0.125	0.124	0.140	0	0	0	0
13	Pb (mg/kg)	0.241	0.280	0.312	0.062	0.034	0.432	0.302	0.024	0.034	0.202	0.310
14	Cr (mg/kg)	0.556	N.D	1.854	1.844	1.145	1.983	1.974	N.D	N.D	0.246	0.236

Table 02. *(average, standard deviation, min, max)heavy metyal concentrations of sediments of badin coastal area.*

Heavy metal	Hg	Ni	Cd	Zn	Cu	Fe	Mn	Co	Pb	Cr
AVERAGE	0.2032	1.455	BDL	2.789	0.3050	71.175	7.375	0.0548	0.216	0.7054
ST DEV	0.173	1.449	BDL	0.865	0.3355	1.610	1.848	0.0634	0.114	0.6949
MIN	0.0428	0.0256	BDL	1.2	0.013	64.62	3.28	0.062	0.024	0.562
MAX	0.8333	4.528	BDL	4.46	1.86	73.4	9.134	0.144	0.435	1.983

References

[1] US EPA 2007.Sediment Toxicity Identification Evaluation (TIE) Phases I, II, and III. Guidance Document US EPA Washington, DC20460.

[2] Sekabira, K.Oryem Origa,H.Basamba, T. A. Mutumba,G. Kakudidi,E.2010. Assessment of heavy metal pollution in the urban stream sediments and its tributaries.Int. J. Environ. Sci. Tech., 7 (3), 435-446.

[3] Aprile, F.M. and Bouvy, M.2008. Distribution and enrichment of heavy metals in sediments at the Tapacura river basin north eastern Brazil. J. Aquat. Sci. Technol., 12(1), 1-8.

[4] Iwuoha, G.N. Osuji, L.C. Horsfall, M. Jnr. 2012.Index Model Analysis Approach to Heavy Metal Pollution Assessment in Sediments of Nworie and Otamiri Rivers in Imo State of Nigeria. Research Journal of Chemical Sciences Vol. 2(8), 1-8.

[5] IUCN 2006. District Vision–Badin. A Framework for Sustainable Development.

[6] SIDA 2013. Sindh Water Sector Improvement Project-I.Regional master plan of Indus delta and coastal zone.Revival of Natural Dhoras Sindh Water. Sector Improvement Plan for Left Bank of Indus, Delta and Coastal Zone.

[7] LBOD 2005.Review of the Performance of the Left Bank Out fall Drain Stage 1 (KPOD, DPOD, Tidal Link and Cholri Weir) Report of the World Bank International Panel of Experts South Asia Region. Agriculture and Rural Development.

[8] EIA 2012. Environmental Impact Assessment for Exploration Activities in Badin Concession – Protected Area Final report.

[9] Gupta, P.K. 2011. Methods in Environmental Analysis Water, Soil and Air.2nd Edition Agribios India.

[10] Burrough, P.A. 1986. Principles of Geographic Information System for land Resources Assessment Monograph on Soils and Resources, Survey No.2. Clarendon Press, Oxford, England.

[11] Facchinelli A. , Sacchi E. , Mallen L.,2001. Multivariate statistical and GIS-based approach to identify heavy metal sources in soils. J. Environ. Pollut.114, 313–324.

[12] Chen C.W., Chen C.F., Dong C.D., 2012. Distribution and Accumulation of Mercury in Sediments of Kaohsiung River Mouth, Taiwan APCBEE Procedia 1 153 – 158.

[13] OgbeibuA.E., OmoigberaleM.O., EzenwaI.M., EzizaJ.O., IgweJ.O.,2014.Using Pollution Load Index and Geoaccumulation Index for the Assessment of Heavy Metal Pollution and Sediment Quality of the Benin River, Nigeria. Natural Environment, 2(1), 1-9.

[14] LuC.X., ChengJ., 2011.Speciation of Heavy Metals in the Sediments from Different Eutrophic Lakes of China. Procedia Engineering, 18, 318-323.

[15] ChenC.W. Chen C.F., Dong C.D. 2012.Copper Contamination in the Sediments of Salt River Mouth, Taiwan Energy Procedia, 16,901-906.

Micro Physical Model for Glaciogenic Particles in Clouds for Precipitation Enhancement

Tamara Tulaikova[1, *], **Svetlana Amirova**[2], **Alexandre Michtchenko**[3]

[1]Wave Research Centre at Prokhorov General Physics Institute, Moscow, Russia
[2]Brody School of Medicine, East Carolina University, Greenville, North Carolina, USA
[3]National Polytechnic Institute, IPN-SEPI-ESIME, Zacatenco, Mexico, D.F.

Email address:
tulaikova@gmail.com (T. Tulaikova), mitchen@mail.ru (A. Michtchenko)

Abstract: The detailed microphysical model is presented for the cases of injection the glaciogenic particles inside natural clouds; nowadays glaciogen aerosols are solid CO_2 or liquid N_2. The model includes calculation for quantity of ice crystals that are forming in the overcooled areas, and effect for water droplets grow in a far zone near glaciogen. The comparison with common *AgI* is presented and discussed.

Keywords: Atmosphere, Clouds, Precipitation Enhancement, Glaciogens

1. Introduction

Concentrations of the greenhouse gases are tightly increasing together with average planet temperature [1], so broader usage of active methods for atmosphere purification and precipitation enhancement seems appropriate mechanism for climate normalization. The developments of methods for the stimulation of precipitation inside natural clouds are based on the fact that a typical cloud can contain thousands and up to millions of tons of water. In addition to it, the sum of CO_2 and water vapor in atmosphere put together the 95% in a mass of greenhouse gases now. Today's atmospheric carbon dioxide levels are the highest in comparison with last times in human history; the CO_2 concentration was not be confined higher then 280 ppm in 100 years ago. Incoming pollution in atmosphere are studied carefully today [2, 3]. The CO_2 input exceeds natural capacities for a process, so we have fast evident accumulation of CO_2 in a free atmosphere. In view of the changes to the climate apart from global warming, we have more hot and arid summers last years, the record-setting hurricane seasons in ocean and etc. The development of technical methods for free atmosphere purification is very important today's world and mitigation approach is necessary in order to combat the imminent dangers related to climate changes. Rains in industrial regions from the middle of last Century to today have led to soil deterioration up to pH = 3 - 4 [4]. Therefore, different systems that are designed to work in precipitation regimes are necessary for atmosphere, soil and climate recovery. The idea of weather modification by precipitation enhancement was generated some time ago by Langmuir [5], but now is the time for the development of the similar ideas aimed at the man-made purification of the free atmosphere; one effective way is further wide development of the methods of precipitation enhancement. Resent approach [6] incorporates the possibility of stepwise CO_2 purification in areas of the free atmosphere by spraying of alkaline compounds inside the clouds via an airplane. The alkali causes significantly increases of the CO_2 solubility in rain droplets during their gravitational fall to provide the effective carbon transport to the ground, because concentrations of carbon ions in water are $[HCO_3^-]$ and $[CO_3^{2-}]$, they increased in 10 and 100 times accordingly by each unit of *pH*. This future technology can really compensate for annual carbon emission by method application at 0.1% of planet surface [7, 8]. Motivation for rains usage, control and modification is obvious.

Today methods for precipitation enhancements have been developed in different countries around the world with wide practice [9, 10]. The most popular current technologies for precipitation stimulation are those that involve the sprinkling of hygroscopic or glaciogenic particles [11, 12]. At present time, the hygroscopic particles should be imported by the airplane into special lowest part of natural convective clouds usually, where rising air flows exist with high velocity only in power cloud, which is difficult in reality. And besides, great number of such experiments with hygroscopic particles didn't bring any results, especially in seaside areas and hot seasons [13]. New glaciogen substances are necessary for development.

The new microphysical model is proposed here for the cases of injection the glaciogenic particles in natural clouds. The solid CO_2 or liquid N_2 are used as glaciogenic substances in clouds to enhance precipitation [14, 15]. The methods for precipitation enhancement into 'cold' clouds were developed in a cold climate in Moscow during last years [16], and series of experiments are performed in St Petersburg [17], so the glasiogenic substances are commonly used in Russia. Outlined in this paper is the temperature analysis in vicinity or further zones around flying glasiogenic particles. The influence for cloud medium is considered. Comparison of both considered glaciogens and *AgI* substance is performed.

2. Temperature Analysis

To intensify the precipitation enhancement in a clouds the glaciogen initiators are used. Such initiators include solid carbon dioxide or drops of liquid nitrogen with the temperature -79°C or -196°C correspondingly. Each glaciogen particle is evaporated inside the cloud that locally lowers the temperature around itself. In supercooled region near particles with temperature T ~ -40°C and lower a homogeneous condensation begins and facilitates formation of a great number of ice crystals, *N*. The smallest ice crystals scattered in different directions with their rapid condensation growth due to initial water reserve of the cloud. At the same time, ice crystals collide with other droplets while moving, this process lowers the temperature of the cloud droplets with increase their condensation grow. Let's consider the area of the temperature field around the moving particle. An initial glaciogenic particle has a radius x_0 with a few mm typically for CO_2, but it is decreases due to further evaporation. The surrounding spherical cooled volume located around, its radius is x_1, for details of this system refer to Figure 1. There T_0 is the particle surface temperature, and T_1 is surrounding temperature. The radii for glaciogen particle are x_0 = 2,3 or 4 mm in practice, but typical cloud droplets radii are r = 1 – 10 μm. Firstly, the stationary influence to Cloud in cooling area with $T_1 < T < T_0$ is considered, note the droplet radii are $r << x_0$. The stationary heat equation in cylindrical coordinates are considered to describe the temperature fields in cooling area $x_0 < x < x_1$:

$$\frac{d^2T}{dx^2} + \frac{1}{x} \cdot \frac{dT}{dx} = 0 \qquad (1)$$

The boundary conditions are:

$$T(x_0) = T_0; \; T(x_1) = T_1 \qquad (2)$$

The solution together with boundary equations describes the temperature fields in cooling area $x_0 < x < x_1$:

$$T(x) = C_1 \ln x + C_2$$
$$C_1 = \frac{T_0 - T_1}{\ln(x_0 / x_1)} \qquad (3)$$
$$C_2 = T_0 - (T_0 - T_1)\frac{\ln(x_0)}{\ln(x_0 / x_1)}$$

The temperature was calculated in cooling zone around glasiogen particles of solid CO_2 or liquid N_2 then their radii are x_0 = 4, 3 or 2 mm. Results of calculations according to stationary model (1-3) are presented at Fig.2a. The calculations accuracy is the temperature difference up to level of $T(x_1) - T_1$ = 0.5°C. The radial cooling zone could be the x_1= 50 mm around glaciogens with radius x_0 = 4 mm. All distances, *x*, are measured in radial direction from the center of coordinates according to the scheme of Figure 1.

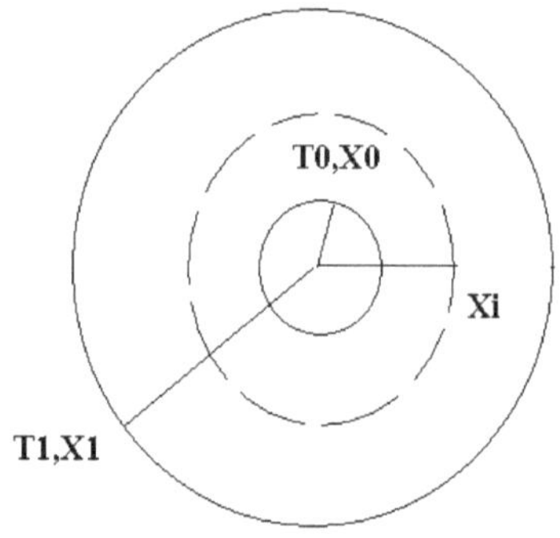

Figure 1. *The scheme of the cooled volume around glaciogen particle.*

However, this stationary model can't be applied or it has specific cases for realization, there is a proof below. Firstly, high speed, *V*, of falling glaciogenic particle provide a small time for action in surrounding, this fact conflicts to the stationary model utilization correctly. Let's consider this topic more carefully. The particle falling velocity *V* is determined usually by equality of gravity force $F_g = mg = 4\pi\rho_p x_0^3 g/3$ with the force of air resistance F_a. The fall velocity can be here approximated roughly using the stationary speed of droplet gravitation sedimentation, where g is the gravitational acceleration according to the following formulas:

$$F_g = F_a, mg = \rho_a \pi x_0^2 V^2 C_{\mathrm{Re}} / 2 \qquad (4)$$

here particle mass is *m*, *g* is acceleration of gravity, the ρ_p, ρ_a, are the density of solid (or liquid) particle and air respectively, and coefficient, C_{Re}, depends of Reynolds number, C_{Re} = 0.6 - 1. The fall velocity $V(x_0)$ can be approximated roughly using the stationary speed of droplet gravitation sedimentation according to the following formulas:

$$V(x_0) \approx \sqrt{2x_0 g \rho_p / \rho_a} \qquad (5)$$

The minimal acting time, t_1, in cooling zone is equal to the diameter of this zone ($2x_0$) divided to velocity from (5) of the flying particle. The time could be defined as follows $t_1 \approx 2x_0 / \left[\sqrt{g\rho_p/\rho_a} \cdot \sqrt{2x_0}\right] = 10^{-2}\sqrt{2x_0}$, there x_0 is measured in meters, so t_1 = 6 - 9 ms. Our physical model propose the plume after flying glaciogen particle that is formed by vaporized supercooled molecules with the same low temperature. It is necessary to introduce the time, t_2, for the plume living and action. The time t_2 determines the diffusion time for glaciogenic molecules motion from the zone of plum that has the same radius as x_0 according to our assumption. The diffusion motion of supercooled molecules from the center of plum is performed according to known diffusion/thermal solutions for length [20], as followed:

$$x_0 \approx \sqrt{D \cdot t_2} \tag{6}$$

The complete acted time is the sum:

$$\begin{aligned} t &= t_1 + t_2 \\ t_1 &\approx x_0^{0.5}\left(2g\rho_p/\rho_a\right)^{0.5} \\ t_2 &\approx x_0^2/\chi \approx x_o^2/D \end{aligned} \tag{7}$$

The plum in the central area has the same temperature as glaciogenic particle's temperature, T_0 = -196 ^{0}C or -79 ^{0}C. The time of action with account of supercooled molecules plume was calculated as t_2 ~ 0.2 0.8 sec then glaciogen particle radii are x_0 = 2 – 4 mm. The χ is a thermal diffusivity, $\chi \approx$ 0.2 cm^2/s. The model of the plume is illustrated by the falling water droplets in the study [21], and is confirmed by realistic numerical values in further calculations at present work. Let's return to the result of stationary model of cooling zone at Fig. 2a to show its further development. The time, $t = t_{st}$, for implementation of the model (1-3) should be in agreement with equation (6) where thermal wave length is x_1, so $t_{st} \approx x_1^2/D$ and t_{st} > 100 seconds. As a result, we get necessary time too long in comparison with [21] for stationary model implementation. The comparison of necessary time with acted time in equation (6) gives $t_{st} \neq t$, so the stationary modal can't be used. Physically, it means that cooled evaporated molecules from glaciogen cannot reach far cooled distance at Fig.2a.

The interpretation for zone of influence of the moving particle should be changed. Non-stationary heat equation in cylindrical coordinates and its solution with the same boundary conditions (2) have the forms:

$$\frac{\partial T}{\partial t} = \chi\left(\frac{\partial^2 T}{\partial x^2} + \frac{2}{x}\frac{\partial T}{\partial x}\right)$$

$$T(x,t) = T_1 - T_0 \cdot erfc\left[\frac{x - x_0}{2\sqrt{\chi t}}\right] \tag{8}$$

There χ is a thermal diffusivity. The complementary error function for the argument, $A(x)=(x-x_0)/(2\sqrt{\chi t})$, could be calculated as simplest approximation of function $erfc(A(x))$ ~ $\exp\left(-A^2(x)\right)$ with the similar results to the next one. The next possible presentation [19] was used here to get more accuracy in far cooled zones up to x_1, as follow:

$$erfc(X) \approx \pi^{-1/2}\left(\frac{x - x_0}{2\sqrt{\chi t}}\right)^{-1} \exp\left[-\frac{(x - x_0)^2}{4\chi t}\right] \tag{9}$$

The acting time t was calculated according to (7) here. Result calculations of the cooling zone according to non-stationary model (8-9) are presented at Figure 2b.

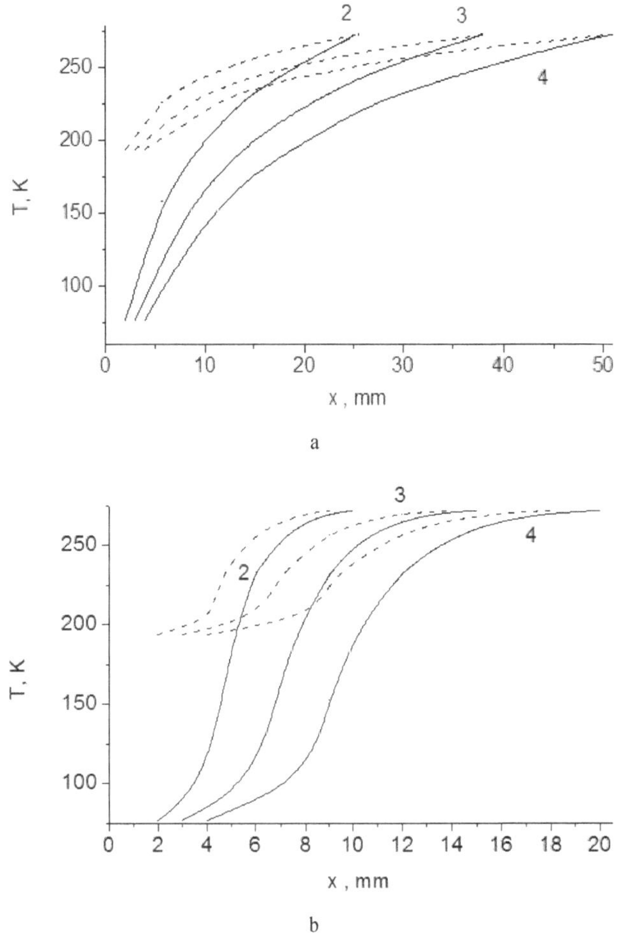

Figure 2. *Calculations of the temperature in cooling zone of the glasiogen particles of CO_2 (dashed lines) and N_2 (solid lines); the graphs titles 2, 3 or 4 correspond to appropriate radius of glaciogen particle x_0. The (a) corresponds to stationary model (1-3), but the (b) corresponds to non-stationary model (7-9).*

As can be seen from Fig.2b the temperature decreases in surrounding media at the distance from the center of axe at Figure 1 as follows. The distance values for complete cooled zone are x_1 = 19; 14 or 9 mm approximately near CO_2 solid particle with its radii x_0 = 4; 3 or 2 mm accordingly. Similar calculations were done around N_2 glaciogens with a radii x_0 = 4; 3 or 2 mm, the zones of influence are x_1 = 20; 15 or 10 mm accordingly. The limit distances with temperature T_i = -40°C were calculated to separate two cooling regimes into surrounding media according to considered mechanisms of cooling influence. We suppose that if $T_i < T <$ T$_0$ the homogeneous condensation predominates, but then $T_i > T > T_1$ the host water droplets in cloud will increased in their sizes. The supercooled zone is located near glaciogen particle then $x_i < x < x_0$, but far cooled zone has radial location at distance $x_i > x > x_1$. Calculations were done with taking into account x_0, the results limit distances are x_i^{CO2} = 9.50; 7.15 or 4.76 mm for CO_2 glaciogen with x_0 = 4; 3 or 2 mm. The zones limits are x_i^{N2} = 12.05; 9.05 or 6.02 mm for liquid N_2 with its droplet radii x_0 = 2, 3, 4 mm correspondingly. All mentioned values of distances x_0, x_i, x_1 are measured from the center of coordinates according to the scheme of Figure 1.

Note that smallest radius x_0 = 2 mm is more real for liquid nitrogen droplets for realization in practice. Considered zone's boundary x_i around glaciogen particle is shown by dashed circle in scheme Fig. 1.

Let's estimate the dynamics of falling glaciogenic particles. Using formulas (5) and (15) for glaciogen evaporation, we estimated the living time of falling glaciogenic particle t_f up to its evaporation. The values are $t_f \approx$ 180 sec for CO_2 particles, and $t_f \approx$ 28 sec for N_2 liquid substance according to Figure 3. The current falling velocity and complete living way could be estimated from Fig. 3.

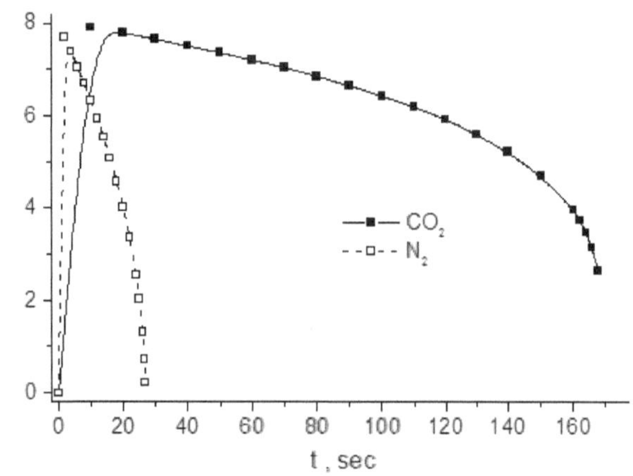

Figure 3. The speed and time for the falling evaporated CO_2 particle or a N_2 droplet taking into account its evaporation by Langmuir equation (15).

3. Resulting Number of Ice Crystals for Different Glaciogens

Firstly we consider the supercooled zone near glaciogen particles. Let's calculate the number of these crystals, N. The water droplets in clouds can be in supercooled state without its crystallization up to the low temperatures ~ -40°C [12], only a narrow central supercooled zone near glasiogenic particle has sue condition for fast homogeneous crystallizations. Water droplets with small radii $r < r_k$ should have sublimation and ones with bigger sizes have time for their condensation grow. The predominant mechanism here is a fast homogeneous condensation of water vapor with its fast crystallization into a large number of ice crystals. The most authors believe the start of homogeneous condensation correspond to oversaturation level $s(T) \approx 3$ - 6. The topic was studied in a science literature with details [22-26]. The critical radius [18] for possibility to grow for water droplets can be calculated using the following formula taking into account:

$$r_k(T) = 2\sigma M / [\rho_w E T \ln(s(T))] \qquad (10)$$

where σ is a water's surface tension, water molar mass is M, the E is a universal gas constant We assumed that droplets with $r = r_k$ have fast crystallization in supercooled area.

Let's calculate the water in this area to build the number of ice crystals in supercooled zone. The initial typical value for supersaturation in local medium was considered as typical cloud value s_0 = 0.01, for example. It provides the water vapor amount W_v as follows:

$$W_v = \frac{P_0(1+s_0)}{P_{atm}} \cdot \frac{N_L}{N_A} M \qquad (11)$$

Taking into account real pressure of water vapor/air and combination of Loschmidt, $N_L = 2.68 \cdot 10^{25}$, and Avogadro, $N_A = 6.022 \cdot 10^{23}$, constants, we obtain the resultant water mass in a water evaporated molecules $W_v \approx 5$ g/m^3. The total fast evaporation of initial cloud droplets was suggested in supercooled zone, so liquid water content W was added in further estimations, for example W = 1g/m^3. The total water content to produce ice crystals will be $W_w = W_v + W$. Other vapor reorganization in medium was neglected here due to long necessary diffusion time. So the crystals number can be calculated by dividing of water vapor amount given by equation (11) to the ice crystals with critical radius $r_k(T)$ from (10).

The supercooled area x_i near glaciogenic particle was calculated as a function of glaciogen initial radius x_0 and can be obtained from Figure 2b then T < 233 K. To get the rate of crystals production near glaciogen particle we use the time of diffusion motion, t_i, of molecules to form cooling zone. This time is $t_0 \approx (x_0^2/D)$ obtained from equation (6). To begin with, the total number of ice crystals that are produced per second, N/t_i in s^{-1}, could be calculated as a sum taken across all of supercooled zone inside glaciogen and at its temperature T = -79°C for CO_2 or T = -196°C for N_2, as follows:

$$N_{x0} = \frac{W_w 4\pi x_0^3 / 3}{t_0 \rho_w 4\pi r_{k,x0}^3 / 3} \qquad (12)$$

The calculated values for N_{x0} are presented at Table 1 in row 1 for glaciogenic particles with radius x_0 = 4 mm for solid CO_2 particle and for liquid N_2 droplet.

At the next step we incorporate wider supercooled zone x_i with the temperature $T_i < T < T_0$ near glaciogen particle taking into account data at Figure 2b with x_i being a radius around glaciogen particles with appropriate local temperature T = -40°C = 233 K. The cooling zone for homogeneous condensation here was considered inside radial zone $x_i < x < x_0$ and during time $t_i = (x_i - x_0)^2 / D$. Taking into account mentioned larger zone for homogeneous condensation we obtain the following equation:

$$N_{xi} \approx \frac{W_w x_i^3}{2\rho_w t_i}\left[r_{k,x0}^{-3} + r_{k,xi}^{-3}\right] \qquad (13)$$

Calculations of this number N_i of the ice crystals are performed according to the formula (13) and presented for at rows 2-4 of the Table 1 for a glaciogen radii x_0 = 4; 3 or 2 mm correspondingly.

Table 1. *Calculations (12-14) of producing of ice crystals numbers, N, near glaciogens.*

Cristals numbers, N/s ; N	*Solid* CO_2	*Liquid* N_2
1) N_{x0}, s^{-1}; x_0 = 4mm (eq.12)	$1.6 \cdot 10^{17}$	$2.5 \cdot 10^{18}$
2) N_{xi}, s^{-1}; x_i then x_0 = 4mm,(13)	$6.2 \cdot 10^{17}$	$8.3 \cdot 10^{18}$
3) N_{xi}, s^{-1}; x_i then x_0 = 3mm	$4.7 \cdot 10^{17}$	$6.2 \cdot 10^{18}$
4) N_{xi}, s^{-1}; x_i then x_0 = 2mm	$3.2 \cdot 10^{17}$	$4.2 \cdot 10^{18}$
5) N_f ; x_1 then x_0 = 4mm,(14)	$1.1 \cdot 10^{20}$	$2.3 \cdot 10^{20}$
6) N_f/g, g^{-1}	$4.2 \cdot 10^{20}$	$8.7 \cdot 10^{20}$

The total number of ice crystals, N_f, in a trajectory of gravitation fall of glaciogenic particle can be estimated by multiplication of N_{xi} to fly time t_f. The complete fly time presented at Fig. 3, so the complete number of ice crystals N_f can be calculated according to following formula:

$$N_f = t_f N_{xi} \qquad (14)$$

The value N_f is listed at the 5th column of the Table 1 then particle radii are equal to 4 mm. The values in the last row of the Table 1 were obtained by dividing the previous data (5 row) to the particle initial weight $4\pi\rho r_0^3 \rho_p/3$, so data at 6th row represent the complete crystals number N_f per 1 gram of glaciogen, N_f/g.

The results of Table 1 suggest that the particles of liquid nitrogen are more effective, but they act only inside smaller cloud layer due to fast evaporation. For a physical description, the further collisions of ice crystals are useful. The moving ice crystals will collide with the water droplets with radius r inside cloud, this collision can cause fast crystallization in overcooled droplets with resulting fast droplet condensation grow.

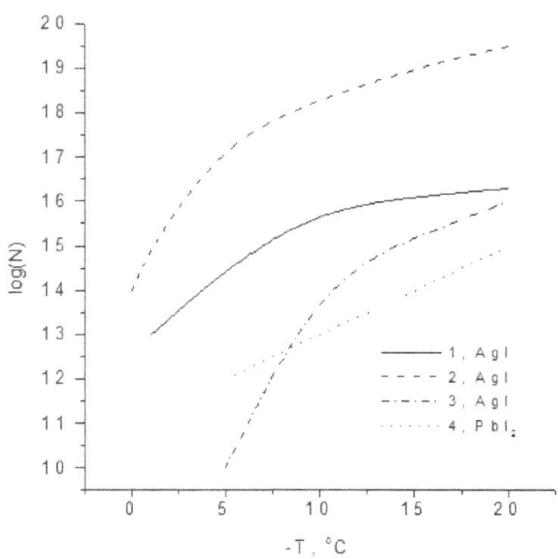

Figure 4. *Theoretical limit of ice crystals, N per 1 gram of AgI or* PbI_2 *[26]. The graph 1 correspond to m = 1, ε = 0; the graph 2 differs from (1) by cooled AgI; the graph 3 differs from (1) by m = 0.996, ε = 0.014; the graph (4) presents the best experimental data for* PbI_2.

Let's compare obtained data from calculations for both CO_2 and N_2 with activity of widespread traditional glaciogens for precipitation enhancement, the AgI and PbI_2. In the monograph [26] the resultant ice forming properties of silver and lead iodide aerosols are investigated in details. Particular attention is given to determine of aerosol ice-forming activity limit, appropriate formulas and calculations were presented for these cases. The theoretical results for number of produced ice crystals are presented for AgI using at Fig.4. The theory predicts that AgI glaciogen activity depends strongly on wetting parameter m and the magnitude of deformation obtained at crystallization, ε.

Theoretical values are presented with comparison to work [26] which gives similar values from the 12 experimental studies from different groups. The best results were obtained in laboratory chambers at temperature T = -5°C ÷ -15°C, so values differ from theoretical limit within 10%. Note, that the best graph (2) was calculated for cooled molecules of AgI such conditions were possible in experimental chambers only, because in reality the AgI is a part of artillery shell with high temperature as a result. In experiments the AgI or PbI_2 reagents were mixed with explosive substance in the central part (5 – 10%) of the artillery shell. The artillery shells with reagent in the central part blow up in a cloud in experiments, and reagent evaporates with formation of mono molecules or clusters [27, 28].

4. Water Droplets Increment in far Cooled Zone

Note that for considered here glaciogens (CO_2 and N_2) there are additional cooling effects for surrounding media in a extended zone, but AgI or PbI_2 do not exhibit such effect due to principal difference of action. Firstly, large numbers of ice crystals are forming. The other effect is the additional fast grow of water droplets in extended zones surrounding moving glaciogen particle, there is $T_1 < T < T_i$. The The appropriate cooling far zone is created by glaciogen particle, this zone changes in time during particle evaporation with appropriate changes on its size and fall velocity, see Fig.2b. ice crystals cannot appear here because supersaturation is low, but host cloud droplets that are located here will have fast condensation grow due to additional temperature decrease $T(x)$. The condensation increment of the radius of usual water droplets trapped in far zone on the basis of the ratio of the Langmuir [18] as follows:

$$d_w^2 \approx \frac{8DMp_b|s(T)|}{ET(t)\rho_w} t + d_w^0 \qquad (15)$$

here new diameter of water droplet is $d_w = 2r$, its initial diameter is d_w^0, the water density is ρ_w, water molar mass is M, the E is a universal gas constant, the saturated water vapor pressure is p_b. The supersaturation near water droplet due to decreased temperature is $s(T)$ given by equation (16). The supersaturation function depends on surrounding temperature as follows:

$$s(T) = \ln(p(T)/p_b) \quad (16)$$

The pressure $p(T)$ of saturated water vapor near water droplet increases according to low temperature in cooled zone, it can be determined on the basis of the relation [18]:

$$\ln[p(T)] = A - B/T(x) \quad (17)$$

here the coefficients are A=21.18 and B=5367 for CGC system. Then a water droplet is located in the zone of influence $x_1<x<x_i$ at the scheme of Fig.1, a supersaturation $s(T)$ near water droplet has strong increase. An appropriate super saturation $s(T)$ near water droplet can be presented as follows:

$$s(x) \approx B\left(T_x^{-1} - T_1^{-1}\right) \quad (18)$$

Maximal value is $s_{max} \approx 8$ near CO_2 glaciogen particle then T_{x0} = -79°C, or $s_{max} \approx 50$ near N_2 then T_{x0} = -196°C. Next we determine the variable temperature in a field of influence of glaciogen particle during movement according to equation (9). Resulting relations for temperature (9) and associated supersaturation function from equation (18) were introduced according to increment formula (15) to get water droplet increments. Alternative scenario is using temperature and distance of cold extended zone near glaciogen directly at the Figure 2b. We use the equation (7) for affecting time t taking into account of supercooled molecules after flight of glaciogen particle. The equations were used for approximate calculations, then the initial radius in (15) correspond to the typical cloud droplet size, $d_w^0 = 10$ μm, but further calculation indicates a low sensitivity to initial droplet size. The resulting form for a water droplet increment is as follow:

$$r_w \approx 0.5\sqrt{d_w^2 - d_0^2} \quad (19)$$

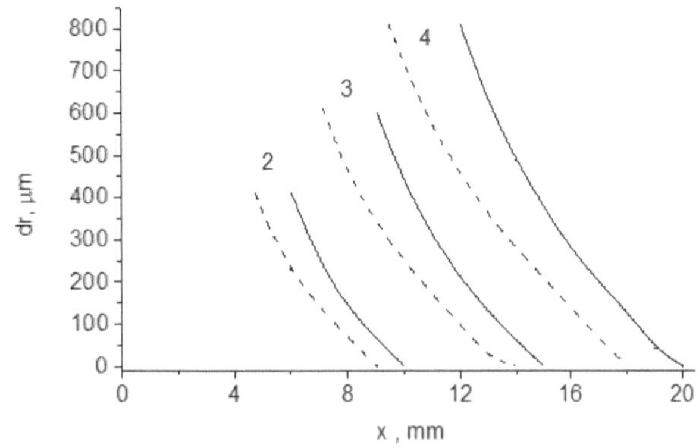

Figure 5. *The increment of water droplets in cooled zone of glaciogen, then 0 >T> -40℃. The graph enumeration 2; 3 or 4 correspond to the glaciogen's radius x_0 = 2; 3 or 4mm accordingly. The solid graphs correspond to N_2 influence, the dashed ones to CO_2.*

The corresponding results are given at Figure 5 and show great values for increment of water droplets in extended zone near glaciogen particles flying in clouds. One can see from Figure 5 the significant influence of a cooling reagent in a distant zone onto observed water droplets. In particular, if initial radius of glaciogen is x_0 = 4 mm, the radius of water droplets increases up to r = 810 μm in the vicinity of that zone. The influence of a distant zone on the environment of a cloud is enormous, considering the big extent of a zone influence is 20 or 18 mm for liquid nitrogen or solid ice respectively.

Conclusion

The solid CO_2 or liquid N_2 glaciogen particles can be applied in cold or warm clouds. Appropriate models and formulas for complete actions are presented and justified by corresponding calculations describing utilization of solid CO_2 or liquid N_2 glaciogens. These substances are no expansive, they exhibit ecological safety and high efficiency in utilization. The comparison with common *AgI* reagent was considered for efficiency of ice crystal production.

Acknowledgements

The authors are grateful for useful discussions to the authors of many experimental technologies for precipitation enhancement for clouds in Russia Dr. Gennady P. Beryulev. The authors thanks to Dr. Bagrat G. Danelyan, who are the chief of the Department of cloud physics and active effects of Moscow Central Aero Observatory now.

References

[1] IPCC: Summary for Policymakers, In: "Climate Change 2014, Mitigation of Climate Change. Contribution of Working Group III to the Fifth Assessment Report of the Intergovernmental Panel on Climate Change", Edenhofer, O., R. Pichs-Madruga, Y. Sokona, E. Farahani, S. Kadner, K. Seyboth, A. Adler, I. Baum, S. Brunner, P. Eickemeier, B. Kriemann, J. Savolainen, S. Schlömer, C. von Stechow, T. Zwickel and J. C. Minx (eds). Cambridge University Press, Cambridge, United Kingdom and New York, NY, USA, 2014.

[2] A. Kleidon, Y. Malhi and P. M. Cox. "Maximum entropy production in environmental and ecological systems Introduction". Philos. T R Soc. B, 365(1545), 2010, pp.1297-1302.

[3] T. W. Choularton., M. W. Gallagher, K. N. Bower, P. J. Connolly, M. J. Flynn, Z. Ulanowski. "Calibration of the cloud particles imager probes using calibration beads and ice crystal analogs the depth of field". Journal of atmospheric and oceanic technology, 2007, 24, p. 1860-1879.

[4] Yu. A., Izrael, I. M. Nazarov, A. J. Pressman, F. J. Rovinsky, A. G. Ryaboshapko, L. M. Filippova. "Acid rain". Hydrometeoisdat, Russia, 1989.

[5] I. Langmuir, and V. J. Schaefer. "Improved methods of conditioning surfaces for adsorption". J. Am. Chem. Soc., 1937, v.59, pp.1762-1763.

[6] T. Tulaikova, S. Amirova. "The effective possibility for atmosphere CO_2 purification", Lap-Lambert, Germany, 2012.

[7] T. Tulaikova, S. Amirova. "The method for effective CO_2 purification in the atmosphere". Global Journal of Science Frontier Research. Vol.15-H, No.1, Version 1.0, 2015, pp. 1-9.

[8] Svetlana Amirova, Tamara Tulaikova. "One possibility for atmosphere CO_2 purification to get climate recovery". Science Discovery in SciencePG. Special Issue: New Technical Ideas for Climate Recovery. Vol. 3, No. 1-2, 2015, pp.1-6.

[9] Ch. Daly, W. P. Gibson, G. H. Taylor, M. K. Doggett and L. J. Smith. "Observation bias in daily precipitation measurements at United States cooperative network station"s. *BAMS*, 2007, v.88, pp.899-912.

[10] K.-M. Lau, Ramanathan V., et all. "The joint aerosol-monsoon experiment. A new challenge for monsoon climate research". 2008, BAMS, v.89, pp.369-383.

[11] S. M. Shmeter and G. P. Beryulev. "Efficiency of cloud and precipitation modification with hygroscopic aerosols". Rus. Meteorology and Hydrology, 2005, v.2, pp.43-60.

[12] A. S. Drofa et al. "Formation of cloud microstructure: the role of hygroscopic particles". Izvestiya. Atmospheric and oceanic physics, 2006, v.42, pp.355-366.

[13] Bruintjes, R.T., 1999: A review of cloud seeding experiments to enhance precipitation and some new prospects. BAMS, 80, 805-820.

[14] Arnett S. Dennis. Weather Modification by Cloud Seeding. International Geophysics Series. Academic Press, New York, London, 1980., Vol 24, 275 p

[15] L. G. Kachurin. Physical basis of impact on atmospheric processes. L., Gidrometeoizdat, 1973.

[16] A. A. Chernikov, Beryulev S. P., Corneev V. P. "Method of active influence on convective clouds". Patent of Russia No. 2005121252/12, Central Aerological Observatory.

[17] B. M. Vorobyev, Zinchenko A. V. "Physics of clouds and active effects". A. I. Voeikov GGO print, St Peterburg, Russia, vol.517, 1988.

[18] Parker C Reist. "Introduction to aerosol science". Macmillan publishing company, New York, 278p. (Reist P. Aerosols. M.: Mir, 1987)

[19] M. Abramowitz, I Stegun. "Mathematical ed., the Handbook of special functions". M.: Nauka, 1979, 830p.

[20] H. S. Carlslaw, J. C. Jaeger. Conduction of heat in solids. Oxford at the Clarendon press, 1964.

[21] M. K. Makuachiv. "Moving droplet Evaporation'. In "Physics of clouds and weather modification". St. Petersburg, Gidrometeoizdat, 1996, No. 89, pp. 1-18.

[22] C. A. Jeffery, P. H. Austin. "Homogeneous nucleation of supercooled water. Results from a new equation of state". Journal of Geophysical Research. 1997, Vol.102, p. 25.26925.279

[23] V. Khvorostyanov, Curry, J. A. "The Theory of Ice Nucleation by Heterogeneous Freezing of Deliquescent Mixed CCN". Part I: "Critical Radius, Energy, and Nucleation Rate". J. Atmos. Sci. 2004, Vol. 61, p. 26762691.

[24] B. O Kramer, Hubner, H. Vortisch, Woste L., Leisner T. "Homogeneous nucleation rates of supercooled water measured in single levitated microdroplets". Journal of Chemical Physics, 1999, Vol. 111, No.14, p. 65216527.

[25] H. R Pruppacher, Klett, J. D. "Microphysics of Clouds and Precipitation". Springer Publications. New York, 1997

[26] Nora A. Plaude, "Study of the ice-forming properties of aerosols silver iodide and iodide of wine". Proceedings of CGO, Moscow, 1967, v 80, pp.5-88.

[27] V. G. Khorguani, A. Zhikharev S., Myakon'kiy G. B. "Ice forming activity of aerosols AgI an PbI2 obtained by explosion". Izvestia of Academy of Sciences USSR. Atmospheric and oceanic physics. 1974, v. X, pp.100 103.

[28] M. K. Jakamokhov and Abshaev A. M. "Dispersion crystallizing reagents by explosion- III Complex Brownian coagulation and the formation of the ice-forming particles". Rus. Meteorology and hydrology, 2012, No. 9, pp. 58-66.

Physicochemical Assay of Water of Kashmir Lakes

Muhammad Aslam Mirza[1, *], Muhammad Aziz Choudhary[1], Muhammad Yar Khuhawar[2], Rafee Arain[2], Asima Naz[1], Kauser Yasmeen[3]

[1]Department of Chemistry, Mirpur University of Science and Technology (MUST), Mirpur AJK, Pakistan
[2]Institute of Advanced Research Studies in Chemical Sciences, University of Sindh, Jamshoro, Pakistan
[3]Department of Chemistry, Federal Urdu University of Arts, Science and Technology, Karachi, Pakistan

Email address:
ma_mirza64@yahoo.com (M. A. Mirza)

Abstract: The lakes of Kashmir are generally shallow and situated in flood plans of the river Jhelum. In most of the lakes, thermal stratification is uncommon, and stable stratification occurs only in deeper lakes such as Manabal and Mangla lakes having depth within the range 12-30 m. Anthropogenic contamination of lakes water is one of the concerns that have received attention at local, regional and global levels due to their toxicological impacts and importance in ecosystems. This preliminary study was designed to investigate the comparative physicochemical profile of the water of Mangla, Baghsar and Benjosa lakes of Kashmir. Twenty six parameters including alkalinity, salinity, pH, electrical conductivity (EC), total dissolved solids (TDS), dissolved oxygen (DO), biological oxygen demand (BOD), chemical oxygen demand (COD), total hardness (TH), chloride, sulphate, nitrite, nitrate, phosphate, Na, K, Ca, Mg, Fe, Pb, Cd, Ni, Co, Zn, and Cu were recorded. Suitability of lakes water for irrigation was checked by using US Salinity diagram, permeability index (PI) and Wilcox diagram. Cluster analysis was carried out to investigate the similarity levels among the selected parameters. Piper's diagram, sodium percentage and residual sodium carbonate (RSC) data served as information tool to understand the ecological status of the water of lakes. COD and BOD patterns in Mangla and Baghsar lakes indicated dual anoxic conditions of the water-rock soil interface and anthropogenic impact while Banjosa showed aerobic activity at the interface. Increasing rate of tourist traffic and greater habitational influence seems to be a major factor contributing towards raising trophic levels of these lakes.

Keywords: Kashmir Lakes, Mangla, Baghsar, Banjosa, Physicochemical Study of Lakes

1. Introduction

In the broad sense, water quality is determined by numerous biological, physical, and chemical parameters that affect the cachet of water for a meticulous use. In recent years attempts in the country have been made to utilize estuaries, reservoirs and other water bodies for fish culture practices by which the protein uptake for our masses could be increased. Freshwater resources are used to meet the day-to-day requirements of human settlement, for production of hydropower and most important of all, for agricultural, fisheries and industrial needs. Pakistan has one of the most extensive canal systems in the world. It must be recognized that in under developed and densely populated countries, which have the most pollution problems, the importance of healthy freshwater environment has increased. All the aforementioned activities are expected to have a significant impact on the water quality. The study of limnological dynamics is of great significance in removing the constrictions, which hinder production of inland fish [1-3]. The seasonal variations in physico-chemical factors have a profound effect on the distribution and population density of both fauna and flora [4-9]. In recent years ecosystem of surface water in Kashmir has been disturbed by anthropogenic activities. Free mixing of detergents, grease, oil and animal waste may be considered responsible for the increasing levels of contaminants in the water bodies [10, 11]. Bioavailability of essential elements enhances the productivity of water that shifts the lake water towards eutrophication [12-17]. The abundance of phytoplankton and zooplankton in the freshwater bodies is greatly regulated by the physicochemical factors [18-19]. The quality of lake water may be altered by anthropogenic activity, geological nature of the catchments area and yearly rate of the rainfall [20].

Over its lifetime, a lake progresses from a more oligotrophic to a more eutrophic state by collecting essentional nutrients such as phosphorus, nitrogen and sedimentation [21]. Nutrients from agricultural areas, storm water runoff, urban development, fertilized yards and gardens, failing septic systems, land clearing, municipal and industrial wastewater, runoff from construction projects, and recreational activities contribute to accelerated enrichment or eutrophication. Land development, construction and agricultural activities near lakes and streams, or farming steep slopes leave soils vulnerable to erosion and accelerate sedimentation [22]. Like algae, aquatic plants (macrophytes) are a vital part of the lake because they provide cover, habitat and sometimes food for fish, the organisms that fish eat, and other wildlife [23]. Lakes can be contaminated by toxic substances including industrial chemicals such as PCBs (polychlorinated biphenyls), metals, and solvents; pesticides from agricultural runoff; urban storm water runoff containing petroleum hydrocarbons, metals and pesticides; and air-deposited chemicals [24]. Toxic contamination may cause dramatic impacts such as fish kills. Less obvious impacts include decreased reproduction or slower growth rates in fish and reductions in invertebrate diversity [25]. One long-term danger of toxic contamination is the bioaccumulation or build-up of toxic substances--mercury, for example--in fish flesh. The toxic effects may be passed on to humans eating the fish [26].

1.1. Baghsar Lake

Baghsar Lake lies at the latitude 32.58° and longitude 74.8° and is situated 975 m above the sea level in Samahni valley of District Bhimber in Azad Jammu & Kashmir (AJ&K) Pakistan[14]. Baghsar lake is a tourist spot and lies on the border of mesotrophic to eutrophic nature. It is bounded to Mirpur District in the northwest; and shares border with Indian administrated Kashmir in the northeast and is linked to Pakistan in the west and south. The area is very rich in archaeological remnants. The lake is nearly 1 Km long sheet of water that soothes the senses of the visitors. Its average depth is 3-4 meter and width 15 to 115 meters. The lake is located just in the core of Baghsar village surrounded by mountains. Main source of water for this lake is rainfall; however, water oozes out from its bed too. The average yearly rainfall is 1500 mm [14].

1.2. Banjosa Lake

Banjosa a site of immense scenic beauty, houses a Lake, situated at an altitude of 1981 m above sea level at about 20 Km from Rawalakot (elevation1615 m), the headquarter of district Poonch. Water inflows of this lake are ice melts, spring ooze, and precipitation. The average yearly rainfall is 1500 mm. The quality of Lake Water may be changed by anthropogenic activity, geological nature of the catchment's areas and yearly degree of precipitation. The lake is periodically drained and recharged to maintain the quality of its water. Khan and Zutshi [27, 28] reported primary productivity and trophic status of Naranbagh Lake and limnological studies on Dal Lake of Kashmir. Leghari et al [29], reported limnological studies of Tatta Pani Hot springs and River Poonch at Tatta village of District Poonch, Kashmir. Mirza et al [9-11] demonstrated the quality of waters of springs, torrents, wells, ponds, snow melt and precipitation in the targeted area. The targeted area of Kashmir stretches between latitude 33°-36° and longitude 73°-75° and the altitude rises sharply from 360 m to 6325 m above sea level. There are the hot plains of the Bhimber and Mirpur districts and the coldest lands such as Rawalakot, Leepa and Trarkhal. Weather conditions vary across landscape because of the lofty mountains like the Pirpanjal that check the moisture-laden air from entering the valley. In summers, the outer plains and the outer hills receive rainfall from monsoon winds while in winters, winds from the Mediterranean cause snowfall and rainfall in the Valleys of Kashmir. The moisture-laden winds cause rainfall in the forests on the hills pushing the temperature down in summer.

1.3. Mangla Lake

Mangla Lake is the 16th largest artificial lake in the world. It was built from 1961 to 1967 across the Jhelum River. It lies at about 108 Km south-east of Islamabad in Mirpur District of Azad Kashmir, Pakistan. The main edifice of the lake consists of 2 spillways, 5 power-cum-irrigation tunnels, 4 embankment lakes and a 1,000 MW power station. The main dam is 10, 3,100 m long and 138 m high with a reservoir of 97,250 Km^2. Since its first sequestration in 1967, sedimentation has increased to the extent of 1.39 Km^3, and at present the gross storage capacity has declined to 5.86 Km^3 from the actual design of 7.25 Km^3. The live capacity has declined to 5.65 Km^3 from 6.59 Km^3. This implies a reduction of 19.22% in the capacity of the lake.

1.4. Geology

All the selected lakes are situated in mountainous area except Mangla Lake that is placed in fairly plan area. Ecological study of the area was carried out by Meddlicott, Lydekker and Mddlemiss [30-32]. Siwalik group of rocks is found mostly in the western part of the Poonch District along with the eastern bank of the Jhelum River. These rocks crop out between Azad Pattan and Kohala. An irregular eastern contact runs from Hajeera through Rawalkot to Dhirkot. Wadia divided the Siwaliks of Kahmir into Lower Palandri, Middle Mang and Upper Sand Rock Stage [33]. Chaudhry and Ashraf reported in Kotli area, the rocks consist of granite and crystalline groups [34]. The rocks of the study area are mainly made of chlorite, muscovite mica granite, sand stone, schists, quartzite, pyrite and calcite. The fractured mica schists inter-bedded with thin bends of granite gneiss, quartzite, and non-foliated granitic rocks act as a barrier to the ground water flow, forming springs. A permeable recharge zone for groundwater is provided by fractured mica schists covered with thin layer of top soil (0.5- 1.5 m) in this area [35, 36].

2. Experimental

2.1. Sampling Strategy

Prior to efficient sampling the lakes were observed for their locations and stratified for their sampling. Each stratum was sampled for predetermined number of sub samples. The sub samples were mixed to form composite samples. The samples were also *in situ* tested for basic physico-chemical parameters as temperature, pH, DO, TDS and EC. The sampling mode was simple grab method. High density polythene bottles (pre-washed) were suspended, at predetermined depths through a Secchi disc, from the surface and stopper removed by a sharp jerk through an attached line. The Lakes were navigated by boat. For elemental analysis each sample was acidified to pH < 2; for organic matter, fixed with $HgCl_2$. Separate samples were taken for specific analysis.

For sediment sample collection, clean, free of rust and chrome plated hand probe and spade were used. Simple random procedure was adopted for the collection of samples. With this system each soil core was selected separately, randomly and independently of previously drawn units. Individual soil cores taken from 10 locations for each composite sample were mixed thoroughly in clean plastic containers. A sample of the soil mixture was removed as the composite sample and placed in a soil sample bag, lined with plastic. This composite sample consisted of a mixture of the individual cores.

2.2. Measurements

Hardness, chloride and alkalinity were determined by titration with standard EDTA, silver nitrate and hydrochloric acid. Conductivity, salinity and total dissolved solids (TDS) were evaluated with Orion 115 conductivity meter. The pH was recorded with Orion 420 A pH meter. The dissolved oxygen (DO) in the samples was determined by Wrinkler method (APHA, 1989). Chemical oxygen demand (COD) was estimated by micro-dichromate oxidation method. Kjeldahl nitrogen was determined by using standard procedure [37]. Orthophosphate, nitrate and nitrite were determined by spectrophotometery. Orthophosphate was determined by reducing phosphomolybdic acid formed with ascorbic acid to molybdenum blue. Total phosphate was estimated by persulphate acid hydrolysis, followed by determination as for orthophosphate. Nitrate was determined after derivatization with brucine sulphate. Nitrite was estimated using N-naphthyl ethylenediamine as derivatizing reagent as reported [37]. Sulphate was determined by turbidimetery as $BaSO_4$ using Hitachi 220 spectrophotometer. A method developed by the International Humic Substance Society (IHSS) was used for the extraction of humic substances from soil samples (Stevenson, 1982). The metal ions Na, K, Ca, Mg, Fe, Pb, Cu, Zn, Ni, Cd, and Co were determined with Varian Spectr AA-20 atomic absorption spectrometer with standard burner head and air acetylene flame at the conditions recommended by the manufacturer. The analysis was carried out in triplicate with integration time 3 sec. and delay time 3 sec. Na, K, Ca and Mg were determined after appropriate dilution. Sample (350 mL) containing nitric acid (1 mL) was heated gently at 90-95 °C and was concentrated to about 5-8 mL. The solution was transferred to volumetric flask and final volume was adjusted to 15 mL. The solution was analyzed for the contents of Fe, Pb, Cu, Zn, Ni, Cd, and Co by air acetylene flame atomic absorption spectrometer.

3. Results and Discussion

3.1. Chemical Characteristics of Lake Waters; pH, EC, TDS, TH, DO, BOD & COD

The water in Mangla Lake (Fig.1) is slightly alkaline (pH 7.5) in summer and autumn and slightly acidic (pH 6.1) in winter. Baghsar and Banjosa lake waters are acidic in summer (pH 6.5 & pH 6.6) and alkaline in autumn and winter (pH 7.09, 7.1 & pH 7.13; 7.99). Over a hydrological year the inter season variance for Mangla is 0.567, Baghsar 0.117 and Banjosa 0.567. TDS, EC, TH, values also concomitantly decrease or increase with a corresponding change in pH. These trends indicate moderate photosynthesis activity and chemical weathering of lithologic limestone and dilution by precipitation or concentration through evatranspiration. Likewise, lower DO, BOD, COD values in all the lakes negate inputs from anthropogenic activity but indicate deeper anaerobic activity at water-bed rock interface (Fig.1).

3.2. Nitrogen, Phosphorus, Hardness, Alkalinity and Salinity

Nitrite- N, nitrate-N and Kjeldahl-N were determined in samples of Banjosa, Mangla and Bhaghsar Lakes; values observed were found in the range 0.001-0.007 mg/L, 2.66-3.24 mg/L and 0.19-5.89 mg/L respectively. Higher concentrations of Kjeldahl-N may be attributed to industrial and domestic sewage of Mirpur city in Mangla Lake and biomass degradation in Banjosa Lake. Banjosa, Mangla and Bhaghsar Lakes indicated orthophosphate and acid hydrolysable phosphate within the range 0.06-0.16 mg/L and 0.07-0.20 mg/L respectively reflecting the deficiency of phosphate into earth crust of the targeted area. Hardness, alkalinity and salinity were observed within the range 131-226 mg/L, 113-211 mg/L and 0.1-0.2 mg/L for all the three Lakes (Table 1).

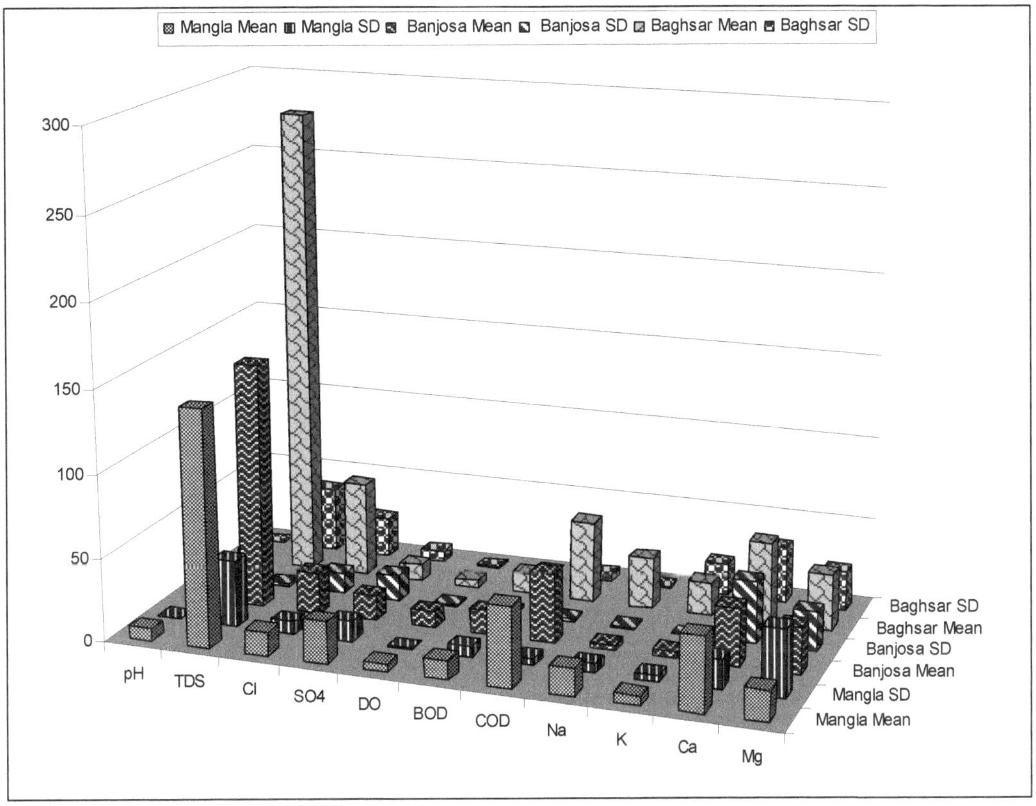

Fig. 1. Comparative study of pH, TDS, Cl⁻, SO₄, DO, BOD, COD, Na, K, Ca, and Mg contents in Lakes of Kashmir.

Table 1. Results of Physicochemical Parameters of Lakes of AJ&K.

Parameters	Mangla Lake	Banjosa Lake	Baghsar Lake
Salinity g/L	0.10±0.041	0.10±0.011	0.20±0.015
M. Alkalinity mg/L	113±18.61	135±27.51	211±46.61
Hardness as $CaC0_3$ mg/L	131±126.21	172±12.61	226±36.61
Nitrite –N mg/L	0.004±0.001	0.001±0.0004	0.007±0.002
Nitrate- N mg/L	3.24±1.24	2.82±1.05	2.66±1.02
K- Nitrogen mg/L	5.89±2.11	1.35+0.31	0.19±0.06
Orthophosphate mg/L	0.15±0.03	0.16±0.02	0.06±0.008
Total Phosphate mg/L	0.20±0.046	0.20±0.015	0.07±0.04

3.3. Major Elemental Distributions

(i). Cations: Ca emerges as the major cationic load contributor in all three lakes; Mangla 52%, Banjosa 49.7 % and Baghsar 36.5 %. Second in donor rank are Mg and Na contributing about 25 % to Mangla and Baghsar lake loads while for Banjosa 40 % is contributed by Mg and fewer than 5 % by Na. The order of cation distribution for Mangla lake water is Ca-Mg-Na-K for Banjosa Ca-Mg-K-Na, and that for Baghsar is Ca-Mg-Na-K (Pipers plot). Na is higher for Mangla (19.4 %) and Baghsar (23.2 %) where most of the anthropogenic activity is centered and both are drained by rivers carrying surface and mountain run offs (Table 2).

(ii). Anions: Anionic concentration is dominated by HCO_3/CO_3, followed by chlorides, sulphates and trace nitrates. This trend pertains to all three lakes. The metal speciation is anticipated as Ca-HCO_3/CO_3 > Mg-HCO_3/CO_3 > Na- HCO_3, Na-Cl > K-Cl > Ca-SO_4 > Mg-SO_4. Insignificant presence is also indicated for NO_2-N, NO_3-N; Mangla Lake indicates 4.2 mg/L Banjosa 1.35 mg/L and Baghsar 0.57 mg/L. Though the Mangla and the Baghsar receive considerable domestic waste water but K-Nitrogen data demonstrates substantial dilution by precipitation. The overall quantity of carbonate alkaline earth type and Cl-group (nominal anthropogenic activity) and is maintained by precipitation dilution (Table 2).

Table 2. Results of physicochemical parameters of Lakes of AJ&K.

Parameters	Mangla Lake Mean (SD)	Banjosa Lake Mean (SD)	Baghsar Lake Mean (SD)
pH 25°C	7.03 ±0.75	7.25±0.73	6.89±0.34
EC µS/cm	223±26.5	232±13.4	459±31.8
TDS mg/L	142±40.06	148±1.41	283.2±38.97
Cl^- mg/L	14.29±7.64	24.63±12.65	57.13±24.2
SO_4 mg/L	25.40±12.02	15.32±15.38	10.12±7.04
DO mg/L	4.64±0.62	10.06±1.41	4.39±0.73
BOD mg/L	11.52±6.60	14.80±1.41	12.6±1.65
COD mg/L	48.00±4.00	41.33±1.53	49.00±4.24
Na mg/L	16.79±5.78	3.09±0.99	30.94±1.08
K mg/L	5.83±3.70	4.01±1.42	19.53±19.9
Ca mg/L	45.01±17.61	34.52±37.47	48.54±34.45
Mg mg/L	18.77±40.01	27.80±24.15	33.94±23.8

3.4. Trace Metals

Trace metals constitute only 0.93 % of TDS in Baghsar, 0.75 % of TDS in Mangla and 1.2 % of TDS in Banjosa. The ratio of trace metals for Baghsar:Mangla is 0.405, Mangla : Banjosa is 0.601 and Banjosa : Mangla is 0.67.

Pb and Co dominate the minor elemental presence. Pb contributing 34 % in Mangla.39.9 % in Baghsar and to 50 % in Banjosa, while Co is 25.3 % in Baghsar, 73.5 % in Mangla and 4.4 % in Banjosa. They form of 5 % COD in Baghsar, 2.2 % in Mangla and 4.4 % Banjosa. All the preceding comparisons of lakes bring about a number inferences: only about 1% of TDS is made up of trace metals of which lake wise Pb is from 34-50 % and that for Co the change is from 5-73.5 %.

The trace metals constitute only 0.93 % of TDS in Baghsar, 0.75 % of TDS in Mangla and 1.2 % of TDS in Banjosa. Pb and Co dominate trace elemental presence; Pb contributes 34 % in Mangla; 39.9 % in Baghsar; 50 % of TDS in Banjosa while Co is 25.3 % in Baghsar; 73.5 % Mangla and 4.4 % of TDS in Banjosa. There is also a marginal change in Cd Concentration (Table 3).

Table 3. *Trace metals data of lakes of Kashmir.*

Parameters	Mangla Lake	Banjosa Lake	Baghsar Lake
Fe mg/L	0.008±0.002	0.027±0.012	0.018±0.005
Pb mg/L	0.037±0.003	0.090±0.014	0.087±0.013
Cu mg/L	0.001±0.0004	0.001±0.0005	0.058±0.010
Zn mg/L	0.023±0.007	0.036±0.008	0.012±0.003
Ni mg/L	0.031±0.015	0.020±0.011	0.017±0.006
Cd mg/L	0.007±0.001	0.004±0.002	0.005±0.002
Co mg/L	0.079±0.031	0.069±0.041	0.067±0.021

3.5. Soil Organic Matter

Baghsar Lake soil analysis revealed the presence of total organic matter (TOM), humic acid (HA), fulvic acid (FA) and non-humic substances (NHS) 0.210 g, 31 %, 33 %, and 36 % respectively. The results obtained from Banjosa lake soil were 0.228 g TOM, 34 % HA, 36 % FA and 30 % NHS whereas, Mangla lake soil indicated TOM 0.102 g, HA 35 %, FA 24 %, and NHS 41 %. Moisture content in Baghsar, Banjosa and Mangla soil samples was observed 27 %, 25 % and 19 % respectively. Based on the percentage ratio of HA and FA, the soils of Baghsar, Mangla and Banjosa were categorized as gray forest, sodpodzolic and gray forest respectively after Kononova [36].

4. Water Suitability for Irrigation

4.1. Hazards of Alkalinity and Salinity

The Lake water samples were also tested regarding their suitability as irrigation water. Two parameters TDS, EC were specifically monitored as these are lineally correlated and are a measure of salinity hazard to crops. TDS values of all samples indicate that lake water is of fresh nature (Table 4) while EC based classification places 2 lakes in excellent and 1 in good class (Table 5).

Since it reduces osmotic activity of plants and interferes with the absorption of nutrients [37, 38], sodium adsorption ratio (SAR) was specifically calculated to determine the suitability of lake water for irrigation [39]. The SAR values range from 0.55-4.82 with an average value of 2.78 during one hydrological year. As lake water samples of the study area fall in the low sodium class A1 (Table 6) (Fig. 2), it implies that no alkali hazard is anticipated to the crops. If the SAR value is greater than 6-9, the irrigation will cause permeability problems on shrinking and swelling types of clays [40, 41].

Table 4. *Nature of lake water based on TDS values.*

TDS mg/L	Nature of water	Representing lakes	No. of lakes
< 1000	Fresh water	Baghsar, Mangla, Banjosa	3
1000 - 10000	Brackish water	Nill	Nill
10000 - 100000	Saline water	Nill	Nill
> 100000	Brine water	Nill	Nill

Table 5. *Classification of irrigation (lake) water based on electrical conductivity (EC).*

EC(μS/cm)	Water class	Hydrological year (2004-2005) Representing lakes	Total no. of lakes
< 250	Excellent	Mangla, Banjosa	2
250-750	Good	Baghsar	1
750-2000	Permissible	Nil	Nil
2000-3000	Doubtful	Nil	Nil
>3000	Unsuitable	Nil	Nil

Table 6. *Classification of lake based on alkalinity hazard.*

SAR	Alkalinity hazard	Water class	Hydrological year (2004-2005) Representing lakes	Total no. of lakes
< 10	A1	Excellent	Baghsar, Mangla, Banjosa	3
10 – 18	A2	Good	Nil	Nil
18 - 26	A3	Doubtful	Nil	Nil
> 26	A4	Unsuitable	Nil	Nil

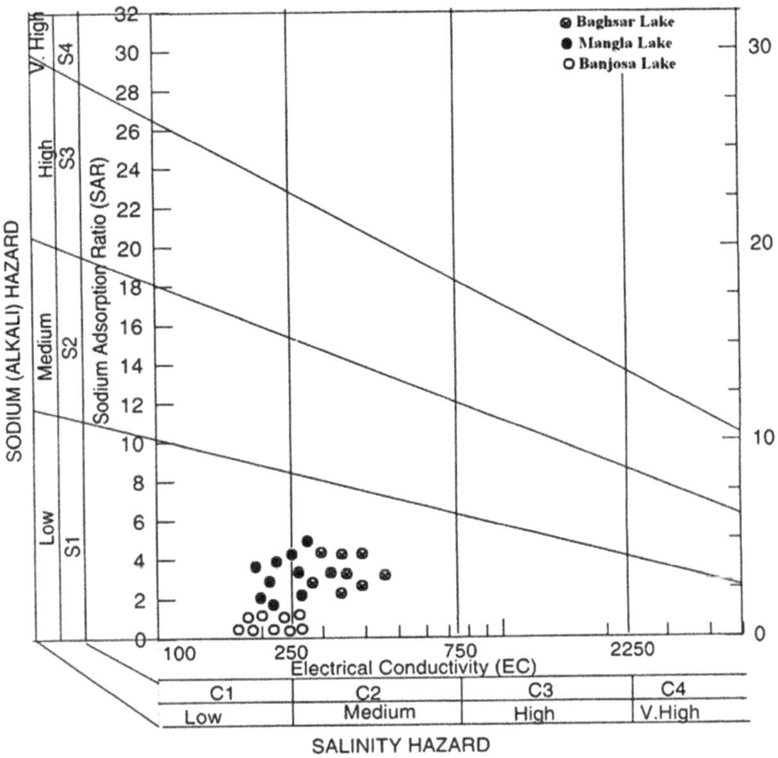

Fig. 2. Salinity and alkalinity hazard in lake water in US salinity diagram.

4.2. Sodium Concentration (Percentage)

The sodium percentage (Na %) was also calculated. The sodium % (Table 7) indicates that the lake water is excellent (Mangla) to good (Baghsar, Banjosa) for irrigationand meets USDA standards [42] (Fig.3).

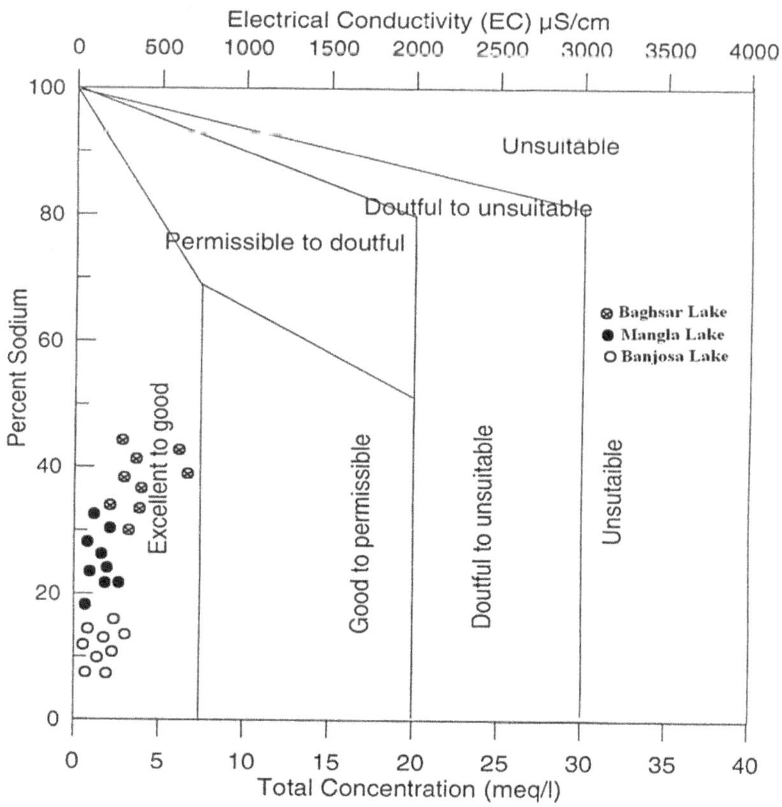

Fig. 3. Suitability of lakes water for irrigation in Wilcox diagram.

Table 7. *Quality of irrigation water based on % Na.*

% Na	Water class	Hydrological year (2004-2005) Representing lakes	Total no. of lakes
< 20	Excellent	Mangla	1
20-40	Good	Baghsar, Banjosa	2
40-60	Permissible	Nil	Nil
60-80	Doubtful	Nil	Nil
>80	Unsuitable	Nil	Nil

4.3. Permeability Index (PI)

If the concentration of sodium ions is high in irrigation water, they get absorbed by clay particles, displacing Ca^{2+} + Mg^{2+} ions reducing permeabilityand eventually resulting in poor internal drainage [41,43,44].

The Permeability Index (PI) was calculated and for samples in the present work PI ranges from 22.5 % to 40.07 % in a hydrological year. The average value is 32.2 % that comes under class–II of Donen's chart making it good for cropping. While % Na categorizes 1 lake (Mangla) as excellent and 2 lakes (Baghsar, Banjosa) as good, none of them as doubtful (Table 8) (Fig.4).

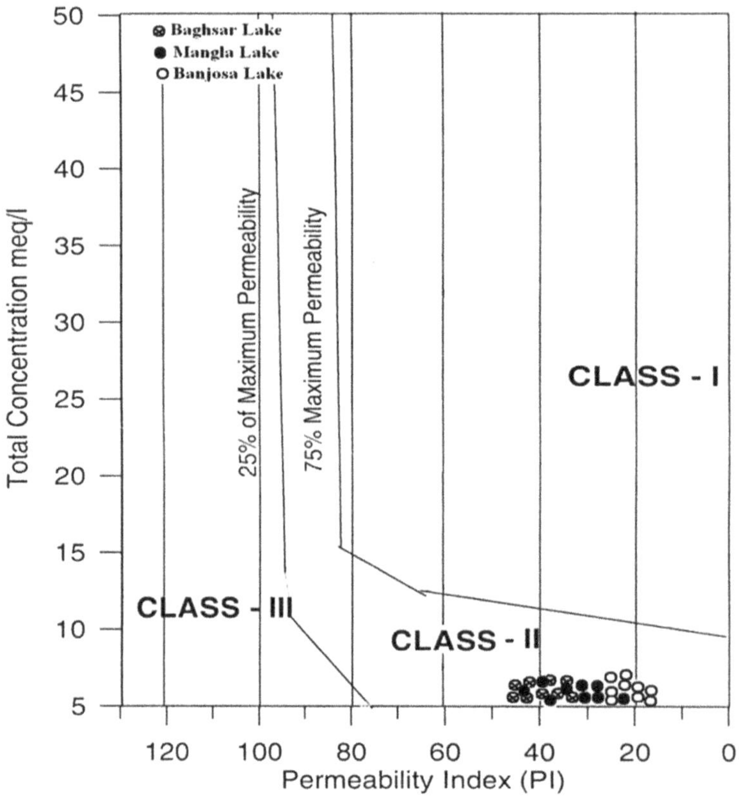

Fig. 4. *Suitability of lakes water for irrigation based on PI.*

The application of these two parameters (Na % & PI) indicate that permeability index renders only none sample as unsuitable for irrigation and the % Na puts no sample as border line indicating that the concentration of sodium is not high enough to render the water unsuitable for irrigation. Total hardness also indicates that hardness of water is mainly due to carbonates and bicarbonates of calcium and magnesium, 1 sample moderate and 2 hard (Table 8).

Table 8. *Hardness based classification of the lake water.*

Total hardness as CaCO3 mg/L	Water class	Representing lakes
< 75	soft	Nil
75-150	Moderately hard	Mangla
150-300	Hard	Baghsar, Banjosa
> 300	Very hard	Nil

4.4. Residual Sodium Carbonate (RSC)

Residual Sodium Carbonate test places the samples in good category of lake water giving it a value of 0.1-0.851 i.e. <1.25 for 3 lakes; none of the lakes fall in doubtful and unsuitable category [45] (Table 9).

Table 9. *Classification of irrigation water based on residual sodium carbonate.*

RSC (meq/L)	Water quality	Hydrological year (2008-2009) Representing lakes	Total no. of lakes
< 1.25	Good	Baghsar, Mangla, Banjosa	3
1.25-2.5	Doubtful	Nil	Nil
> 2.5	Unsuitable	Nil	Nil

4.5. Hydrochemical Profile

In Piper's diagram, major cation-anion distribution indicates Ca as the predominant ion Ca-HCO_3 followed by Mg-HCO_3. The order being Ca -Mg - HCO_3 > Na-K-HCO_3; for chlorides the distribution is mixed Ca-Na-Mg-K-Cl. The nature of water (Mangla Lake) is more of Ca-Mg- Cl/NO_3 type because of anthropogenic activity from Mirpur city (Fig. 5). The SO_4 ions show a little presence because of a fewer industries and lake of sulphur deposits in the area.

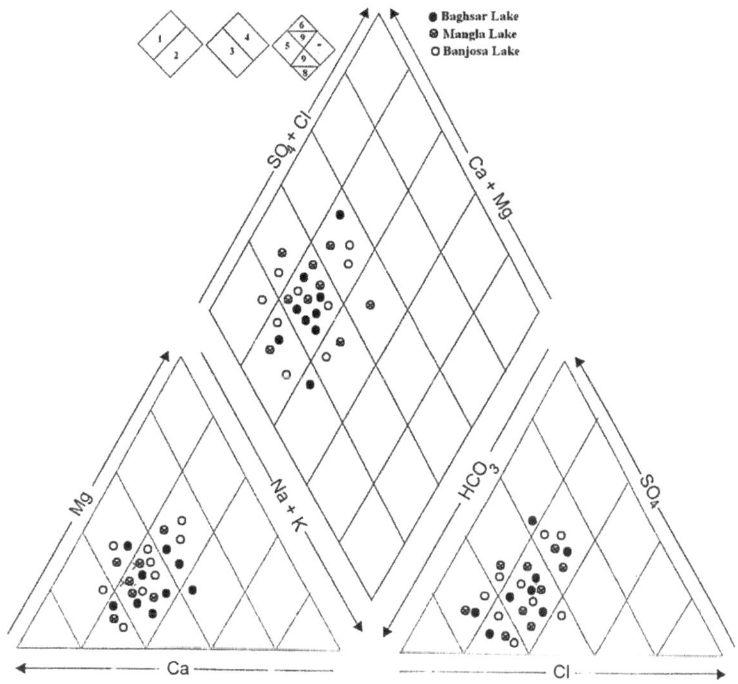

Fig. 5. Hydrochemical profile of lakes water in Piper diagram.

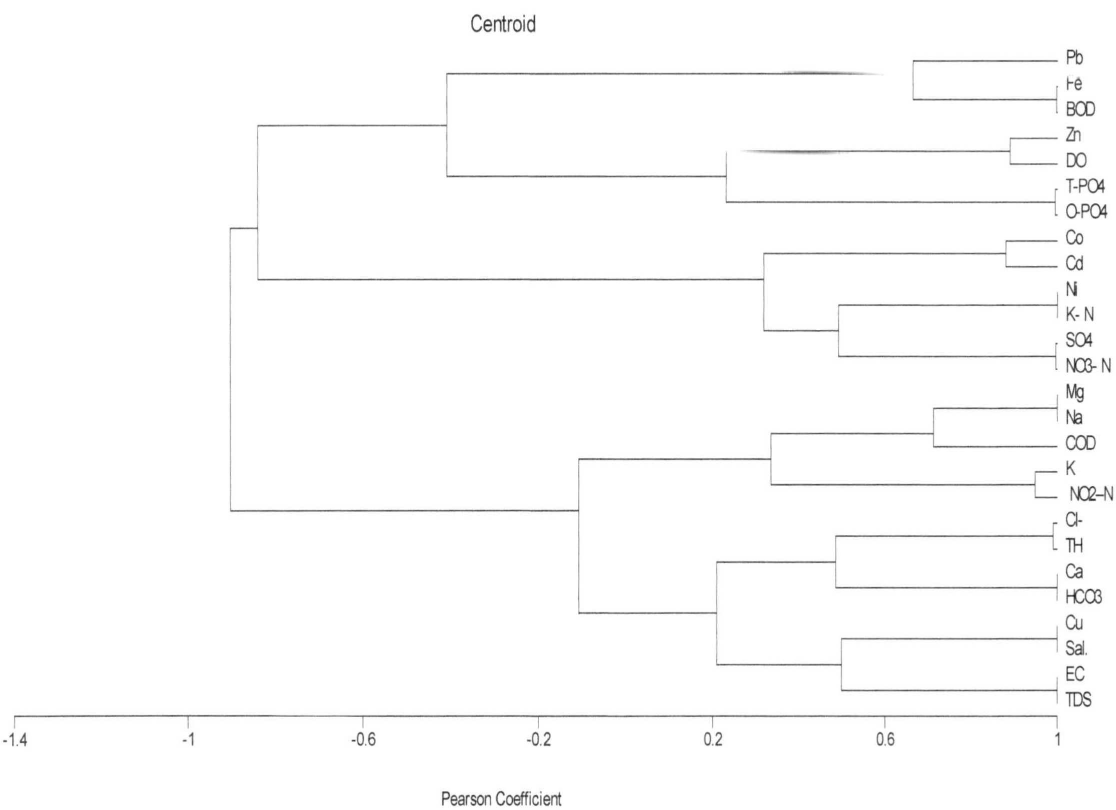

Fig. 6. A dandrogram showing cluster analysis of lakes water in Kashmir.

4.6. Cluster Analysis of Lakes Water

Cluster analysis of 26 parameters including major and trace metals in the water samples was carried out in order to investigate the multivariate distribution as shown in Fig. 6. Five clusters were observed for altogether 26 parameters in water samples; close association of Pb, Fe and BOD with the Zn, DO, O-PO_4 and T-PO_4 in the water shown by first two clusters implied that most of the inputs were associated with the anthropogenic activities with special reference to the Baghsar and Mangla Lakes which are situated within thickly populated areas. Third cluster comprising of Co–Cd, Ni-Kjeldahl nitrogen, and sulfate-nitrate nitrogen showed similar trend of these parameters in the hydrolitho-environment. The anthropogenic contributions were shown by the forth cluster consisting of Na, Mg, COD, K and NO_2-N which might be contributed by the agricultural activities as well as dry deposition of the suspended particulates. Fifth cluster showed close association among six parameters subdivided in to two clusters with three parameters each (Cl^--TH-Ca-HCO_3 and Cu-Sal-EC-TDS) showing predominant contributions of geological formations.

4.7. Correlation Coefficient Matrix (r) of selected Metals

The data on correlation matrix between selected metals is shown in table 10 and the significant r-values at $p<0.05$ are made bold. Very strong correlations were observed between the following elemental pairs: Mg-Na (r=1.00), Co-Ni (r=0.999) whereas K-Na (r=0.890), Ca-K (r=0.911), Mg-K (r=0.884), Pb-Fe (r=0.904), Cu-Na (r=0.779), Cu-K (r=0.979), Cu-Ca (r=0.975), Cu-Mg (r=0.772), Cd-Ni (r=0.859) and Co-Cd (r=0.882) showed strong correlation depicting lithogenically linked interactions or common anthropogenic contamination sources. The metal pairs with negative r-values such as Zn-Na (r=-0.994), Zn-K (r = -0.993) etc representing the mutual exchange of the respective metals with soil/water matrix. The remaining pairs indicated either positive or negative correlation showing their independent variations in the studied matrix.

Table 10. *Correlation coefficient matrix of 14 physico chemical parameters of Kashmir lakes.*

	Na	K	Ca	Mg	Fe	Pb	Cu	Zn	Ni	Cd
K	0.890									
	0.302									
Ca	0.622	0.911								
	0.573	0.271								
Mg	1.000	0.884	0.613							
	0.007	0.309	0.580							
Fe	-0.603	-0.172	0.250	-0.612						
	0.588	0.890	0.839	0.581						
Pb	-0.203	0.266	0.641	-0.214	0.904					
	0.870	0.828	0.557	0.863	0.282					
Cu	0.779	0.979	0.975	0.772	0.030	0.456				
	0.431	0.130	0.142	0.439	0.981	0.699				
Zn	-0.994	-0.933	-0.701	-0.993	0.515	0.098	-0.841			
	0.067	0.234	0.506	0.075	0.655	0.937	0.364			
Ni	-0.051	-0.501	-0.814	-0.039	-0.766	-0.968	-0.666	0.156		
	0.968	0.666	0.395	0.975	0.444	0.163	0.536	0.900		
Cd	0.468	0.014	-0.401	0.479	-0.987	-0.960	-0.189	-0.372	0.859	
	0.690	0.991	0.737	0.682	0.102	0.180	0.879	0.757	0.343	
Co	-0.002	-0.458	-0.785	0.009	-0.796	-0.979	-0.629	0.108	0.999	0.882
	0.999	0.697	0.426	0.994	0.413	0.132	0.567	0.931	0.031	0.312

4.8. Statistical Dimension and Inferences

Data on a few selected parameters was treated to express statistical confidence and inferences drawn thereof, by calculating standard deviation of inter seasonal and inter lake variation in pH, TDS, Cl^-, SO_4^{2-}, Ca^{2+} Na^+, Mg^{2+}, K^+, DO, BOD, COD.

The parameters selected were to reflect:

(i) The in situ environment of the lake.

(ii) Biogeochemical processes occurring deeper in the bed-rock soil, at the water-bed rock interface and, vertically along the water matrix.

(iii) Alterations in the geochemical composition of water due to so called natural pollution and /or anthropogenic pollution.

(iv) Inter-lake comparison.

4.9. The inferences

pH varies from slightly acidic to slightly alkaline indicating moderate levels of photosynthesis regulating CO_2 availability and Ca-CO_3, Ca-SO_4 & Mg-CO_3, Mg-SO_4 equilibrium facilitated by mineralization and cation exchange by organic acids mostly humic and fulvic acids.

Standard deviation (SD) for SO_4 is also randomly distributed indicating chemical weathering of rock salt and its mobility to slightly acidic aqueous phase.

Interlake and inter-season variation of DO (Table 2) show

better precision for Mangla and Baghsar while lesser for Banjosa, BOD is more precise for Banjosa and Baghsar but less for Mangla. COD also repeats BOD pattern, Mangla and Baghsar boast of dual anoxic-conditions of the water-rock soil interface and anthropogenic impact while Banjosa shows aerobic activity at the interface and also deeper in the soil. These trends are of preliminary and tentative nature until more frequent intera-seasonal sampling is done to ascertain intra-seasonal variation and also a better understanding of the nature of biogeochemical processes operating in the water system, and quantify anthropogenic impact and its considerable attenuation by dilution.

A discerning look at SD data (Table 2) brings up a pattern where the SD values for almost all the variables match for any two of the three lakes implying thereby that a combination of any two of the three lakes (i) do not differ in situ environment; (ii) are of similar rock soil composition; (iii) receive same degree of anthropogenic pollution and (iv) undergo equal levels of physical and chemical weathering.

5. Conclusions

Preliminary study of biogeochemical processes of three lakes and their intra and inter lake biogeochemical processes operating in their hydro and lithospheres indicate the existence of sound eco equilibrium between these processes. The natural system is nevertheless severely strained by anthropogenic activity involving variegated land use for urban settlements and industrialization, adding sewage and effluents, into these lakes.

The physicochemical study of some parameters of the lakes show high SD symptomatic of systematic error suggesting for more closely spaced sample taking over the seasons so as to precise the data and carry out more significant data tests and also enhance our ability to monitor and predict the 'straying' of any of the processes and, take remedial measures.

References

[1] M. N. González-Alcaraz, C. H. Miguel, & J. Álvarez-Rogel, When liming and revegetation contribute to the mobilisation of metals: Learning lessons for the phytomanagement of metal-polluted wetlands. Journal of Environmental Management, 15, 72–80, 2013.

[2] J. M. Arocena, J. M. van Mourik, & Á. Faz Cano. Granular soil structure indicates reclamation of degraded to productive soils: A case study in Southeast Spain. Canadian Journal of Soil Science, 92, 243–251, 2012.

[3] F. Degryse, E. Smolders, & R. D. Parker. Partitioning of metals (Cd, Co., Cu, Ni, Pb, Zn) in soils: concepts, methodologies, prediction and applications—A review. European Journal of Soil Science, 60, 590–612, 2009.

[4] C. M. Laluraj, & G. Gopinath. Assessment of seasonal variation of groundwater quality of Pheretic aquifer-A river basin system. Environmental Monitoring and Assessment, 117, 45-57, 2006.

[5] S. Mahboob. Seasonal changes in planktonic life and water chemistry of Ajmal fish farm, Samundri road, Faisalabad. M.Sc. Thesis, Agri. Univ., Faisalabad, 1986.

[6] S. Mahboob, A. N. Sheri, M. B. Sial, & M. Javed. Influence of physico-chemical factors on the dray weight of planktonic biomass in a commercial fish farm. Pakistan Journal of Agricultural Sciences, 25, 271–278, 1988b.

[7] S. Mahboob, A. N. Sheri, M. B. Sial, M. Javed, & M. Afzal. Seasonal changes in physico-chemistry and planktonic life of a commercial fish farm. Pakistan Journal of Agricultural Sciences, 25, 22–27, 1989a.

[8] S. Mahboob. Influence of fertilizer and artificial feed on the growth performance in composite culture of major, common and some Chinese carps. Ph. D. Thesis, Agriculture. University, Faisalabad, 1992.

[9] S. Mahboob, & A. N. Sheri. Effect of fertilizer and artificial feed on the seasonal distribution of planktonic life in fish ponds. Pakistan Journal of Biological Sciences 5, 461–263, 2001.

[10] M. Latif. Comparison of Physico-chemical and biological aspects of two fish ponds fertilized with broiler droppings and N: P: K (25:25:0). M.Sc. Thesis, Agriculture. University, Faisalabad, 1990.

[11] M. Hassan. Effect of pond fertilization by broiler droppings on growth performance and meat quality of major carps. Ph.D. Thesis, Agriculture. University, Faisalabad, 1998.

[12] M. A. Mirza, M. Y. Khuhawar, & R. Arain. Quality and distribution of lead in underground water Kashmir. Asian Journal of Chemistry, 20, 5915-5923, 2008.

[13] M. A. Mirza, M. Y. Khuhawar, & R. Arain, Quality of spring water in the catchment areas of Indus River. Asian Journal of Chemistry, 19, 5279-5304, 2007.

[14] M. A. Mirza, G. M. Arain, & M. Y. Khuhawar. Comparative physicochemical study of five main Ponds of District Bhimber Azad Jammu & Kashmir. Journal of Chemical Society of Pakistan, 28, 430-434, 2006.

[15] M. A. Mirza, G. M. Arain, & M.Y. Khuhawar. Hydrochemistry of Snowmelt in Sudhanoti & Poonch Districts of Azad Kashmir Pakistan. Sindh University Research Journal (Science Series), 38, 25-32, 2006.

[16] M. A. Mirza, G. M. Arain, & M. Y. Khuhawar. Chemistry of Torrents Water in the State of Azad Jammu & Kashmir, Pakistan. The Nucleus, 44, 25-31, 2007.

[17] M. A. Mirza, M. Y. Khuhawar, & A. J. Kandhro. Quality of Rain Water and its Suitability for Drinking and Agricultural use in District Bhimber Azad Kashmir, Pakistan. Pakistan Journal of Analytical Chemistry, 6, 72-79, 2005.

[18] H. A. Muhlhauser, A. Herpic, N. Mladinic, P. N. Montecino, & S. Cabrera. Water quality and limnological features of a high altitude Andean lake, Chungard in northern Chile. Rev. Chil. Hist. Nat., 68, 369–289. 1995.

[19] C. D. Jersabek, & R. Schabetsberger. Limnological aspects of an alpine Karst lake with extreme changes in water level. Limnologica, 26, 1–13, 1996.

[20] http://www.ajk.gov.pk/main/index.html

[21] J. D. Hem. Study and Interpretation of the Chemical Characteristics of Natural Water, USGS Water Supply, p. 2254, 1985.

[22] G. D. Cooke, E. B. Welch, S.A. Peterson & P. R. Newroth. Restoration and Management of Lakes and Reservoirs. 2nd Ed. Lewis Publishers, Boca Raton, FL. 548 pp. 1993.

[23] M. V. Gibbons, H. L. Gibbons, M. D. Sytsma. A Citizens' Manual for Developing Integrated Aquatic Vegetation Management Plans. First Ed. Prepared for Washington Department of Ecology Water Quality Program, 1994.

[24] U. S. Environmental Protection. Agency Volunteer Lake Monitoring: A Methods Manual. EPA 440/4-91-002. 1991.

[25] H. L. Edmiston, V. B. Meyers. Florida Lakes: A Description of Lakes, Their Processes, and Means of Protection. Water Quality Management and Restoration, Florida Department of Environmental Regulation, Tallahassee, Fl. 29 pp. 1993.

[26] North American Lake Management Society. NALMS Management Guide for Lakes and Reservoirs. Alchua, FL. 42 pp, 1989.

[27] M. A. Khan, & D. P. Zutshi. Primary productivity and trophic status of a Kashmir Himalayan lake. Hydrobiologia 68, 3–8, 1980.

[28] D. P. Zutshi, K. K. Vass, Limnological studies on Dal Lake Sirinagar Indian. Natl. Sci. Acad B, Biol. Sci. 48, 234-241, 1982.

[29] S. M. Leghari, A. Bano, M. Y. Khuhawar, & T. M. Jahangir. Limnological Study of Tatta Pani Hot Spring and River Punch at Tatta Pani Vilalge, District Punch, Azad Kashmir, Pakistan. Scientific Khyber, 13, 73-85, 2000.

[30] C. S. Meddlicott. Notes on the Sub-Himalayan series in the Jammu. (Jammoo) Hills. Rec. Geol. Surv. (India). 9, 49-57, 1876.

[31] R. Lydekker. The Geology of Kashmir and Chamba territories and the British district of Kaghan. Mem. Geol. Surv. India, 22, 1-334.1883.

[32] C. C. Middlemiss. The geology of Hazara and the Black Mountains. Mem. Geol. Surv. India. 26, 1-302, 1896.

[33] D. N. Wadia. North Punjab and Kashmir. Rec. Geol. Surv. India, 62,152-156, 1928.

[34] M. Ashraf, M. N. Chaudhry, & K. A. Qureshi. Stratigraphy of Kotli area of Azad Kashmir and its correlation with standard type area of Pakistan. Kashmir Jour. Geol.1, 19-30, 1983.

[35] K. Kumar, D. S. Rawat, R. Joshi. Chemistry of spring water in Almora, Central Himalaya, India. Environ. Geol. 31(3/4), 150-156, 1997.

[36] APHA, Standard Methods for Examination of Water and Wastewater, American Public Health Association, Washington DC, edn. 17, 1497. 1989.

[37] F. J. Stevenson. Humus Chemistry, Willey Interscience Publications, New, 1982.

[38] M. M. Kononova. Soil organic Matter: Its Nature, Its role in Soil Fertility, 2nd Ed., Pergammon, Oxford, 1966.

[39] A. Saleh, F. Al-Ruwaih, M. Shehata. Hydrogeochemical processes operating within the main aquifers of Kuwait. J. Arid Environ. 42, 195-209,1999.

[40] G. Gilly, G. Corrao, & S. Favilli. Concentrations of nitrates in drinking water and incidence of gastric carcinomas. First descriptive study of the Piemonate Region, Italy. Science of the Total Environment, 34, 35–48, 1984.

[41] K. R. Karanth. Groundwater assessment, development and management. Tata McGraw Hill New Delhi, pp720, 1987.

[42] H. M. Ragunath. Groundwater. Wiley Eastern Ltd, New Delhi. pp563. 1987.

[43] Collins, A. Jenkins, The impact of agricultural land use on stream chem-istry in the middle Hills of the Hima-layas, Napal. J Hydrol 185,71–86,1996.

[44] C. J. Pilbeam, P. J. Gregory, R. C. Munankarmy, & B. P. Tripathi. Leaching of nitrate from cropped rainfed terraces in the mid-hills of Nepal. Nutrient Cycling in Agroecosystems, 69, 221-232, 2004.

Simulation of Potassium Transport in Carbonate Aquifer

Omar Chaalal[1], Ahmed Murad[2], Ahmed M. Soliman[3], Rafiq Islam[5], Ismail A. El Haty[3], D. Hank[4]

[1]Chemical and Petroleum Engineering Department, College of Engineering, United Arab Emirates University, Al-Ain, United Arab Emirates
[2]Geology Department, College of Science, United Arab Emirates University, Al-Ain, United Arab Emirates
[3]Chemistry Department, College of Science, United Arab Emirates University, Al-Ain, United Arab Emirates
[4]Nuclear Fuel Technology Department, Hot Labs. Centre, Atomic Energy Authority., Cairo, Egypt
[5]National Polytechnic School of Algiers, Algeria

Email address:
omar.chaalal@uaeu.ac.ae (O. Chaalal), ahmed.murad@uaeu.ac.ae (A. Murad), a.soliman@uaeu.ac.ae (A. M. Soliman), ismailelhaty@uaeu.ac.ae (I. A. El Haty), d.Hank@hotmail.com (D. Hank)

Abstract: *Sophisticated* prediction of aquifer performance requires numerical simulation. To date, no comprehensive simulation has been reported on groundwater modeling. Most available simulators are not applicable for fractured aquifer, and do not account for contaminant leaching and degradation, particularly in the vadose zone. Consequently, studying contaminant transport in a fractured or vuggy formation offers a forthidable challenge. This paper addresses the problem of contaminant transport in carbonate aquifer, in the presence of fractures. Most of aquifers in UAE are of limestone or carbonate origins. A series of experiments was conducted using potassium nitrate as the contaminant. *Dynamic* adsorption and desorption tests were carried out using both homogeneous and fractured formation models. Initial modeling and experiments were carried out for a range of initial concentration values. The concentration at the outlet was measured with the Flame Ionization technique. A numerical model was developed using the surface excess theory, combined with a non-Fickian dispersion coefficient. Numerical results agreed favorably with experimental results. It was found that the non-Fickian model was necessary for modeling fracture flow results and with this version, there was no need to use the dual porosity/dual permeability formulation. Strong dependence of adsorption on initial concentration was observed and was justified with the numerical model.

Keywords: Carbonate Aquifer, Modling, A non-Fickian Dispersion Coefficient, Potassium Nitrate, Dual Porosity and Numerical Simulation

1. Introduction

In the past three decades, fractured rock domains have received increasing attention by the researchers from a number of disciplines, including hydrogeology, petroleum engineering, and environmental engineering. The importance of fractures is particularly enhanced when one deals with carbonate formations, both in water flow and petroleum production, because most of these formations are known to be fractured. The subject has been investigated by the petroleum engineers in connection with multiphase flow, because many important petroleum reservoirs are in fractured rock formations. Of special interest are reservoirs composed of fractured porous rocks in which the matrix blocks, surrounded by the network of fractures are porous. The permeability of such blocks are often rather low, but the porosity and, hence, the storage capacity for fluids are very high.

Environmental engineers have been investigating the problem in connection with geological isolation of radioactive waste and remediation of ground water. Hydrologists deal with fractured formations because numerous deep aquifers are fractured. Chemical engineers mainly deal with chemical transport in limestone, but not necessarily in its fractured form (Arsic et al., 1991; Couturier et al., 1993).

In Canada, Switzerland, and Sweden large national research programs are underway to study radio-nuclide transport in crystalline rocks. Finland also intends to site their repository in similar rocks and has a research program directed toward this end. Other countries, including Japan, France, Spain, the United Kingdom, and the USA have or have had research programs oriented towards crystalline rocks. In addition, an international field of research program is in progress in the Stripa mine, Sweden. The research is aimed at understanding the processes that determine the flow paths and the flow rates of water in the fractured rock masses.

Solutes that are dissolved in water will be carried by the moving water, but various other mechanisms are involved in the process of solute transportation. Small molecules or ions diffuse in a concentration gradient and can move from one 'stream tube' to another.

Different water volumes move with different velocities and may mix at more or less regular intervals. The dissolved species may also experience kinetic effects caused by physical processes. One of such processes that may have a large impact for solute transport in fractured rock is the diffusion in and out of zones in which the water is moving so slowly that it can, for all practical purposes, be assumed to be stagnant. Such stagnant zones can be expected in fractures with uneven surfaces and with fracture-filling materials. In rocks with a connected matrix can be very much larger than the mobile water volume in the fracture. Because the stagnant zones are reached by diffusion, the volume of stagnant water accessed depends on the residence time of the flowing water.

Different approaches to solute transport in fractured media have been attempted. These include the analysis of a single fracture (e.g. Grisak and Pickens, 1981; Tang et al., 1981; Sudicky and Frind, 1984). While some researchers considered the porous medium as comprised of equally-spaced fractures (Sudicky and Frind, 1984), others have used deterministic study of flow in an irregular network of fractures contained in an impermeable rock. Yet others (Schwartz and Smith, 1988) considered the same problem with a stochastic model. Recently, Choi et al. (1997) used a dual-permeability, dual-porosity approach to model fluid flow through a fractured formation. They built their model using non-Darcian flow through the fracture while allowing Darcian flow through the matrix. Even though this work was devoted to advancing the hydrodynamic aspect of fluid flow, no species transport was considered. Adsorption and desorption are important phenomena in chemical transport. A Langmuir isotherm has been used most commonly to model adsorption at the solid/liquid interface. The widespread use of the Langmuir model arises from its ability to represent the general shape of many adsorption isotherms. However, more recently, the surface excess theory has been proven to be more theoretically sound and better apt in modeling adsorption (Islam and Chakma, 1991).

Furthermore, Sarwar and Islam (1997) demonstrated that the surface excess theory is also capable of representing chemical transport through fractured formations. However, non-Fickian dispersion had to be introduced, along with non-equilibrium adsorption and desorption rates. They considered the transport several chemicals in silica sand formation and observed very good agreement between experimental and numerical results, even for the regime with non-uniform adsorption and desorption profiles. Recently Zargham et al (2011) have described the adsorption behavior in a system of solid phase carbonate and of an aqueous phase of surfactant. They found that the rate of adsorption dependents on availability of surfactant in the system and the adsorption of surfactant increased with increasing surfactant concentration. This research comprises of laboratory flow experiments and numerical modeling of KNO3 transport in carbonate formations. Such flow is of relevance to agricultural applications for which run-offs from fertilizers constitute major contamination problems for groundwater. Also, in petroleum engineering applications, mathematical models of chemical transport are of importance, especially for the case of fractured carbonate formations.

2. Experimental Setup and Procedure

A glass column was used for all experimental tests. Because only hydrostatic pressure was used to maintain flow through the porous medium, little pressure was anticipated in the column and a glass column was sufficient. Natural carbonate rocks were collected from the outcrops of Al-Ain (near Jabel Hafeet). These samples were crushed in order to facilitate packing of this material in the column. The porosity of the crushed rock was approximately 15%. This was considered to be the result of adequate packing. The packed column was connected to an inlet (for injection of the contaminated water) and was released to the outside at atmospheric pressure. A two-folded aluminum foil was placed parallel to the axis of the cylinder to model a single fracture. The foil contains approximately 50 holes (performed with a 0.05 mm needle) per cm^2, in order to assure cross flow as well as fracture flow. This was the procedure previously implemented by Sarwar and Islam (1997).

The Nitrate used was a solution of KNO_3. This form of the potassium is considered to be difficult to remove from a water stream. Two different concentrations of nitrite, 50 and 150 ppm, were used in this study. Standard solutions were prepared using laboratory-grade KNO_3. Deionized water was used for all the experiments, as a solvent (adsorption tests) as well as a washing agent (desorption tests).

2.1. Procedure

In order to observe both adsorption and desorption, contaminated water was injected in each sample and was followed by fresh water after a high number of pore volumes of fluid injection. During this process, samples were collected at the outlet and were analyzed for potassium content in the effluent. The concentration at the outlet was measured with a Digital Flame Analyzer (Model 2655000, as supplied by Cole Palmer Instrument Co., Chicago, USA). This instrument was made available by the Water and Electricity Laboratory of Al-Ain in UAE.

2.2. Theory

Adsorption of a chemical is considered to take place from

the monomer phase. Surface excess is the measure of adsorption from liquid mixtures that describes the difference between the actual amount of component *i* adsorbed and the amount of component *i* that would be present in the adsorbed phase if the composition and the density of that phase were same as those of the phase in equilibrium. For a binary system, the surface excess of component *i* is given by:

$$n_i^e = n^o(x_i^o - x_i), \quad (1)$$

where *i*=1 denotes chemical (*i*=1) and *i*=2 denotes the bulk phase carrying the chemical. If it is assumed that the liquid phase is directly in contact with the solid, the above equation assumes the following form

$$n_i^e = n'(x_i' - x_i), \quad (2)$$

Using the monomer adsorption model, the surface excess can be expressed as

$$\frac{1}{n'} = \frac{x_1'}{m_1} + \frac{x_2'}{m_2} \quad (3)$$

where m_1 and m_2 represent the amounts of monolayer coverage for the chemical and the bulk phase, respectively. At this point the concept of selectivity can be introduced. Selectivity, *S*, is defined as follows:

$$S = \frac{x_1'/x_2'}{x_1/x_2} \quad (4)$$

Even though the monolayer model is based on the assumption of monolayer adsorption, it has been taken for granted that the monolayer model is also suitable for multi-layered adsorption (Basu et al. 2006; Song and Islam, 1994). With selectivity, the expression of surface excess can be written as

$$n_a^e = \frac{m_1 x_1 x_2 (S-1)}{Sx_1 + (m_1/m_2)x_2} \quad (5)$$

The amount of chemical adsorbed is given by

$$n_1' = \frac{m_1 x_1 S}{Sx_1 + (m_1/m_2)x_2} \quad (6)$$

Finally, the kinetic term is calculated as:

$$\frac{\partial n_1^e}{\partial i} = k_i(n_1^e - n_1^{ea}) \quad (7)$$

where $k_i = k_1$ for desorption. This term introduces the possibility of desorption of a chemical in a dynamic displacement system.

3. Governing Equation

The one-dimensional mass transfer equation that includes the surface excess theory to describe adsorption in the solid/liquid interface is given as (Song and Islam, 1994):

$$\frac{\lambda q \rho_1}{A\phi}\frac{\partial^2 C}{\partial z^2} - \frac{q\rho_1}{A\phi}\frac{\partial C}{\partial z} = \frac{1-\phi}{\phi}\rho_r n_o \frac{\partial C}{\partial i} + \frac{1-\phi}{\phi}\rho_r\left(\frac{\partial n_1^{ea}}{\partial i}\right) \quad (8)$$

4. Numerical Solution Method and Accuracy

In order to discretize Eq. (8) the numerical scheme of Barakat and Clark (1966) was used. This scheme has been proven to be powerful in solving diffusive convection equation in the past. Numerical simulation was carried out in a linear grid system of 100 grid blocks. This number was selected after a series of trial runs for optimum accuracy. Time was discretized with a Δt of 50 seconds. This value was found to be appropriate considering that the spatial grid blocks were rather small. The following boundary and initial conditions were used:

$$C(z,0) = 0, \text{ for } 0 \le z \le L$$

$C(0,t) = C_o + \lambda(\partial C/\partial z)$, for the adsorption test. For desorption, C_o is set to zero.

$$\left.\frac{\partial C}{\partial z}\right|_{z=L} = 0, \text{ for } t>0$$

Table 1. *Input Data*

C_o	50, 100 ppm	λ	Variable, see Table 2
K	11.3 mD	m_1/m_2	15
n_o	200 mg/g	S	25
Q	10 ml/min	m_1	$0.05n_o$
Φ	0.15	k_1	2/hr
ρ_i	1 cm^3/ml	k_2	Variable, see Table 2
ρ_r	2.8 g/ml	PV	22 cm^3

Numerical simulation runs were conducted in order to obtain best match between experimental and numerical results. The best match data are reported in Table 1. Note that the dispersion parameter was found to be a function of concentration. This aspect will be discussed later.

5. Experimental Results and Discussion

Eight experimental runs were conducted with KNO_3 as the chemical with two different concentrations (50 and 150 ppm). Description of each run is given in Table 2. For each concentration, four runs were conducted in absence or presence of fracture and adsorption and desorption. Following is a discussion of various experimental runs.

Figure 1 shows the results of Runs 1 and 2 (adsorption and desorption with 50 ppm solution). These two runs were conducted on the same porous material. Adsorption test was simply followed by the desorption test that involved the injection of de-ionized water. As can be seen from Fig. 1, the outlet concentration rose to the same level as that of the inlet only after the injection of 10 pore volumes (PV) of water. This value is relatively higher as compared to the case using silica sand, for which saturation (to a dimensionless concentration of unity) was reached within two pore volumes of chemical injection (Sarwar and Islam, 1997). Indeed, carbonate formations exhibit much higher adsorptive capacity than do silica sands. Crushed carbonate has much greater surface area than silica sand. This is reflected in the fact that the permeability of crushed carbonate is several hundred times smaller than that of silica sand. The outlet concentration is measured in terms of potassium content. It was not possible in this study to determine the fate of NO_2^- in this study. It might be interesting to find the partition of NO_2^- in the solid and the liquid phases. The desorption trail in Fig. 1 shows much faster decline in outlet concentration than that of the adsorption curve. This was in contrast to silica sand for which slower desorption was observed (Sarwar and Islam, 1994). However, only limited amount of chemical was injected in their case and most likely nothing more than a monomer layer was allowed to be formed.

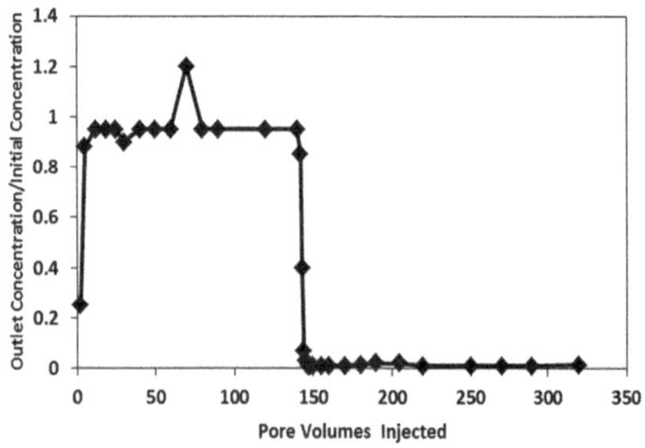

Fig. 1. Chemical adsorption and desorption in absence of fracture (initial concentration =50 ppm).

Figure 2 shows the adsorption and desorption profiles of Runs 3 and 4, respectively. These runs were conducted with 50 ppm nitrite solution during the adsorption phase. However, a single fracture was added to the system. This resulted in a smaller pressure drop across the porous medium in order to maintain the same flow rate as the homogeneous case (Fig. 1). However, the chemical transport was not particularly affected by the presence of the fracture. The saturation (the point of reaching concentration ratio of unity) took place at approximately the same time as that of the homogeneous case. It is conceivable that fractures will play a greater role at higher flow rates. This aspect of the study is currently underway.

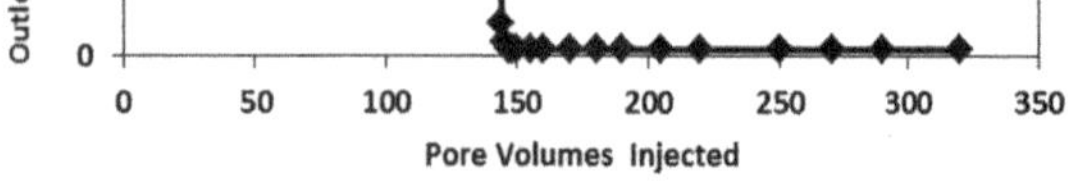

Fig. 2. Chemical adsorption and desorption in presence of fracture (initial concentration =50 ppm).

Figure 3 shows adsorption and desorption results of Runs 3 and 4 that were conducted with 150 ppm solution. The adsorption appears to reach steady state at a much slower pace than the one reached by lower concentration cases. Adsorption indeed depends on the concentration gradient that is likely to be higher in this case than in the case of lower absolute concentration. This effect, however, cannot be observed unless a dynamic test is performed as done in this study. The effect of concentration on desorption is less pronounced.

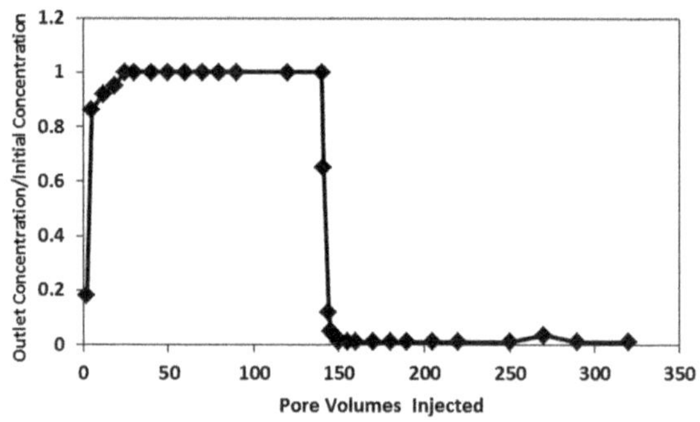

Fig. 3. Chemical adsorption and desorption in absence of fracture (initial concentration =150 ppm).

Figure 4 shows the adsorption and desorption profiles for the 150 ppm in presence of a fracture. For this case, the effect of fracture is clearly shown. The whole adsorption curve is moved to the left, in effect showing that chemical transport remain faster than the homogeneous case. Also, fractures play a greater role in the presence of higher concentration of the contaminant. This contribution of fractures can be modeled by non-Fickian dispersion (Sarwar and Islam, 1994). It is also conceivable that the adsorption and desorption rates are also affected. This will be discussed in the following section on numerical modeling.

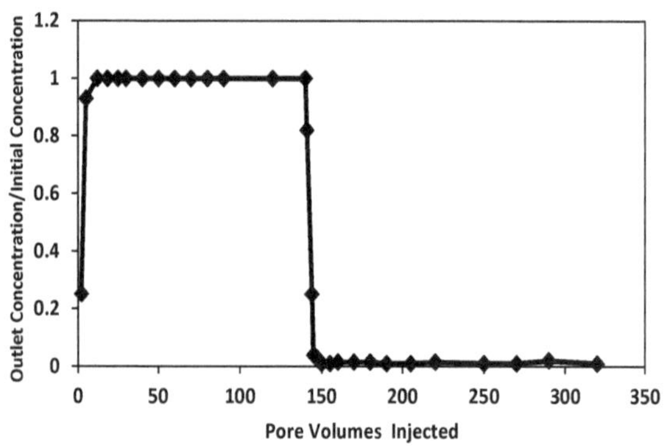

Fig. 4. Chemical adsorption and desorption in presence of fracture (initial concentration =150 ppm).

6. Discussion on Mathematical Modeling

Numerical simulation runs were conducted individually for each section of adsorption and desorption separately. This helped determine the adsorption and desorption rates independently. Following is a discussion on the mathematical modeling results.

Figure 5 compares numerical and experimental results of Run 1 (adsorption with 50 ppm solution, in absence of fracture). The agreement between experimental and numerical results is excellent. Resulting best-fit parameters are listed in Table 1. Numerical modeling of such a process is further consolidated if the breakthrough values could be monitored. Unfortunately, breakthrough values were not monitored in the experimental study.

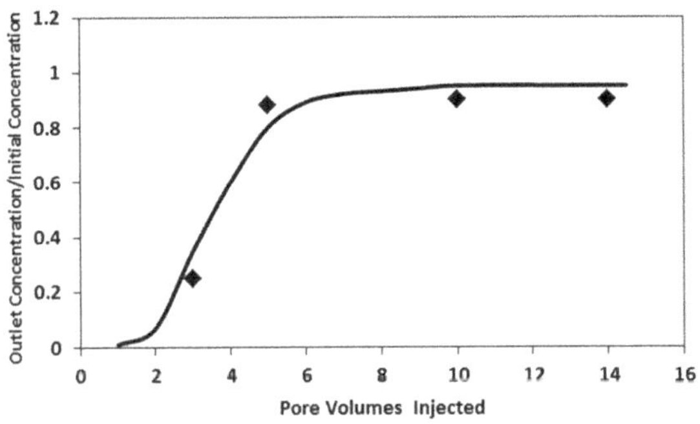

Fig. 5. Comparison of experimental and numerical results of an initial concentration of 50 ppm (no fracture).

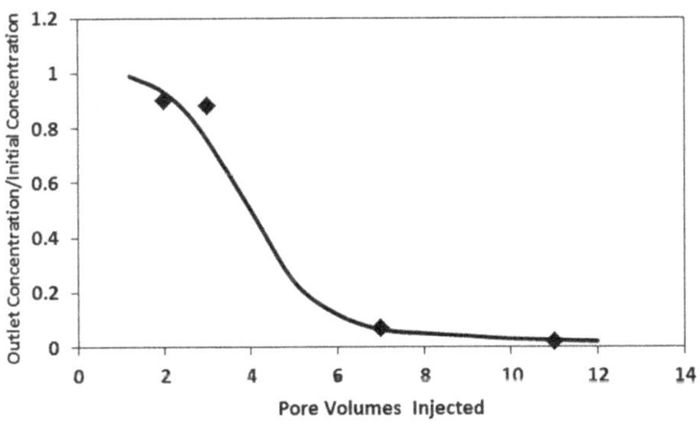

Fig. 6. Comparison of experimental and numerical results of an initial concentration of 50 ppm (no fracture).

The experimental and numerical results of Run 2 (desorption with 50 ppm solution in absence of a fracture) are shown in Fig. 6. A much lower desorption rate compared to adsorption rate had to be used in order to match the experimental desorption data. It commonly found that the adsorption rate is higher than the desorption rate. This, in fact, forms the basis for irreversible thermodynamics and the surface excess theory. In order to obtain a better match between experimental and numerical results, the dispersion parameter used in the desorption phase was somewhat smaller than that of the adsorption phase. It is expected that the dispersion of fresh water into contaminated water (desorption test) be different than the dispersion of contaminated water into fresh water (adsorption test). This process is further complicated by the presence of tightly packed crushed carbonate rocks. No other variable was adjusted in order obtain the match between experimental and numerical results. Also, these numbers are significantly different from those of silica sand (Sarwar and Islam, 1994). Note, however, that the concentration of nitrate used by Sarwar and Islam was much lower than the one used in this study.

Figure 7 compares experimental and numerical results of Run 3. For this run, an adsorption test was performed using a concentration of 150 ppm, in absence of fractures.

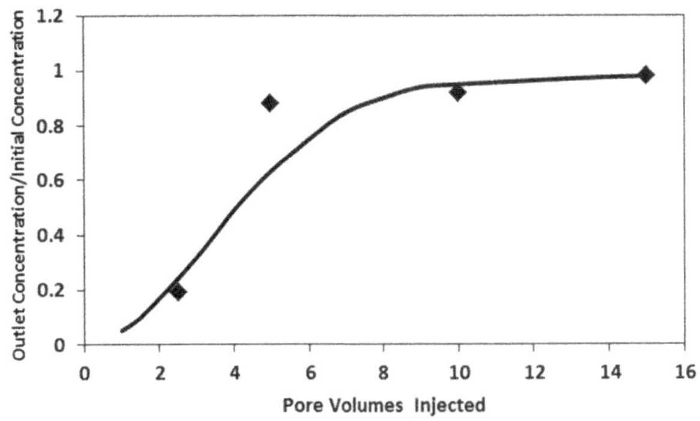

Fig. 7. Comparison of experimental and numerical results of an initial concentration of 150 ppm (no fracture).

For this numerical run no parameter was adjusted. Note that the agreement between experimental and numerical results is remarkably good, especially considering that no parameter was adjusted. In order to obtain a good match between experimental and numerical results of the desorption test, the dispersion rate was first modified. This alone did not improve the agreement that showed elongated desorption trail. In order to remedy this discrepancy, the rate of desorption was modified (see Table 2). It is not unusual to find that the desorption rate, be a function of concentration (Sarwar and Islam, 1994). It is particularly important to realize that at higher concentrations, the assumption of monolayer may not applicable, in which case desorption will depend on the concentration, rather than on the surface properties of the solid. Figure 8 shows that the agreement between experimental and numerical results is greatly improved by modifying the desorption rate.

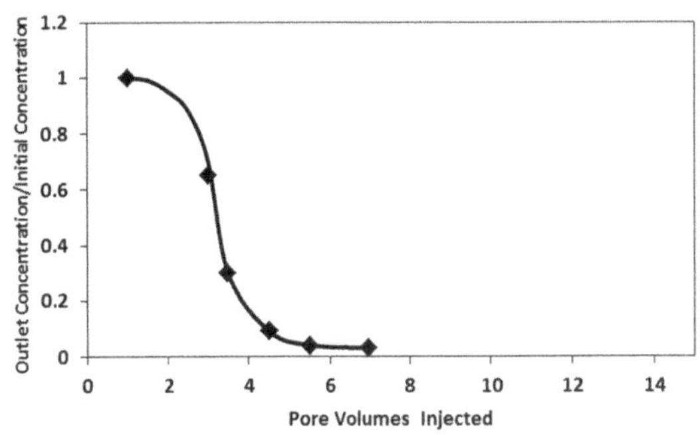

Fig. 8. Comparison of experimental and numerical results of an initial concentration of 150 ppm (no fracture).

Figure 9 compares experimental and numerical results of Run 5 for which 50 ppm solution was used to conduct the adsorption test in presence of a single fracture. Excellent agreement between experimental and numerical results is obtained only after modifying the dispersion rate to one third of that of the previous adsorption tests. The presence of fracture invokes an increase in permeability, and hence creates a channel for chemical transport. This can result in decreased adsorption (due to bypassing of chemicals) or a decrease in dispersion (as convection is affected by an increase in permeability). Note that the adsorption or desorption rates were not changed due to the presence of the fracture. An excellent agreement between experimental and numerical results can be seen in Fig. 10. The agreement is particularly good during the quick change in outlet concentration for during which the dynamic phenomena are most active. Considering that only the dispersion parameter was adjusted for this desorption case, the agreement between experimental and numerical results is impressive.

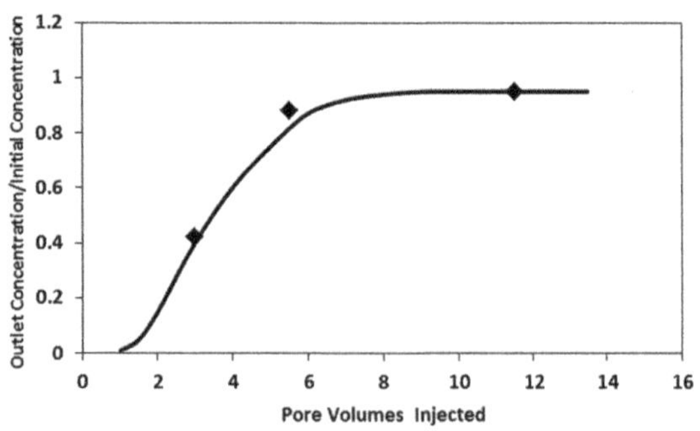

Fig. 9. Comparison of experimental and numerical results of an initial concentration of 50 ppm (with fracture).

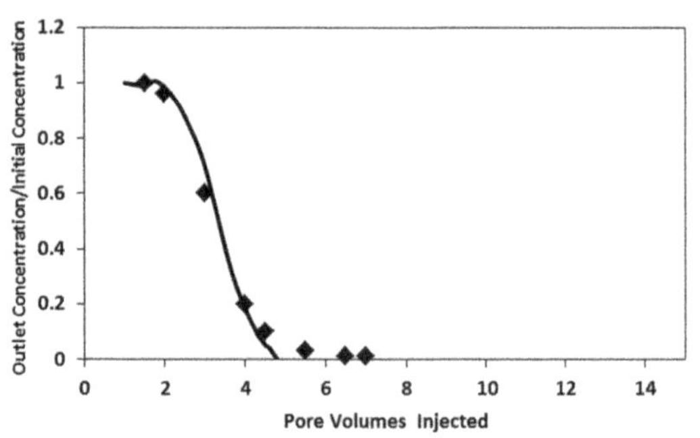

Fig. 10. Comparison of experimental and numerical results of an initial concentration of 50 ppm (with fracture).

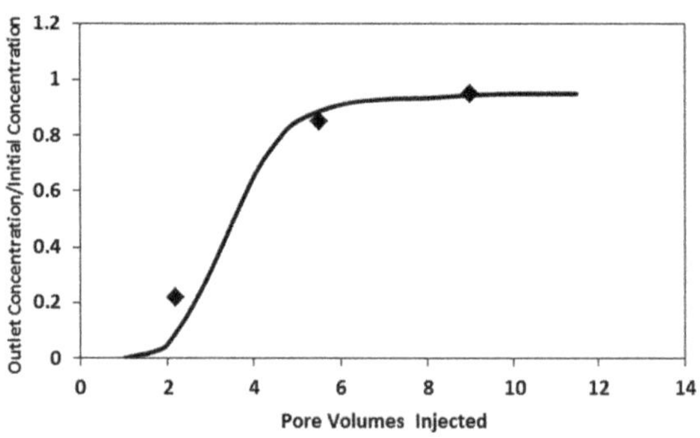

Fig. 11. Comparison of experimental and numerical results of an initial concentration of 150 ppm (with fracture).

Figure 11 shows the comparison between experimental and numerical for Run 7 that was conducted with a 150 ppm solution using a fractured formation. Figure 12 shows the same for the desorption case. Note that the agreement between experimental and numerical results is obtained without making any adjustment to the adjustable parameters (see Table 2).

***Table 2.** Experimental Runs*

Run no.	Λ	K_2	Description of runs
1	3	0.03	50 ppm, adsorption test, no fracture
2	1.1	0.03	50 ppm, desorption test, no fracture
3	3	0.03	150 ppm, adsorption test, no fracture
4	0.25	0.01	150 ppm, desorption test, no fracture
5	1	0.03	50 ppm, adsorption test, with fracture
6	0.1	0.03	50 ppm, desorption test, with fracture
7	1	0.03	150 ppm, adsorption test, with fracture
8	0.1	0.01	150 ppm, desorption test, no fracture

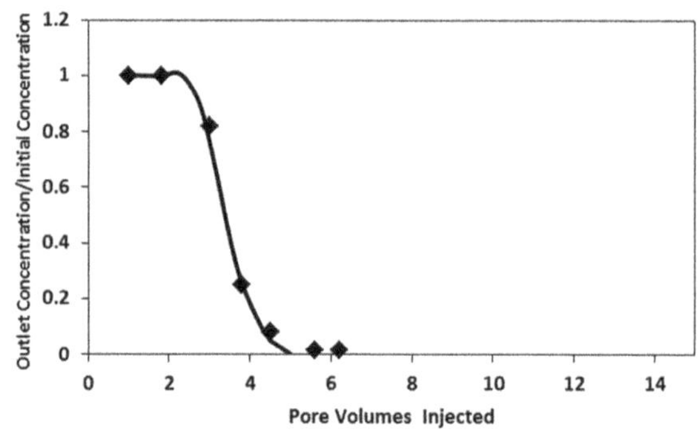

Fig. 12. Comparison of experimental and numerical results of an initial concentration of 150 ppm (with fracture).

7. Conclusions

A series of experimental and numerical runs was conducted in order to describe nitrite flow in a carbonate formation with or without the presence of a linear fracture. Experimental results show that the kinematics of adsorption and desorption depend on the concentration of the solute, leading to the use of a non-Fickian dispersion model. This model is coupled with the surface excess theory and showed excellent agreement with experimental results with minimum number of adjustable parameters.

Acknowledgments

Partial funding of this research was provided by the United Arab Emirates University [College of Science (21S065-COS/IRG-09/13)]

Nomenclature

A= cross sectional area of the core, cm^2

C= concentration of the chemical, %wt

C_o= concentration of the chemical , %wt

k_i= kinetic constant of adsorption (i=1) or desorption (i=2), 1/hr

k= absolute permeability of the porous medium, mD

L= length of the porous bed, cm

m_i= monolayer coverage of component i per unit of adsorbent, mg/g

$n°$= amount of liquid per unit mass of adsorbent, mg/g

n'= amount adsorbed per unit mass of adsorbent, mg/g

n_i'= amount of component i adsorbed per unit mass of adsorbent, mg/g

$n^e{}_i$ = equilibrium surface excess of component i per unit mass of adsorbent, mg/g

n_i^{ea}= actual surface excess of component i per unit mass of adsorbent, mg/g

PV= pore volume, $A\phi L$, cm^3

q= flow rate, ml/hr

S= selectivity, dimensionless

t= time, s

u= local speed, cm/s

x_i= bulk phase mass fraction of component i

x_i'= adsorbed phase mass fraction of component i

z= distance along porous bed, cm

a= entrainment rate constant, 1/cm

λ= dispersion parameter, cm

ϕ= porosity of the rock

ρ_l= liquid density, mg/ml

ρ_r = rock density, g/m1

References

[1] Arsic, B., Oka, S., and Radovanovic, M., 1991, "Characterization of Limestones for SO2 Absorption in Fluidized Bed Combustion", FBC Technology and the Environmental Challenge, Hilger, London, p. 171.

[2] Barakat, H.Z. and Clark, J.A., 1966, "On the Solution of the Diffusion Equations by Numerical Methods", ASME Trans. Heat Transfer, vol. 88, 83.

[3] Basu, A. , Mustafiz, S., Islam, M. R., Bjordalen, N., Rahman M. S. and Chaalal, O. 2006, "A Comprehensive Approach for Modeling Sorption of Lead and Cobalt Ions through Fish Scales as an Adsorbent" Chem. Eng. Comm., 193:580–605.

[4] Choi, E.S., Cheema, T.J., and Islam, M.R., 1997, "A New dual porosity/dual permeability model with non-Darcian flow through fractures", J. Petroleum Science and Engineering, vol. 17, 331-344.

[5] Couturier, M.F., Karidio, I., and Steward, F.R., 1993, "Study on the Rate of Breakage of Various Canadian Limestones in a Circulating Transport Reactor", Circulating Fluidized Bed Technology IV, A.A. Avidan, ed., Amer. Inst. Chem. Eng., New York, p. 672.

[6] Islam, M.R., and Chakma, A., 1991, "Mathematical Modelling of Enhanced Oil Recovery by Alkali Solutions in the Presence of Cosurfactant and Polymer", J. Pet. Sci. Eng., vol. 5, 105-126.

[7] Sarwar, M., and Islam, M.R., 1997, "A Non-Ficician Surface Excess Model for Chemical Transport Through Fractured Porous Media", Chem. Eng. Comm., vol. 160, 1-34.

[8] Schwartz, F.W., and Smith, L., 1988, "A Continuum Approach for Modeling Mass Transport in Fractured Media", Water Resources Res., vol. 24 (8), 1360-1372.

[9] Song, F.Y. and Islam, M.R., 1994, "Effect of Salinity and Rock Type on Sorption Behavior of Surfactants as Applied in Cleaning of Petroleum Contaminants", J. Pet. Sci. Eng., vol. 10 (4), 321-336.

[10] Sudicky, E.A., and Frind, E.O., 1982, "Contaminant Transport in Fractured Porous Media: Analytical Solution for a Single Fracture", Water Resources Res., vol. 21 , 1677-1683.

[11] Tang, D.H., Frind, E.O., and Sudicky, E.A., 1981, "Contaminant Transport in Fractured Porous Media: Analytical Solution for a System of Parallel Fractures", Water Resources Res., vol. 17 (3), 555-564.

[12] Zargham Salari, Mohammad Ali Ahmadi, Riaz Kharrat and Abbas Abbaszadeh Shahri 2011 "Experimental Studies of Cationic Surfactant Adsorption onto Carbonate Rocks" Australian Journal of Basic and Applied Sciences, 5(12): 808-813.

Permissions

All chapters in this book were first published in AJEP, by Science Publishing Group; hereby published with permission under the Creative Commons Attribution License or equivalent. Every chapter published in this book has been scrutinized by our experts. Their significance has been extensively debated. The topics covered herein carry significant findings which will fuel the growth of the discipline. They may even be implemented as practical applications or may be referred to as a beginning point for another development.

The contributors of this book come from diverse backgrounds, making this book a truly international effort. This book will bring forth new frontiers with its revolutionizing research information and detailed analysis of the nascent developments around the world.

We would like to thank all the contributing authors for lending their expertise to make the book truly unique. They have played a crucial role in the development of this book. Without their invaluable contributions this book wouldn't have been possible. They have made vital efforts to compile up to date information on the varied aspects of this subject to make this book a valuable addition to the collection of many professionals and students.

This book was conceptualized with the vision of imparting up-to-date information and advanced data in this field. To ensure the same, a matchless editorial board was set up. Every individual on the board went through rigorous rounds of assessment to prove their worth. After which they invested a large part of their time researching and compiling the most relevant data for our readers.

The editorial board has been involved in producing this book since its inception. They have spent rigorous hours researching and exploring the diverse topics which have resulted in the successful publishing of this book. They have passed on their knowledge of decades through this book. To expedite this challenging task, the publisher supported the team at every step. A small team of assistant editors was also appointed to further simplify the editing procedure and attain best results for the readers.

Apart from the editorial board, the designing team has also invested a significant amount of their time in understanding the subject and creating the most relevant covers. They scrutinized every image to scout for the most suitable representation of the subject and create an appropriate cover for the book.

The publishing team has been an ardent support to the editorial, designing and production team. Their endless efforts to recruit the best for this project, has resulted in the accomplishment of this book. They are a veteran in the field of academics and their pool of knowledge is as vast as their experience in printing. Their expertise and guidance has proved useful at every step. Their uncompromising quality standards have made this book an exceptional effort. Their encouragement from time to time has been an inspiration for everyone.

The publisher and the editorial board hope that this book will prove to be a valuable piece of knowledge for researchers, students, practitioners and scholars across the globe.

List of Contributors

Nweke Chinenyenwa Nkeiruka and Nwabanne Joseph Tagbo
Department of Chemical Engineering, Nnamdi Azikiwe University, PMB 5025, Awka, Nigeria

Dejen Yemane Tekle and Azeb Gebresilassie Tesema
Mekelle University, College of Health Science, Department of Public Health, Mekelle, Ethiopia

Abebe Beyene Hailu
Jimma University, College of Public Health and Medical Sciences, Department of Environmental Health Science and Technology, Jimma, Ethiopia

Taffere Addis Wassie
Addis Ababa University, Ethiopian Institute of Water Resource, Addis Ababa, Ethiopia

Amin Kalantarifard, Eun-Song Byeon, Jo Gwanggon and Go Su Yang
Department of Environmental Engineering, Chonbuk National University, Jeonju, South Korea

Shumpei Kitazaki, Kei Nakagawa and Hiroshi Asakura
Graduate School of Fisheries Science and Environmental Studies, Nagasaki University, Nagasaki, Japan

Tomonori Kindaichi
Graduate School of Engineering, Hiroshima University, Hiroshima University, Higashihiroshima, Japan

Anup Prakash Upadhyay and Arvind Bijalwan
Indian Institute of Forest management, Nehru Nagar, Bhopal, Madhya Pradesh, India

Ibiang Young Bassey, Usanga Daniel Archibong and Williams Uduak Ita
Department of Genetics and Biotechnology, University of Calabar, PMB 1115, Calabar, Nigeria

Ekanem Bassey Effiong
Department of Genetics and Biotechnology, University of Calabar, PMB 1115, Calabar, Nigeria

Department of Science Technology, Akwa Ibom State Polytechnic Ikot Osurua, PMB 1200, Ikot Ekpene, Nigeria

Ricardo Díaz Martín
Chairman building Engineering department, CEU San Pablo University, Madrid, Campus de Monte príncipe

Gaston Sanglier Contreras
Associate professor building Engineering department, CEU San Pablo University, Madrid, Campus de Monte príncipe

Adib Guardiola Mouhaffel
CEU San Pablo University, Madrid, Campus de Monte príncipe

Abdul Jabbar Kandhro
Department of Community Medicine, Peoples University of Medical & Health Sciences for Women, Nawabshah, Shaheed Benazirabad

Ali Muhammad Rind, Abdul Aziz Mastoi, Khalida Faryal Almani and Muhammad Sohaib Rajpout
Center for Environmental Sciences, University of Sindh, Jamshoro, Sindh, Pakistan

Sujo Meghwar and Muhib Ali Laghari
Department of Geography, University of Sindh, Jamshoro, Sindh, Pakistan

Tetiana Lutchyn
National Technical University of Ukraine "Kyiv Polytechnic Institute", Institute for Energy Saving and Energy Management, Kyiv, Ukraine

Kamal O. Abdullah, Salar Z. Muhammed and Adil M. Hussein
Physic Dep., Faculty of Science and Science Education, University of Sulaimani, Kurdistan region, Iraq

Waleed Hussein Ali Hussein
Department of Architectural Engineering, Faculty of Engineering, Fayoum University, Fayoum, Egypt

Biswajit Mukhopadhyay and Buddhadev Bhattacherjee
Faculty: Civil Engineering; Institute: Kolkata Metropolitan Water & Sanitation Authority, City: Kolkata, India

Col Islam-ul-Haque
Chairman Eco-Logical Sustainability Through Environment Protection Services, Inc, (Eco-STEPS), Islamabad, Pakistan

Asim Saleem
Country WASH advisor, Plan International Pakistan, Islamabad, Pakistan

Mohamed Cassim Mohamed Zakeel and Prathiba Aruni Weerasinghe
Department of Plant Sciences, Faculty of Agriculture, Rajarata University of Sri Lanka, Puliyankulama, Anuradhapura, Sri Lanka

Banumu Arachchige Dilhani Ganga Kumari
Postgraduate Institute of Agriculture, University of Peradeniya, Peradeniya, Sri Lanka

Hetti Arachchige Mangalika Wickremasinghe
Department of Agricultural Biology, Faculty of Agriculture, University of Peradeniya, Peradeniya, Sri Lanka

Muluken Nega Bazeze
College of Agriculture and Natural Resources, Dilla University, Dilla, Ethiopia

Teshome Soromessa and Eyale Bayable
Center for Environmental Science, College of Natural Science, Addis Ababa University, Addis Ababa, Ethiopia

Adhena Ayaliew Werkneh, Belay Zimbelachew Medhanit, Angaw Kelemework Abay and Jemal Yimer Damte
Department of Chemistry, College of Natural Science, Jigjiga University, Jigjiga, Ethiopia

Abdunnaser Mohamed Etorki
Department of Chemistry, University of Tripoli, Tripoli, Libya

Ibrahim Salem Shaban
Department of Environmental Science and Engineering, Libyan Academy, Tripoli, Libya

Tchadanaye New Mahamat
Department of Chemistry, Faculty of Exact and Applied Sciences, University of N'djamena, N'djamena, Chad

Noumi Guy Bertrand
Department of Chemistry, Faculty of Science, University of Ngaoundere, Ngaoundere, Cameroon

Paloumi Yabe
Department of Chemistry, Faculty of Exact and Applied Sciences, University of N'djamena, N'djamena, Chad

Department of Chemistry, Faculty of Science, University of Ngaoundere, Ngaoundere, Cameroon

Cristina Damian, Ana Olteanu, Mircea Oroian, Ana Leahu and Sorina Ropciuc
Stefan cel Mare University of Suceava, Faculty of Food Engineering, 13 Universitatii, Suceava, Romania

Margaret Tita Awah, Gus Djibril Kouankap Nono, Pierre Wotchoko and Mispa Ayuk Tabot
Department of Geology, HTTC, The University of Bamenda, Bambili ; Bamenda, Cameroon

Veronique Kamgang Kabeyene
Department of Earth Sciences, Faculty of Sciences, University of Yaoundé 1, Yaoundé, Cameroon

Alice Magha
Department of Geology, HTTC, The University of Bamenda, Bambili ; Bamenda, Cameroon

Department of Earth Sciences, Faculty of Sciences, University of Yaoundé 1, Yaoundé, Cameroon

A. Tesfay Gebretsadik
Department of Industrial Chemistry, Addis Ababa Science and Technology University, Addis Ababa, Ethiopia

Tarekegn Berhanu
Strengthening of Agricultural Pesticide Residue Analysis System Project, JICA, Addis Ababa, Ethiopia

Belete Kefarge
Department of Chemistry, Jigjiga University, College of Natural and Computational Science, Jigjiga, Ethiopia

John Safari, Vicent Gowele and James Lwelamira
Institute of Rural Development Planning (IRDP), Dodoma, Tanzania

Dibal Hyeladi, Daspan Raymond and Lar Uriah
Department of Geology University of Jos, Jos, Nigeria

Shem Auta Dahilo
Federal College of Education, Akwanga, Nasarawa State, Nigeria

Ayele Behaylu
Department of Geography and Environmental Studies, Collage of Social Science and Humanities, Adigrat University, Adigrat, Ethiopia

Fitra Suzanti
Biology Education Study Program, University of Riau, Riau, Indonesia

Retna Astuti Kuswardani
Faculty of Agriculture, University of Medan Area, Medan, Indonesia

Suci Rahayu
Faculty of Mathematics and Natural Sciences, University of North Sumatra, Medan, Indonesia

Agus Susanto
Palm Research Center of Marihat, North Sumatra, Indonesia

M. A. Misdaq and J. Ouguidi
Nuclear Physics and Techniques Laboratory, Faculty of Sciences Semlalia, University of Cadi Ayyad, Marrakech, Morocco (URAC-15 Research Unit Associated to the CNRST, Rabat, Morocco)

A. Matrane
Nuclear Physics and Techniques Laboratory, Faculty of Sciences Semlalia, University of Cadi Ayyad, Marrakech, Morocco (URAC-15 Research Unit Associated to the CNRST, Rabat, Morocco)

Nuclear Medicine Service, Mohamed VI University Hospital Centre, Faculty of Medicine and Pharmacy, University of Cadi Ayyad, Marrakech, Morocco

Mohammad Amin Qureshi, Ghulam Murtaza Mastoi, Mushtaque Ahmed Baloch, Muhammad Ali Bhatti, Muhammad Waseem Junejo, Rashid Ali Malukhani and Imran Ali Abbasi
Centre for Environmental Sciences, University of Sindh, Jamshoro, Sindh, Pakistan

Sarfaraz Ali Mallah
M. A. Kazi Institute of Chemistry University of Sindh Jamshoro, Sindh, Pakistan

Tamara Tulaikova
Wave Research Centre at Prokhorov General Physics Institute, Moscow, Russia

Svetlana Amirova
Brody School of Medicine, East Carolina University, Greenville, North Carolina, USA

Alexandre Michtchenko
National Polytechnic Institute, IPN-SEPI-ESIME, Zacatenco, Mexico, D.F.

Muhammad Aslam Mirza, Muhammad Aziz Choudhary and Asima Naz
Department of Chemistry, Mirpur University of Science and Technology (MUST), Mirpur AJK, Pakistan

Muhammad Yar Khuhawar and Rafee Arain
Institute of Advanced Research Studies in Chemical Sciences, University of Sindh, Jamshoro, Pakistan

Kauser Yasmeen
Department of Chemistry, Federal Urdu University of Arts, Science and Technology, Karachi, Pakistan

Omar Chaalal
Chemical and Petroleum Engineering Department, College of Engineering, United Arab Emirates University, Al-Ain, United Arab Emirates

Ahmed Murad
Geology Department, College of Science, United Arab Emirates University, Al-Ain, United Arab Emirates

Ahmed M. Soliman and Ismail A. El Haty
Chemistry Department, College of Science, United Arab Emirates University, Al-Ain, United Arab Emirates

D. Hank
Nuclear Fuel Technology Department, Hot Labs. Centre, Atomic Energy Authority., Cairo, Egypt

Rafiq Islam
National Polytechnic School of Algiers, Algeria

Index